XXX International Conference on Photonic, Electronic, and Atomic Collisions (ICPEAC 2017)

AA006209

Journal of Physics Conference Series Volume 875

Cairns, Australia
26 July - 1 August 2017

Editors:

Alexander Bray
Sean Hodgman
Joshua Machacek
Anatoli Kheifets

ISBN: 978-1-5108-4729-3
ISSN: 1742-6588

Printed from e-media with permission by:

Curran Associates, Inc.
57 Morehouse Lane
Red Hook, NY 12571

Some format issues inherent in the e-media version may also appear in this print version.

Printed by Curran Associates, Inc. (2017)

For permission requests, please contact the Institute of Physics
at the address below.

Institute of Physics
Dirac House, Temple Back
Bristol BS1 6BE UK

Phone: 44 1 17 929 7481
Fax: 44 1 17 920 0979

techtracking@iop.org

Additional copies of this publication are available from:

Curran Associates, Inc.
57 Morehouse Lane
Red Hook, NY 12571 USA
Phone: 845-758-0400
Fax: 845-758-2633
Email: curran@proceedings.com
Web: www.proceedings.com

Table of contents

Volume 875

XXX International Conference on Photonic, Electronic, and Atomic Collisions (ICPEAC2017)

26 July to 1 August 2017, Cairns, Australia

Accepted papers received: 5 July 2017
Published online: 30 August 2017

Strong field-ultrafast

Structure-spectroscopy

Photons – Molecule

Weak field

Strong field-ultrafast

Cold - Storage - Cluster

Photons – Condensed Matter/Other

Leptons – Atom/Ion

Leptons – Molecule

Leptons – Condensed Matter

Heavy Particles – Cold

Heavy Particles – Atom/Ion

Heavy Particles – Molecule

Heavy Particles – Condensed Matter

Other Processes

The 30th International Conference on Photonic, Electronic and Atomic Collisions (XXX ICPEAC) was held at the Cairns Convention Centre, Queensland, Australia on 26 July - 1 August, 2017, organized by the Australian National University (ANU, Canberra), Curtin University (Perth, Western Australia) and Griffith University (Queensland). Held biannually, ICPEAC is one of the largest conferences on the atomic and molecular physics calendar, and covers the recent progress in photonic, electronic, and atomic collisions with matter. With a history dating back to 1958, the ICPEAC has its second coming to the Sunshine State of Queensland, the first of which being the 1991 Australian ICPEAC in Brisbane. With the selection of Cairns as the venue for the ICPEAC XXX the local organizers began their preparations six years ahead in anticipation of the event.

The conference brought together 500 participants from 50 countries and attracted 680 contributed papers for presentation in poster sessions. Among the latter, 311 are presented in issues 2-12 of this volume of the Journal of Physics Conference Series. In addition, five plenary lectures by Linda Young, Clifford Surko, Henrik Cederquist, Tilman Pfau and Reinhard Kienberger have been presented along with 62 progress reports and 26 special reports. The latter "hot topics" were selected by the ICPEAC Program Committee from the contributed papers. Detailed write-ups of most of the invited lectures and reports are presented in issue 1 of this volume, constituting a comprehensive and tangible record of the meeting.

Preceeding the conference, three distinguished scientists: Barry Schneider (NIST, USA), Chris Greene (Purdue University, USA) and Thomas Pfeifer (Max-Plank Institute, Germany) presented their tutorial lectures specifically designed for students and early career researchers.

During the conference, the 2017 IUPAP Young Scientist Prize was awarded to Johannes Feist (Universidad Autónoma de Madrid, Spain). The Sheldon Datz Prize for an Outstanding Young Scientist Attending ICPEAC was awarded to Dermot Green (Queen's University Belfast, UK).

The Local Organizing Committee would like to express its gratitude to Tourism and Events Queensland (TEQ), the International Union for Pure and Applied Physics (IUPAP), all the conference sponsors for their financial support. With particular emphasis we thank the group of volunteers from the ANU and Curtin University for their efforts in orchestrating the event.

Finally, the Local Organizing Committee would like to thank all the ICPEAC participants for their support and contributions to the conference.

The Editors,
Alexander Bray, Sean Hodgman, Joshua Machacek, Anatoli Kheifets
(Australian National University, Canberra, Australia)

IOP Conf. Series: Journal of Physics: Conf. Series **875** (2017) 001002 doi:10.1088/1742-6596/875/1/001002

Peer review statement

All papers published in this volume of *Journal of Physics: Conference Series* have been peer reviewed through processes administered by the proceedings Editors. Reviews were conducted by expert referees to the professional and scientific standards expected of a proceedings journal published by IOP Publishing.

Autoionization of very-high-n strontium Rydberg atoms

X. Zhang* , G. Fields*, F.B. Dunning*[1], S. Yoshida†, and J. Burgdörfer†

* Department of Physics & Astronomy, Rice University, Houston, Texas, USA
† Institute for Theoretical Physics, Vienna University of Technology, Vienna, Austria, EU

Synopsis The autoionization of high n, n~280-430, strontium Rydberg states through excitation of the 5s $^2S_{1/2}$→ 5p $^2P_{1/2}$ transition in the core ion is investigated. The autoionization rates decrease rapidly as L is increased paving the way for production of long-lived two-electron-excited planetary atoms.

The ability to create localized wave packets in Rydberg atoms whose behavior mimics that of a classical particle suggests the possible engineering of atoms that contain two excited electrons, each moving along quasi-classical orbits. While such atoms, termed planetary atoms, have been the subject of much theoretical interest,[1] their practical realization is a challenge because interactions between the excited electrons can lead to rapid autoionization. In the present work we explore autoionization using very-high n strontium Rydberg atoms, and the techniques that can be used to control it.

Strontium atoms in a collimated beam are excited to selected n^1P_1 or n^1F_3 states and autoionization is induced by exciting the 5s $^2S_{1/2}$→5p $^2P_{1/2}$ transition in the core ion using a 100ns-long pulse of 422 nm laser radiation. (Use of a short pulse minimizes population of the long-lived 4d $^2D_{3/2}$ state which can also induce autoionization) The number of surviving Rydberg atoms is determined by field ionization.

Measurements with both nP and nF states in which the frequency of the 422 nm laser is scanned reveal broad autoionization features whose widths decrease with increasing n and, in the limit of very high n, approach that of the transition in the Sr^+ core ion.

The effect of increasing the angular momentum, L, of the excited Rydberg electron is examined either by initially exciting selected extreme Stark states in a dc electric field or by sudden application of a "pump field" to n^1P_1 atoms which creates a Stark wavepacket that undergoes Stark precession resulting in periodic changes in $|L|$. [2] Even modest increases in L are found to result in a dramatic decrease in the autoionization loss. This is illustrated in Fig. 1 which shows the fraction of n=320 $^1P_1$ atoms that survive autoionization versus the duration of the pump pulse.

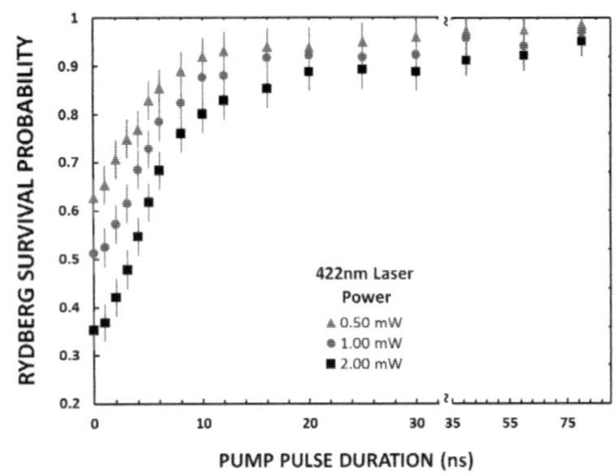

Figure 1. Rydberg atom survival probability versus pump pulse duration.

Pulse durations of ~10-15 ns are sufficient to populate states with $|L|$~20-30.

The present work indicates that by placing one electron in an outer localized very-high-n, high-ℓ, ℓ~n, near-circular Bohr-like orbit, it should be possible, with further excitation of the inner electron, to produce long-lived two-electron-excited planetary atoms.

Research supported by the NSF, The Robert A. Welch Foundation, the FWF (Austria), and the SFB-NextLite.

References

[1] G. Tanner, K. Richter, and J.-M. Rost 2000 *Rev. Mod. Phys.* **72**, 497
[2] B. Wyker, S. Ye, T.J. McKinney, F.B. Dunning, S. Yoshida, C.O. Reinhold, and J. Burgdörfer 2011 *J. Phys. B :At. Mol. Opt. Phys* **44**, 184002

[1] E-mail: fbd@rice.edu

IOP Conf. Series: Journal of Physics: Conf. Series **875** (2017) 022002 doi:10.1088/1742-6596/875/3/022002

Photoionization of neutral iron from ground and metastable states

Oleg Zatsarinny [1], **Luis Fernandez-Menchero, and Klaus Bartschat**

Department of Physics and Astronomy, Drake University, Des Moines, Iowa 50311, USA

Synopsis The B-spline R-matrix method is used to investigate the photoionization of neutral iron from the ground and metastable states in the energy region from the ionization thresholds to 2 Ry. The sensitivity of the results is checked by comparing the predictions obtained in different approximations. Inclusion of all terms from the $3d^64p$ and $3d^54s4p$ configurations considerably changes the low-energy resonance structure.

Neutral Fe and its ions play important roles in many aspects of astrophysics. Within fundamental atomic physics, the large number of possible terms resulting from the approximately half-open 3d shell, in combination with the near degeneracy of the $3d^x$, $3d^{x-1}4s$, $3d^{x-2}4s^2$, and other configurations for the neutral elements and the lowly-charged ions, result in very complex spectra that are strongly influenced by configuration interaction. Therefore, the calculations become extremely cumbersome and the spectra very rich in structure and difficult to analyze.

Accurate photoionization cross sections are required for the analysis and interpretation of spectroscopic observations. Thus far, the only data available for the photoionization of Fe I are limited to transitions between *LS* states. The most extensive calculations, presented by Bautista [1], involved 52 *LS* terms of the Fe II target. This is only a small part of the Fe II spectrum, which raises the question of the convergence of results.

The purpose of the present work was to perform elaborate and extensive calculations for the photoionization of neutral iron by using highly accurate target wave functions and by including fine-structure effects in the close-coupling expansions directly. The calculations were performed with the advanced BSR code [2], which employs the R-matrix method in a B-spline basis. To represent the target states, we used term-dependent non-orthogonal orbital sets. This allowed us to generate a more accurate description of the Fe I target states than those previously employed.

The target expansion included all states of the $3d^64s$, $3d^7$, $3d^64p$, $3d^54s^2$, and $3d^54s4p$ configurations. This adds up to 261 *LS* terms or 716 fine-structure levels. The convergence of the results was checked by comparing data obtained in different approximations. As an example, Fig. 1 compares the predicted photoionization cross section of the ground state of Fe I in two *LS* approximations. BSR-39 includes all relevant terms of the $3d^64s$, $3d^7$, and $3d^54s^2$ configurations, as well as the low-lying states of the $3d^64p$ and $3d^54s4p$ configurations as in [1], whereas BSR-112 also includes all terms of the $3d^64p$ and $3d^54s4p$ configurations. Note that both the resonance structure and the background cross sections change considerably, with the BSR-39 results being only in qualitative agreement with the cross sections obtained in [1]. This is due to the different target representations.

Ultimately, the present work aims to provide systematic data for photoionization of the fine-structure levels. Further details of the calculation and some preliminary results for ground-state and metastable photoionization in the Breit-Pauli approximation will be presented at the conference.

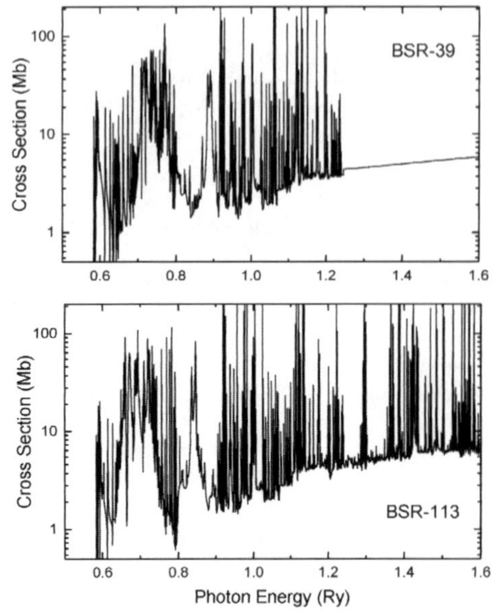

Figure 1. Comparison of the photoionization cross section of the ground state $(3d^64s^2)^5D$ of Fe I between two scattering models, BSR-39 and BSR-112.

This work was supported by the US National Science Foundation under grants PHY-1403245 and PHY-1520970, and the XSEDE allocation PHY-090031.

References

[1] M.A. Bautista 1997 *Astron. &Astrophys. Suppl. Ser.* 122 167

[2] O. Zatsarinny 2006 *Comp. Phys. Commun.* 174 273

[1] E-mail: oleg.zatsarinny@drake.edu

Low-energy outer-shell photodetachment of the negative ion of aluminum

Oleg Zatsarinny[* 1], **Klaus Bartschat**[*], **Elizabeth Nagy**[†], **Sergey Gedeon**[†], **Viktor Gedeon**[†], **and Vladimir Lazur**[†]

[*] Department of Physics and Astronomy, Drake University, Des Moines, Iowa 50311, USA
[†] Department of Theoretical Physics, Uzhgorod State University, Uzhgorod, 88000, Ukraine

Synopsis The photodetachment of the negative ion of Al is investigated by employing the *B*-spline *R*-matrix method for photon energies ranging from threshold to 12 eV. Several prominent resonance features are predicted, thereby providing new challenges in the study of this highly correlated process.

The photodetachment of Al^- investigated in the present work is part of our systematic study of the photodetachment of negative ions from the IIIB group of the periodic system [1,2]. Negative ions in this group have the $(ns^2np^2)^3P$ ground state, and photoexcitation of the ns electron leads to prominent $nsnp^3$ resonances. The accurate description of these resonances is a major challenge for theory. As shown in our recent study of the photodetachment of B^- [1] and Ga^- [2], their interpretation is not trivial and has led to different assignments in various previous calculations.

The calculations were performed with the B-spline R-matrix code [3]. The multi-configuration Hartree-Fock method with non-orthogonal orbitals in connection with B-spline expansions was employed for an accurate representation of the target wave functions. The scattering model contained the lowest 10 bound states of aluminum up to the $3s^26s$ state, plus the 3s-excited states $(3s3p^2)^2S$ and $(3s3p^2)^2P$ that lie in the target continuum. The latter states were included to allow for a proper description of the 3s photodetachment.

Figure 1 exhibits our results for the total photodetachment cross section of Al^- from its ground $(3s^23p^2)^3P$ state, with affinity of 432 meV. The cross section shows a complicated energy dependence due to different partial-wave and resonance contributions. Details will be presented at the conference. In particular, the minimum at 3.87 eV is due to the $(3s3p^3)^3D^o$ window resonance, whereas the sharp peak at 3.69 eV is due to the $(3s^24s4p)^3P^o$ resonance. This finding is in close agreement with the experimental measurements [4].

The Al^- ion has an excited $(3s^23p^2)^1D$ bound state with a binding energy of 0.109 eV. The total photodetachment cross sections for the ^1D initial state is presented in Fig. 2. The cross sections exhibit prominent resonance-like structures, but not all peaks can be related to particular resonances: only the maximum around 6 eV is due to the combined contribution of two $(4s4p^3)$ ^1P and ^1D negative-ion states. All other structures reflect near-threshold maxima caused by the opening of new channels.

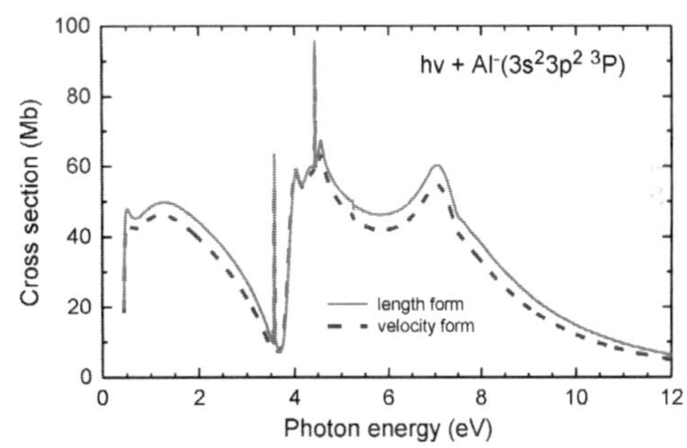

Figure 1. Total photodetachment cross sections of Al^- from the ground $(3s^23p^2)^3P$ state.

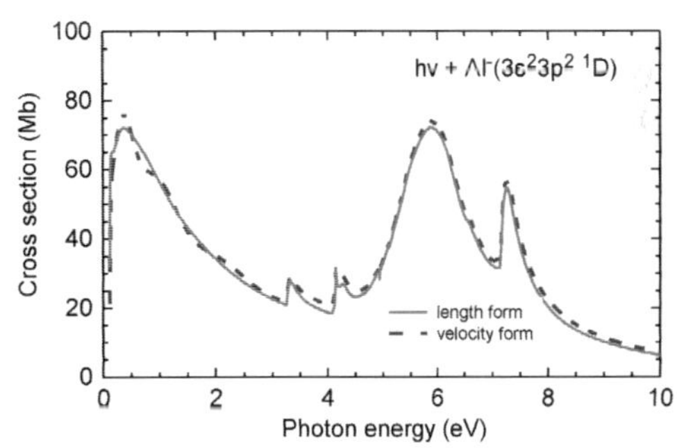

Figure 2. Total photodetachment cross sections of Al^- from the excited $(3s^23p^2)^1D$ state.

This work was supported by the US National Science Foundation under grants PHY-1403245 and PHY-1520970, and the XSEDE allocation PHY-090031.

References

[1] K. Wang *et al.*, 2016 *Eur. Phys. J. D* **70** 72
[2] K. Wang *et al.*, 2016 *Phys. Rev. A* **94** 023402
[3] O. Zatsarinny 2006 *Comp. Phys. Commun.* **174** 273
[4] B. J. Davies *et al.* 1997 *Phys. Rev. A* **56** 378

[1] E-mail: oleg.zatsarinny@drake.edu

IOP Conf. Series: Journal of Physics: Conf. Series **875** (2017) 022004 doi:10.1088/1742-6596/875/3/022004

Multi-electron processes in K-shell double and triple photodetachment of oxygen anions

S. Schippers*[1], R. Beerwerth†,‡, L. Abrok§, S. Bari♭,♮, M. Martins△, S. Ricz§, J. Viefhaus♮, S. Fritzsche†,‡, A. Müller◇

* I. Physikalisches Institut, Justus-Liebig-Universität Gießen, 35392 Giessen, Germany
† Helmholtz-Institut Jena, 07743 Jena, Germany
‡ Theoretisch-Physikalisches Institut, Friedrich-Schiller-Universität Jena, 07743 Jena, Germany
§ Institute of Nuclear Research of the Hungarian Academy of Sciences, 4026 Debrecen, Hungary
♭ European XFEL GmbH, 22869 Schenefeld, Germany
♯ DESY Photon Science, FS-SCS, 22607 Hamburg, Germany
♮ DESY Photon Science, FS-PE, 22607 Hamburg, Germany
△ Institut für Experimentalphysik, Universität Hamburg, 22761 Hamburg, Germany
◇ Institut für Atom- und Molekülphysik, Justus-Liebig-Universität Gießen, 35392 Giessen, Germany

Synopsis Absolute cross sections for double and triple photodetachment of O^- ions have been measured at photon energies around the threshold for *K*-shell ionization using the photon-ion merged-beams technique at a synchrotron light-source. In addition, corresponding large-scale *ab-initio* calculations were carried out at a level of complexity that has never been invoked before. From the comparison between experiment and theory it becomes apparent that the inclusion of multi-electron processes such as double shake-up in the theoretical calculations is crucial for the explanation of the experimental findings.

Absolute cross sections of double and triple photodetachment of O^- ions were measured in the experimental photon energy range of 524–543 eV that comprised the threshold for K-shell ionization [1]. For the experimental measurements, the photon-ion merged-beams technique [2] was employed using the permanently installed end station PIPE [3] at the "Variable Polarization XUV Beamline" (P04) [4] of the synchrotron light source PETRA III at DESY in Hamburg, Germany. Using resolving powers of up to 13000, the position, strength and width of the below-threshold $1s2s^2 2p^6\ ^2S$ resonance (Fig. 1) as well as the positions of the $1s2s^2 2p^5\ ^3P$ and $1s2s^2 2p^5\ ^1P$ thresholds for K-shell ionization were determined with high-precision. In addition, systematically enlarged multi-configuration Dirac-Fock calculations have been performed for the resonant detachment cross sections by utilizing the MCDF method as implemented in the GRASP [5] and RATIP [6] codes. Results from these *ab initio* computations agree very well with the measurements for the resonance width and branching fractions for double and triple detachment, if *double* shake-up (and -down) of valence electrons and the rearrangement of the electron density is taken into account. For the absolute cross sections, however, a previously found discrepancy between measurements and theory is confirmed [1].

This research was carried out at the light source PETRA III at DESY, a member of the Helmholtz Association (HGF). We would like to thank G. Hartmann, F. Scholz, and J. Seltmann for assistance in using beamline P04 as well as S. Klumpp for his continuous support of the PIPE setup. Funding by BMBF (contracts 05K10RG1, 05K10GUB, 05K16RG1, 05K16GUC, and 05K16SJA) is gratefully acknowledged.

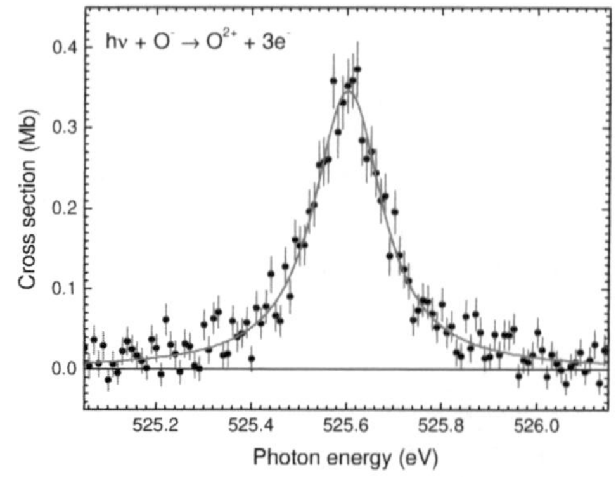

Figure 1. High-resolution measurement (symbols) of the $1s2s^2 2p^6\ ^2S$ resonance in the triple detachment channel [1]. The resonance width has been determined from a Voigt line-profile fit (line) to be 164 ± 14 meV.

References

[1] S. Schippers *et al.* 2016 *Phys. Rev. A* **94** 041401(R)
[2] S. Schippers *et al.* 2016 *Contemp. Phys.* **57** 215
[3] S. Schippers *et al.* 2014 *J. Phys. B* **47** 115602
[4] J. Viefhaus *et al.* 2013 *Nucl. Instrum. Meth. A* **710** 151
[5] P. Jönsson *et al.* 2007 *Comput. Phys. Commun.* **177** 597
[6] S. Fitzsche 2012 *Comput. Phys. Commun.* **183** 1525

[1]E-mail: stefan.schippers@physik.uni-giessen.de

IOP Conf. Series: Journal of Physics: Conf. Series **875** (2017) 022005 doi:10.1088/1742-6596/875/3/022005

Photoionization of atomic fluorine

Jiayu Sun[*1], **Jinlei Liu**[*], **Jianhua Wu**[*2], **Jianmin Yuan**[*†], **Zengxiu Zhao**[*]

[*] Department of Physics, College of Science, National University of Defense Technology, Changsha 410073, People's Republic of China
[†] IFSA Collaborative Innovation Center, Shanghai Jiao Tong University, Shanghai 200240, People's Republic of China

Synopsis We use *R*-matrix method to study the near threshold (17.5eV-20eV) photoionization of neutral fluorine atom and obtain total photoionization cross sections in ^3P to ^1D range. The theoretical calculation results fit the experimental results better after we consider core-valence electron correlation.

Photoionization cross sections of atoms or ions are widely used in calculations of plasma opacity in both astrophysics and radiation transport of thermal plasma. As for fluorine atom, people first observed abundant autoionization resonance structures in experiments. Then the high-quality experimental data inspired people to study photoionization of atomic fluorine theoretically. Most recently Saha employed MCHF method for detailed investigation of the autoionization resonances between ^3P and ^1D thresholds.

In the present study, we aim to use the nonrelativistic close-coupling *R*-matrix [1] approach with consideration of core-valence interaction to calculate the photoionization cross sections from ^3P to ^1D threshold of fluorine. And comparison is made with the experimental results. Eleven single electron orbits($1s$,$2s$,$2p$,$3s$,$3p$,$3d$,$4s$,$4p$,$4d$,$4f$,$5g$) are included to form the configuration interaction wave function of the positive fluorine using CIV3 programm. To discuss the effects of core-valence electron correlation, we chose three different cases of electron excitation, which are listed in Table 1. Case A stands for the situation that all the electrons on $1s$ and $2s$ shells are kept frozen and only two electrons can be excited from $2p$ shell; case B that only one electron on $2s$ shell can be excited; case C that two electrons on $2s$ shell can be both excited.

Table 1. Target energy and ionization energy.

State	A	B	C	MCHF	Expt.
^3P	0.0	0.0	0.0	0.0	0.0
^1D	2.652	2.624	2.612	2.710	2.588
^1S	6.361	7.248	5.649	5.560	5.569
$^2P^o$	17.51	16.62	17.27	17.25	17.44

Comparing these three different cases, it is clear that the energy levels fit the experimental results much better after considering the core-valence electron correlation effects.

Between the energy range from 17.5eV to 20eV, fluorine atom is mainly ionized into $F^+(^3P)$ state according to Table 1. In Fig. 1, the total cross sections of ^3P state are displayed in both length and velocity gauges, compared with MCHF results and experimental results.

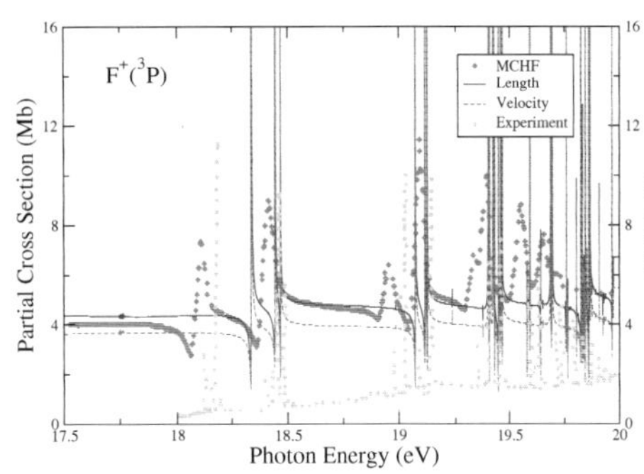

Figure 1. Total photoionization cross sections of fluorine between the ^3P and ^1D thresholds. MCHF calculations of Saha: violet diamond. Experimental data of Caldwell *et al.*: green dot.

Smaller step length is applied in RMATRX1 [1] program as well as the consideration of configuration interaction. So that we distinguish the resonance states that are overlapped together in MCHF calculation results. It is shown that our theoretical result has a good agreement with the experiment, with more and sharper autoionization resonance peaks.

References

[1] Keith A. Berrington *et al.* 1995 *COMPUT. PHYS. COMMUN.* **092** 000290

[1]E-mail: jaimesun2756@163.com
[2]E-mail: wujh@nudt.edu.cn

Relativistic Effects in Photoionization of Outer *ns* Subshells of Heavy Atoms

David A. Keating* [1], **Pranawa C. Deshmukh**[+2], **Steven T. Manson**[*3]

[*] Department of Physics and Astronomy, Georgia State University, Atlanta, GA 30303, USA
[+] Department of Physics, Indian Institute of Technology Tirupati Tirupati 517506, India

Synopsis Dynamic changes due to relativity are very prevalent effects in the photoionization of heavy atoms. Presented is an exploration of the evolution of these dynamic changes as the nuclear charge, Z, increases and how they affect the cross sections.

At high enough Z, relativistic effects become important contributors to the quantitative and qualitative nature of atomic properties. It is important to identify and study these relativistic effects as they evolve with increasing Z in order to fully understand atomic properties. The most prevalent relativistic effect is the creation of spin-orbit doublet channels and dynamic changes, i.e., changes in the wave functions, which also shifts the thresholds. The *ns* subshells are chosen because they do not have a spin-orbit doublet, which simplifies the study of the dynamic effects.

A theoretical study of the 6s subshell of mercury (Z = 80), radon (Z = 86), radium (Z = 88) and nobelium (Z = 102) and the 7s subshell of radium and nobelium have been performed using the relativistic random phase approximation (RRPA) methodology [1]. In order to determine which features in the photoionization cross section are due to relativistic effects, calculations are performed using the (nonrelativistic) random phase approximation with exchange (RPAE) methodology [2]. The RPA methodologies were chosen because of their ability to exclude channels from the calculations. This is important because interchannel coupling can obscure and diminish the dynamic effects. Calculations are performed with and without interchannel coupling for comparisons.

Without the inclusion of interchannel coupling there is a very clear difference between the RRPA and RPAE results, the absence of a Cooper minimum from the RRPA curves, which is clearly present in the RPAE results. As Z increases the Cooper minimum, in the RPAE, falls into the discrete spectrum, as can be seen in top plot of Figure 1(A). Studying the descent of the minimum into the discrete spectrum reveals that the Cooper minimum is clearly missing in the RRPA results and not just hidden in the autoionization region. Interchannel coupling pulls some of the Cooper minimum back out of the discrete, in the RPAE curve (see B plot of Figure 1), and induces one due to the coupling with the 5d subshell in the RRPA curve. The influence of the 5d subshell completely reshapes the two methodologies results into very similar cross sections. The major differences in the (B) bottom plot of Figure 1 are due to the differences between their 5d cross sections.

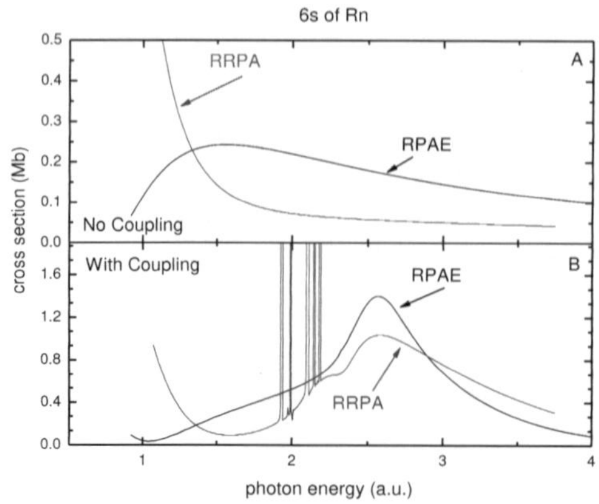

Figure 1 Photoionization cross section of the 6s subshell of radon with (A) and without (B) interchannel coupling included.

References

[1] W. R. Johnson and C. D. Lin, Phys. Rev. A **20**, 964 (1979)
[2] M. Ya. Amusia, Atomic Photoeffect (Plenum, NY, 1990)

[1] E-mail: dkeating2@studnet.gsu.edu
[*] E-mail: pcd@iittp.ac.in
3 E-mail: smanson@gsu.edu

Calculations of electron shakeoff probabilities using pseudostates

Takeshi Mukoyama* [1] **and Nobuyuki Tamura**† [2]

* Institute for Nuclear Research of the Hungarian Academy of Sciences (ATOMKI), Bem tér 18/c, H-4026 Debrecen, Hungary
† Graduate School of Science, Chiba University, 1-33 Yayoi-cho, Inage, Chiba 263-8522, Japan

Synopsis Electron shakeoff probabilities accompanying inner-shell vacancy production are calculated in the screened hydrogenic model using pseudostates. The results are compared with the values calculated with continuum wave functions.

When an inner-shell vacancy is created, other electrons in the same atom experience sudden change in the central potential and have a probability to be ejected from the atom. This process is called *shakeoff* and has been extensively studied both theoretically and experimentally.

In the sudden approximation, the shakeoff probability can be given in terms of imperfect overlap between the .initial- and final-state wave functions. However, the final state corresponds to the continuum state and calculations involving continuum wave functions are sometimes complicated and tedious. It is interesting to calculate electron shakeoff probabilities with a simpler approach.

In the present work, we use the pseudostate method and calculate the shakeoff probability in the screened hydrogenic model. For this model, the shakeoff probabilities with continuum wave functions are known in the simple analytical forms [1]. We consider a Schrödinger equation for hydrogenic potential and expand the eigenfunction $\varphi(r)$ in terms of a set of basis functions:

$$\varphi(r) = \sum_{i=1}^{N} c_i \chi_i(r),$$

where N is the number of basis set, c_i is the expansion coefficient and $\chi_i(r)$ is the Slater-type orbital (STO). With suitable choice of N and parameters of STO's, the atomic Hamiltonian is diagonalized and the energy eigenvalues of the hydrogenic atom are well reproduced. In addition to these bound states, we obtain states with positive discrete energy, i.e. *pseudostates*. The eigenfunctions corresponding to pseudostates in the final state are used to calculate the overlap integral with the initial bound-state wave function. The basis set consists of ten STO's both for $l = 0$ and $l = 1$. The screening constants are obtained according to the Slater's recipe.

Figure 1 shows the comparison of the pseudostate wave function of Ne for the energy eigenvalue $E = 4.83$ a.u. and $l = 0$ with the corresponding continuum wave function. The former is normalized to the latter at the first peak. It is clear that both wave functions have similar shape.

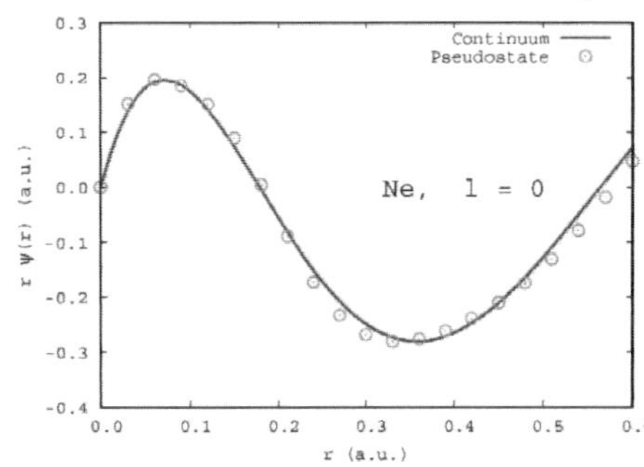

Figure 1. Comparison of the pseudostate wave function of Ne for $E = 4.83$ a.u. and $l = 0$ with the corresponding continuum wave function.

The shakeoff probability is calculated as the sum of overlap integrals for the pseudostates. The results for Ne accompanying K-shell vacancy production is shown in Table 1 and compared with the values obtained with continuum wave functions [1].

Table 1. Electron shakeoff probabilities for Ne accompanying K-shell vacancy production (%).

	1s	2s	2p
Pseudostate	0.028	1.18	1.35
Continuum	0.029	1.12	1.41

The present results indicate that the pseudostate approach is useful to calculate the electron shakeoff probabilities accompanying inner-shell vacancy production.

References

[1] T. Mukoyama, 2005 *X-Ray Spectrom.* **34**, 64.

[1] E-mail: mukoyama@atomki.mta.hu
[2] E-mail: tamtam@chiba-u.jp

IOP Conf. Series: Journal of Physics: Conf. Series **875** (2017) 022008 doi:10.1088/1742-6596/875/3/022008

Distorted wave photoionization cross sections for use in NLTE model atmospheres: Ni – Ni^{10+}

S. P. Preval[*1], **N. R. Badnell**[*2]

[*] Department of Physics, University of Strathclyde, Glasgow, G4 0NG, United Kingdom

Synopsis The Opacity Project is a highly successful venture in which energy levels, oscillator strengths, and photoionization cross sections were calculated for astrophysically important elements up to iron. However, metals heavier than iron are being discovered in stellar atmospheres. The photoionization cross sections for elements and ions not covered by the Opacity Project are usually approximated with hydrogenic formulae. We present a series of valence and inner-shell distorted wave photoioniza-tion cross section calculations for use in model atmosphere calculations covering Ni - Ni^{10+}. We discuss current progress, and future work.

Stellar atmosphere models can only ever be as good as the atomic data supplied to them. The Opacity Project has served as a gold standard for atomic data for use in opacity calculations and bespoke stellar atmosphere modelling. The Opacity Project has calculated energy levels, oscillator strengths, and photoionization cross sections for all astrophysically relevant ions up to iron. Beyond iron, the photoionisation cross sections are usually approximated with hydrogenic formulae, or extrapolated. There do exist distorted wave calculations by [1], however, these are totals only and are not partially resolved. In the case of white dwarf stars, hydrogenic cross sections are a poor approximation. [2] considered the effect on measured atmospheric abundances for the hot white dwarf G191-B2B when using model atmospheres calculated with hydrogenic and realistic nickel PI data. It was shown that the abundances and the flux distribution in the extreme ultraviolet wavelengths changed dramatically, with flux attenuation being 80% greater in the realistic case compared to the hydrogenic case. This result alone is motivation to continue the work of the Opacity Project and calculate PI data for heavier metals. To this end, we present our progress in calculating valence and inner-shell PI data for Ni - Ni^{10+}.

The cross sections were calculated using the distorted wave code AUTOSTRUCTURE [3] We treat PI from different complexes separately, splitting the calculation into K-shell, L-shell, and M-shell respectively. This is a valid approximation to make, as mixing between levels arising from K, L, and M-Shell PI is small. The K and L-Shell cross sections were calculated with term resolution, and the M-Shell cross sections were calculated with level resolution using kappa-averaged relativistic wavefunctions. An example of the data calculated is given in Figure 1, where we have plotted the total direct PI cross sections for Ni10+, while also showing the individual contributions from the K, L, and M shells.

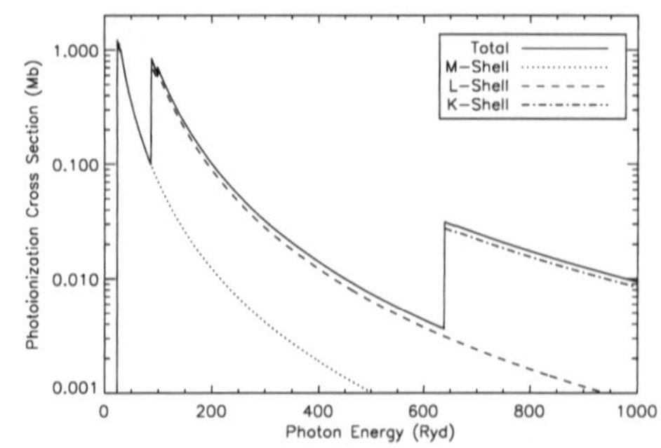

Figure 1. Total direct photoionization cross section of Ni^{10+} (solid line), along with contributions from M-Shell (dotted line), L-Shell (dashed line), and K-Shell (dot dashed line) photoionization.

Currently, direct PI cross sections have been calculated for Ni-Ni^{10+}. We intend to calculate the resonant contributions to the total PI cross section. In addition, we will assess the effect of this atomic data on white dwarf model atmospheres, and in particular, changes to atmospheric structure, temperature, and atmospheric abundances.

References

[1] Witthoeft M C, Bautista M A, Garcia J, Kallman T R, Mendoza C, Palmeri P, Quinet P, 2011, *ApJS*, **196**, 7

[2] Preval S P, Barstow M A, Badnell N R, Hubeny I, Holberg J B, 2017, *MNRAS*, **465**, 269

[3] Badnell N R, 2011, *Comput. Phys. Commun.*, **182**, 1528

[1]E-mail: simon.preval@strath.ac.uk
[2]E-mail: badnell@phys.strath.ac.uk

Photoionization of polarized Ne II in the region of autoionizing states

E.V. Gryzlova[*1], **A.N. Grum-Grzhimailo**[*2], **S.I. Strakhova**[*], **S.M. Burkov**[†], **K. Ueda**[‡], **M. Meyer**[§], **G. Sansone**[¶]

[*] Skobeltsyn Institute of Nuclear Physics, Lomonosov Moscow State University, Moscow 119991, Russia
[†] Pacific National University, Khabarovsk 680035, Russia
[‡] Institute of Multidisciplinary Research for Advanced Materials, Tohoku University, Sendai 980-8577, Japan
[§] European XFEL GmbH, 22869 Schenefeld, Germany
[¶] Physikalisches Institut, Universität Freiburg, 79106 Freiburg, Germany

Synopsis A detailed theoretical study on photoionization of polarized positive neon ion including its autoionizing states is presented, with reference to recent experiments at the free-electron laser FERMI. The feasibility for a complete photoionization experiment on positive ions is demonstrated.

By using intense linearly and/or circularly polarized extreme ultraviolet pulses of a free electron laser (FEL), the sequential two-photon double ionization (2PDI) has been realized in noble gas atoms (see reviews [1, 2]). Within the two-step model, the second step of the 2PDI represents the photoionization of the polarized positively charged ion, which can be studied by means of angle-resolved electron spectroscopy at FELs. Since the residual single-charged ion produced after the first step photoionization is generally polarized, a unique opportunity is opened to study photoionization of positively charged ions with an unprecedented level of detail, i.e. up to realizations of a complete experiment. Another poorly explored field is ionic autoionizing states, which may be now excited by photons from the polarized ground ionic state in the sequential 2PDI.

We calculated cross sections and angular distributions of photoelectrons produced upon ionization of $Ne^+(2p^5)$ in the region of the autoionizing states with dominant configuration $Ne^{+*}(2s2p^53p)$. The theoretical results are compared to recent recent measurements at FERMI. Two theoretical approaches were used: (a) perturbation theory for the both photoionization steps with multiconfiguration intermediate-coupling Hartree-Fock wavefunctions, combined with diagonalization method for treating the autoionizing states of Ne^{++} and (b) perturbation theory for the first-step ionization and R-matrix calculations for the second-step ionization from polarized neon ion.

Using both circularly and linearly polarized FEL beams, the complete photoionization experiment on positively charged ions becomes possible, enabling us to find, within a certain theoretical model, any observable from the measured photoionization amplitudes. Furthermore, photoelectron angular distributions from the second-step ionization help to complete the experiment also on the first-step ionization. As an illustration of the theoretical work, Fig. 1 shows electron clouds reconstructed from the measured ionization amplitudes for both ionization steps.

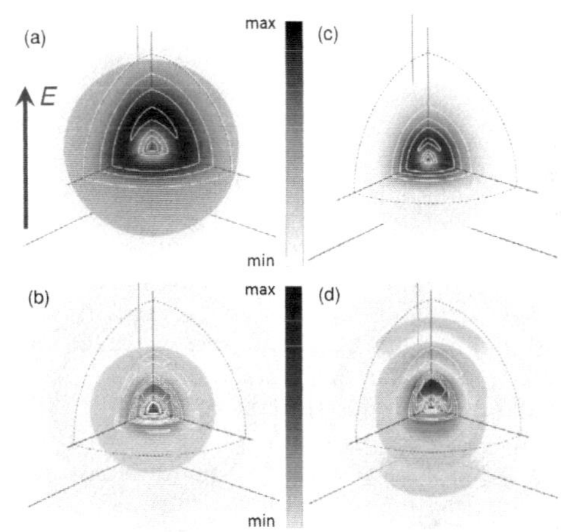

Figure 1. Inner structure of the spacial electron density of the residual ions $Ne^+(2p^5\,^2P)$ (a) and $Ne^{++}(2p^4\,^3P)$ (c) and the corresponding first-step (b) and second-step (d) photoelectrons. The photon energy of 56.5 eV is in the region of the autoionizing resonance $Ne^{+*}(2s2p^53p)$. The dashed line indicates the radius 5 a.u. (linear scale). The electric vector $\vec{E}$ of the linearly polarized FEL photons is indicated

Details of the calculations and their comparison with the experiment will be presented at the conference.

References

[1] M. Braune *et al.* 2016 *J. Mod. Opt.* **63** 324

[2] A.N. Grum-Grzhimailo *et al.* 2016 *J. Mod. Opt.* **63** 334

[1]E-mail: gryzlova@gmail.com
[2]E-mail: grum@sinp.msu.ru

The influence of ambient gas on laser produced plasmas and the crater morphology of laser-ablated Al

D. X. Sun[1], **Y. Q. Wang, M. G. Su, Q. Min, S. Q. Cao and C. Z. Dong**[2]

Key laboratory of Atomic and Molecular Physics & Functional Materials of Gansu Province, College of Physics and Engineering, Northwest Normal University, Lanzhou 730070, Gansu, China

Synopsis In this work, the influence of the ambient gas (N_2, O_2, Ar and He) on the plasma emission signal, plasma lifetime, electron number density and temperature, as well as the crater morphology of laser-ablated Al has been studied experimentally.

A laser pulse focused to a sufficiently high irradiance can produce plasma useful for spectrochemical analysis (generally called laser induced breakdown spectroscopy, LIBS). LIBS is generally carried out in the atmospheric pressure in air. Therefore, the laser ablation of solid targets in the presence of a background gas has received increased attention due to its importance in pulsed laser deposition, nanoparticle formation and growth, laser micromachining.

Compared with expansion into a vacuum, the interaction of the plume with an ambient gas is a far more complex gas dynamic process due to the rise of new physical processes involved such as deceleration, thermalization of the ablated species, interpenetration, recombination, formation of shock waves and clustering.

In this work, the influences of the background gas on the emission signal, mass ablation rate, the plasma temperature and the electron number density have been investigated in experiment. The effect of the ambient atmosphere (N_2, O_2, Ar and He) on the plasma parameters and on the emission intensity is found to be maximum in the case of Ar environment.

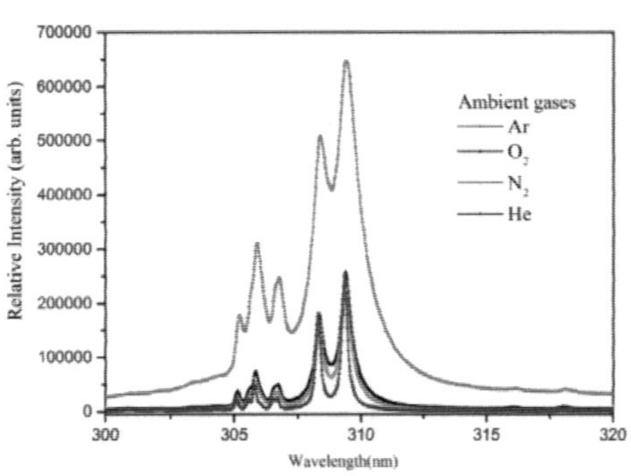

Figure 1. Emission signal of Al plasma detected under different ambient gas, (delay time 300 ns, gate width 200 ns).

We observed that the spectral lines of Al are more intensive, as shown in Fig. 1. Excitation temperature is higher. In addition to, the plasma lifetime is longer in case of Ar condition. However, the influence is in contrast to the case of He condition. The difference of the plasma parameters reveal that the hotter and denser plasmas are formed in an argon atmosphere rather than in helium or others.

Fig. 2 shows the craters after 10 pulses ablation under different surrounding gas. The morphology of crater is quite different, which dependent on the thermalization and shockwave propagation in surrounding gas. Moreover, the ablated efficiency can be obtained from these craters, and the ablated mass under helium is more than that under other gases.

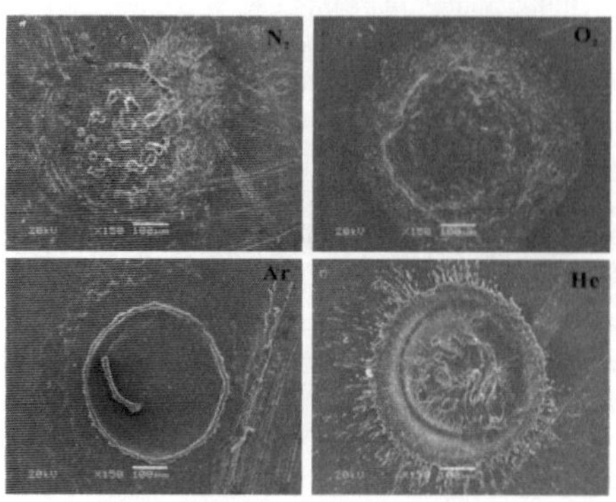

Figure 2. SEM images of crater morphology of laser ablated Al under different ambient gas (10 pulses, 60 mJ/pulse).

This work is supported by the National Natural Science Foundation of China (Grants Nos. 11564037, U1332206, and 11364037).

References

[1] S. Bashir *et al.* 2012 *Appl. Phys.* A **107** 203

[1] E-mail: sundx@nwnu.edu.cn

[2] E-mail: dongcz@nwnu.edu.cn

IOP Conf. Series: Journal of Physics: Conf. Series **875** (2017) 022011 doi:10.1088/1742-6596/875/3/022011

Resonance induced population transfer of Fe XVII ions in plasma environment

Chensheng Wu*‡, Xiang Gao* [1]

*Beijing Computational Science Research Center, Beijing 100193, China
‡Department of Engineering Physics, Tsinghua University, Beijing 100084, China

Synopsis We resolve the discrepancy between the existing theoretical oscillator strengths and the recent X-ray free electron laser (XFEL) experiment for the intensity ratio between two of the strongest emission lines from Ne-like Fe XVII (Fe^{16+}) ion. Our study shows that, by considering the dynamic resonance induced population transfer between Fe XVII and Fe XVI (Fe^{15+}) ions, which coexist in the XFEL experiment, we are able to successfully resolve this difference in theory and ex-periment. Further experimental works are suggested for a more detailed understanding of the dynamic resonance processes for ions under the plasma environment.

The x-ray emission lines of the Ne-like Fe^{16+} ion have been observed in a variety of astrophysical objects, including the Sun, stellar coronae, elliptical galaxies, and supernova remnants. Two of the most intense lines of the Ne-like Fe^{16+} ion are the $2p^53d\ ^1P_1 \rightarrow 2p^6\ ^1S_0$ dipole emission line (3C) at 15.01Å and the $2p^53d\ ^3D_1 \rightarrow 2p^6\ ^1S_0$ inter-combination line at 15.26 Å (3D). However, the diagnostic utility of these two lines has been limited by the fact that although extensive studies have been carried out, discrepancies between the theoretical estimates and the measurements from astrophysical and laboratory sources persist. In a recent benchmark experiment [1], the Fe^{16+} ion were first generated in an electron beam ion trap (EBIT) and then photo-excited by the X-ray free electron laser (XFEL). This experiment was designed to allow a direct comparison of experimental and theoretical results, excluding the effects of electron collisions. The measured weighted 3C/3D ratio was 2.61±0.23, which is significantly lower than the most elaborated theoretical values at 3.4 or higher [1]. A more recent detailed atomic structure calculation suggests that other physical processes beyond those included in the atomic structure calculation of an isolated atomic system may be responsible for this discrepancy [2,3].

With an extended large scale full relativistic configuration interaction calculation including the quantum electrodynamics (QED) term such as the Breit interaction, the excitation energies from the current calculation are less than 0.02% from the NIST data and the f_{3C}/f_{3D} ratio is 3.567±0.003. We first came to the conclusion that the reliability of the theoretical results for an isolated Fe XVII ion from the present calculation and other earlier atomic structure calculations may not be the cause for the discrepancy between the experimentally observed intensity ratio and the theoretical estimates. Our detailed analysis on the influence from the outside plasma environment suggests that such effect is not the likely cause for the difference in f_{3C}/f_{3D} ratio either, although our calculation does suggest a substantial decrease in theoretical oscillator strength ratio that overlaps with the experimental intensity ratio if the plasma density is substantially higher than the ones in the XFEL experiment. Since it is known that the energies of the Fe XVII 3D line and the Fe XVI C line are both close to 812 eV and the pulse length of the XFEL experiment is sufficiently longer than the time scale of the autoionization process transferring the upper state of the Fe XVI ion to the ground state of Fe XVII ion, we decided to focus our investigation on the dynamic resonance induced population transfer from the Fe XVI to the Fe XVII ions. By solving the relevant rate equations, we are able to generate finally the theoretically simulated spectra as well as the intensity ratio in agreement with the experimental spectra from the XFEL experiment.

References

[1] S. Bernitt et. al. 2012 Nature **492**, 225
[2] N. S. Oreshkina, et. al. 2014 Phys. Rev. Lett. **113**, 143001
[3] S. D. Loch, et. al. 2015 Astrophys. J. Lett. **801**, L13

[1] E-mail: xgao@csrc.ac.cn

IOP Conf. Series: Journal of Physics: Conf. Series **875** (2017) 022012 doi:10.1088/1742-6596/875/3/022012

Correlation dynamics in double photoionization of excited helium atom by a single ultrashort XUV pulse

Aihua Liu [1], **Dajun Ding**

Institute of Atomic and Molecular Physics, Jilin University, Changchun 130012, China

Uwe Thumm [2]

Department of Physics, Kansas State University, Manhattan, Kansas 66506, USA

Synopsis We study single- and two-photon double ionization of helium by short XUV pulses by numerically solving the time-dependent Schrödinger equation in full dimensionality within a finite element discrete variable representation scheme. We discuss the joint energy and angular distributions, identify sequential and non-sequential contributions in the double ionization by ultrashort pulse, and track the dynamics of the ionization process in distinct double ionization regimes.

Sequential and non-sequential double ionization of helium atoms has been well studied for decades in electron-momentum resolved spectra [1, 2 and refs. therein]. However with the availability and application of ultrashort subfemtosecond XUV pulses, the distinction of correlated and sequential double ionization has become blurred and requests new approaches and quantitative measures [1].

Solving the time-dependent Schrödinger equation *ab initio* with our recently developed implementation of a fully dimensional finite-element discrete-variable representation scheme [2] for few-photon double ionization of the helium atom, we investigate the single XUV photon double ionization of helium excited states [for example, He($1s2s^1S$) and He($1s2p^1P$)]. We studied electronic correlation based on joint photoelectron energy distributions and joint angular distributions. Our discussion of angular distributions reveals two different pathways to double ionization: (1) the inner K-shell electron absorbs one photon and knocks out the L-shell electron (This process favors 136° angular difference between emitted photoelectrons) or (2) the L-shell electron absorbs one photon and knocks out the K-shell electron (This process occurs predominantly at an angular difference of 65°).

Figure 1 displays the joint energy distribution of helium single-photon double ionization from the $1s2s^1S$ by 78 eV XUV photons. The uneven distribution is different from the single photon double ionization from ground state of helium atoms, even though both initial states have ^1S symmetry.

[1]E-mail: aihualiu@jlu.edu.cn
[2]E-mail: thumm@phys.ksu.edu

This work is supported by the U.S. NSF, U.S. DoE and the National Basic Research Program of China (973 program, Grant No. 2013CB922200) and National Science Foundation of China (Grant No. 11534004).

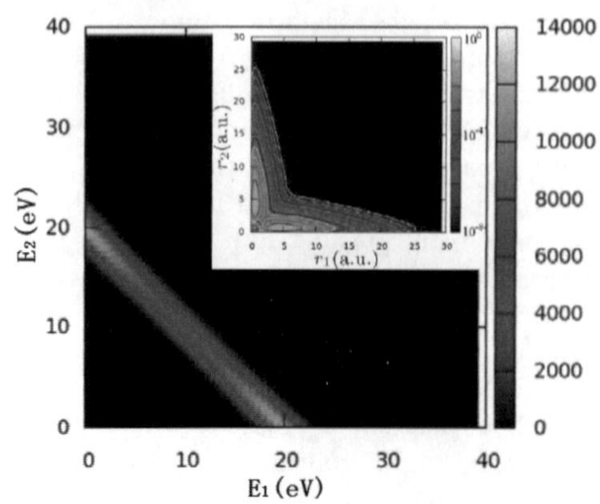

Figure 1. (Color online) Normalized joint photoelectron energies distributions for the double ionization of He($1s2s^1S$) by a single 78 eV photon. The XUV pulse has a peak intensity of $5\times10^{14}W/cm^2$ and pulse length of 2.12 fs (40 o.c.). The inset shows the undistorted excited initial state. The color bar is in arbitrary units.

References

[1] A. Liu and U. Thumm 2015 *Phys. Rev. Lett.* **115**, 183002.

[2] A. Liu and U. Thumm 2014 *Phys. Rev. A* **89**, 063423.

Impact of fullerene polarizability on Wigner time delay in photodetachment of fullerene anions C_N^-

V. K. Dolmatov[1]

Department of Physics and Earth Science, University of North Alabama, Florence, Alabama, 35632, U.S.A.

Synopsis The initial insight into the impact of fullerene polarizability on photodetachment time delay from fullerene anions C_N^- is exposed. The impact is demonstrated to be significant.

Time delay in photoemission of electrons upon photoionization of atoms [1] has become an *ad hoc* topic of many modern studies. To the best of the author's knowledge, no study has been performed with regard to photodetachment time delay from fullerene anions C_N^-. Yet, because fullerenes possess large polarizability, it is interesting to elucidate the impact of polarizability on corresponding photodetachment time delay. It is the aim of the present work to expose the effect of the target polarization on photodetachment time delay from highly polarizable C_N^-s.

To meet the goal, photodetachment time delays from fullerene anions with $N = 60$ and 240 are chosen as case studies. The attached electron is assumed to be a 2p-electron. In view of the complexity of the given many-electron problem, a reasonable simplification of the problem is desirable. In the present work, a popular modelling of a fullerene cage C_N by an attractive spherical potential $U_c(r)$ of certain depth U_0, inner radius r_0 and thickness Δ ([2] and references therein) is utilized. A fullerene anion C_N^- is then formed by the binding of an external electron by this potential into a $n\ell$-state. To account for the impact of the polarization potential V_{pol} of a fullerene on the motion of a released electron, V_{pol} is approximated by a static dipole polarization potential: $V_{pol}(r) \approx -\alpha/[2(r^2+b^2)^2]$ [3]. Here, α is the static polarizability of C_N and b is a parameter of the order of r_0. Thus, a released from C_N^- electron moves in the field of an effective potential $U_{eff}(r) = U_c(r) + V_{pol}(r)$. Then, to determine the sought Wigner time delay $\tau_{n\ell}$, one solves corresponding time-independent Schrödinger equations, finds the electron's initial-state and final-state wavefunctions, calculates the effective photodetachment amplitude $D_{n\ell}$ and its phase $\varphi_{n\ell}$, and completes the study by the calculation of $\tau_{n\ell}$ in accordance with its definition, as an electron-energy derivative of the phase: $\tau_{n\ell} = d\varphi_{n\ell}(E)/dE$ (in atomic units, a.u.).

For illustration, time delays τ_{2p} upon 2p-photodetachment of $C_{60}^-(2p)$, calculated with and without accounting for fullerene polarizability, are depicted in Fig. 1. The latter is self-explanatory, and details of a strong impact of polarizability of C_{60} on time delay are evident.

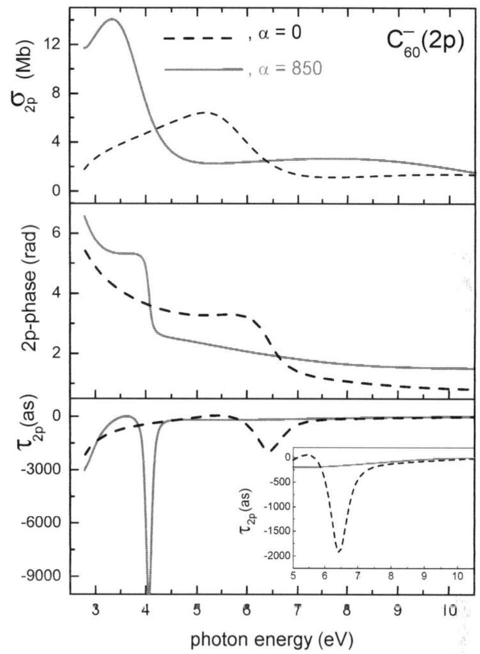

Figure 1. Calculated photodetachment cross section σ_{2p}, phase φ_{2p} and time delay τ_{2p} upon photodetachment of $C_{60}^-(2p)$ with (solid lines) and without (dashed lines) account for C_{60} polarizability α. Parameters: $\alpha \approx 850$, $U_0 \approx -0.26$, $r_0 \approx 5.26$ and $\Delta \approx 2.91$ a.u. [3].

The help of A. Edwards and C. Lane in performed calculations and the travel grant support from COAS, UNA are gratefully acknowledged.

References

[1] M. Schultze *et al.* 2010 *Science* **328** 1658

[2] V. K. Dolmatov *et al.* 2004 *Radiat. Phys. Chem.* **70** 417

[3] V. K. Dolmatov *et al.* 2017 *Phys. Rev. A* **95** 012709

[1]E-mail: vkdolmatov@una.edu

IOP Conf. Series: Journal of Physics: Conf. Series **875** (2017) 022014 doi:10.1088/1742-6596/875/3/022014

Electron Matter-Wave Vortices in Double Photoionization of Helium by Attosecond Pulses

J.M. Ngoko Djiokap*[1], A.V. Meremianin†, N.L. Manakov†, S.X. Hu‡, L.B. Madsen§, and Anthony F. Starace*[2]

* Department of Physics and Astronomy, University of Nebraska, Lincoln, Nebraska 68588-0299, USA
† Department of Physics, Voronezh State University, Voronezh 394006, Russia
‡ Laboratory for Laser Energetics, University of Rochester, Rochester, New York 14623-1299, USA
§ Department of Physics and Astronomy, Aarhus University, DK-8000 Aarhus C, Denmark

Double photoionization of helium by a pair of time-delayed, elliptically-polarized attosecond pulses is shown to produce two-electron momentum distributions that exibit *two*-start spiral vortex structures. These structures originate from Ramsey interference of the created pair of two-electron wavepackets, each carrying a total angular momentum of unity. The predicted vortex-shaped patterns occur for any energy partitioning between electrons, and are exquisitely sensitive to the time delay between the two pulses, their relative phase, their ellipticity and handedness. Moreover, these kinds of vortices occur for both in-plane (cf. Figs. 1, 2) and out-of-plane detection geometries; however, they only occur for fixed angular separation $\hat{\mathbf{p}}_1 \cdot \hat{\mathbf{p}}_2$ between the electron momenta.

We solve *ab initio* the seven-dimensional two-electron, time-dependent Schrödinger equation and analyze the results using perturbation theory. Such vortices are general phenomena, as similar patterns have been predicted for both atomic [1] and molecular [2] single-electron ionization processes. Recently, similar Fermat spiral vortices were observed in an experiment [3] involving femtosecond multiphoton ionization of potassium. Those results are consistent with our theoretical predictions for both photoionization [1] and multiphoton ionization [4] of helium.

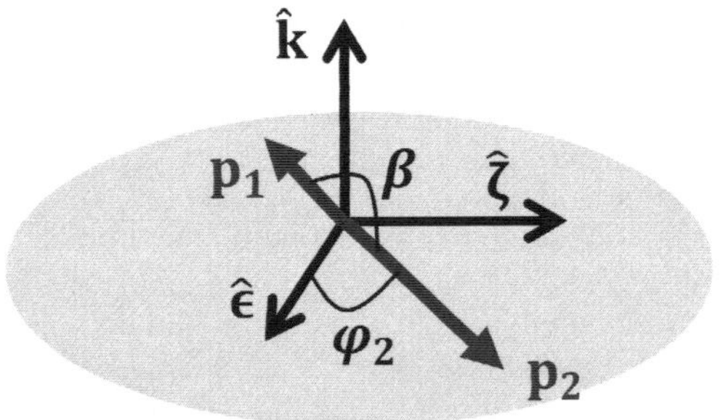

Figure 1. In-plane detection geometry in which $\mathbf{p}_1, \mathbf{p}_2$ are ejected back-to-back (BTB) with unequal energy sharing (UES). This plane is defined by the major ($\hat{\epsilon}$) and minor ($\hat{\zeta}$) axes of the polarization ellipse; it is orthogonal to the laser propagation direction, $\hat{\mathbf{k}}$.

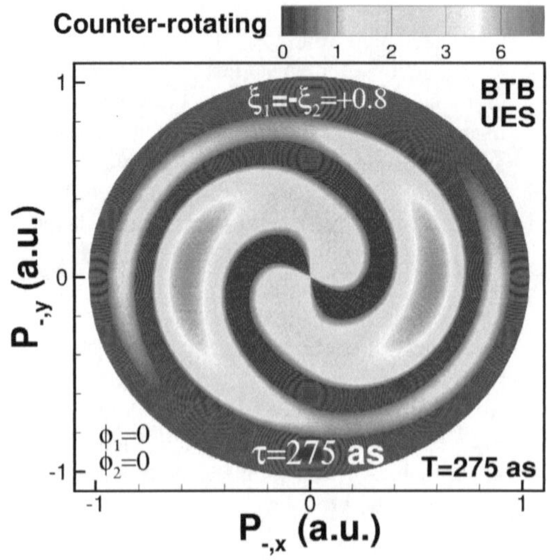

Figure 2. Fermat spiral vortex pattern in double ionization of He by a pair of oppositely elliptically polarized attosecond pulses (with degrees of circular polarization $\xi_1 = -\xi_2 = +0.8$) for the detection geometry in Fig. 1. The vortex pattern is exhibited by the $\mathbf{P}_-$ distribution in the polarization plane, where $\mathbf{P}_- = (\mathbf{p}_1 - \mathbf{p}_2)/2$ defines the relative momentum of the ionized pair of electrons. The two pulses have the same carrier frequency of 90 eV; intensity of 50 TW/cm^2; carrier-envelope phase $\phi_1 = \phi_2 = 0$; and are delayed in time by $\tau = T$, where $T = 275$ as is the total duration of the six-cycle pulses.

Research supported in part by the U.S. Department of Energy, Office of Science, Basic Energy Sciences, Grant No. DE-FG02-96ER14646.

References

[1] J.M. Ngoko Djiokap *et al.* 2015 *Phys. Rev. Lett.* **115** 113004

[2] K.-J. Yuan *et al.* 2016 *Phys. Rev. A* **93** 053425

[3] D. Pengel *et al.* 2017 *Phys. Rev. Lett.* **118** 053003

[4] J.M. Ngoko Djiokap *et al.* 2016 *Phys. Rev. A* **94** 013408

[1]E-mail: marcelngoko@unl.edu
[2]E-mail: astarace1@unl.edu

Attosecond time delays in the valence photoionization of xenon and iodine at energies degenerate with core emissions

Maia Magrakvelidze*[1], Himadri Chakraborty[† 2]

*Department of Physics, University of Mary Washington, Fredericksburg, Virginia 22401, USA
†Department of Natural Sciences, D.L. Hubbard Center for Innovation and Entrepreneurship, Northwest Missouri State University, Maryville, Missouri 64468, USA

Synopsis We perform time-dependent local density functional calculations of the quantum phase and Wigner-Smith time delays in the valence photoionization of Xe versus I atoms at higher energies where coupling with the core emission becomes important. Results for the 5*s* emission are presented.

Study of photoemission dynamics in real time using coherent pump-probe experiments is a key for exploring electron correlations in matter. Such experiments may involve the conventional streaking or the standard reconstruction of attosecond beating by interference of two-photon transitions (RABITT) measurements. Since the additional delay introduced by the probe pulse from the Coulomb-laser coupling can be estimated independently and subtracted from the measured result, the RABITT approach is consistent with the Wigner-Smith (WS) route to determine emission time that involves energy differential of the photoamplitude phase [1].

We compute these quantum phases and resulting WS delays in the photoionization of valence electrons of Xe and I atoms, using a scheme of time-dependent local density approximation (TDLDA) [2]. Analysis of the results with the goal to temporally access details of electron correlations has been carried out.

To identify two spectral features in the valence emission as induced due to its dynamical correlation with the degenerate core (4*d*) channel, we plot the 5*s* cross sections of Xe and I in the top panel of Figure 1. Note these features in each of the curves as (i) a strong shape resonance at lower energies and (ii) a Cooper minimum (CM) at higher energies.

The bottom panel shows the 5*s* time delays for the atoms in the same energy range. Strong negative delays near the resonance indicate accelerated emissions of the electron for both Xe and I. This corroborates to the general expectation of a collective-type resonance environment in which electrons feel transient repulsion as the emission process becomes favored [3]. In contrast, an opposite temporal trend is displayed near the CM region where the delays appear positive. This slower emergence near CM dovetails with the time-behavior earlier seen in the Ar emission near a spectral minimum [2]. Also note that for Xe our result closely agrees with the streaking measurement for the delay of 4*d* relative to 5*s* [4].

The work presents first results for an open-shell atom like I, although modeled in a spherical frame. Robust similarities between Xe and I, we hope, will encourage attosecond measurements on these atoms.

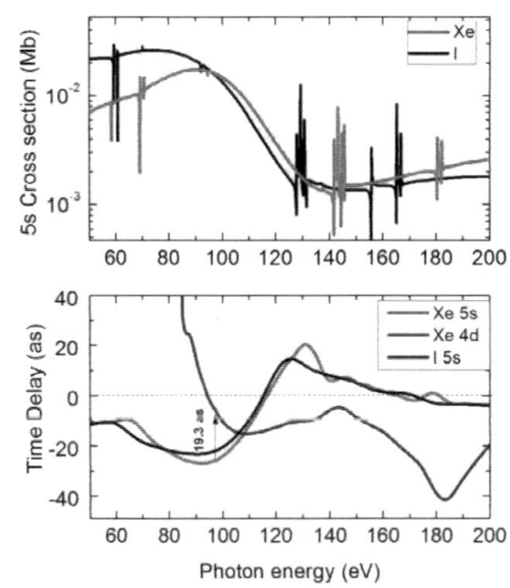

Figure 1. Xe and I 5*s* cross sections (top). WS time delays for Xe and I 5*s*, and Xe 4*d* (bottom). The arrow indicates the measurement from [4].

Supported by the U.S. National Science Foundation research Grant No. PHY1413799.

References

[1] E.P. Wigner, 1955 Phys. Rev. **98**, 145.
[2] M. Magrakvelidze *et al*, 2015 *Phys. Rev. A* **91**, 063415.
[3] M. Magrakvelidze *et al*, 2016 *Phys. Rev. A* **94**, 013429.
[4] A. Verhoef *et al*, CLEO:OSA Tech. Digest 2013 QF2C.4.

[1] E-mail: mmagrakv@umw.edu

[2] E-mail: himadri@nwmissouri.edu

IOP Conf. Series: Journal of Physics: Conf. Series **875** (2017) 022016 doi:10.1088/1742-6596/875/3/022016

Localizing High-Lying Rydberg Wave Packets with Orthogonally-Polarized Two-Color Laser Fields

Seyedreza Larimian*, **Ji-Wei Geng**†, **Stefan Roither***, **Daniil Kartashov***, **Li Zhang***, **Mu-Xue Wang**†, **Qihuang Gong**†, **Liang-You Peng**†, **Christoph Lemell**‡, **Shuhei Yoshida**‡, **Joachim Burgdörfer**‡, **Andrius Baltuška***, **Markus Kitzler***, **Xinhua Xie***,§1

*Photonics Institute, Technische Universität Wien, A-1040 Vienna, Austria
†State Key Laboratory for Mesoscopic Physics and Collaborative Innovation Center of Quantum Matter, School of Physics, Peking University, Beijing 100871, China
‡Institute for Theoretical Physics, Technische Universität Wien, A-1040 Vienna, Austria
§Institute of Theoretical Chemistry, University of Vienna, A-1010 Vienna, Austria

Synopsis We demonstrate controlling over the localization of high-lying Rydberg wave packets in argon atoms with orthogonally-polarized two-color laser fields. Experiments and accompanying semiclassical simulations show clear evidences on asymmetric localization of high-lying Rydberg electrons in argon atoms after the interaction with two-color laser fields.

Highly excited Rydberg atoms and molecules, in comparison with normal atoms and molecules, have unique properties and can be exploited in the studies of the quantum phenomena on human-sized level and the transition from the quantum to the classic world [1]. In a strong laser pulse, valence electrons of an atom or a molecule can be detached through tunnelling or barrier suppression ionization. After conclusion of the pulse, some of the released electrons may be recaptured by the ionic Coulomb potential and populate highly excited Rydberg states [2]. Recently, we demonstrated the detection of such states with the electron-ion coincidence spectroscopy [3].

In this submission, we report on the control of the formation of spatially localized high-lying Rydberg wave packets by waveform controlled orthogonally-polarized two-color (OTC) fields in argon atoms. With the reaction microscope, we measured ionization signals of high-lying Rydberg states of argon atoms induced by a weak dc field (1.5 V/cm) and the black-body radiation as a function of the relative phase between the two-color laser fields. We found that the weak dc field ionization yields of high-lying Rydberg argon atoms oscillate over the relative phase with a period of 2π, while the photoionization signal by the black-body radiation (BBR) with a period of π, as shown in Fig. 1. By performing semi-classical simulations, we show that the populated high-lying Rydberg electron wave packets can be asymmetrically localized around the nucleus after the interaction of an argon atom with a cycle-shaped OTC field. This leads to the 2π periodicity of the weak dc field ionization signal from the high-lying Rydberg states.

Our work provides an effective way to control the population and the localization of high-lying Rydberg wave packets in atoms and molecules, and may find potential applications in quantum information and chemistry.

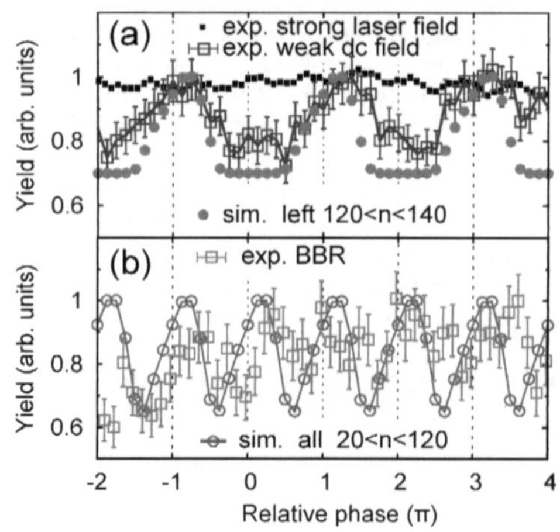

Figure 1. (a) The normalized dc field ionization yield from high-lying Rydberg states and the strong field ionization yield as a function of the relative phase are presented as open blue and filled black squares. Simulated population of corresponding Rydberg states ($120 < n < 140$) at the left (negative) side is indicated by the red circles. (b) The normalized BBR photoionization yield and the simulated population of Rydberg states with $20 < n < 120$ are indicated by green squares and grey circles), respectively.

References

[1] T. F. Gallagher 2005 *Rydberg atoms* (Cambridge University Press).

[2] T. Nubbemeyer, *et al.* 2008 *Phys. Rev. Lett.* **101**, 233001

[3] S. Larimian, *et al.* 2016 *Phys. Rev. A* **94**, 033401

1E-mail: xinhua.xie@univie.ac.at

Sideband and streaking regimes in two-color photoionization by twisted X-waves in strong infrared fields

B. Böning[†1], **W. Paufler**[†] **and S. Fritzsche**[†,♯]

[†] Theoretisch-Physikalisches Institut, Friedrich-Schiller-Universität Jena, D-07743 Jena, Germany
[♯] Helmholtz-Institut Jena, Fröbelstieg 3, D-07743 Jena, Germany

Synopsis The photoionization by short twisted XUV pulses (X-waves) is investigated in the presence of an intense infrared beam within the S-matrix formalism. It is demonstrated how sidebands are affected by the twist of the ionizing pulse as the pulse duration increases.

During the past decade, the generation of attosecond laser pulses enabled experimental studies of electron dynamics in atoms. In particular, the photoionization by XUV pulses in the presence of an additional strong low-frequency laser field has attracted much interest. Such a setup allows, for example, to extract timing information about the ionization process and to characterize the participating laser pulses. Here, we explore this photoionization if the plane-wave XUV pulse is replaced by a twisted X-wave that carries additional orbital angular momentum (OAM).

Detailed studies of the excitation and ionization by twisted light have revealed that selection rules of both bound-bound and bound-free transitions are modified [1, 2]. However, the presence of an additional plane-wave infrared field also affects the continuum states accessible to the photoelectron and hence leads to the formation of sidebands in the recorded spectra [3]. A question of particular interest is how these sidebands are modified and how the streaking regime arises if the duration τ_X of the twisted pulse is shortened to the order of the infrared cycle length T_{IR}. For plane-wave pulses, this transition is displayed in Fig. 1. For $\tau_X < T_{IR}$, the photoelectrons appear at a definite time in the infrared field and are streaked to the same energy. The energy-differential ionization probability therefore exhibits a single peak in the streaking regime (red curve). If τ_X is increased, the interference of photoelectrons entering the infrared field at different times leads to the formation of sidebands ($\tau_X > T_{IR}$, blue curve). It was recently demonstrated that the angular distribution of photoelectrons in the transition regime provides information about the initial atomic state [4].

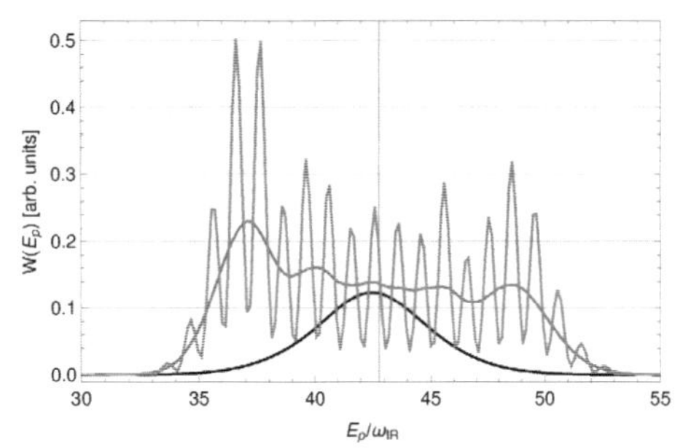

Figure 1. Energy-differential ionization probability $W(E_p)$ for three durations τ_X of the ionizing XUV pulse with photon energy $\omega_X = 80$ eV. Red curve: sideband regime for $\tau_X = 7$ fs; black curve: streaking regime for $\tau_X = 1$ fs; blue curve: transition for intermediate $\tau_X = 3$ fs. The central black line marks the position of the XUV photoline without infrared field. Further laser parameters are $\lambda_{IR} = 800$ nm, $I_{IR} = 10^{13}$ W/cm^2 and $I_X = 10^9$ W/cm^2. Black and blue curves are magnified by a factor of 2.

In this contribution, we discuss how the two-color ionization process is affected by the OAM of the X-wave. More specifically, we investigate the streaking and sideband regimes and the transition from one to the other with increasing X-wave pulse duration.

References

[1] O. Matula *et al* 2013 *J. Phys. B* **46** 205002

[2] M. Scholz–Marggraf *et al* 2014 *Phys. Rev. A* **90** 013425

[3] D. Seipt *et al* 2015 *Phys. Rev. A* **94** 053420

[4] A. Kazansky *et al* 2010 *Phys. Rev. A* **82** 033420

[1]E-mail: birger.boening@uni-jena.de

IOP Conf. Series: Journal of Physics: Conf. Series **875** (2017) 022018 doi:10.1088/1742-6596/875/3/022018

Intracycle interference in laser assisted XUV atomic hydrogen ionization

A. A. Gramajo *[1], **R. Della Picca** *, **R. C. Garibotti** *, **and D. G. Arbó** †

*Centro Atómico Bariloche, Av. Exequiel Bustillo 9500, 8400 Bariloche, Argentina
† Institute for Astronomy and Space Physics - IAFE (UBA-Conicet), Buenos Aires, Argentina

Synopsis We present a theoretical study of ionization of the hydrogen atom due to an XUV pulse in the presence of an IR laser with both electric fields linearly polarized in the same direction. By means of a very simple semiclassical model we explain the photoelectron (PE) spectrum as the interplay of *intra-* and *intercycle* interferences.

Laser assisted photoemission (LAPE) corresponds to two-color multiphoton ionization, where one of the fields has high frequency compared to the other. Depending on the pulse durations, two regimes can be distinguished: (i) the XUV pulse is much shorter than the IR period, i.e., streak camera [1], and (ii) the XUV pulse is longer than the laser period [2], with the ensuing formation sidebands (SBs) due to the absorption/emission of additional laser photons.

In this work we focus on the second regime for atomic hydrogen photoionizaton. Within the semiclassical model [3], the electron spectrum can be explained in terms of *intracycle* interferences stemming from trajectories within the same optical cycle and *intercycle* interference stemming from trajectories released at different cycles. The PE espectra are characterized by peaks at energy values given by

$$E_n = n\omega_L + \omega_X - I_p - U_p \tag{1}$$

modulated by the intracycle interference pattern. In Eq. (1), $\omega_{X(L)}$ is the frecuency of XUV(IR) pulse, I_p is the ionization potential, U_p the ponderomotive energy and n is the number of absorbed / emitted IR photons.

From the Fig. 1(a) for emission parallel to the polarization axis, the *intercycle* interference arises as the coherent superposition of the electron trajectories born at different optical cycles giving rise to SBs equally spaced by ω_L. Besides, the *intracycle* interference arises as the coherent superposition of the two electron trajectories released in the same optical cycle and giving rise to the modulation of the SBs. For emission perpendicular to the polarization axis, the intracycle interference pattern can also be thought as the interference between electron trajectories born at each half cycles of the same optical cycle (*intrahalfcycle*) and interference between trajectories from the first and the second half cycle of the laser (*interhalfcycle*). The destructive *interhalfcycle* interference explains the characteristic SBs separated by twice the IR photon energy, as we can see in Fig. 1(b). Our calculations are in good agreement with quantum results within the strong field approximations and the numerical solution of the time-dependent Schrödinger equation [3].

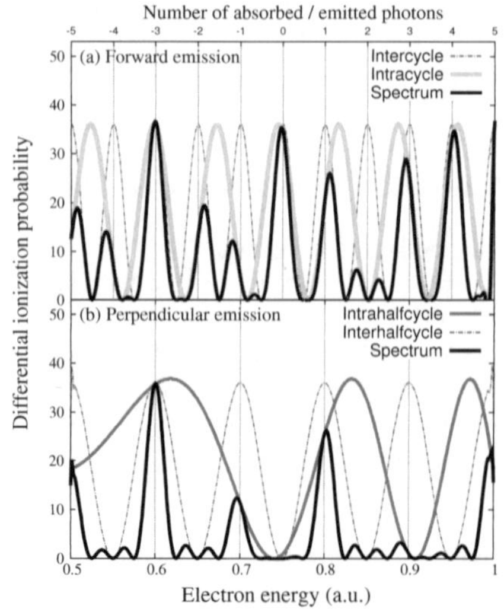

Figure 1. Photoelectron spectra in (a) the forward and (b) the perpendicular directions (black line). The color line depicts the intra- and intercycle interferences. Vertical lines exhibits the positions of the SBs according to Eq. (1). The IR laser amplitude is $F_{L0} = 0.05$ and $\omega_L = 0.05$. $\omega_X = 1.5$ and duration of the XUV is two IR cycles

References

[1] J. Itatani *et al*, *Phys. Rev. Lett.* **88**, 173903 (2002).
[2] V. Véniard *et al*, *Phys. Rev. Lett.* **74**, 4161 (1995).
[3] A. A. Gramajo *et al*, *Phys. Rev. A* **94**, 053404 (2016).

[1]E-mail: gramajo.anaalicia@cab.cnea.gov.ar

IOP Conf. Series: Journal of Physics: Conf. Series **875** (2017) 022019 doi:10.1088/1742-6596/875/3/022019

Non-perturbative semiclassical model for strong-field ionization

N. I. Shvetsov-Shilovski [*, 1], **M. Lein**[*], **L. B. Madsen** [†], **E. Räsänen**[‡], **C. Lemell** [§], **J. Burgdörfer** [§], **D. G. Arbó** [¶, 2], **and K. Tőkési** [&,$]

[*] Leibniz Universität Hannover, Hannover, Germany, EU
[†] Department of Physics and Astronomy, Aarhus University, Århus C, Denmark, EU
[‡] Tampere University of Technology, Tampere, Finland, EU
[§] Vienna University of Technology, Vienna, Austria, EU
[¶] Institute for Astronomy and Space Physics (UBA-Conicet), Buenos Aires, Argentina
[&] Institute for Nuclear Research, Hungarian Academy of Sciences, Debrecen, Hungary, EU
[$] ELI-ALPS, ELI-HU Non-profit Ltd., Dugonics tér 13, H-6720 Szeged, Hungary, EU

Synopsis We present a non-perturbative semiclassical model for strong-field ionization that accounts for path interferences of tunnel-ionized electrons in the ionic potential within the framework of a classical trajectory Monte-Carlo representation of the phase-space dynamics.

Above-threshold ionization, high harmonic generation, and non-sequential double ionization are the most well-known examples of strong-field physics concerning with highly nonlinear phenomena originating from the interaction of strong laser radiation with atoms and molecules. In their original formulation, semiclassical models do not take into account the effect of the Coulomb potential of the parent ion on the electron motion after ionization. In turn, the Coulomb-corrected strong field approximation (CCSFA) and the quantum trajectory Monte Carlo method (QTMC) invoke first-order perturbation theory to include the Coulomb potential.

We formulate a semiclassical two-step (SCTS) model based on the theory of semiclassical time-dependent propagators [1] that accounts for the Coulomb potential beyond the perturbation theory. We derive a semiclassical expression for the transition amplitude for strong-field ionization that differs from the QTMC [2] and CCSFA [3] models improving the agreement with quantum simulations.

The figure 1 shows the photoelectron momentum distribution for H atom. Our SCTS result qualitatively resembles the results of the direct numerical solution of the time dependent Schrödinger equation (TDSE). However, we found that the low-energy spectrum shows marked deviations in the bouquet-shaped interference structure, where the number of radial nodal lines is controlled by the dominant partial-wave angular momentum [4]. Whereas the SCTS model closely matches the nodal pattern of the TDSE, the QTMC model yields fewer nodal lines, which is a direct consequence of the underestimate of the Coulomb interactions in the QTMC treatment of the interference phase.

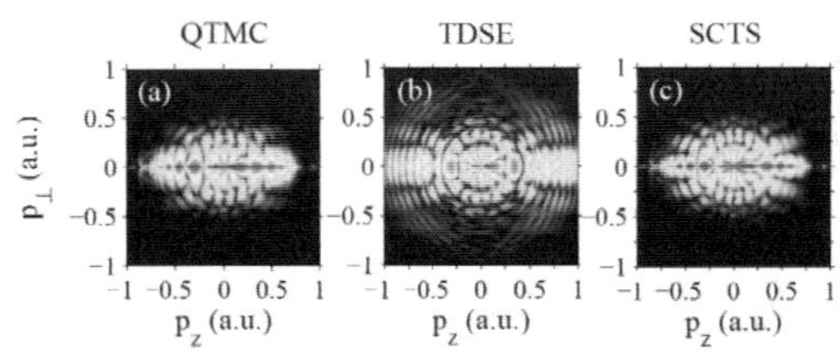

Figure 1. Vectorial momentum distributions for the H atom ionized by a laser pulse with a duration of $n = 8$ cycles, wavelength of $\lambda = 800$ nm, and peak intensity of $I = 9 \times 10^{13}$ W/cm^2 obtained from (a) the QTMC model, (b) solution of the TDSE, and (c) the present SCTS model. The laser field is linearly polarized along the z-axis.

Work supported by the European Community's FP7 through the CRONOS Project No. 280879, The Academy of Finland Project No. 267686, the Nordic Innovation Top-Level Research Initiative Project No. P-13053, the COST Action CM-1204, the STSM Grant COST Action CM-1204, ERC-StG (Project No. 277767, TDMET), the VKR Center of Excellence QUSCOPE, FWF-SFB049 NextLite, CONICET Argentina PIP0386, ANPCyT Argentina PICT-2014-2363, and UBACyT0617BA.

References

[1] D. J. Tannor, *Introduction to Quantum Mechanics. A Time-Dependent Perspective* (University Science Books, Sausalito, 2007).
[2] N. I. Shvetsov-Shilovski *et al.*, Phys. Rev. A **94**, 013415 (2016).
[3] M. Li *et al.*, Phys. Rev. Lett. **112**, 113002 (2014).
[4] D. G. Arbó *et al.*, Phys. Rev. Lett. **96**, 143003 (2006).

[1] E-mail: n79@narod.ru
[2] E-mail: diego@iafe.uba.ar

IOP Conf. Series: Journal of Physics: Conf. Series **875** (2017) 022020 doi:10.1088/1742-6596/875/3/022020

Intensity-dependent shift in transverse electron momentum distribution for strong field ionization

Nida Haram*, Han Xu*, Atia-tul-Noor*, Satya Sainadh Undurti*, Igor Ivanov§, Igor Litvinyuk*, R.T. Sang*

*Centre for Quantum Dynamics and Australian Attosecond Science Facility, Griffith University, Nathan, Queensland 4111, Australia

§Centre for Relativistic Laser Science, Institute for Basic Science, Gwangju-500-712, Republic of Korea

Synopsis We present an intensity dependent study of transverse electron momentum distribution obtained from strong field ionization of argon using linearly polarized few-cycle (6 fs) laser pulses in the intensity range 10^{15}W/cm^2-10^{16}W/cm^2.

In order to explain the correlated many-particle dynamic in atoms and molecules, Cold Target Recoil Ion Momentum Spectroscopy (COLTRIMS) provides a kinematically complete picture of fragmentation process with high momentum resolution and detection efficiency [1,2]. In this project, we are interested in investigating the influence of laser intensity on transverse electron momentum distribution (TEMD) with COLTRIMS.

TEMD describes the photoelectron momenta in the direction perpendicular to the polarization plane of the laser pulse. The effect of varying ellipticity on TEMD in strong field atomic ionisation was studied by our group earlier [3]. In this work we investigate the effect of photon momentum as a function of laser intensity on the TEMD by examining the peak shifts with reference to zero transverse momentum i.e. $P_\perp=0$ using linearly polarized few-cycle laser pulses. Figure 1 shows the projection of three dimensional electron momentum distribution onto a two-dimensional imaging detector. The detector plane contains the axis of gas jet (x) and laser propagation direction (y), whereas (z) corresponds to time-of-flight axis.

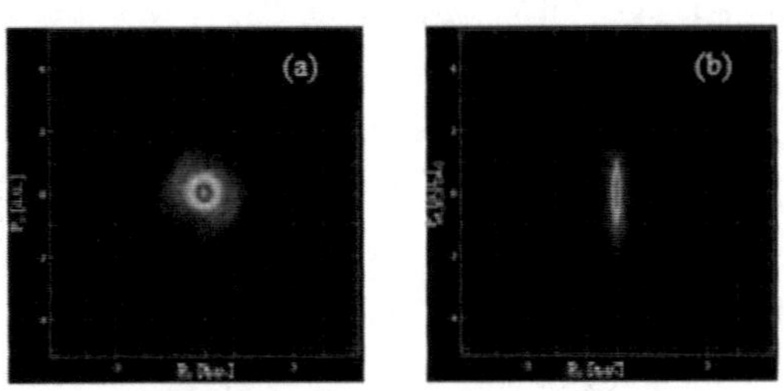

Figure 1. Electron momentum spectra in (a) xy and (b) xz plane at 800nm, 1.7x10^{15}W/cm^2

For a linearly polarized laser beam this distribution exhibits a cusp-like peak at zero momentum due to Coulomb focusing effect at $P_\perp=0$ — a long-range Coulomb interaction between emitted electron and its parents ion [4,5]. The shifts of the central peaks have been observed experimentally at intensities on the order of 10^{13}W/cm^2 in both directions; parallel as well as anti-parallel to the photon momentum direction; due to the complex contribution of Coulomb attraction from ion and magnetic Lorentz force [5,6].

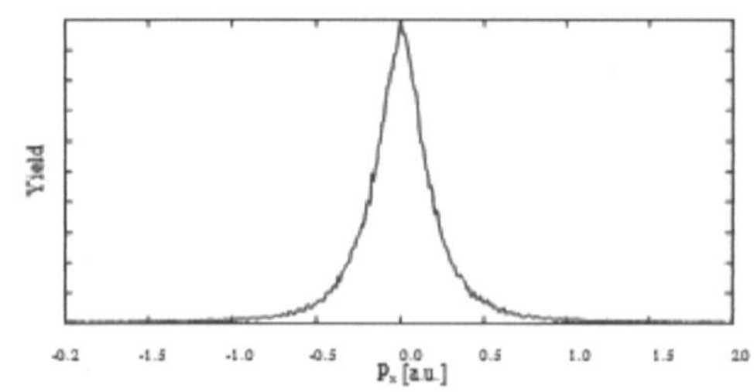

Figure 2. Projection of electron momentum distribution on laser propagation direction at 1.7x10^{15}W/cm^2

Experimentally, we observed that the peak of transverse electron momentum distribution along beam propagation direction is shifting with increased intensity. We compare our measurements with *ab initio* fully relativistic simulations of Ar ionisation. This study might help to resolve the questions related to time-resolved atomic and molecular holography, as the influence of photon momentum is noticeable in the holographic patterns [7].

References

[1] Ullrich J et al 2003 Rep. Prog. Phys. 66 1463
[2] Dörner R et al 2000 Phys. Rep. 330 95
[3] Ivanov I A et al 2016 Sci. Rep. 6 19002
[4] Ivanov I A 2015 Phys. Rev. A. 92 63417
[5] Ludwig A et al 2014 Phys. Rev. Lett. 24 243001
[6] Rudenko A et al 2005 J. Phys. B 38 L191
[7] Chelkowski S et al 2015 Phys. Rev. A 92 51401(R)

[1] E-mail: R.Sang@griffith.edu.au

IOP Conf. Series: Journal of Physics: Conf. Series **875** (2017) 022021 doi:10.1088/1742-6596/875/3/022021

Photoionization Time delay in atomic Barium

Aarthi Ganesan* [1], **Pranawa C Deshmukh**[†#2], **Anatoli S Kheifets**[‡], **Valeriy K Dolmatov**[§], **and Steven T Manson**[¶]

[*] Department of Physics, CPGS, Jain University, Bengaluru 560011, India
[†] Department of Physics, Indian Institute of Technology Tirupati, Tirupati 517506, India
[#] Department of Physics, Indian Institute of Science Education and Research, Tirupati 517507, India
[‡] Research School of Physics and Engineering, The Australian National University, Canberra ACT 0200, Australia
[§] Department of Physics and Earth Science, University of Northern Alabama, Florence, Alabama 35632, USA
[¶] Department of Physics and Astronomy, Georgia State University, Atlanta, GA 30301, USA

Synopsis Wigner-Eisenbud-Smith (WES) photoionization time delay is calculated for atomic barium 5s subshell using Relativistic Random Phase Approximation (RRPA), RRPA *with* relaxation (RRPA-R) and the Relativistic Multi-Configuration Tamm-Dancoff (RMCTD) techniques. Electron correlations are treated differently in different many body methods; the WES time delay is sensitive to these differences. The present study is aimed at determining the suitability of various many-body methods for the study of the WES time-delay.

The time-delay in electron ejection into the continuum after photon absorption is of much interest [1-5]. Photoionization time-delay provides valuable dynamical information about electron correlation and relativistic effects in a many-electron system.

In the present work photoionization time delay is calculated for the first time for atomic barium. The high Z value of atomic barium requires careful treatment of relativistic and electron correlation effects. In the present work, the Wigner-Eisenbud-Smith (WES) time delay [6] for Ba5s subshell is calculated in the region of second Cooper minimum [7] using the (i) RRPA [8], (ii) and the RRPA-R [9] and (iii) the RMCTD approximation [10].

The results from the RRPA method are shown in the adjacent figure. The photoionization cross section of Ba5s undergoes a 'correlational Cooper minimum' well above the ionization threshold [7] (Fig. 1). The photoionization time delay in the relativistic $5s \rightarrow p_{1/2}$ and $5s \rightarrow p_{3/2}$ channels shows a dip ('time advancement') in the region of the Cooper minimum (Fig. 1). The time-delay calculated from RRPA-R and RMCTD are also determined (not shown) to understand the influence of different electron correlations from different many-body theories on the attosecond time delay.

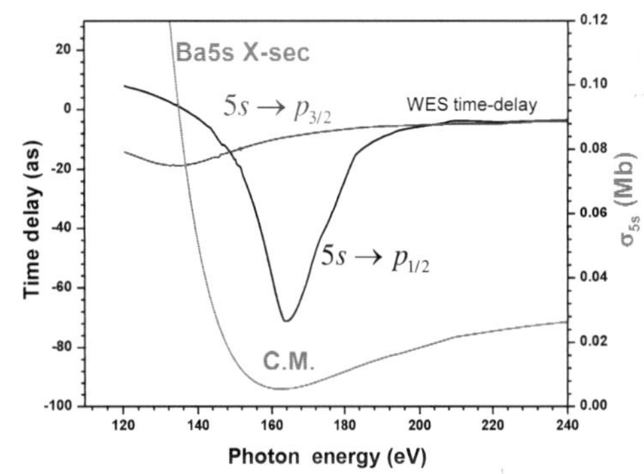

Figure 1: RRPA results for photoionization time delay in Ba5s channels in the region of second Cooper minimum. Black line: Time-delay in $5s \rightarrow p_{1/2}$ channel; Blue line: Time-delay in $5s \rightarrow p_{3/2}$ channel; Red line: cross-section.

References

[1] M. Schultze *et al.* 2010 *Science* **328** 1658
[2] A. S. Kheifets and I. A. Ivanov 2010 *Phys. Rev. Lett.* **105** 233002
[3] A. S. Kheifets 2013 *Phys. Rev. A* **87** 063404
[4] P. C. Deshmukh *et al.* 2014 *Phys. Rev. A* **89** 053424
[5] S. Saha *et al.* 2014 *Phys. Rev. A* **90** 053406
[6] E.P. Wigner 1955 *Phys. Rev.* **98** 145
[7] G. Aarthi *et al.* 2013 *J. Phys B* **46** 185002
[8] W. R. Johnson and C. D. Lin 1979 *Phys. Rev. A* **20** 964
[9] V. Radojevic *et al.* 1989 *Phys. Rev. A* **40** 727
[10] V.Radojevic and W. R. Johnson 1985 *Phys. Rev. A* **31** 2991

[1] E-mail: aarthihari07@gmail.com
[#2] E-mail: pcd@iittp. ac.in

IOP Conf. Series: Journal of Physics: Conf. Series **875** (2017) 022022 doi:10.1088/1742-6596/875/3/022022

Confinement effects and angular dependence of Wigner-Eisenbud-Smith time delay

A. Mandal[*,1], **P. C. Deshmukh**[†,#,2], **V. K. Dolmatov**[‡,3], **A. Kheifets**[§,4], **and S. T. Manson**[¶,5]

[*] Department of Physics, Indian Institute of Technology Madras, Chennai, 600036, India
[†] Department of Physics, Indian Institute of Technology Tirupati, Tirupati, 517506, India
[#] Department of Physics, Indian Institute of Science Education and Research Tirupati, Tirupati 517507, India
[‡] Department of Physics and Earth Science, University of North Alabama, Florence, AL 35632, USA
[§] Research School of Physics and Engineering, The Australian National University, Canberra ACT 0200, Australia
[¶] Department of Physics and Astronomy, Georgia State University, Atlanta, 30303, USA

Synopsis The effect of confinement on the angle dependent Wigner-Eisenbud-Smith time delay in photoionization of atomic xenon is studied. A finite spherical annular well potential is used to model the nearly spherical confinement due to fullerene C_{60} molecular cage. The confinement is seen to have a significant effect on the time delay.

Estimation of Wigner-Eisenbud-Smith (WES) time delay in atomic photo-transitions has received much attention in recent years [1]. Some recent studies reveal the effect of confinement on the time delay [2]. The interference of different photoionization channels play an important role in determining the time delay and also determines the angle dependence of WES time delay (except for *ns* photoionization) [3].

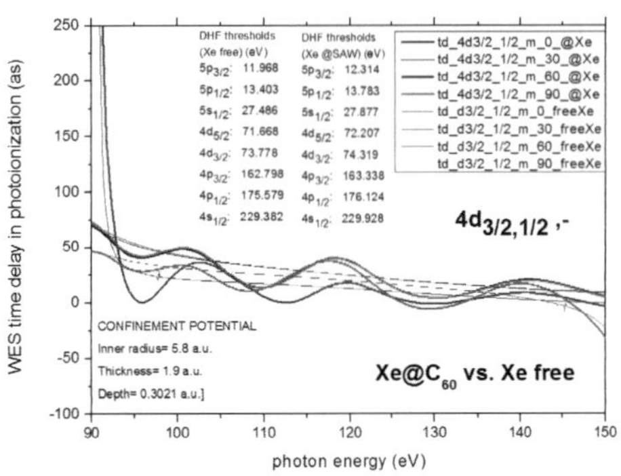

Figure 1. WES time delay for dipole transitions from $4d_{3/2,1/2}$ to the spin *down* final state for confined Xe (bold lines); thin lines are for free Xe.

Based on the formalism of [4] and [3] we study the WES time delay for free and confined Xe atom for the photoionization from $4d_{3/2}$ and $4d_{5/2}$ subshells at various angles. Illustrative results are shown in Fig. 1 where we have plotted the WES time delay in the dipole photoionization channel from the $4d_{3/2,1/2}$ level to the final state with $\nu = -1/2$ (spin down). Since measuring the initial m-states and the polarization of the emerging photoelectron is difficult, the weighted average over all the initial states with different m quantum numbers and sum over photoelectron spin states has been calculated and is shown in Fig. 2. The results of both the individual channels and the weighted average show significant effects of the confinement on the oscillation of the time delay about the results for the free atom. More details concerning the individual photoionization amplitudes could be obtained if the photoelectron spin states were detected.

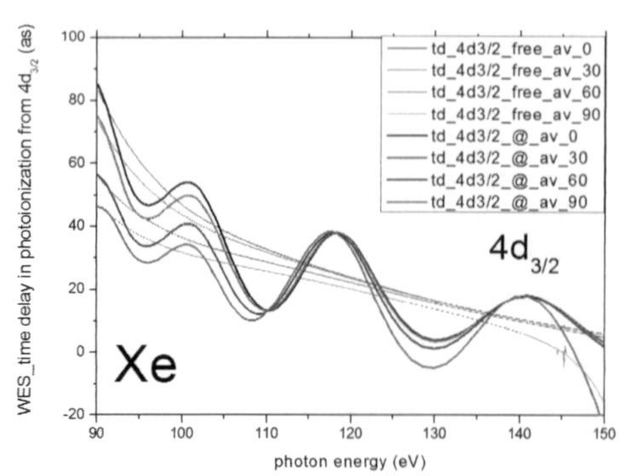

Figure 2. $4d_{3/2}$ time delay, averaged over initial and summed over final states. Bold lines are the WES time delay for confined Xe and thin lines are for free Xe.

References

[1] R. Pazourek *et al.* 2015 *Rev. Mod. Phys.* **87** 765

[2] P. C. Deshmukh *et al.* 2014 *Phys. Rev. A* **89** 053424

[3] A. Kheifets *et al.* 2016 *Phys. Rev. A* **94** 013423

[4] W. R. Johnson *et al.* 1979 *Phys. Rev. A* **20** 964

[1]E-mail: amankur@physics.iitm.ac.in
[2]E-mail: pcd@iittp.ac.in
[3]E-mail: vkdolmatov@una.edu
[4]E-mail: a.kheifets@anu.edu.au
[5]E-mail: smanson@gsu.edu

IOP Conf. Series: Journal of Physics: Conf. Series **875** (2017) 022023 doi:10.1088/1742-6596/875/3/022023

Polarization effects in above-threshold ionization of Mg with a mid-infrared strong laser field

Huipeng Kang[*,§ 1], **SongPo Xu**[*,†], **YanLan Wang**[*,†], **XuanYang Lai**[*], **Thomas Pfeifer**[‡], **XiaoJun Liu**[* 2], **Jing Chen**[¶ 3], **Ya Cheng**[&], **and ZhiZhan Xu**[&]

[*] State Key Laboratory of Magnetic Resonance and Atomic and Molecular Physics, Wuhan Institute of Physics and Mathematics, Chinese Academy of Sciences, Wuhan 430071 , China
[†] University of Chinese Academy of Sciences, Beijing 100049, China
[‡] Max-Planck-Institut für Kernphysik, 69117 Heidelberg, Germany
[§] Institut für Kernphysik, Goethe Universität Frankfurt, 60438 Frankfurt am Main, Germany
[¶] HEDPS, Center for Applied Physics and Technology, Peking University, Beijing 100084, China and Institute of Applied Physics and computational Mathematics, P. O. Box 8009, Beijing 100088, China
[&] State Key Laboratory of High Field Laser Physics, Shanghai Institute of Optics and Fine Mechanics, Chinese Academy of Sciences, P. O. Box 800-211, Shanghai 201800, China

Synopsis Using a semiclassical approach, we find an enhancement of low-energy structure in above-threshold ionization spectra of Magnesium by intense, mid-infrared laser pulses. By performing electron trajectory and recollision-time distribution analysis, we demonstrate this enhancement results from a focusing of electrons by laser induced dynamical polarization.

Along with the advances in generation of tunable long-wavelength intense laser pulses, experimental studies revealed novel features of above-threshold ionization (ATI) electrons deeply in the tunneling ionization regime [1]. One is the surprising bump appearing in the low-energy part of photoelectron energy spectra (PES), coined as "low-energy structure" (LES) [1, 2]. It's well known now that the LES can be attributed to the Coulomb focusing of outgoing electrons by the parent ion. However, so far our understanding relies on the single-active electron (SAE) approximation. In this contribution, we theoretically investigate the ATI of Magnesium with mid-infrared (~2000 nm) strong laser fields using a semiclassical approach [3], with the purpose of studying multielectron effects beyond the SAE picture.

Figure 1(a) displays our calculated PES of Mg at an intensity of 3×10^{13} W/cm^2. In addtion, we also calculated the corresponding spectrum based on SAE approximation in Fig. 1(b). It is noteworthy that the relative yield of the LES is clearly enhanced when including the multielectron effects, irrespective of the decreased absolute counts. In Figs. 1(c) and 1(d), we present the normalized angular distributions of the electrons that belong to the energy range of LES in Figs. 1(a) and 1(b), respectively. The calculated results show that the distribution becomes narrow with the multielectron effects taken into account, implying additional ionic focusing effect on LES other than Coulomb focusing effect.

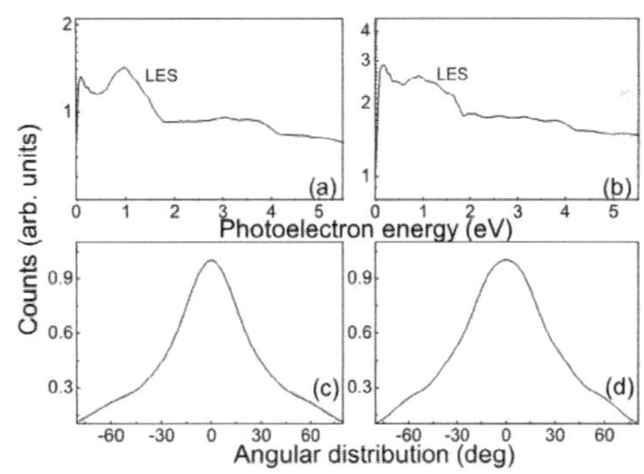

Figure 1. (a) Calculated PES of Mg at an intensity of 3×10^{13} W/cm^2 and a wavelength of 2000 nm. (b) Corresponding spectrum based on SAE approximation. The laser parameters are the same as in (a). (c), (d) Normalized angular distributions of electrons that contribute to the energy range of LES in (a) and (b), respectively.

To uncover the underlying mechanism, we analyze the recollision-time distribution and electron trajectory for the SAE and the multielectron case. We find the confining electrons by the dynamical dipole potential facilitate the recollision process, and thus the production of LES.

References

[1] C. I. Blaga *et al.* 2009 *Nat. Phys.* **5** 335
[2] W. Quan *et al.* 2009 *Phys. Rev. Lett.* **103** 093001
[3] N. I. Shvetsov-Shilovski *et al.* 2012 *Phys. Rev. A* **85** 023428

[1] E-mail: kang@atom.uni-frankfurt.de
[2] E-mail: xjliu@wipm.ac.cn
[3] E-mail: chenjing@iapcm.ac.cn

IOP Conf. Series: Journal of Physics: Conf. Series **875** (2017) 022024 doi:10.1088/1742-6596/875/3/022024

Exploration of spectral and temporal fine structures of high-order harmonic generation[1]

Peng-Cheng Li*2, Xiong-Yuan Lei*, Xiao-Xin Zhou*3, Shih-I Chu†4

* College of Physics and Electronic Engineering, Northwest Normal University, Lanzhou 730070, China
† Department of Chemistry, University of Kansas, Lawrence, Kansas 66045, USA

Synopsis We present an *ab initio* precision investigation of the time-frequency characteristics of high-order harmonic generation (HHG) calculated by solving accurately and efficiently the time-dependent Schrödinger equation by means of the time-dependent generalized pseudospectral method. We extend a new synchrosqueezing transform (SST) to obtain the time-frequency spectra of the HHG and the dynamical phase associated with the dipole-emission time profile. The SST time-frequency analysis allows us to explore the in depth dynamics and contributions of the multi-rescattering trajectories and resonant trajectories in HHG. It also exhibits the subtle details of the spectral and temporal fine structures of the HHG, providing novel insights regarding the dynamical origin of the HHG in below- and above-threshold regimes.

In the last several years, the study of high-order harmonic generation (HHG) is one of the most rapidly developing topics in strong-field physics [1]. Particularly, the HHG has been experimentally demonstrated to be a promising method to produce a unique source of coherent light in the extreme ultraviolet (XUV) and soft x-ray regions of the spectra, as well as for the generation of ultrashort attosecond laser pulse. The production of attosecond pulse represents a very important tool for the detection and control of the electronic dynamic behaviors [2]. Although the semiclassical three-step model can obtain the simple physical picture of the HHG, to account for the dynamical origin of HHG and attosecond pulse generation, it is necessary and valuable to also explore the spectral and temporal structures of the HHG from an *ab initio* simulations. This can facilitate the further experimental development of metrology for optimal control of attosecond pulse generation.

Recently, we have extended the synchrosqueezing transform (SST) method newly developed to explore the dynamical origin of the near- and below-threshold HHG simulated from an *ab initio* calculations [3] . By comparing with several other time-frequency transform methods, such as wavelet transform, the Gabor transform, and the Wigner-Ville transform, we find that the SST allows us to probe more deeply the delicate spectral characteristics in HHG, particularly for near- and below-threshold harmonics.

Figure 1 we present the SST time-frequency analysis of the HHG. For the comparison, the semiclassical simulation is also shown. It is clear that not only the semiclassical trajectories of the first returning are matched very well, but also the multi-rescattering trajectories are in good agreement with the SST time-frequency spectra.

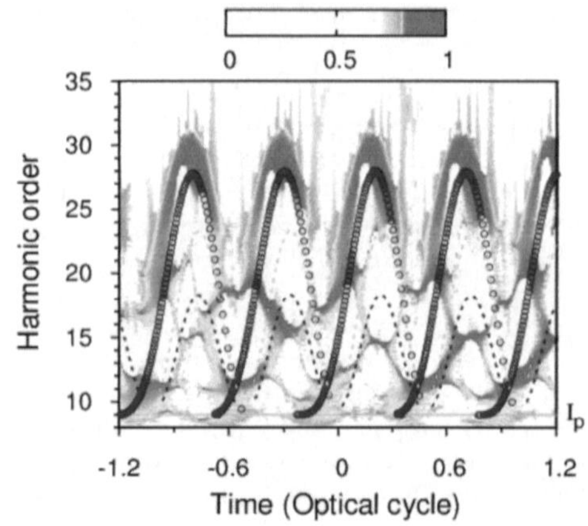

Figure 1. SST time-frequency spectra of above-threshold HHG in a 800-nm, 40 optical cycles, cosine-squared laser pulse with peak intensity $I = 1.5 \times 10^{14}$W/cm^2. The blue circles and the dashed lines indicate the semiclassical trajectories of the first returning and multiple returning, respectively, and the green vertical line indicates the corresponding ionization threshold marked by I_p.

This work is partially supported by DOE. P. C. Li. and X. X. Zhou is partially supported by National Natural Science Foundation.

References

[1] F. Krausz *et al.* 2015 *Rev. Mod. Phys.* **81** 163-234
[2] C. Ott *et al.* 2014 *Nature* **516** 374-378
[3] P. C. Li *et al.* 2015 *Nature Communications* **6** 7178

[2]E-mail: lipc@nwnu.edu.cn
[3]E-mail: zhouxx@nwnu.edu.cn
[4]E-mail: sichu@ku.edu

Study of angular dependence of photoionization time delay in $(n-1)d$ channels for Zn, Cd and Hg

S. Banerjee[*1], **P. C. Deshmukh**[†,#2], **A. Kheifets**[‡3], **V. K. Dolmatov**[§4], **and S. T. Manson**[¶5],

[*] Department of Physics, Indian Institute of Technology Madras, Chennai, 600036, India
[†] Department of Physics, Indian Institute of Technology Tirupati, Tirupati, 517506, India
[#] Department of Physics, Indian Institute of Science Education and Research, Tirupati, 517507, India
[‡] Research School of Physics and Engineering, The Australian National University, Canberra ACT 0200, Australia
[§] Department of Earth Science, University of North Alabama, Florence, AL 35632, USA
[¶] Department of Physics and Astronomy, Georgia State University, Atlanta, 30303, USA

Synopsis The angular dependence of Wigner-Eisenbud-Smith photoionization time delay has been investigated for the $nd \to \varepsilon f$ dipole channels in atomic Zn, Cd and Hg using the relativistic random phase approximation (RRPA). The studies have been carried out also in the nonrelativistic limit. This technique enables identification of the relativistic effects on the angular dependence of photoionization time delay.

Photoionization of $(n-1)d$ subshells of atomic Zn, Cd and Hg (which all have outer electron configuration $(n-1)d^{10}ns^2$) has been studied previously using the relativistic random phase approximation (RRPA) formalism [1, 2]. The Wigner-Eisenbud-Smith (WES) photoionization time delay [3, 4, 5, 6] results for photoionization of atomic Zn and Cd are also reported [7]. These calculations are extended in the present study to (a) include Hg, for which relativistic effects are expected to be significant, being a high-Z atom with similar electronic configuration as Zn and Cd (b) carry out the studies in the RRPA and also in the nonrelativistic limit of the RRPA (NRL-RRPA) in order to pinpoint the effects of relativistic interactions, and (c) study angular dependence of the WES time-delay. The angle-resolved studies of the WES time delay have been carried out using the formulation described in Reference [8].

The results show the evolving importance of relativistic interactions, with increasing Z, on both the total WES time delay as well as the angular distribution of the time delay. Also, the interplay between relativistic and many-body interactions is highlighted. Results in non-relativistic limit are not shown in the figures.

There is of course no Cooper minimum in the $3d$ photoionization of Zn, but significant effects are seen at the Cooper minimum in the case of Cd and Hg, with the relativistic effects most dominant for Hg.

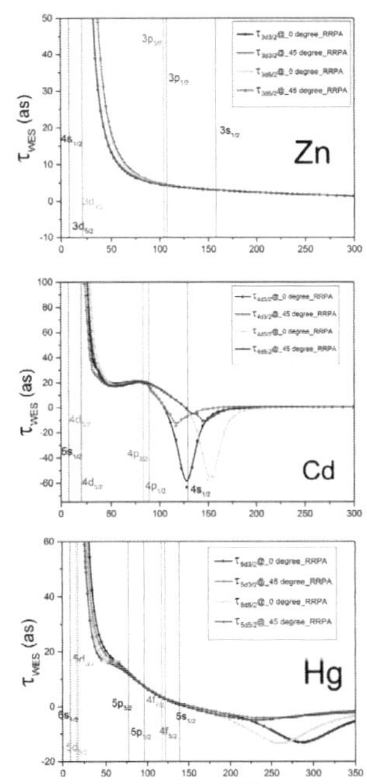

Figure 1. WES $(n-1)d$ photoionization time-delay in Zn, Cd, Hg

References

[1] W. R. Johnson *et al.* 1982 *Phys. Rev. A* **25** 337

[2] W. R. Johnson and C. D. Lin 1979 *Phys. Rev. A* **20** 964

[3] M. Schultze *et al.* 2010 *Science* **328** 1658

[4] L. E. Eisenbud 1948 *Ph. D. thesis, Princeton Univ.*

[5] E. P. Wigner 1955 *Phys. Rev.* **98** 145

[6] F. T. Smith *et al.* 1960 *Phys. Rev.* **118** 349

[7] S. Saha *et al.* 2015 *J. Phys. Conf. Ser* **635** 092061

[8] A. Kheifets *et al.* 2016 *Phys. Rev. A* **94**(1) 013423

[1]E-mail: souravbanrj444@gmail.com
[2]E-mail: pcdeshmukh@gmail.com
[3]E-mail: a.kheifets@anu.edu.au
[4]E-mail: vkdolmatov@una.edu
[5]E-mail: smanson@gsu.edu

Improving harmonic conversion efficiency via optimizing macroscopic properties of the gas medium

Xiaoyong Li* [1], **Guoli Wang**†, **Juan Fan***, **Jinyu Ma***, **and Yawen Jiang***

* Experimental Center, Northwest University for Nationalities, Lanzhou, 730030, China
† College of Physics and Electronic Engineering, Northwest Normal University, Lanzhou, 730070, China

Synopsis We study the effects of length and pressure of gas jet on the conversion efficiency of macroscopic harmonics. Our results show that an optimal flux for certain range of harmonics can be obtained by optimizing macroscopic properties of the gas jet.

High-order harmonic generation (HHG) has been considered as one of the best methods to obtain an ultra-short coherent light source in the extreme ultraviolet (XUV) and soft x-ray regions. It can also produce attosecond pulse, which provides a powerful tool to reveal ultrafast dynamics of the electron in the atoms and molecules.

However, for HHG applications, there are still some drawbacks[1]. One of the main disadvantages is its low conversion efficiency. In essence, the conversion efficiency depends not only on the parameters of laser field, but also on the macroscopic properties of the gas jet[2]. So the HHG conversion efficiency is not necessarily enhanced monotonously by increasing the pressure and the length of the gas jet.

In this paper, we want to find out whether there is a highest conversion efficiency for certain range of harmonics. Figures 1 and 2 show the conversion efficiency for harmonics of Ne atoms in the plateau region (H30-H45) and cut off region (H45-H55), respectively. It turns out that the conversion efficiency goes up as the pressure is increasing, but it goes down when it climb over the top as the pressure is getting higher for each length of the gas jet. For a fixed length of the gas jet, there is also optimal conversion efficiency for each gas jet pressure. Additionally, for the best conversion efficiency of different harmonic regions, the optimal pressure and length of the gas jet are different.

By means of removing the possible factor (for example, absorption and dispersion) terms in our theoretical model, we find out that the absorption of HHG is the most important factors under the condition of our simulation.

Our study indicates that the conversion efficiency of HHG for certain range can be improved by selecting proper pressure and length of the gas jet.

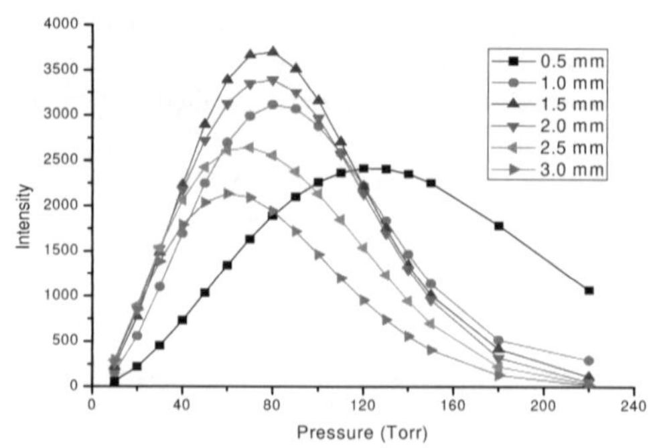

Figure 1. The conversion efficiency of harmonics H30-H45 as a function of the length and pressure of the gas jet.

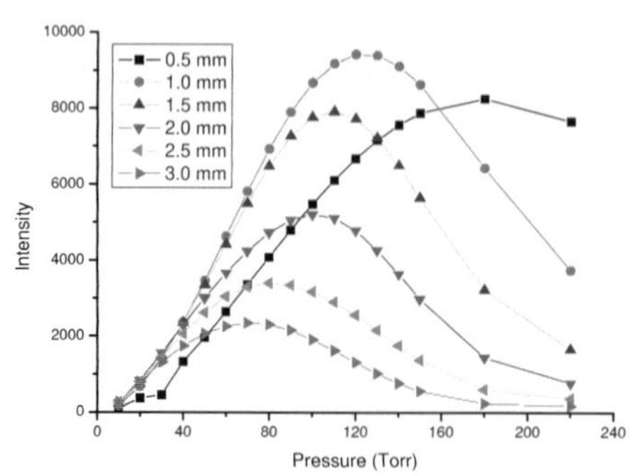

Figure 2. The conversion efficiency of H45-H55 as a function of the length and pressure of the gas jet.

Acknowledgment This work was supported by the National Natural Science Foundation of China under Grant No. 11564033, 11364038, 11264036; and the Fundamental Research Funds for the Central Universities under Grant No. 31920170032, 31920150023 .

References

[1] C. Winterfeldt *et al. Rev. Mod. Phys.* **80** 117.
[2] Y. Tamaki *et al. Phys. Rev. A* **62** 063802.

1 E-mail: lxycock@163.com

IOP Conf. Series: Journal of Physics: Conf. Series **875** (2017) 022027 doi:10.1088/1742-6596/875/3/022027

Ionization of atoms by few-cycle laser pulses: spatial and temporal interference effects

Sándor Borbély[*1], **Attila Tóth**[†], **and Ladislau Nagy**[*]

[*] Faculty of Physics, Babeş-Bolyai University, Kogălniceanu Street 1, 400084 Cluj-Napoca, Romania, EU
[†] ELI-ALPS, ELI-HU Non-Profit Ltd, H-6720 Szeged, Dugonics tér 13, Hungary, EU

Synopsis The intra-pulse interference effects appearing during the ionization of atoms by few-cycle laser pulses is studied. These are the result of coherent superposition of electronic wave packets, which follow different spatio-temporal paths. With the help of a wave function analysis tool, the spatial and temporal interference patterns are disentangled and studied separately.

The momentum distribution of continuum electrons created during the ionization of atoms by few-cycle laser pulses is modulated by interference effects, which are the results of superposition between electronic wave packets following different spatial and temporal paths [1]. The spatial interference, is one from the numerous possible interference scenarios [1], and occurs when wave packets emitted at the same time (i.e., during the same quarter-pulse cycle), but following different spatial paths are coherently added in the continuum (spatial interference). The formed radial interference pattern (see Fig. 1) is the result of the interference between the direct (i.e., unscattered) and the scattered wave packets [2, 3]. By considering the direct wave packet as reference, while the scattered wave packet as a signal wave, the spatial interference pattern can be interpreted as the holographic mapping (HM) of the target atom's state [2].

In our previous theoretical studies [3, 4, 5] we have used specially tailored two-cycle laser pulses, where the large portion of ionization occurred during a single half-cycle. As a result of this, in the ionization spectrum, the dominant interference pattern was the HM, which allowed us the systematic study of it's properties. The high photon energy and high pulse intensity (see the example in Fig. 1) used in our theoretical studies is not feasible in experiments, thus in the present work we have modified our laser pulse parameters (lowering the photon energy and field intensity, adding more optical cycles) in order to converge towards experimentally available laser pulses.

As a result of these modifications the interference pattern became more complicated, composed by the superposition of both spatial (HM) and temporal interference patterns. In order to perform a systematic study on these interference patterns, we have developed a wave function analysis tool, which allows the disentanglement of the different interference patterns based on the numerical solution of the time-dependent Schrödinger equation (TDSE).

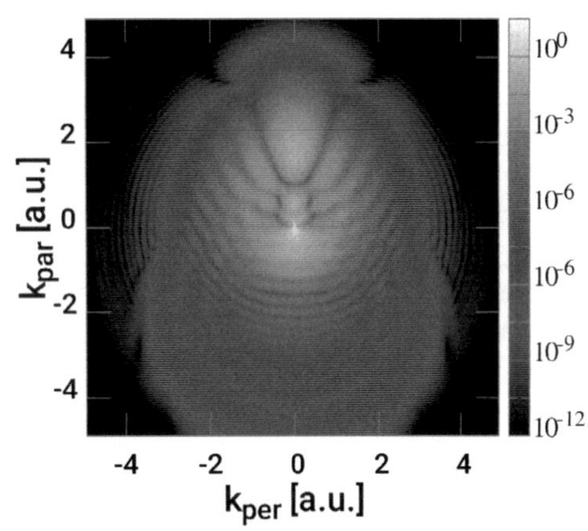

Figure 1. The distribution of the continuum electrons as a function momentum components parallel and perpendicular to the laser polarization vector obtained for H target. The parameters of the driving laser pulse are: $\omega = 0.4445$ a.u. (photon energy), $\tau = 28.27$ a.u. (pulse duration), $E_0 = 1$ a.u.(pulse strength).

During the numerical solution of the TDSE, we have separated and stored in partial wave functions the electronic continuum wave packets created during each half-cycle of the laser pulse. By calculating the ionization spectrum separately for each partial wave function we have obtained the pure HM pattern created by each half-cycle of the driving laser field. The final electron spectrum is obtained by adding coherently the partial ionization amplitudes. During this coherent summing the temporal interference effects emerge.

[1]E-mail: sandor.borbely@phys.ubbcluj.ro

References

[1] X.-B Bian *et al* 2011 *Phys. Rev. A* **84** 043420.
[2] Y. Huismans *et al* 2011 *Nature* **331** 61.
[3] S. Borbély *et al* 2013 *Phys. Rev. A* **87** 013405.
[4] S. Borbély *et al* 2014 *Eur. Phys. J. D* **68** 339.
[5] S. Borbély *et al* 2014 *J. Phys. Conf. Ser.* **635** 092085.

IOP Conf. Series: Journal of Physics: Conf. Series **875** (2017) 022028 doi:10.1088/1742-6596/875/3/022028

The influence of gas-jet position on the macroscopic high-order harmonic generation

Guoli Wang[1], Penghao Zhao, Xiaoxin Zhou[2]

College of Physics and Electronic Engineering, Northwest Normal University, Lanzhou 730070, China

Synopsis We simulate the macroscopic Xe high-order harmonic spectra generated by strong mid-IR 1500-nm laser pulse. Our simulations confirm the experimental findings that the intensity of high-energy harmonics can be enhanced effectively by putting gas jet correctly to a position before laser focus.

High-order harmonic generation (HHG), which appears when atoms or molecules are exposed in intense laser field, has become a widespread and intensive research field in recent years, due to many important applications. To obtain the high-photon-energy harmonics, mid-infrared driving laser pulses are used more widely according to cutoff law. However, the limitation is that the harmonic yield decreases very quickly with the increase of laser wavelength.

As a matter of fact, to obtain strong harmonic yield, many factors can be controlled. In this work we perform a theoretical simulation on harmonic spectra dependence on gas jet position. Figure 1 shows the comparison between our simulation and a recent experiment [1]. In the simulation, the single-atom response is calculated by the quantitative rescattering theory [2], and the macroscopic spectra are obtained by solving three-dimensional Maxwell equations. The Xe gas target if of 1.5 mm length considering the density distribution to be Lorentzian. The laser pulse at the entrance of gas jet is assumed as Gaussian beam in both time and space, with the wavelength of 1500 nm and pulse duration of 20 fs. The laser peak intensity is 2.53×10^{14} W/cm^2, and the beam waist is 55 μm (Rayleigh range is 6.3 mm). The generated harmonics are collected by a 1-mm diameter pinhole placed 150 mm downstream of the interaction region. We can see that the main experimental features are well reproduced in the simulation. When the gas jet is put behind the laser focus (z=4 mm), the harmonic yield decreases very quickly when increasing the photon energy, which is a typical phenomenon for the high ionized medium due to bad phase matching for the higher-energy photons. However, when the gas jet is placed before the laser focus (z=-4 mm), an enhancement about one order of magnitude for harmonics around 65 eV is observed. Vozzi et al have demonstrated the importance of propagation effects of laser pulse in the gas medium for such enhancement of high-energy harmonics in the Ref. [1]. Moreover, our simulations show that this phenomenon is also related to the collecting configuration of experiment, i.e, the size and position of pinhole (the propagation distance of harmonics in free space), because the spatial distributions of harmonic emission from different positioned gas jet at the far field are very different.

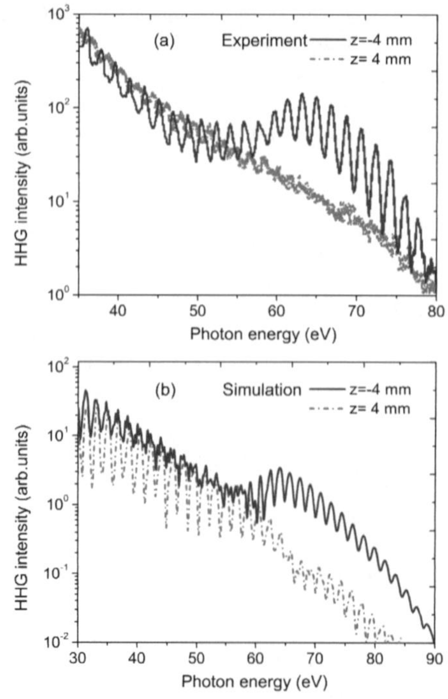

Figure 1. Comparison of the simulated and experimental [1] Xe HHG spectra. Here, the value of z is the entrance position of gas jet with respect to the laser focus: positive value means that the jet is placed behind the laser focus.

Project supported by the National Natural Science Foundation of China (Nos. 11364038, 11465016).

References

[1] C. Vozzi *et al.* 2011 *New J. Phys.* **13** 073003

[2] C. D. Lin *et al.* 2010 *J. Phys. B* **43** 122001

[1]E-mail: wanggl@nwnu.edu.cn
[2]E-mail: zhouxx@nwnu.edu.cn

IOP Conf. Series: Journal of Physics: Conf. Series **875** (2017) 022029 doi:10.1088/1742-6596/875/3/022029

Circular Dichroism in the Multi-Photon Ionization of Oriented Helium Ions

M. Ilchen[1,2,*1], N. Douguet[3], T. Mazza[1], A. J. Rafipoor[1], C. Callegari[4], P. Finetti[4], O. Plekan[4], K. C. Prince[4,5,6], A. Demidovich[4], C. Grazioli[4], L. Avaldi[7], P. Bolognesi[7], M. Coreno[7], M. Di Fraia[8], M. Devetta[9], Y. Ovcharenko[1], S. Düsterer[10], K. Ueda[11], K. Bartschat[3], A. N. Grum-Grzhimailo[1,12], A. V. Bozhevolnov[13], A. K. Kazansky[14,15,16], N. M. Kabachnik[1,12,16], and M. Meyer[1,*2]

[1] European XFEL GmbH, Holzkoppel 4, 22869 Schenefeld Hamburg, Germany
[2] PULSE at Stanford, 2575 Sand Hill Road, Menlo Park, 94025 California, USA
[3]Department of Physics and Astronomy, Drake University, Des Moines, Iowa 50311, USA
[4]Elettra-Sincrotrone Trieste, 34149 Basovizza, Italy
[5]Istituto Officina dei Materiali, Consiglio Nazionale delle Ricerche, Area Science Park, 34149 Trieste, Italy
[6]Molecular Model Discovery Laboratory, Department of Chemistry and Biotechnology, Swinburne University of Technology, Melbourne, Victoria 3122, Australia
[7]CNR Istituto Struttura della Materia, Via del Fosso del Cavaliere, 100-00133 Roma, Italy
[8]Department of Physics, University of Trieste, 34128 Trieste, Italy
[9]Istituto di fotonica e nanotecnologie CNR-IFN, Milano, Italy
[10]Deutsches Elektronen-Synchrotron (DESY), Notkestrasse 85, 22603 Hamburg, Germany
[11]Institute of Multidisciplinary Research for Advanced Materials, Tohoku University, Sendai 980-8577, Japan
[12]Skobeltsyn Institute of Nuclear Physics, Lomonosov Moscow State University, Moscow 119991, Russia
[13]Sankt Petersburg State University, Universitetskaya nab. 7/9, Sankt Petersburg 199164, Russia
[14]Departamento de Fisica de Materiales, UPV/EHU, 25018 San Sebastian/Donostia, Spain
[15]IKERBASQUE, Basque Foundation for Science, 48011 Bilbao, Spain
[16]Donostia International Physics Center (DIPC), 25018 San Sebastian/Donostia, Spain

Synopsis Intense, circularly polarized extreme-ultraviolet (XUV) and near-infrared (NIR) laser pulses are combined to double-ionize atomic helium via the oriented intermediate $He^+(3p)$ resonance state. A strong and NIR-intensity dependent circular dichroism is accounted to a helicity dependent AC-Stark shift.

The dichroic interaction of circularly polarized X-rays with matter is a multi-disciplinary field of science. With the advent of circularly polarized free-electron lasers (FELs) [1, 2], the research field of circular dichroism phenomena in non-linear and ultrafast physics has been extended to the soft X-ray regime. Using high intensity, narrow bandwidth XUV pulses from FERMI in Italy with a photon energy of 48.4 eV, we were able to first ionize and subsequently orient ionic helium in the 3p (m=+1) magnetic substate. From there, the population of this state was controlled by a helicity-dependent AC Stark shift generated by an overlapped near-infrared laser. The measured circular dichroism of electrons emitted via multi-photon ionization of the ionic 3p state is intensity dependent to a surprisingly strong extent, therefore allowing for an easily controllable and polarization selective transparency of such resonances [3]. The experimental results depicted in Fig. 1 are in excellent agreement with two independent calculations based on the Time Dependent Schrödinger Equation (TDSE) [4, 5]. Potential applications in the general context of chirality research at FELs will be presented.

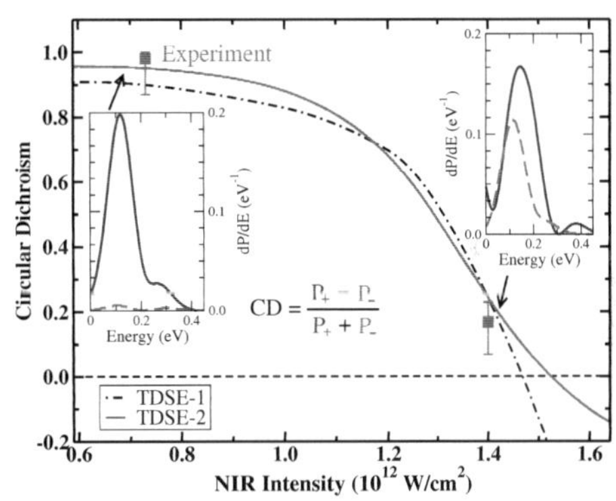

Figure 1. Circular dichroism for electrons ejected from the oriented $He^+(3p)$ state after multi-photon absorption (E_{kin} ≈200 meV) as function of the NIR peak intensity [3].

References

[1] E. Allaria, *et al.*, *Phys. Rev. X* **4**, 041040 (2014).
[2] A. A. Lutman, *et al.*, *Nat. Phot.* **10**, 1038 (2016).
[3] M. Ilchen, *et al.*, *Phys. Rev. Lett.* **118**, 013002 (2017).
[4] A. K. Kazansky and N. M. Kabachnik *J. Phys. B* **40**, 2163 (2007).
[5] N. Douguet *et al.*, *Phys. Rev. A* **93**, 033402 (2016).

[1]E-mail: markus.ilchen@xfel.eu
[2]E-mail: michael.meyer@xfel.eu

IOP Conf. Series: Journal of Physics: Conf. Series **875** (2017) 022030 doi:10.1088/1742-6596/875/3/022030

Few-cycle strong-field ionization of atomic hydrogen with elliptically polarized infrared light

N. Douguet[†1], **K. Bartschat**[†2]

[†] Department of Physics and Astronomy, Drake University, Des Moines, Iowa, 50311, USA

Synopsis The strong-field ionization of atomic hydrogen induced by elliptically polarized light is studied in detail. In particular, we investigate the ionization spectra for various pulse characteristics, consider the circular dichroism from an excited oriented electronic state, and finally analyze the tunneling regime for few-cycle pulses.

The utilization of pulses with elliptical or circular polarization is currently of great interest due to many important applications, such as photoelectron circular dichroism for the investigation of chiral molecules, or in attoclock experiments to time-resolve the dynamics of light absorption by atoms and molecules. Even for the simplest case of atomic hydrogen, the ultrafast electron dynamics in an intense few-cycle elliptically polarized optical pulse represents a complex problem, whose complete description remains a challenge.

In this work, we present a detailed theoretical and computational investigation of the strong-field ionization of atomic hydrogen by elliptically polarized light in the long-wavelength regime (800 nm). Solving the time-dependent Schrödinger equation, we analyze the ionization spectra at various peak intensities up to 4×10^{14} W/cm^2 for a variety of pulse ellipticities. The calculations are performed with the length and velocity forms of the electric dipole operator. In particular, we compare the extreme cases of circularly and linearly (studied in [1]) polarized light, considering important features such as the partial-wave population over time and the maximum angular momentum needed to achieve convergence.

Starting from an oriented atomic state, we also consider the dynamics (see Figure 1) responsible for the circular dichroism [2], from the multiphoton to the tunneling regime and a variety of pulse lengths. A model based on the strong-field approximation is employed in an attempt to predict the variation of the dichroism as a function of the laser peak intensity.

Finally, we analyze the asymptotic electron momentum distribution and its offset angle with respect to the starting direction of a circularly polarized pulse to shed some light on the electron dynamics in the tunneling regime and to assist in the design and understanding of attoclock experiments.

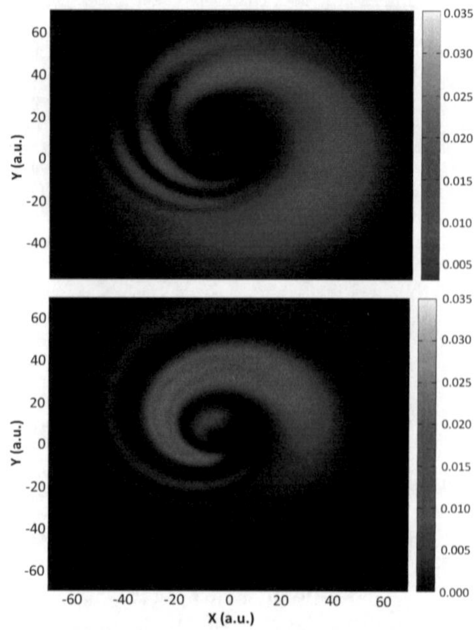

Figure 1. Electron wave-packet probability ($t = 5$ fs) in the photoionization of hydrogen initially in an oriented 2p state, induced by a 4-cycle 770 nm circularly polarized infrared pulse with peak intensity 4×10^{13} W/cm^2, for the initial state co-rotating (top panel) or counter-rotating (lower panel) with the field.

This work is supported by the United States National Science Foundation, Grant No. PHY-1430245 and the XSEDE allocation No. PHY-090031.

References

[1] A. N. Grum-Grzhimailo, B. Abeln, K. Bartschat, D. Weflen, and T. Urness, Phys. Rev. A 2010 **81** 043408

[2] M. Ilchen *et al.*, Phys. Rev. Lett. 2017 **118** 013002

[1]E-mail: nicolas.douguet@drake.edu
[2]E-mail: klaus.bartschat@drake.edu

Relativistic ionization probabilities of hydrogenlike ions exposed to intense laser pulses

I. V. Ivanova[*], **A. Saenz**[‡], **A. I. Bondarev**[*†], **I. A. Maltsev**[*†], **V. M. Shabaev**[*1], **D. A. Telnov**[*]

[*] Department of Physics, St. Petersburg State University, Universitetskaya nab. 7/9, 199034 St. Petersburg, Russia
[‡] AG Moderne Optik, Institut für Physik, Humboldt-Universität zu Berlin, Newtonstrasse 15, D-12 489 Berlin, Germany
[†] SSC RF ITEP of NRC "Kurchatov Institute", Bolshaya Cheremushkinskaya 25, 117218 Moscow, Russia

Synopsis The relativistic intense-field ionization of heavy hydrogenlike ions is investigated. The investigation is performed by numerical solving of the time-dependent Dirac equation. The interaction of intense laser pulses with ions is considered beyond the dipole approximation. As an example, the ionization probabilities of the hydrogenlike tin ion exposed to a strong laser field are calculated.

In the nearest future, the development of experimental facilities will allow for carrying out investigations of heavy atoms and ions exposed to extremely intense laser fields. High photon energies and intensities of laser pulses enable to study relativistic effects of an ion-laser interaction. Novel facilities are able to produce short pulses with a peak intensity of 10^{21} W/cm^2 or even higher. For example, the High-Intensity Laser Ion-Trap Experiment (HILITE) is under construction at GSI (Darmstadt, Germany) and is planned to be performed by the SPARC collaboration [1]. The rapid development of laser technologies has triggered the theoretical investigations of one-electron ions exposed to extremely intense laser pulses [2, 3, 4, 5, 6].

Previously, we developed a method for treatment of hydrogenlike ions exposed to an intense laser field [7]. In that method, the interaction with the laser field was described in the electric dipole approximation. In the present work, the method has been extended beyond the dipole approximation. The vector potential of the laser field is represented by the multipole expansion, and dependence on the spatial coordinates is expressed through the spherical Bessel functions (SBF).

Using this method, we have calculated the ionization probability of the hydrogenlike tin ion (Z = 50) initially in the ground state and exposed to a 20-cycle pulse with the carrier frequency 100 keV and peak intensity 1.56×10^{22} W/cm^2. Fig. 1 shows the partial ionization probabilities into the states with different orbital quantum numbers l. In the figure, four curves are shown in which different values of L_{max} (1 to 4) are adopted. Here, L_{max} is the maximal order of SBF retained in the expansion of the vector potential. The curves show how convergence of the partial ionization probabilities with larger l is reached when the non-dipole terms of higher orders are included in the expansion. Additionally, the calculations of the total ionization probability are performed within the dipole approximation (DIP) and beyond it (BYD); the following results are obtained: $P_{DIP} = 3.1 \times 10^{-5}$ and $P_{BYD} = 3.9 \times 10^{-5}$. The 21% difference between the results demonstrates the importance of the non-dipole effects for the laser field frequency and intensity regime used in the calculations.

This work was supported by BMBF (Grant No. 05P15KHFAA), by SPbSU-DFG (Grant No. 11.65.41.2017), by RFBR (Grant No. 16-02-00233), and by the G-RISC (Project No. P-2016b-6).

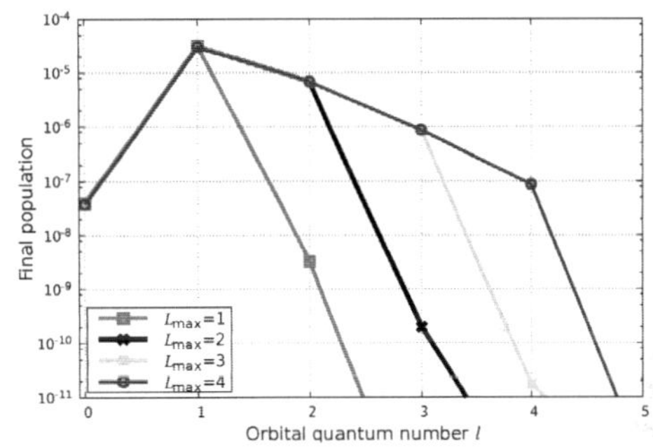

Figure 1. Final population of different continuum l-states for the hydrogenlike tin ion exposed to the intense laser pulse. L_{max} is the maximal order of SBF included in the calculations.

References

[1] S. Ringleb *et al.* 2015 *Phys. Scr.* **T166** 014067

[2] S. Selstø *et al.* 2009 *Phys. Rev. A* **79** 043418

[3] M.S. Pindzola *et al.* 2012 *Phys. Rev. A* **85** 032701

[4] Y.V. Vanne and A. Saenz 2012 *Phys. Rev. A* **85** 033411

[5] E.B. Rozenbaum 2014 *Phys. Rev. A* **89** 012514

[6] A.S. Simonsen *et al.* 2016 *Phys. Rev. A* **93** 053411

[7] I.V. Ivanova *et al.* 2015 *J. Phys. Conf. Ser.* **635** 092040

[1]E-mail: v.shabaev@spbu.ru

IOP Conf. Series: Journal of Physics: Conf. Series **875** (2017) 022032 doi:10.1088/1742-6596/875/3/022032

Time-Dependent Two-Particle Reduced Density Matrix Theory: Application to High-Harmonic Generation

Fabian Lackner[*1], **Iva Březinová**[*2], **Takeshi Sato**[†‡], **Kenichi L. Ishikawa**[†‡], **Joachim Burgdörfer**[*]

[*] Institute for Theoretical Physics, Vienna University of Technology, Wiedner Hauptstraße 8-10/136, 1040 Vienna, Austria, EU

[†] Photon Science Center, School of Engineering, The University of Tokyo,7-3-1 Hongo, Bunkyo-ku, Tokyo 113-8656, Japan

[‡] Department of Nuclear Engineering and Management, School of Engineering, The University of Tokyo, 7-3-1 Hongo, Bunkyo-ku, Tokyo 113-8656, Japan

Synopsis We follow a new approach for calculating high-harmonic spectra for multi-electron atoms and molecules by propagating the two-particle reduced density matrix. Calculated results are in very good agreement with state-of-the-art many-body wavefunction-based benchmark calculations.

High-harmonic generation (HHG) is one of the fundamental processes in strong field physics whose applications range from attosecond metrology [1], tunable table-top XUV/Soft X-ray sources [2] to high precision spectroscopy [3] and orbital imaging [4]. On the atomic level the theoretical description of HHG is challaging because of the multi-electron nature of the underlying process. While simple models such as the single-active-electron approximation (SAE) or time-dependent Hartree-Fock (TDHF) are well suited to describe qualitative features of HHG, advanced theories capable of correctly treating electron correlation are needed for a quantitative description [5]. However, with increasing system size conventional wavefunction-based methods such as the multi-configurational time-dependent Hartree-Fock (MCTDHF) method soon become unfeasible due to their exponential scaling with particle number. One way to overcome this exponential barrier is to abandon a wavefunction-based description and propagate time-dependent reduced densities instead. Considerable success along this line has been achieved by time-dependent density functional theory (TDDFT) [6]. However, accurate calculations in TDDFT face the difficulty of unknown exchange-correlation functionals that are difficult to improve systematically. We propose to go beyond the limitations of TDDFT by propagating the time-dependent two-particle reduced density matrix (TD-2RDM). As a hybrid between the electron density and the many-body wavefunction the 2RDM fully includes two-particle correlations. We have implemented the TD-2RDM method to describe high-harmonic generation of fully three-dimensional multi-electron atoms [7]. To obtain accurate results for the electronic response we use an advanced closure scheme [8] that is suited to conserve all constants of motion associated with symmetries of the Hamiltonian. We benchmark the performance of the TD-2RDM method by comparing it to a state of the art MCTDHF calculation [9] as well as to TDDFT calculations (see Fig.1). We find very good agreement between the TD-2RDM and the MCTDHF method while TDDFT within the local density approximation shows clear deviations indicating that the correct treatment of two-particle correlations is essential to obtain accurate HHG spectra.

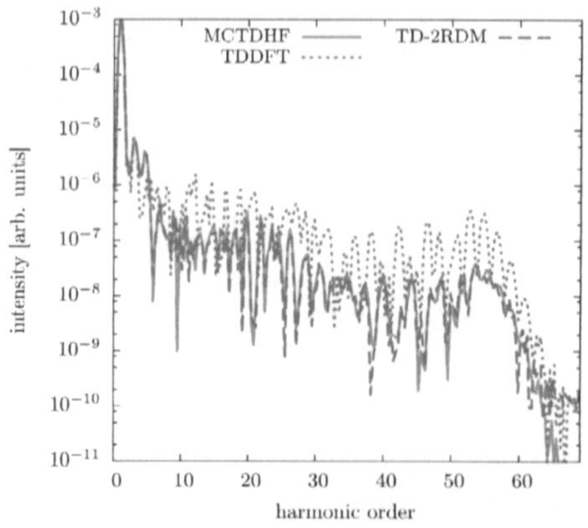

Figure 1. HHG spectrum of beryllium subject to a 3-cycle laser pulse with intensity $I = 4.0 \times 10^{14}$W/cm^2. MCTDHF(—), TD-2RDM(- -), TDDFT(···).

References

[1] M. Hentschel *et al.* 2001 *Nature* **414** 509

[2] Ph. Zeitoun *et al.* 2004 *Nature* **431** 426

[3] A. D. Shiner *et al.* 2001 *Nature Phys.* **7** 464

[4] J. Itatani *et al.* 2004 *Nature* **432** 867

[5] S. Sukiasyan *et al.* 2009 *Phys. Rev. Lett.* **102** 223002

[6] X. M. Tong *et al.* 2015 *Phys. Rev. A* **92** 043422

[7] F. Lackner *et al. Phys. Rev. A* in press

[8] F. Lackner *et al.* 2015 *Phys. Rev. A* **91** 023412

[9] T. Sato *et al.* 2016 *Phys. Rev. A* **94** 023405

[1]E-mail: fabian.lackner@tuwien.ac.at
[2]E-mail: iva.brezinova@tuwien.ac.at

Intracycle interference in ionization of Ar by a laser assisted XUV pulse

D. G. Arbó [*] [1], S. D. López [*], M. Kubin [†], J. Hummert [†], M. J. J. Vrakking [†] and O. Kornilov [†]

[*] Institute for Astronomy and Space Physics IAFE (UBA-Conicet), Buenos Aires, Argentina
[†] Max Born Institute, Berlin, Germany, EU

Synopsis We present a theoretical and experimental study of the subcycle interference in laser assisted XUV ionization of Ar atoms. Averaging over the focal volume happens to blur the intracycle interference, which thus cannot be measured directly. We show that even at these conditions, the intracycle interference can be obtained through the subtraction of two different angle and energy-resolved distributions at slightly different laser intensities.

Intracycle interferences are crucial in the understanding of above-threshold ionization by near infrared lasers [1]. Recently, it was calculated for the case of laser assisted XUV ionization [2,3].

In this work, we show that even in experiments where interferences are missing or hidden by the varying intensities in the focal volume, the intracycle interference evidenced in the modulation of the sidebands emerges when the two angle-resolved and energy-resolved distributions taken at somewhat different laser intensity are subtracted.

The experiments presented here are carried out using recently constructed XUV time delay compensating monochromator beamline [4] and a velocity map imaging spectrometer. This system is capable of producing a wavelength-selected, but short XUV pulse and thus is best suitable for studying laser-assisted XUV ionization processes. In the experiments, Atoms are ionized by 29.5 eV photons in the presence of IR field, which leads to generation of several sidebands.

In Figure 1 we observe the difference of the angle and energy-resolved distribution photoelectron distribution recorded for two different laser intensities. The red color corresponds to positive values and the blue color to negative ones. Near zero values of the difference are in white color. We see the resemblance between the calculated map by numerically solving the time-dependent Schrödinger equation and the measured map. In both, the distributions align around the polarization axis of both IR and XUV fields (0 and 180 deg.).

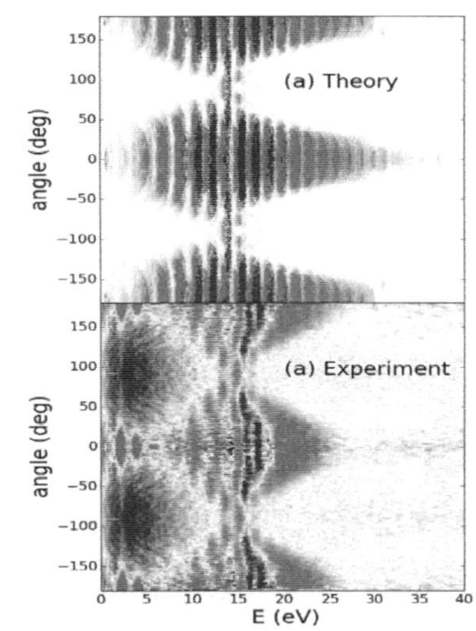

Figure 1. Difference of photoelectron distributions for two somewhat different laser intensities: around I ~ 6 × 10^{13} W/cm^2 for the Ar atom ionized by an XUV pulse due to the 19th harmonic of a laser pulse of $\lambda = 800$ nm. (a) Theory, (b) Experiment. The red color corresponds to positive values and the blue to negative. The laser field is linearly polarized along the z-axis.

This work has been supported by the Argentine-Germany collaboration Conicet-DAAD 2015, CONICET (Argentina PIP0386), ANPCyT (Argentina PICT-2014-2363).

References

[1] D. G. Arbó *et al.*, Phys. Rev. A **81**, 021403(R) (2010).
[2] A. K. Kazansky and N. M. Kabachnik, Journal of Physics B **43**, 035601 (2010).
[3] A. A. Gramajo *et al.*, Phys. Rev. A **94**, 053404 (2016).
[4] M. Eckstein et al, J. Phys. Chem. Lett. **6**, 419 (2015) ; M. Eckstein et al, arXiv:1604.02650 [physics.ins-det]

[1] E-mail: diego@iafe.uba.ar

IOP Conf. Series: Journal of Physics: Conf. Series **875** (2017) 022035 doi:10.1088/1742-6596/875/3/022035

Manipulating dynamical interference in photoionization processes by inhomogeneous strong-field

Hicham Agueny [1] **and Jan Petter Hansen**

Department of Physics and Technology, Allegt. 55, University of Bergen, N-5007 Bergen, Norway

Synopsis We investigate dynamic interference in photoionization processes by intense spatially inhomogeneous fields. Simulations from numerical solutions of the three-dimensional (3D) time-dependent Schrödinger equation show that the phenomenon manifests by a multiple peak structure in the momentum distribution of the ejected electron. Its sensitivity to the inhomogeneity strength of the laser field offers the possibility to manipulate the mechanism.

Ultrafast laser pulses can easily exceed atomic field strengths of 10^9 V/cm, opening new doors in the investigation and control of ultrafast processes. Laser-induced Stark shift is one such process. The latter leads to an energy shift of atomic energy levels in an oscillating electric field, which affects electronic transitions. We refer to the energy level shifts as dynamical Stark effect [1]. Because of the non-perturbative nature of the mechanism, it becomes an increasingly important tool for controlling and measuring quantum systems [2].

In this context, subcycle dynamical Stark effect induced by a few-cycle near infrared laser field was investigated and measured using attosecond transient absorption spectroscopy [3]. Furthermore, a strong dynamic interference induced by intense high-frequency laser pulses and caused by the dynamical Stark effect in the continuum has been observed by Demekhin and Cederbaum [4].

Because of the importance of the effect to induce quantum features as we gain more understanding of ultrafast processes, we will address its impact on the photoionization process by intense spatially inhomogeneous fields. Simulations from numerical solutions of the 3D-time-dependent Schrödinger equation exhibit interference patterns in the momentum distribution of the ejected electron (see Fig. 1).

During the conference, we will present results for the above-mentioned process for both linear and circular polarizations, and show how the inhomogeneity strength influences the dynamical interference. The study will be addressed for the hydrogen atom and will be extended to both diatomic and triatomic systems. Dependence of the dynamical interference on the pulse ellipticity will also be addressed.

The phenomenon will be discussed by examining the momentum-space geometry which is a basic principle for the attosecond streak camera [5].

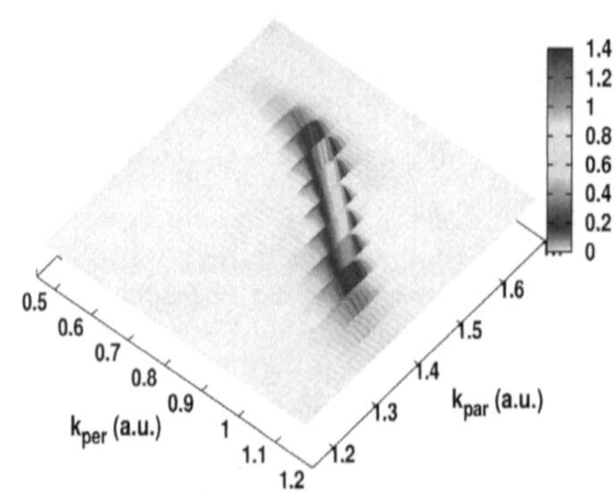

Figure 1. Map of the ejected electron momentum distribution on the k_{par}-k_{per} plane (parallel and perpendicular momentum components). The process is studied for hydrogen atom exposed to Gaussian-shaped pulses of 18 fs duration, carrier frequency of 53.60 eV and peak intensity of 7 10^{16} W/cm^2 (homogeneous field and linear polarization case).

References

[1] M. V. Fedorov, *Atomic and Free Electrons in a Strong Light Field* (World Scientific, River Edge, 1997)

[2] B. J. Sussman *et al* 2006, *Science* **69** 012701

[3] M. Chini *et al* 2012 *Phys. Rev. Lett.* **109** 073601

[4] P. V. Demekhin and L. S. Cederbaum 2012 *Phys. Rev. Lett.* **108** 0253001

[5] M. Hentschel *et al* 2001 *Nature (London)* **414** 509

[1]E-mail: hicham.agueny@uib.no

IOP Conf. Series: Journal of Physics: Conf. Series **875** (2017) 022036 doi:10.1088/1742-6596/875/3/022036

Coherent control of the photoelectron angular distribution in photoionization of neon

N. Douguet[†1], **A. N. Grum-Grzhimailo**[*2], **E. V. Gryzlova**[*3], **E. I. Staroselskaya**[*4], **K. Bartschat**[†5]

[†] Department of Physics and Astronomy, Drake University, Des Moines, Iowa, 50311, USA
[*] Skobeltsyn Institute of Nuclear Physics, Lomonosov Moscow State University, Moscow 119991, Russia

Synopsis We investigate bichromatic photoionization of neon induced by linearly polarized XUV femtosecond pulses. The fundamental frequency is tuned near an optically allowed transition to create an interference between one-photon and two-photon ionization pathways. The asymmetry in the photoelectron angular distribution is analyzed for a variety of pulse characteristics.

Light-induced coherent control of the photoelectron angular distribution (PAD) in neon was recently achieved using the Free-Electron Laser (FEL) at FERMI [1]. This experimental tour de force promises a rich field of possibilities in the control of matter and has prompted new research in this exciting field. In order to gain a better understanding of these processes, we investigated two-pathway interferences in the ionization of neon induced by the fundamental and second harmonic of a femtosecond XUV pulse, i.e., the $\omega+2\omega$ process. We consider the cases for which either $(2p^53s)\,^1P_1$ [2] or $(2p^54s)\,^1P_1$ are chosen as intermediate states to enhance the two-photon ionization probability.

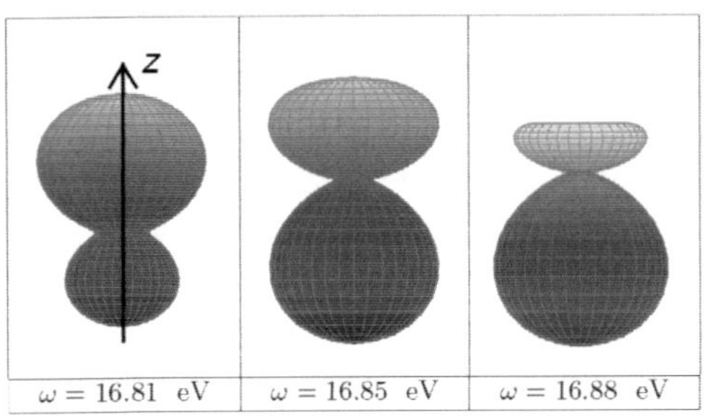

Figure 1. PAD at three fundamental frequencies near the $(2p^53s)\,^1P_1$ resonance for a 250-cycle bichromatic XUV pulse with fundamental peak intensity 10^{12}W/cm^2. The relative intensity of the second harmonic is equal to 1% of the fundamental. The *Z*-axis represents the direction of light polarization.

Using a time-dependent approach supported by a perturbative formalism, we analyze the effects on the PAD when varying the fundamental frequency (see Fig. 1), the intensity ratio between the harmonics, and the carrier envelope phase. The role of the pulse length, fundamental intensity, and experimental averaging is also discussed in detail. Our results are compared with new experimental data [3]. We also discuss the additional degree of freedom provided by adding an infrared field [4], thereby leading to above-threshold ionization (ATI) processes (Fig. 2). We compare the PADs of the sidebands obtained in time-dependent calculations with those from a model based on the strong-field approximation (SFA) using Coulomb-Volkov wavefunctions.

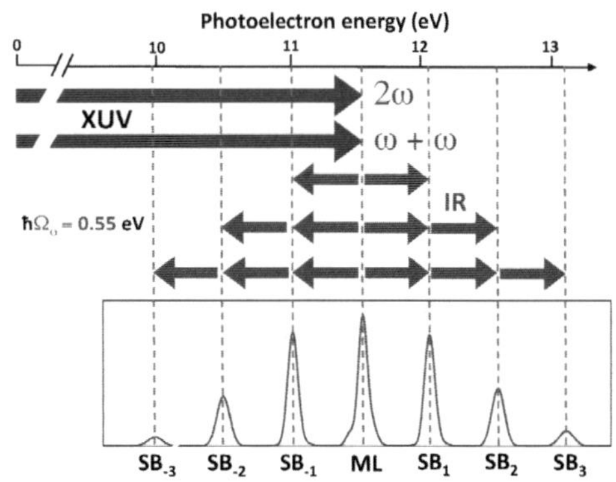

Figure 2. Photoionization scheme and associated spectrum for the $\omega+2\omega$ process with an additional optical field. The blue arrows indicate XUV photon absorption while the red arrows show the minimum number of photons absorbed or emitted to produce the different sidebands (SBs) around the mainline (ML).

This work is supported by the United States National Science Foundation, Grant No. PHY-1430245 and the XSEDE allocation No. PHY-090031.

References

[1] K. C. Prince *et al.* 2016 *Nature Phot.* **10** 176
[2] N. Douguet *et al.* 2017, Europ. J. Phys. D **71**, in press, arXiv:1611.01878
[3] G. Sansone *et al.* 2017, private communcation
[4] N. Douguet, A. N. Grum-Grzhimailo, K. Bartschat 2017 *Phys. Rev. A* **95** 013407

[1]E-mail: nicolas.douguet@drake.edu
[2]E-mail: algrgr1492@yahoo.com
[3]E-mail: gryzlova@gmail.com
[4]E-mail: kate_7-7@mail.ru
[5]E-mail: klaus.bartschat@drake.edu

Coherent control of the THz radiation in an inhomogenous plasma channel

Xiao-Bo Zhang, Xin Qiao, Li-Hong Cheng, Ai-Xia Zhang and Ju-Kui Xue*

Key Laboratory of Atomic & Molecular Physics and Functional Materials of Gansu Province, College of Physics and Electronics Engineering, Northwest Normal University, Lanzhou 730070, China

Synopsis The effective control on the intensity, polarization and the beam width of the THz radiation based on laser-plasma interaction is realized in a magnetized rippled density plasma channel.

Teraherta (THz) radiation based on laser-plasma interaction is of widespread interest for a variety of promising applications in the physics, biology and medicine [1]. Due to the material breakdown and low conversion efficiency of the non-plasma based schemes, the laser-plasma interaction as a key method for the THz radiation has been observed and discussed widely. In addition to frequency and amplitude of the THz radiation, how to control the polarization and the beam width of the THz radiation is the key issue to the applications of the THz science [2, 3, 4]. In previous works, the polarization of the THz radiation is controlled mainly by adjusting the location of the β-barium borate (BBO) crystal [3]. For the beam width control of the THz radiation, some rough methods (limiting the detection area with an aperture, the different parabolic metal mirror) are suggested [2]. Particularly, the control of the beam width of THz radiation via the laser-plasma interaction has never been discussed. So it's still an opening subject to design an efficient scheme for coherently manipulation the polarization and the beam width of the THz radiation.

In this paper, based on laser-plasma interaction, we propose an efficient new scheme for the control of the THz radiation generation by beating of two color lasers in a a magnetized plasma channel. The effects of the magnetic field and the rippled density plasma channel on the THz radiation are discussed in detail. We find that the coherent control of the polarization of the THz radiation is realized completely by adjusting the direction of the magnetic field, while the intensity and the beam width of the THz radiation can be controlled by changing the plasma channel generated by the super-Gaussian lasers and the strength of the magnetic field effectively. Particularly, the intensity of the THz radiation can forms a platform across the channel and the beam of the THz radiation shows a focussing feature, which can be controlled by adjusting the rippled density plasma channel and the strength of the magnetic filed. Figure 1 depicts the key results of the THz field (MV/cm) controlled by the plasma channel and the magnetic field direction. Such powerful and simple THz control techniques pave the way for promising THz applications.

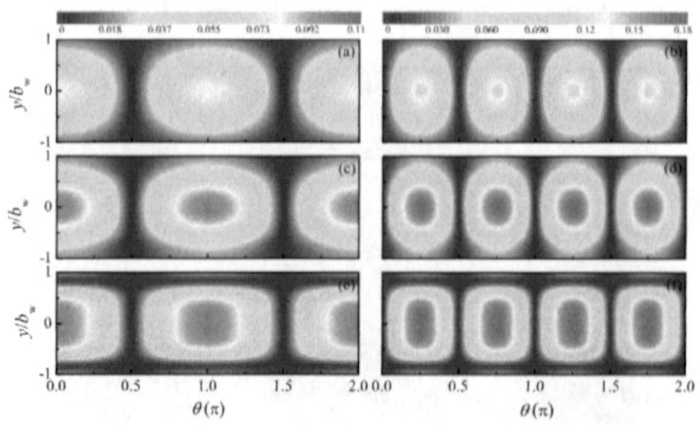

Figure 1. (Color online) Variation of E_{0y} (the left column) and E_{0x} (MV/cm) (the right column) against the plasma channel and the angle of the magnetic field θ.

This work is supported by the National Natural Science Foundation of China under Grants No. 11274255, 11305132, and 11475027.

References

[1] M. Tonouchi 2007 *Nat. Photonics* **1** 97

[2] S. L. Chen *et al* 2014 *Nat. Photonics* **8** 537

[3] Z. Zhang *et al* 2016 *Phys. Rev. Lett.* **117** 243901

[4] W. M. Wang *et al* 2015 *Phys. Rev. Lett.* **114** 253901

*E-mail: xuejk@nwnu.edu.cn

IOP Conf. Series: Journal of Physics: Conf. Series **875** (2017) 022038 doi:10.1088/1742-6596/875/3/022038

Laser-electron interaction in plasma channel with dispersion effect

Li-Hong Cheng and Ju-Kui Xue*

Key Laboratory of Atomic & Molecular Physics and Functional Materials of Gansu Province, College of Physics and Electronics Engineering, Northwest Normal University, Lanzhou 730070, China

Synopsis Laser-electron interaction in a plasma channel is presented, taking into the action of superluminosity of the laser phase velocity induced by the dispersion of the plasma and the channel.

Electron acceleration in laser-plasma physics is gaining growing attention. Beams of electron at GeV have been produced by laser wakefield acceleration [1]. However, wakefield acceleration typically requires very expensive and multi-terawatt laser systems. Unlike wakefield acceleration, direct laser acceleration is a linear process with no threshold intensity, and it offers an attractive alternative for producing energetic electrons. Direct laser acceleration has been demonstrated in gases, vacuum[2], and preformed plasma channels [3, 4].

However, there are many open questions, both from the basic principle and the understanding in underling process, needed to be further discussed in the production of highly energetic electron in laser-plasma interaction. Particularly, each direct acceleration scheme has to focus on overcoming one of the underlying assumptions of the Lawson-Woodward theorem to achieve net energy gain. Furthermore, previous work [4] neglected the response of laser in ion channel, by using the assume that the laser pulse propagates as the light velocity in a plasma channel with no electromagnetic dispersion. Generally, the presence of the plasma and plasma channelling could independently lead the laser pulse being subject to wave dispersion.

To solve the critical problem (electrons gain nonzero net energy in laser-plasma interaction), we investigate the dynamics of electron irradiated by a linearly polarized laser pulse in a plasma channel, and introduce the dispersive nature of electromagnetic waves in plasma channel via the specification of an arbitrary phase velocity. The coupled effects of laser phase velocity and plasma channel on electron acceleration are discussed in detail. The superluminosity of the laser results in rich and complex electron dynamics which are depicted in the plane of the phase velocity and plasma charge density in figure 1. For weak superluminosity (regions I and IV), the electron oscillation is mainly depends on the plasma charge density, and the effect of the phase velocity can be neglected. For moderate superluminosity (region II), a cross-over region can exist, where the highly energetic electron could be found and the net energy gain is several times greater than energy gain in vacuum. For strong superluminosity (regions III), although the maximum electron energy is limited, the electron could get adjustable net energy gain from the laser, due to the superluminosity of the laser. The electron oscillations are no longer limited by the charge density threshold and the electron can always get net energy from the laser.

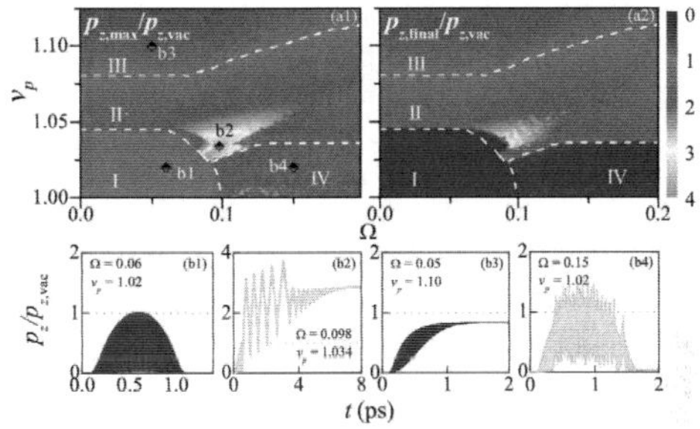

Figure 1. (color online) Longitudinal momentum of electron irradiated by the laser pulse for different phase velocity v_p and plasma charge density Ω.

This work is supported by the National Natural Science Foundation of China under Grants No. 11274255, 11305132, and 11475027.

References

[1] W. P. Leemans *et al* 2006 *Nat. Phys.* **2** 696

[2] T. Plettner *et al* 2005 *Phys. Rev. Lett.* **95** 134801

[3] C. Gahn *et al* 1999 *Phys. Rev. Lett.* **83** 4772

[4] A. V. Arefiev *et al* 2012 *Phys. Rev. Lett.* **108** 145004

*E-mail: xuejk@nwnu.edu.cn

IOP Conf. Series: Journal of Physics: Conf. Series **875** (2017) 022039 doi:10.1088/1742-6596/875/3/022039

Attoclock using Atomic Hydrogen

U Satya Sainadh* [1], Han Xu* [1], X. Wang § [2], Atia-Tul-Noor* [1], William C. Wallace* [1], N. Douguet† [3], Igor Ivanov ‡ [4], Klaus Bartschat† [3], Anatoli Kheifets¶ [5], R.T. Sang * [1], Igor Litvinyuk* [1].

* Australian Attosecond Science facility, Center for Quantum Dynamics, Griffith University, Brisbane QLD , Australia.
§School of Nuclear Science & Technology, Lanzhou University, Lanzhou, 730000, China.
† Department of Physics and Astronomy, Drake University, Des Moines, Iowa 50311, USA.
‡Center for Relativistic Laser Science, Institute for Basic Science, Gwangju 500-712, Republic of Korea.
¶Research School of Physics and Engineering, The Australian National University, Canberra ACT 0200, Australia.

Synopsis We present the results of attosecond angular streaking of atomic Hydrogen using elliptically polarized, 5.5 fs pulses that are centered around 770 nm, within the intensity range of 1.65 to 4 X 10^{14} W/cm^2. We measure angular offsets in the photo-electron momentum distribution relative to the peak of the electric field of light in the polarization plane .We find a strong agreement in these results with the solutions of complete 3D-TDSE simulations. Further, we compute the contribution of coulomb effects (between the ionized electron-parent ion) to the measured angular offsets using Yukawa potential, subtracting which yields the real 'tunneling delays'.

The 'attoclock' technique [1], also known as attosecond angular streaking, employs near-circularly polarised few-cycle femtosecond pulses, wherein the peak electric field is strong enough to suppress the Coulomb barrier and facilitate electron tunnel ionisation. The ionised electrons and ions get streaked in the residual (post-peak) circular field of light mapping the instant at which electrons appear in the continuum to its final momentum. Thus measuring the peak field and photo-electron momenta gives us the time difference between the instant of peak electric field and the instant at which electron appears in the continuum, respectively. The time-difference materialises as an angular offset in the photo-electron momentum distribution in the polarisation plane relative to the polarisation ellipse of the light.

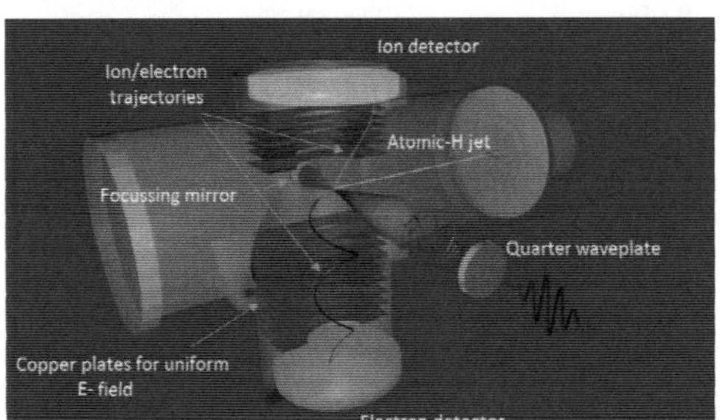

Figure 1: A schematic of the experimental set-up wherein the atomic-H gas jet interacts with the elliptic polarised light inside the COLTRIMS.

Although previous attoclock experiments [1,2] hinted at zero-tunnelling delays, exact theoretical solutions were not available to study the ionisation dynamics in detail. Atomic Hydrogen (H), being the simplest atomic system, can be solved exactly using 3D-TDSE that enables us to benchmark the field and also be used to validate various theoretical models that help us in understanding ionisation dynamics in complex atomic systems.

We performed the attoclock experiment with H using COLTRIMS and 770nm, 6fs pulses at intensities from 0.165-0.39 PW/cm2. The H gas jet source [3] is an RF-discharge tube that dissociates H2 with a dissociation fraction of 60%. We present and compare these results with the full solution of 3D-TDSE codes calculated using exactly the same experimental parameters. Further, using simulations with Yukawa potential we present the contributions of coulomb potential to the angular offsets and finally present an upper bound for 'tunnelling delays'.

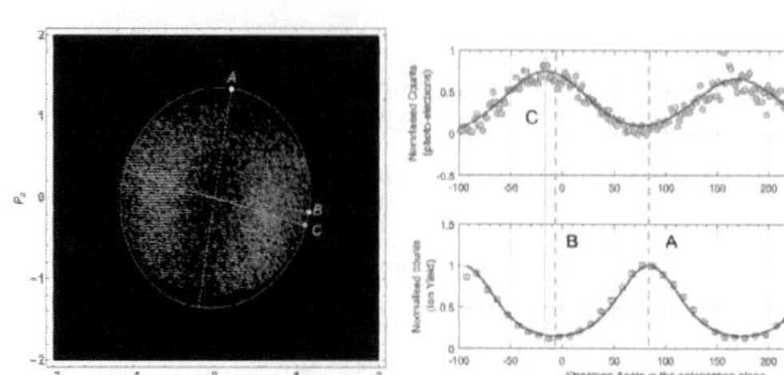

Figure 2: Experimental data of momentum distribution of photoelectrons in the polarisation plane. A corresponds to the peak electric field and B & C are the expected and measured peaks of photoelectron-momentum distribution in the polarisation plane

References
[1] P. Eckle *et.al.* 2008 Nat. Phys. **4**, 565
[2] A.N. Pfeiffer *et.al.* 2013 Phys.**414**, 8491
[3] J.P. Schwonek, 1990 PhD thesis, Massachusetts Institute of Technology, Cambridge, MA.

[1] E-mail: sainadh.undurti@griffithuni.edu.au

IOP Conf. Series: Journal of Physics: Conf. Series **875** (2017) 022040 doi:10.1088/1742-6596/875/3/022040

Control of photoemission delay in resonant two-photon transitions

L. Argenti*†|, Á. Jiménez Galán†‡, R. Taïeb§, J. Caillat§, A. Maquet§, F. Martín†¶

* Dept. Physics & CREOL, University of Central Florida, Orlando, FL (USA).
† Dept. Chemistry, Autonomous University of Madrid, Spain (EU).
‡ Max Born Institute, Berlin, Germany (EU).
§ Sorbonne University & UPMC Université Paris 6, Paris, France (EU).
¶ IMDEA-Nanociencia & IFIMAC, Madrid, Spain (EU).

Synopsis In contrast to one-photon transitions and non-resonant multiphoton transitions, time delay in resonant multi-photon electron emission can exhibit large positive and negative values that have no scattering equivalent, due to the interference of multiple ionization paths.

The photoelectron emission time delay τ associated to one-photon absorption, which coincides with half the Wigner delay τ_W [1, 2] experienced by an electron scattered off the ionic potential, is a fundamental descriptor of the photoelectric effect [3]. Although it is hard to access directly from experiment, it is possible to infer it from the time delay of two-photon transitions, $\tau^{(2)}$, measured with attosecond pump-probe schemes [4, 5], provided that the contribution of the probe stage can be factored out [6]. In absence of resonances, τ can be expressed as the energy derivative of the one-photon ionization amplitude phase, $\tau = \partial_E \arg D_{Eg}$, and, to a good approximation, $\tau = \tau^{(2)} + \tau_{cc}$, where τ_{cc} is associated to the dipole transition between Coulomb functions. Here we show that, the presence of a resonance, the correspondence between τ and $\partial_E \arg D_{Eg}$ is lost. Furthermore, while $\tau^{(2)}$ can still be written as the energy derivative of the two-photon ionization amplitude phase, $\partial_E \arg D^{(2)}_{Eg}$, it does not have any scattering counterpart. Indeed, $\tau^{(2)}$ can be much larger than the lifetime of an intermediate resonance in the two-photon process, or more negative than the lower bound imposed on scattering delays by causality. Finally, we show that $\tau^{(2)}$ is controlled by the frequency of the probe pulse, ω_{IR}, so that by varying ω_{IR}, it is possible to radically alter the photoelectron group delay [7].

To draw our conclusions, we use a general finite-pulse analytical model for resonant multi-photon ionization amplitudes [8, 9], which we have developed and applied successfully to a number of attosecond interferometric experiments and simulations [8, 10, 5]. We also offer selected examples in the case of helium, using parameters obtained from accurate *ab initio* simulations of the photoionization of the helium atom [11].

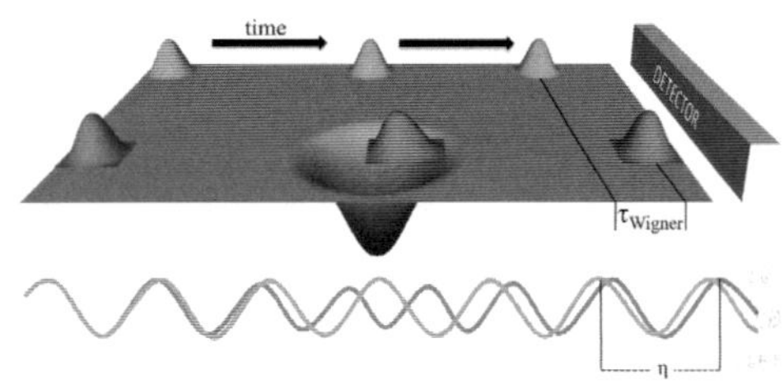

Figure 1. Sketch of the Wigner time delay. As the electron wave packet (green) scatters off the potential, it acquires a phase shift η as compared to a reference free electron (orange). This phase shift is related to the time lapsed between the detection of the scattered electronic wave packet and the detection of the unscattered free electron, called the Wigner time delay.

References

[1] E. P. Wigner 1955 *Phys. Rev.* **98** 145.
[2] F. T. Smith 1960 *Phys. Rev.* **118** 349.
[3] R. Pazourek *et al.* 2015 *Rev. Mod. Phys.* **87** 765.
[4] P. M. Paul *et al.* 2001 *Science* **292** 1689.
[5] V. Gruson *et al.* 2016 *Science* **354** 734.
[6] M. Dahlström *et al.* 2013 *Chem. Phys.* **414** 53.
[7] L. Argenti *et al.* 2017 *submitted*.
[8] Á Jiménez *et al.* 2014 *Phys. Rev. Lett.* **113** 263001.
[9] Á Jiménez *et al.* 2016 *Phys. Rev. Lett.* **93** 023429.
[10] M. Kotur *et al.* 2016 *Nature Commun.* **7** 10566 .
[11] L. Argenti & E. Lindroth 2010 *Phys. Rev. Lett.* **105** 053002.

[1]E-mail: luca.argenti@ucf.edu

IOP Conf. Series: Journal of Physics: Conf. Series **875** (2017) 022041 doi:10.1088/1742-6596/875/3/022041

Inelastic x-ray scattering technique and its application on determining the electronic structures of atoms and molecules

Lin-Fan Zhu * [1]

* Hefei National Laboratory for Physical Sciences at Microscale, Department of Modern Physics, University of Science and Technology of China, Hefei, Anhui 230026, People's Republic of China

Synopsis Non-resonant inelastic x-ray scattering (NRIXS) is a powerful tool to reveal the electronic structures of matter while it has a very low cross section. Recently we have extended the NRIXS to atomic and molecular physics with a low-density target. The electronic structures of the ground/excited states and the photoabsorption cross sections of some atoms and molecules have been determined by NRIXS and the newly-developed dipole (γ,γ) method.

Electronic structures of the ground and excited states of atoms or molecules are of fundamental importance to atomic and molecular physics, and are investigated by the high energy electron scattering method traditionally. However, the electron scattering method has its intrinsic merits and demerits. One of its merits is the large scattering cross sections, and one of its demerits is the strong interaction between the incident electron and target which leads to the invalidity of the first Born approximate (FBA), and that will complicate the explanation of the experimental observations. Furthermore, the differential cross section (DCS) of high-energy electron scattering rapidly diminishes with a rate of q^{-4} as the momentum transfer increases, so it is difficult to measure the dynamic parameters at large momentum transfer. However, NRIXS method has the advantages that the FBA is almost always satisfied, which provides a powerful tool to study the dynamic parameters and the electronic structures of the ground and excited states of atoms and molecules.

The disadvantage of IXS is its very low cross sections, i.e., about $10^{-29}m^2$, and it is the reason that the x-ray scattering technique is extensively used in condensed matter physics rather than in atomic and molecular physics. Recently, with the dramatic progress of the third generation synchrotron radiation and crystal analyzer, it provides the possibility to measure the dynamic parameters and electronic structures of atoms and molecules by using the high-resolution IXS.

Utilizing the high flux of the third generation synchrotron radiation and high density target sealed in a gas cell at about 1 MPa, we recently realized absolute dynamic parameter measurements of atoms and molecules by the IXS [1,2]. The electronic structures of the ground and excited states of some atoms and molecules [3-12] have been determined by the IXS at incident photon energies of about 10 keV and energy resolution of about 70 meV.

Recently, the generalized oscillator strenghts (GOSs) for the valence-shell excitations of acetylene have been measured, and the primary result for R30 is shown in Fig.1.

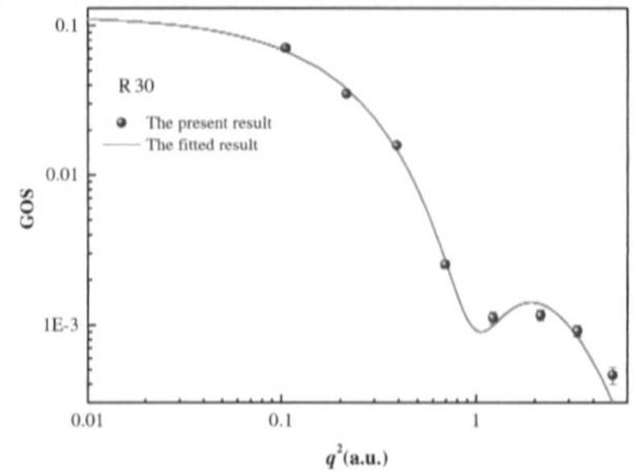

Fig.1 The GOS of R30 state of acetylene measured by the IXS

References

[1] B. P. Xie, L. F. Zhu, K. Yang et.al., 2010 *Phys. Rev. A* **82** 032501.
[2] L. F. Zhu, L. S. Wang, B. P. Xie et.al., 2011 *J. phys. B* **44** 025203.
[3] L. F. Zhu, W. Q. Xu, K. Yang et.al., 2012 *Phys.Rev. A* **85** 030501(R).
[4] K. Xu, K. Yang, Y. W. Liu et al., 2012 *Phys. Rev. A* **86** 022509.
[5] Y. W. Liu, X. X. Mei, X. Kang et al., 2014 *Phys. Rev. A* **89** 014502.
[6] Y. G. Peng, X. Kang, K. Yang et al., 2014 *Phys. Rev. A* **89**, 032512
[7] D. D. Ni, X. Kang, K. Yang et al., 2015 *Phys. Rev. A* **91**, 042501
[8] X. L. Zhao, K. Yang, L. Q. Xu et al.., 2015 *J. Chem. Phys.* **142**, 084301
[9] L. Q. Xu, Y. W. Liu, X. Kang et al., 2015 *Phys. Rep.* **5**, 18350
[10] X. Kang, Y. W. Liu, L. Q. Xu et al., 2015 *AstroPhys. J.* **807**, 96
[11] X. Xu, D. D. Ni, X. Kang et al., 2016*J. Phys. B* **49**, 064010
[12] Y. W. Liu, X. Kang, L. Q. Xu et al., 2016 *AstroPhys. J.* **819**, 142

[1] E-mail: lfzhu@ustc.edu.cn

Photodissociation of sympathetically crystallized CaH^+

Naoki Kimura[*1] **Masatoshi Kajita**[†] **Kunihiro Okada** [*2]

[*] Department of Physics, Sophia University, Tokyo 102-8554, Japan
[†] National Institute of Information and Communications Technology, Tokyo 184-8795, Japan

Synopsis We demonstrated photodissociation of sympathetically crystallized CaH^+ toward rovibrational spectroscopy by UV-UV or IR-UV double resonance. The photodissociation of CaH^+ was successfully confirmed at λ = 283-287 nm.

Rovibrational precision spectroscopy on sympathetically crystallized molecular ions is expected to be an important tool for discussions of fundamental physical constants stabillity [1]. Especially, some precision measurements of vibrational transitions in diatomic hydrides have been proposed toward time variation detection of proton-electron mass ratio β (= m_p / m_e) [2]. The vibrational transition of $X^1\Sigma(v,N) = (0,0) \rightarrow (1,0)$ in CaH^+ is one of the candidates. Recentry, sympathetic cooling and spectroscopic studies of CaH^+ have been demonstrated by several reserach groups [3] [4].

We have constructed a cryogenic linear Paul trap for rotational cooling of Coulomb crystallized CaH^+ [3] [5]. Now we are trying to observe the laser induced fluorescence (LIF) from internally cooled CaH^+. Unfortunately, the transition wavelength to observe the LIF has not been experimentally determined yet. Therefore, as an alternative plan, we started a photodissociation experiment in order to determine the rovibrational constants. Here, we report on the photodissociation experiment of sympathetically crystalized CaH^+ toward rovibrational spectroscopy.

In Fig. (a-1), we show an observed LIF image from a laser cooled Ca^+ crystal. As we demonstrated before, CaH^+ ions can be generated by a laser induced chemical reaction of $Ca^+(^2P_{1/2})$ + H_2 âĘŠ CaH^+ + H. As shown in Fig. (a-2), generated CaH^+ ions are sympathetically crystalized in a Ca^+ crystal. The ion numbers of Ca^+ and CaH^+ are determined by comparing LIF image with simulation images obtained by molecular dynamics simulations [3]. In the photodissociation experiment, we irradiated a UV pulsed laser beam (λ = 283-287 nm) to a mixed Coulomb crystal. Theoretical calculation predicted that the photodissociation via the excited state of $1^1\Pi$ in CaH^+ could occur [6]. As shown in Fig. (a-2) and (a-3), we successfully observed the photodissociation of sympathetically crystalized CaH^+. Fig. (b) shows a decay curve of the number of CaH^+ as a function of the laser irradiation time. A dissociation rate was determined to be $\gamma = 2.3(0.5)\times10^{-2}\ s^{-1}$ using a least-square fitting of a single exponential function to the data in Fig.b.

This work was supported by JPSJ KAKENHI Grant Number 15K13545 from the Ministry of Education, Culture, Sports, Science and Technology (MEXT).

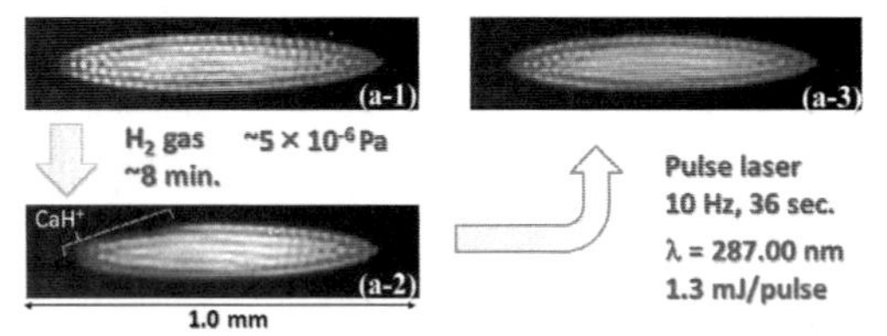

Figure a. CCD images of mixed species Colomb crystals during CaH^+ generation and photodissociaition. The ion numbers of Ca^+ in (a-1) and CaH^+ in (a-2) are 450 and 64, respectively.

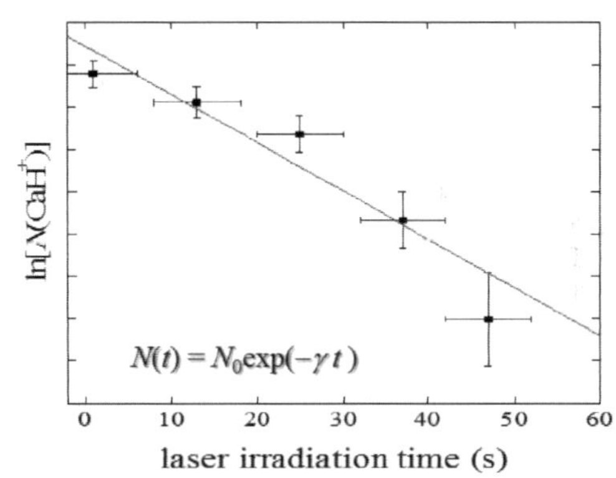

Figure b. A plot of the ion number of CaH^+ as a function of the laser irradiation time.

References

[1] X. Calmet *et al.*, 2002 *Eur. Phys. J. C* **24** 639

[2] M. Kajita *et al.*, 2009 *J. Phys. B* **42** 154022

[3] N. Kimura *et al.*, 2011 *Phys. Rev. A* **83** 033422

[4] N. B. Khanyile *et al.*, 2015 *Nat. Commun.* **6** 7825

[5] K. Okada *et al.*, 2015 *J. Phys. Conf. Ser.* **635** 032060

[6] M. Abe *et al.*, 2012 *Chem. Phys. Lett.* **521** 31-35

[1]E-mail: n-kimura-brl@eagle.sophia.ac.jp
[2]E-mail: okada-k@sophia.ac.jp

IOP Conf. Series: Journal of Physics: Conf. Series **875** (2017) 022043 doi:10.1088/1742-6596/875/3/022043

Accessing the hyperfine splitting in highly charged helium-like ions via angle-resolved x-ray spectroscopic analysis

Z. W. Wu[*,†,1], **A. V. Volotka**[*,‡], **A. Surzhykov**[§,¶] **and S. Fritzsche**[*,♯]

[*] Helmholtz-Institut Jena, D-07743 Jena, Germany
[†] College of Physics and Electronic Engineering, Northwest Normal University, Lanzhou 730070, P.R. China
[‡] Department of Physics, St. Petersburg State University, 198504 St. Petersburg, Russia
[§] Physikalisch-Technische Bundesanstalt, D-38116 Braunschweig, Germany
[¶] Technische Universität Braunschweig, D-38106 Braunschweig, Germany
[♯] Theoretisch-Physikalisches Institut, Friedrich-Schiller-Universität Jena, D-07743 Jena, Germany

Synopsis A new scheme is presented for determining the hyperfine splitting of highly charged ions via angle-resolved spectroscopic analysis. In particular, we propose two measurements on the angular distribution and linear polarization of the emission line $1s2p\,^3P_1, F = I-1, I, I+1 \rightarrow 1s^2\,^1S_0, F_f = I$ of He-like isotopic ions, following the stimulated decay of the $1s2s\,^1S_0, F_i = I$ level. We find that the angular and polarization behavior of the emitted photons strongly depends on the splitting of the overlapping hyperfine $1s2p\,^3P_1, F$ resonances, which could be a promising "signature" for the hyperfine splitting determination.

Hyperfine splitting in highly charged high-Z ions helps to perform accurate tests of QED in strong electromagnetic fields as generated by heavy nuclei. H-like ions as the simplest atomic system have received much attention for such kind of studies. For instance, the first successful measurement was performed for the $1s$ hyperfine splitting of H-like $^{209}_{83}Bi^{82+}$ ions [1].

Helium-like ions are still simple enough to be expected to serve for such QED tests. To the best of our knowledge, however, there have been no such kind of experimental studies up to now, although some theoretical work has been done. When compared to H- or Li-like ions, the ground state of He-like ions does not split into hyperfine levels. As for the excited levels like $1s2p\,^{1,3}P_1, F$, their linewidth is comparable to or even larger than their hyperfine splitting, which are hardly resolved due to their mutual overlap.

For this reason, we here propose a new scheme to measure the hyperfine splitting in He-like ions by analyzing the angular distribution and linear polarization of emitted fluorescence photons. More specifically, we consider the two-step radiative decay

$$\begin{aligned} &1s2s\,^1S_0, F_i = I + \gamma_1(\hbar\omega) \\ &\longrightarrow \quad 1s2p\,^3P_1, F = I-1, I, I+1 \\ &\longrightarrow \quad 1s^2\,^1S_0, F_f = I + \gamma_2 \end{aligned} \tag{1}$$

of He-like ions in order to resolve the hyperfine splitting of the intermediate levels. In this scheme, the initial level $1s2s\,^1S_0, F_i = I$ can be populated via the K-shell ionization of Li-like ions at ion storage rings [2], or alternatively, via the prompt $2s2p\,^1P_1, F = I-1, I, I+1 \rightarrow 1s2s\,^1S_0, F = I$ decay following the resonant electron capture of initially H-like ions into the $2s2p\,^1P_1$ level [3]. The first-step decay may further be stimulated by a laser that has proper intensity and adjustable γ_1 photon energy in order to compete with dominant $2E1$ decay of the initial $1s2s\,^1S_0, F_i = I$ level. Finally, the fluorescence γ_2 photons as emitted in the subsequent spontaneous decay are observed.

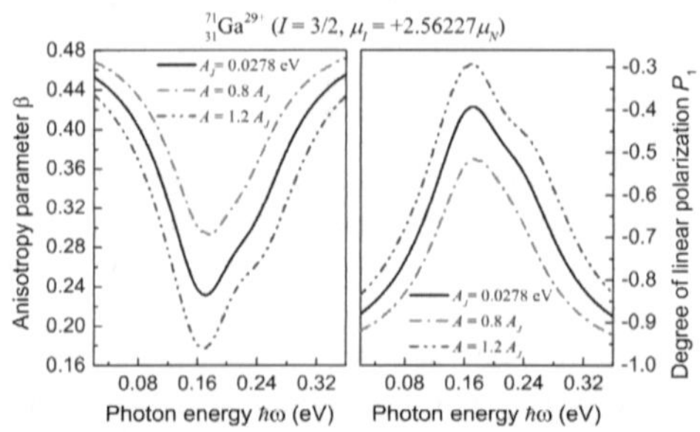

Figure 1. Anisotropy parameter β and linear polarization P_1 of the γ_2 photons for $^{71}_{31}Ga^{29+}$ ions as functions of photon energy of the stimulating laser. Results are given for the calculated hyperfine constant $A_J = 0.0278$ eV as well as for two assumed values $0.8A_J$ and $1.2A_J$.

Figure 1 shows the calculated γ_2 angular distribution and linear polarization for $^{71}_{31}Ga^{29+}$ ions. As seen clearly, these angular and polarization properties depend strongly on the hyperfine structure values. This could be used as a "signature" for determining the associated hyperfine splitting. We therefore suggest that accurate x-ray measurements may indeed serve as a promising tool to measure the hyperfine splitting in highly charged He-like ions.

References

[1] I. Klaft *et al.* 1994 *Phys. Rev. Lett.* **73** 2425.

[2] S. Trotsenko *et al.* 2010 *Phys. Rev. Lett.* **104** 033001

[3] P. H. Mokler *et al.* 1990 *Phys. Rev. Lett.* **65** 3108

[1]E-mail: zhongwen.wu@uni-jena.de

IOP Conf. Series: Journal of Physics: Conf. Series **875** (2017) 022044 doi:10.1088/1742-6596/875/3/022044

Photoionization of Ne^+ ions and Ne atoms near the K edge

A. Müller*[1], D. Bernhardt*, A. Borovik Jr.†, T. Buhr†, J. Hellhund*, K. Holste†, A. L. D. Kilcoyne‡, S. Klumpp§¶, M. Martins¶, S. Ricz£, J. Seltmannℜ, J. Viefhausℜ, S. Schippers†

* Institut für Atom- und Molekülphysik, Justus-Liebig-Universität Gießen, 35392 Giessen, Germany
† I. Physikalisches Institut, Justus-Liebig-Universität Gießen, 35392 Giessen, Germany
‡ Advanced Light Source, Lawrence Berkeley National Laboratory, Berkeley, CA 94720, USA
§ DESY Photon Science, FS-FLASH-D, 22607 Hamburg, Germany
¶ Institut für Experimentalphysik, Universität Hamburg, 22761 Hamburg, Germany
£ Institute of Nuclear Research of the Hungarian Academy of Sciences, Debrecen, 4026 Hungary
ℜ DESY Photon Science, FS-PE, 22607 Hamburg, Germany

Synopsis Absolute cross sections for photoprocesses near the K edge of Ne^+ ions and neutral Ne atoms have been determined at resolving powers up to 27000 facilitating the precise determination of natural widths of the most prominent resonances as well as the associated transition probabilities and branching ratios.

Single, double and triple photoionization of Ne^+ ions and electron production associated with single-photon absorption by a neutral Ne atom have been investigated [1]. Absolute cross sections were measured for Ne^+ employing the **P**hoton-**I**on Spectrometer at **PE**TRA III (PIPE) [2]. Electron emission from neutral Ne in a gas cell was monitored as a function of photon energy. By normalizing the electron-emission spectrum, fine-structure photoabsorption cross sections for Ne atoms were determined. Photon energies for the experiments with Ne^+ ions were between 840 and 930 eV covering the range from the lowest-energy resonances, which are associated with excitation of one single K-shell electron, up to double excitations involving one K- and one L-shell electron, well beyond the K-shell ionization threshold. Photoionization of neutral Ne was investigated at the K edge in the energy range from 863.6 to 870.8 eV. The associated electron production cross section is shown in Fig. 1.

Photon energy bandwidths ΔE were between 32 and 500 meV. Resolving powers up to $E/\Delta E \approx$ 27000 were achieved facilitating the precise determination of natural line widths of the most prominent lines associated with K-vacancy levels [3] in Ne and Ne^+. The uncertainty of the energy scale is estimated to be 0.2 eV. From the data obtained for Ne^+ ions, transition probabilities and branching ratios could be determined for the decay of K-vacancy levels including the fluorescence yield of Ne K_α radiation. For comparison with existing theoretical calculations, astrophysically relevant photoabsorption cross sections were inferred for Ne^+ by summing the measured and inferred partial ionization channels. The experiments revealed the presence of complex Auger-decay mechanisms. The ejection of three electrons from the lowest K-shell-excited $Ne^+(1s2s^22p^6\ ^2S_{1/2})$ level, for example, requires cooperative interaction of at least four electrons.

Funding by BMBF (contracts 05K10RG1, 05K10GUB, 05K16RG1, and 05K16GUC) for setting up the PIPE endstation is gratefully acknowledged. SK acknowledges support from the European Cluster of Advanced Laser Light Sources (EUCALL) project which has received funding from the *European Union's Horizon 2020 research and innovation programme* under grant agreement No 654220.

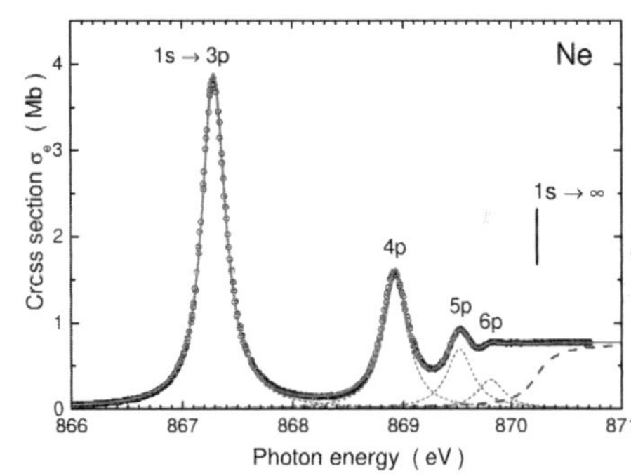

Figure 1. Cross section for the production of electrons subsequent to absorption of a single photon by a neutral neon atom with promotion of a K-shell electron to either $1s2s^22p^6np\ ^1P$ autoionizing levels with $n = 3,4,5,6,...$ or to the continuum. The dotted and dashed lines represent individual contributions to the cross section.

References

[1] A. Müller *et al.* 2017 *Astrophys. J.* **836** 166

[2] S. Schippers *et al.* 2014 *J. Phys. B* **47** 115602

[3] A. Müller 2015 *Phys. Scr.* **90** 054004

[1]E-mail: Alfred.Mueller@iamp.physik.uni-giessen.de

IOP Conf. Series: Journal of Physics: Conf. Series **875** (2017) 022045 doi:10.1088/1742-6596/875/3/022045

Impact of laser polarization on spectra of Eu $4f^76p_{3/2}$ *nl* autoionizing states

C. J. Dai* [1], **X. X. Chang**† [2], **and J. Wang**† [3]

* School of Science, Tianjin University of Technology, Tianjin 300384, China
† Key Laboratory of Display Materials and Photoelectric Devices, Ministry of Education, Tianjin 300384, China

Synopsis Experimental spectra of the Eu $4f\ ^76p_{3/2}$ n*l* autoionizing states are investigated with three-step isolated-core excitation technique, while the impact of laser polarization on the spectra is studied for the first time.

The Eu atom is a very challenging system due to its half-filled 4f shell and much more complex nature in its spectrum. Although some studies on the Eu atom have been reported[1-2], no works on the Eu $4f^76p_{3/2}$ n*l* (*l*=0,2) autoionizing states have been seen. This work will devote to the spectra of these autoionizing states with three dye lasers.

The bound Eu $4f^76sn$*l* Rydberg states are populated from the ground state via the $4f^76s6p$ states by two-step excitation, from which the Eu $4f^76p_{3/2}$ n*l* autoionization spectra are obtained by tuning the wavelength of the third laser around the Eu $4f^76s^+$—$4f^76p^+_{3/2}$ ionic line. Then, different configurations of polarization of three lasers are applied to fulfill the three-step ICE scheme in order to observe their spectral differences. To have a better understanding about the the impact of laser polarization on the spectra, the complex features are explained with the physical principles, such as the selection rules of total angular momentum under different excitation paths, and those of magnetic quantum number under different configurations of polarization of three lasers.

The wavelength of the third laser λ_3 is scanned from 394 nm to 402 nm to cross the ionic line of the $4f^76s^+$—$4f^76p_{3/2}^+$ (J=4) , while the wavelength of the first two lasers, λ_1 and λ_2, are fixed at 564.74 and 529.61 nm. All lasers are linearly polarized, which are set either parallel (π) or perpendicular (σ) to the MCP detector (i.e., direction of electric field to collect the ions), corresponding to the two cases where the selection rule of $|\Delta M|$ = 0 or $|\Delta M|$ = 1 applies. With this method， spectra of several n values for the Eu $4f^76p_{3/2}$ n*l* (*l*=0,2) autoionization Rydberg series have been obtained. As an example, the observed spectrum of Eu $4f^76p_{3/2}6d$ autoionizing state is shown in figure 1, to be compared with a multi-peak fit to the famous Fano-Beaulter line shape.

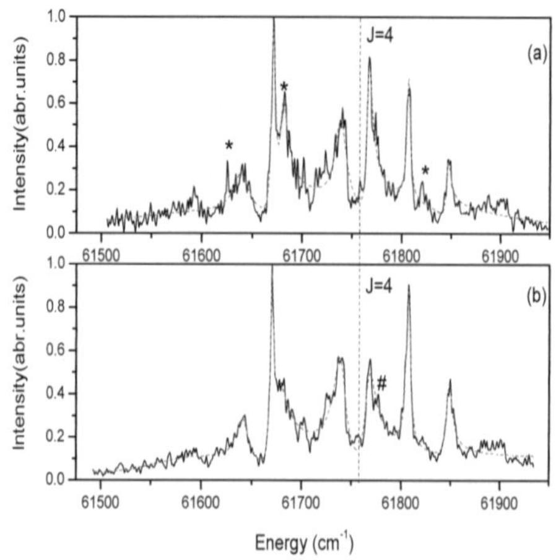

Figure 1. Observed spectrum (black) $4f^76p_{3/2}6d$ autoionizing state with a multi-peak fit (red) ,(a) $|\Delta M|$=0 and (b) $|\Delta M|$=1.

As shown in figure 1, instead of a single broad profile of Eu $4f^76p_{3/2}6d$ autoionizing state, many narrow peaks originated from other autoionization series are superimposed, leading to a structured spectrum. This phenomenon manifests the heavy configuration interaction among the Eu $4f^76p_{3/2}6d$ autoionizing state and $4f^76p_{1/2}$ n*l* and $4f^75d_{1/2,3/2}$ n*l* autoionization series. Some differences can be found in the two spectra, particularly those peaks marked with sign *or # , which demonstrates the impact of different kinds of polarization schemes. The similar effects are also observed on the spectra of the Eu $4f^76p_{3/2}$ ns autoionizing states .

References

[1] J Xie and C J Dai 2011 J. Phys. **B44(1) 015002**.
[2] X Wang and C J Dai 2012 J. Phys. **B45(16) 165001**

[1] E-mail: Daicj@126.com
[2] E-mail: Changxx@126.com

IOP Conf. Series: Journal of Physics: Conf. Series **875** (2017) 022046 doi:10.1088/1742-6596/875/3/022046

Studying Cold Potassium Rydberg Atoms with an AC-MOT

Matthew Harvey*[1], John Agomuo*, Ahmad Sakaamini*, Andrew Murray*[2]

*Photon Science Institute, School of Physics & Astronomy, University of Manchester, Manchester M13 9PL, UK.

Synopsis A new method is described for the detection and study of cold Rydberg atoms. Cold atoms held in an AC-driven magneto optical trap (AC-MOT) are selectively laser excited to a Rydberg state in a stepwise process. Ions or electrons from interactions between Rydberg atoms are then electrostatically extracted and detected via a channel electron multiplier. The apparatus is capable of studying Rydberg atoms from n = 15 to n ~ 220.

A new method for Rydberg atom detection and study is described. A beam of ^{39}K atoms is produced by an oven source and is then slowed by a Zeeman slower before being further cooled and trapped in an alternating current magneto-optical trap (AC-MOT) [1].

In the AC-MOT, the trapping laser polarisation is switched synchronously with an alternating magnetic **B**-field. This allows the trapping fields to be switched off in less than 20 µs, ~300 times faster than for a conventional MOT. Field-free experiments can then take place without any loss of trapping efficiency. Figure 1 shows a simplified block diagram of the experiment.

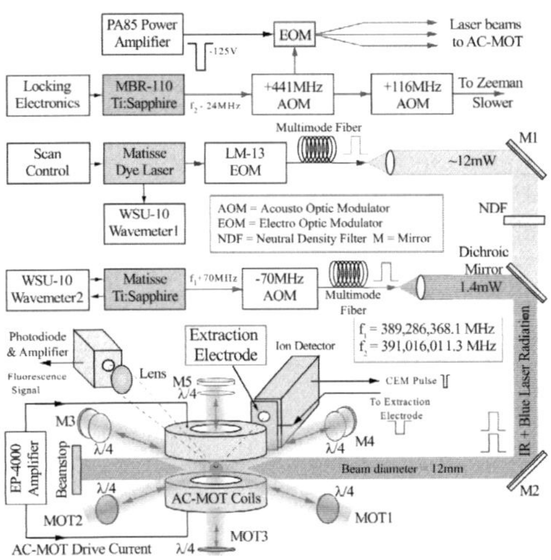

Figure 1. Simplified block diagram of the experiment.

Trapped atoms are selectively laser excited to Rydberg states in a stepwise process. Ions or electrons from interactions between Rydberg atoms are then detected using a threshold penetrating-field detector [2]. The experiments take place in a field-free interaction region that is set by the two MOT coils, and four electrostatic deflector plates in the horizontal plane (not shown). The trapping light is provided by a continuous wave (CW) MBR-110 Ti:Sapphire (Ti:S) laser. A second CW Matisse Ti:S laser is switched into the experiment by an AOM to resonantly excite ground state atoms to the $4^2P_{1/2}$ state. These atoms are further excited to Rydberg states by a UV-pumped, tunable CW Matisse dye laser ($\lambda \approx 418 \rightarrow 470$ nm) that is switched in by an EOM. These lasers are typically switched on for a few µs, after which a penetrating field is switched to deflect ions or electrons into the detector.

The laser frequencies are measured to within 1 MHz by WSU-10 wavemeters calibrated by saturated absorption [3]. This highly sensitive technique allows cold Rydberg atoms from n = 15D → n ~220S to be created and detected. Figure 2 shows the measured spectra for 15D to 196S, collected by scanning the dye laser. The amplitude variation is mainly due to varying dye laser power.

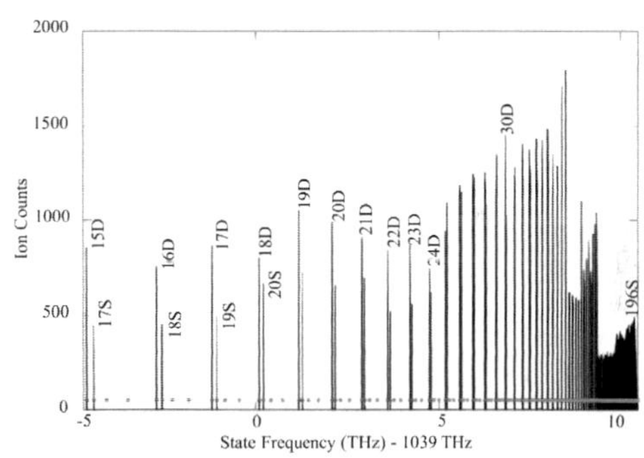

Figure 2. Excitation of Rydberg atoms from n = 15D to n = 196S using CW laser radiation, offset from 1039 THz.

The progress of these studies will be presented at the conference.

References

[1] M Harvey and AJ Murray 2008 *Phys. Rev. Lett.* **101** 173201

[2] S Cvejanovic and FH Read 1974 *J. Phys. B* **7** 1180

[3] TW Hänsch *et al.* 1971 *Phys. Rev. Lett.* **27** 707

[1]E-mail: Matthew.Harvey@manchester.ac.uk
[2]E-mail: Andrew.Murray@manchester.ac.uk

IOP Conf. Series: Journal of Physics: Conf. Series **875** (2017) 022047 doi:10.1088/1742-6596/875/3/022047

Evolution analysis of EUV radiation from laser-produced tin plasmas

M. G. Su*[1], Q. Min*, S. Q. Cao*, D. X. Sun*, P. Hayden†, G. O'Sullivan‡ and C. Z. Dong*[2]

* Key Laboratory of Atomic and Molecular Physics & Functional Material of Gansu Province, College of Physics and Electronic Engineering, Northwest Normal University, Lanzhou 730070, China.

† School of Physical Sciences and National Centre for Plasma Science and Technology, Dublin City University, Glasnevin, Ireland

‡ School of Physics, University College Dublin, Belfield, Dublin, Ireland

Synopsis We present a simplified radiation hydrodynamic model based on the fluid dynamic equations and the radiative transfer equation to rapidly investigate the radiation properties and dynamics in laser-produced tin plasmas. The self-absorption features of EUV spectra measured at an angle of 45° to the direction of plasma expansion have been successfully simulated and explained. The evolution of the plasma temperature has also been evaluated.

One of fundamental aims of extreme ultraviolet (EUV) lithography is to maximize brightness or conversion efficiency of laser energy to radiation at specific wavelengths from laser produced plasmas (LPPs) of specific elements for matching to available multilayer optical systems. For an investigation of EUV radiation of laser-produced tin plasmas, it is crucial to study the related atomic processes and their evolution so as to reliably predict the optimum plasma and experimental conditions [1].

In this work, a simplified radiation hydrodynamics model was developed to explain the complicated self-absorption profiles in EUV spectra of highly-charged tin ions. The origin of self-absorption bands and dips around 13.5 nm in Sn spectra have been successfully simulated and explained, as is indicated in Fig.1.

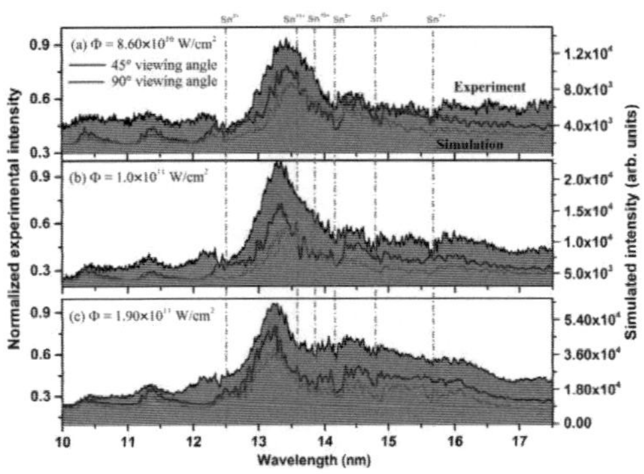

Figure 1. Comparisons of spectral profiles between the experimental and two simulated results for 90° and 45° viewing angles based on the radiation hydrodynamics model.

The plasma temperature is a very important parameter in the characterization of any plasma, since it determines the ion charge and energy distributions. To provide an intuitive description for the plasma expansion process, the contour images of plasma temperature are shown in Fig. 2. It is clearly seen that with increasing power density the distribution in space of the plasma temperature shows a gradual increase at each position considered. The results show an intuitive physical picture of the distributions of temperature within the expanding plume, which are helpful for a more detailed understanding of the spectral features and hydrodynamics evolution for highly charged ions of the mid- and high-Z elements.

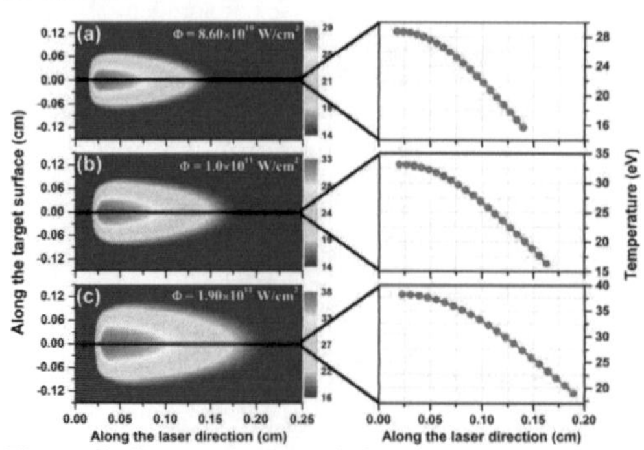

Figure 2. Contour images of plasma temperature in laser-produced tin plasmas at the different pulse power densities. The graphs on the right side show the varieties of plasma temperature along the laser direction.

This work is supported by the National Natural Science Foundation of China (Grants Nos. 11364037, U1332206, and 11564037)

References

[1] V. Y. Banine *et al.* 2011 *J. Phys. D: Appl. Phys.* 44 253001

[1] E-mail: nwnu_sumg@163.com

[2] E-mail: dongcz@nwnu.edu.cn

IOP Conf. Series: Journal of Physics: Conf. Series **875** (2017) 022048 doi:10.1088/1742-6596/875/3/022048

Radiative lifetime of metastable Xe^{3+} measured using an electrostatic ion beam trap

Manabu Saito[*1], **Asahi Chikaoka**[*], **Takuya Majima**[†], **Makoto Imai**[*], **Hidetsugu Tsuchida**[†]

[*] Department of Nuclear Engineering, Kyoto University, Kyoto 615-8540, Japan
[†] Quantum Science and Engineering Center, Kyoto University, Uji 611-0011, Japan

Synopsis We have measured the radiative lifetime of the $^2P_{3/2}$ metastable level in Xe^{3+} using an electrostatic ion beam trap. The lifetime was obtained by measuring the number of photons emitted during the $^2P_{3/2}\rightarrow{}^2D_{3/2}$ transition of Xe^{3+} as a function of the storage time. The lifetime is compared with previously reported theoretical and experimental values.

Measurements of radiative lifetimes for forbidden transitions of ions provide information that would be useful not only for improving the atomic and molecular structure models but also for applications such as the diagnostics of density and temperature in astrophysical and laboratory plasmas. The radiative lifetime of the $^2P_{1/2,3/2}$ metastable levels in the ground-state configuration ($5s^25p^3$) of Xe^{3+} has been studied experimentally and theoretically by several research groups. The experimental lifetime of the $^2P_{1/2}$ state (15.6 ± 0.9 ms [1]) was in good agreement with theoretical values (15.8 ms [1] and 16.4 ms [2]). On the other hand, the experimental lifetime of the $^2P_{3/2}$ state (5.3 ± 0.5 ms [1]) was smaller than the theoretical values (6.6 ms [1] and 6.8 ms [2]) even considering the experimental error. In this work, we measured the radiative lifetime of the $^2P_{3/2}$ metastable state of Xe^{3+} using an electrostatic ion trap of the ion-beam storage type [3, 4]. The trap was placed away from the ion source, which enabled us to prevent the inflow of the ion-source gas into the trap. This minimized the influence of ion-beam losses in the trap due to collisions with residual gases.

Figure 1 shows the experimental arrangement. A Xe^{3+} beam of 3.6 keV was produced using an electron impact ion source. The beam was chopped using an electrostatic chopper and was mass analyzed in a 45° deflecting magnet. The ion bunch was then injected into the trap, and was multiply reflected between the entrance and exit-side electrodes along the trap axis, resulting in the storage of the ion-beam in the trap. The 447 nm photons emitted during the $^2P_{3/2}\rightarrow{}^2D_{3/2}$ transition were measured with an optical system, which is composed of a view port, a bandpass filter (450 ± 12.5 nm), and a photomultiplier tube (PMT) as shown in Fig. 1. The count rate of photons as a function of time was recorded using a multichannel scaler (MCS). Since the count rate of the photons at a given time t is proportional to the number of Xe^{3+} ions in the $^2P_{3/2}$ state at the time t, we can obtain the lifetime of the $^2P_{3/2}$ state from the time spectrum of the photons. The trap was maintained at a pressure of about 1×10^{-7} Pa during measurements.

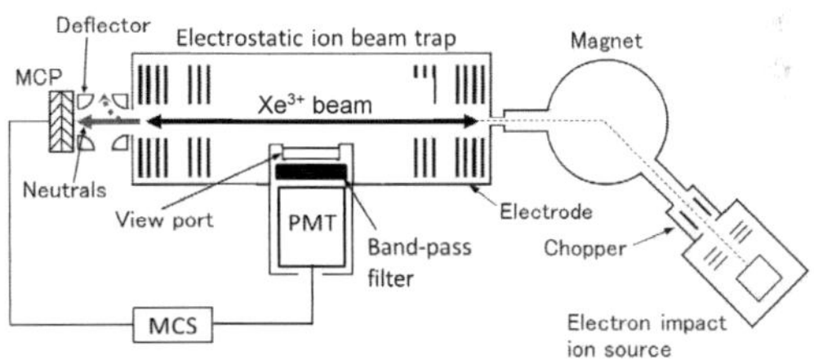

Figure 1. Experimental arrangement

We obtained a lifetime of 6.0 ± 0.7 ms from the time spectrum. However, this value is an apparent lifetime. Since the number of stored Xe^{3+} ions decreases with the trapping time due to electron capture or elastic collision with residual gases, the obtained value has to be corrected for the ion-storage lifetime. For that reason, the number of the stored Xe^{3+} ions as a function of time was monitored through the measurements of Xe^{2+} ions escaping from the trap, which were produced by single electron capture from residual gases in the trap. The preliminary result, corrected for the ion-storage lifetime, is 7.2 ± 0.9 ms. This value is in agreement with the previous theoretical values within the experimental error.

References

[1] A. G. Calamai and C. E. Johnson 1992 Phys. Rev. A **45** 7792
[2] E. Biémont et al. 1995 Astron. Astrophys. Suppl. Ser. **111** 333
[3] K. G. Bhushan et al. 2000 *Phys. Rev.* A **62** 012504
[4] M. Saito et al. 2015 *Phys. Rev.* A **91** 012508

[1] E-mail: saito@nucleng.kyoto-u.ac.jp

Hyperfine stark shifts of ground states of ^{87,85}Rb

X. Wang, J. Jiang[1], L. Y. Xie, D. H. Zhang, C. Z. Dong[2]

Key Laboratory of Atomic and Molecular Physics and Functional Materials of Gansu Province, College of Physics and Electronic Engineering, Northwest Normal University, Lanzhou, 730070, China

Synopsis The hyperfine stark shifts of ground states of ^{87,85}Rb are calculated by using relativistic configuration interaction plus core polarization (RCICP) approach, which agree with other theoretical and experimental results very well. The hyperfine Stark shift of ^{87}Rb is larger than that of ^{85}Rb.

The hyperfine stark shift is closely related to the difference of scalar polarizabilities between the hyperfine states with the same (L, J) but different F quantum numbers. It is essential to evaluate the effect of blackbody radiation (BBR) shift, which are very important to determine the uncertainty of atomic clock. The hyperfine structure(hfs) transition of the ground state of ^{133}Cs serves as a primary frequency standard for the current definition of a second. The similar hfs transition of Rb atom is also considered as secondary microwave frequency standard [1].

The polarizabilities of the ground states of Rb atom are calculated by using relativistic configuration interaction plus core polarization (RCICP) approach. After considering the hyperfine splittings, the energies and transition matrix elements of the hyperfine structure of ^{87,85}Rb are computed. Then the scalar(α_1) and tensor(α_1^T) polarizabilities of the hyperfine ground states of ^{87,85}Rb are determined.

Table 1 gives the differences of scalar and tensor polarizabilities of the hyperfine ground states of ^{87,85}Rb. There are some results of the hyperfine Stark shifts of ^{87,85}Rb [2-6] that are often reported as the Stark shift coefficients k, with units of $(\text{Hz}/(\text{V/m})^2$. This is converted into a.u. by multiplying 0.4018778×10^8. The present hyperfine Stark shifts of ^{87,85}Rb agree with the relativistic configuration interaction plus many-body perturbation (RCI+MBPT) [3], the relativistic linearized coupled cluster single-double with partial triple contributions (RLCCSDT) [4], and the perturbation theory [2] very well. These theoretical results are in good agreement with the experimental values by Mowat *et al.* [5] and Dallal *et al.* [6]. The hyperfine Stark shift of ^{87}Rb is larger than that of ^{85}Rb. The tensor polarizabilities of the hyperfine states do not exceed 10^{-4} a.u. in magnitude. The tensor polarizability of the $F=1$ ground state of ^{87}Rb is positive and that of the $F=2$ ground state of ^{87}Rb is negative. The difference of present tensor polarizabilities of ^{87}Rb is more negative than the experimental value [6]. Similarly, the difference of tensor polarizabilities of ^{87}Rb is also larger than that of ^{85}Rb. It is predictable since ^{85}Rb has the smaller hyperfine splittings than ^{87}Rb.

Table 1. The differences of scalar and tensor dipole polarizabilities (in a.u.) of the hyperfine ground states of ^{87,85}Rb. The notation $a[b]$ means $a\times10^b$.

Method	α_1
^{87}Rb: $\alpha_1(F=2)-\alpha_1(F=1)$	
Present RCICP	0.995[−2]
RCI + MBPT [3]	0.997(8)[−2]
RLCCSDT [4]	0.997(3)[−2]
Perturbation theory [2]	0.972[−2]
Expt. [5]	0.99(24)[−2]
Expt. [6]	0.9967(32)[−2]
^{85}Rb: $\alpha_1(F=3)-\alpha_1(F=2)$	
Present RCICP	4.486[−3]
Perturbation theory [2]	4.311[−3]
Expt. [5]	4.389(96)[−3]
^{87}Rb: $\alpha_1^T(F=2)-\alpha_1^T(F=1)$	
Present RCICP	−1.0409[−4]
Expt. [6]	−0.8841(1045)[−4]
^{85}Rb: $\alpha_1^T(F=3)-\alpha_1^T(F=2)$	
Present RCICP	−6.1115 [−5]

This work is supported by National Natural Science Foundation of China (Grant No. 11564036, U1332206, 11464042, U1330117, U1331122)

References

[1] J. Camparo 2007 *Phys. Today* **60** 33
[2] T. Lee *et al.* 1975 *Phys. Rev. A* **11** 1784
[3] E. J. Angstmann *et al.* 2006 *Phys. Rev. A* **74** 023405
[4] M. S. Safronova *et al.* 2010 *Phys. Rev. A* **82** 022510
[5] J. R. Mowat *et al.* 1972 *Phys. Rev. A* **5** 1059
[6] Y. Dallal *et al.* 2015 *Phys. Rev. Lett.* **115** 183001

[1] E-mail: phyjiang@yeah.net
[2] E-mail: dongcz@nwnu.edu.cn

Magic wavelengths for the $6s^2\,{}^1S_0$ - $6s6p\,{}^3P_1$ transition in ytterbium atom

Zhi-Ming Tang *†, **Yan-Mei Yu** † [1], **Ji-Guang Li** ‡, **Chen-Zhong Dong** * [2]

* Key Laboratory of Atomic and Molecular Physics and Functional Materials of Gansu Province, College of Physics and Electronic Engineering, Northwest Normal University, Lanzhou 730070, P. R. China
† Beijing National Laboratory for Condensed Matter Physics, Institute of Physics, Chinese Academy of Sciences, Beijing 100190, P. R. China
‡ Data Center for High Energy Density Physics, Institute of Applied Physics and Computational Mathematics, Beijing 100088, P. R. China

Synopsis The electric-dipole dynamic polarizabilities of the $6s^2\,{}^1S_0$ and $6s6p\,{}^3P_1$ states of Yb atom are calculated by using the CI+MBPT method. Magic wavelengths for the $6s^2\,{}^1S_0$ - $6s6p\,{}^3P_1$ transition are identified.

The magic wavelengths for a transition are wavelengths of the laser field at which the ac Stark shift of the transition energy is zero so that the transition frequency is insensitive to the laser intensity [1]. Using the optical trap with a magic wavelength could permit coherent control of electronic transitions independent of the atomic center-of-mass motion. Up to now, the magic wavelengths for the clock transition $6s^2\,{}^1S_0$ - $6s6p\,{}^3P_0$ in ytterbium have been well investigated [2]. However, no study on magic wavelengths for the $6s^2\,{}^1S_0$ - $6s6p\,{}^3P_1$ transition in Yb is available. Indeed, the ${}^1S_0$ - ${}^3P_1$ transition in bivalence atom including Yb is also vital in many experiments involving laser cooling and trapping such as the measurement of the atomic electric dipole moment [3].

In this work, the electric-dipole dynamic polarizabilities of the $6s^2\,{}^1S_0$ and $6s6p\,{}^3P_1$ states of Yb atom are calculated by using the configuration interaction plus many-body perturbation theory (CI+MBPT) method [4] and the CI-MBPT program package [5]. The one-electron basis set for the calculation includes 1 - 23s, 2 - 22p, 3 - 22d, 4 - 20f, and 5 - 18g orbitals, where the core and 6 - 9s, 6 - 9p, 5 - 8d orbitals are Dirac-Hartree-Fock (DHF) ones constructed by solving the DHF equations in the V^N, V^{N-1}, V^{N-2} approximations step by step, while all the rest orbitals are virtual ones. The polarizabilities are determined by using the sum-over-states method employing the calculated reduced dipole matrix elements and the experimental transition energies.

The dynamic polarizabilities of $6s^2\,{}^1S_0$ and $6s6p\,{}^3P_1$ states at photon frequencies below 0.115 a.u. are shown in Fig. 1. 'Magic' conditions occur when the polarizabilities of the two states intersect off resonance marked by small circles with numbers on the plot. We predict thirty-six magic wavelengths as is shown in Table 1, while sixteen of them, from No. 7 to No. 22, are located in the visible region.

[1]E-mail: ymyu@aphy.iphy.ac.cn
[2]E-mail: dongcz@nwnu.edu.cn

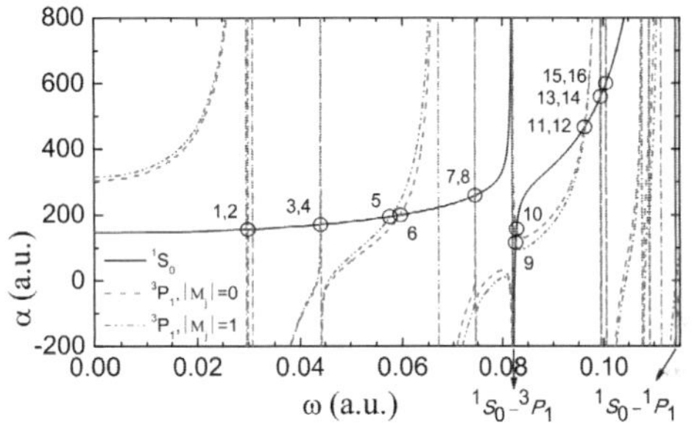

Figure 1. (color online) Dynamic polarizabilities of $6s^2\,{}^1S_0$ and $6s6p\,{}^3P_1$ states of Yb atom.

Table 1. Magic wavelengths for the $6s^2\,{}^1S_0$ - $6s6p\,{}^3P_1$ transition in Yb atom.

No.	λ^*(nm)	No.	λ^*(nm)	No.	λ^*(nm)	No.	λ^*(nm)
1	1526.19	10	550.38	19	417.62	28	379.35
2	1522.80	11	473.76	20	417.60	29	378.06
3	1035.36	12	472.67	21	410.67	30	377.83
4	1035.24	13	458.23	22	410.02	31	369.33
5	789.64	14	458.19	23	387.38	32	369.19
6	766.13	15	453.21	24	387.38	33	364.27
7	611.45	16	453.21	25	379.93	34	364.07
8	611.35	17	423.78	26	379.93	35	335.05
9	551.59	18	423.71	27	379.36	36	334.75

This work is supported by the National Natural Science Foundation of China under Grants No. 91536106 and No. U1332206.

References

[1] H. Katori *et al.* 2003 *Phys. Rev. Lett.* **91** 173005
[2] K. Guo *et al.* 2010 *J. Phys. B: At. Mol. Opt. Phys.* **43** 135004
[3] M. Bishof *et al.* 2016 *Phys. Rev. C* **94** 025501
[4] V. A. Dzuba *et al.* 1996 *Phys. Rev. A* **54** 3948
[5] M. Kozlov *et al.* 2015 *Comput Phys Commun* **195** 199

IOP Conf. Series: Journal of Physics: Conf. Series **875** (2017) 022051 doi:10.1088/1742-6596/875/3/022051

Lifetimes of bound excited states of Pt^-

KC Chartkunchand*[1], M. Kamińska*†, E. K. Anderson*, M. K. Kristiansson*, G. Eklund*, O. M. Hole*, R. F. Nascimento‡, M. Blom*, M. Björkhage*, A. Källberg*, P. Löfgren*, P. Reinhed*, S. Rosén*, A. Simonsson*, R. D. Thomas*, S. Mannervik*, V. T. Davis§, P. A. Neill§, J. S. Thompson§, D. Hanstorp¶, H. Zettergren*, H. Cederquist*, and H. T. Schmidt*[2]

* Department of Physics, Stockholm University, AlbaNova, SE-106 91 Stockholm, Sweden
† Institute of Physics, Jan Kochanowski University, 25-369 Kielce, Poland
‡ Centro Federal de Educação Tecnológica Celso Suckow da Fonseca, Petrópolis, 25620-003 RJ, Brazil
§ Department of Physics, University of Nevada, Reno, Nevada 89557, United States
¶ Department of Physics, University of Gothenburg, SE-412 96 Gothenburg, Sweden

Synopsis Measurements of the radiative lifetimes of the two excited states of the platinum anion Pt^- are presented. Pt^- ions stored in the cryogenic ion storage ring DESIREE were photodetached at different photon wavelengths and the resulting yield of neutral Pt measured as a function of time was recorded. Analysis of the neutral decay curves show a 2.54 ± 0.10 s lifetime for the higher-lying $5d^{10}6s\ ^2S_{1/2}$ excited state and a lifetime in the range of 50–200 ms for the lower-lying $5d^9 6s^2\ ^2D_{3/2}$ excited state. This is the first study to report the lifetime of a bound anion excited state with an electron configuration different from that of the anion ground state.

Anions have long been fascinating systems to study. They are a testament to the subtle nature of electronic structure and dynamics on the atomic scale, with complex electron correlation effects responsible for their very existence. Properties such as bound state lifetimes are crucial in developing a theoretical understanding of these correlation effects. Measurements of these lifetimes provide critical benchmarks for theoretical methods treating these effects in anions and more general atomic systems.

We present studies of Pt^- conducted at the cryogenic electrostatic ion storage ring DESIREE [1, 2]. The storage temperature of 13 K results in pressures on the order of 10^{-14} mbar, significantly reducing ion losses due to collisions with residual gases. Photodetachment due to thermal blackbody radiation is also reduced, an important consideration for loosely-bound systems such as anions. DESIREE's capabilities for studying anion excited state lifetimes were recently demonstrated by lifetime measurements of the $3p^5\ ^2P_{1/2}$ excited state of the sulfur anion [3] and the $3d^9 4s^2\ ^2D_{3/2}$ excited state of the nickel anion [4].

In the present experiment [5], Pt^- anions were stored in DESIREE, with storage times of ~950 s. State-selective photodetchment was used on the stored Pt^- beam: 550 nm photons detached all three bound states of Pt^-, 750 nm photons detached the $5d^9 6s^2\ ^2D_{3/2}$ and $5d^{10}6s\ ^2S_{1/2}$ excited states, and 1360 nm photons detached only the $^2S_{1/2}$ state.

Analysis of decay curves due to photodetachment (Figure 1) yield a radiative lifetime of 2.54 ± 0.10 s for the $^2S_{1/2}$ excited state. Decay to the $5d^9 6s^2\ ^2D_{5/2}$ ground state of Pt^- proceeds via an electric quadrupole (E2) transition, the first of its kind to be observed in an anion. Measurement of the $^2D_{3/2}$ excited state, which decays via a magnetic dipole (M1) transition to the ground state, is complicated by its faster decay compared to the higher-lying $^2S_{1/2}$ state. As a result, only a lifetime in the range of 50–200 ms could be estimated for the $^2D_{3/2}$ excited state.

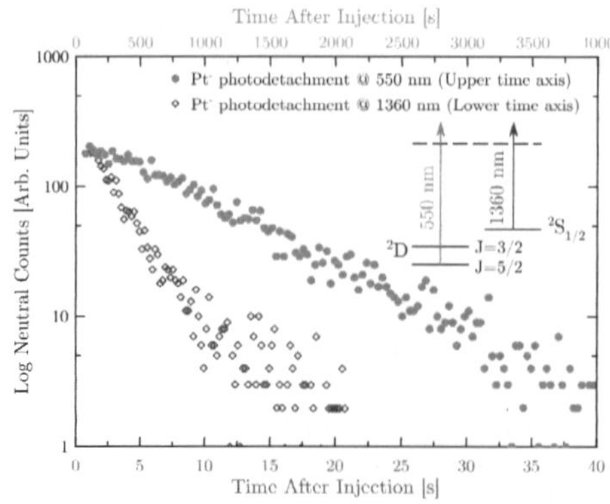

Figure 1. Neutral counts for Pt^- photodetachment at 550 nm and 1360 nm photon wavelengths. The decay curve measured at 550 nm yields the storage time of Pt^- ions in the ring (~950 s).

References

[1] R. D. Thomas *et al.* 2011 *Rev. Sci. Instrum.* **82** 065112

[2] H. Schmidt *et al.* 2013 *Rev. Sci. Instrum.* **84** 055115

[3] E. Bäckström *et al.* 2015 *Phys. Rev. Lett.* **114** 143003

[4] M. Kamińska *et al.* 2016 *Phys. Rev. A* **93** 012512

[5] K. C. Chartkunchand *et al.* 2016 *Phys. Rev. A* **94** 032501

[1]E-mail: kiattichart.chartkunchand@fysik.su.se
[2]E-mail: schmidt@fysik.su.se

IOP Conf. Series: Journal of Physics: Conf. Series **875** (2017) 022054 doi:10.1088/1742-6596/875/3/022054

Lamb shift measurement of antihydrogen for determining the charge radius of antiproton and a stringent test of CPT symmetry

Naofumi Kuroda*[1], David A. Cooke†, Paolo Crivelli†, Hiroyuki Higaki‡, Gianluca Janka†, Yasuyuki Matsuda*, Balint Radics†, Christian Regenfus†

* Institute of Physics, The University of Tokyo, Komaba, Meguro-ku, 153-8902 Tokyo, Japan
† Institute for Particle Physics, ETH Zürich, 8092 Zurich, Switzerland
‡ Graduate School of Advanced Science of Matter, Hiroshima University, Higashi-Hiroshima, 739-8530 Hiroshima, Japan

Synopsis Measurements of Lamb shift in antihydrogen atom have been planned to extract the antiproton charge radius which has never been experimentally investigated. A hyperfine spin filter and a microwave cavity can be employed.

Cold antihydrogen ($\bar{\mathrm{H}}$) atoms have been synthesized recently at CERN to test CPT symmetry and weak equivalence principle. Several groups at CERN have launched experiments of precise spectroscopy for 1S–2S transition and ground-state hyperfine transition, while the Lamb shift of $\bar{\mathrm{H}}$ atoms has not been targeted. We plan a measurement of the $2S_{1/2}$–$2P_{1/2}$ Lamb shift of $\bar{\mathrm{H}}$ at CERN AD-ELENA facility.

Precise spectroscopic study of the Lamb shift in $\bar{\mathrm{H}}$ will yield information on the antiproton ($\bar{p}$) size which has never been investigated experimentally. The most precise direct measurement of the Lamb shift using atomic hydrogen beams achieved 9×10^{-6}[1], which determined the charge radius of the proton at an accuracy of 3%[2]. Any CPT violation could also emerge from a discrepancy between the proton and the antiproton radii [3].

The measurement has been planned to be performed at the GBAR beamline constructing at AD-ELENA, where $\bar{p}$ beams at several keV in kinetic energy are injected into a dense positronium (Ps) target and provide neutral anti-atoms together with positively charged anti-ions by the following charge exchange reactions;

$$\bar{p}+\mathrm{Ps} \rightarrow \bar{\mathrm{H}}+e^-,$$
$$\bar{\mathrm{H}}+\mathrm{Ps} \rightarrow \bar{\mathrm{H}}^+ +e^-.$$

Calculations (eg. [4]) predict $2S$ state $\bar{\mathrm{H}}$ atoms. The idea of beam spectroscopy of the Lamb shift is to use this flux of $2S$ $\bar{\mathrm{H}}$. From a designed beam intensity of AD-ELENA, the line center can be determined with an uncertainty of 100 ppm, which gives antiproton radius at the level of 10% in precision[5]. A microwave at a cavity induces a transition from thus obtained $2S_{1/2}$ of $\bar{\mathrm{H}}$ to short lived $2P_{1/2}$, antihydrogen atoms de-excite. Antihydrogen atoms in $2S$ metastable state are quenched at the end of a spectrometer line. The Lyman α light from the quenched anti-atom is detected by a photo detector.

[1]E-mail: kuroda@phys.c.u-tokyo.ac.jp

The resonance profile should be composed of hyperfine states of antihydrogen atom. Figure 1 shows the Breit-Rabi diagram for antihydrogen $2S_{1/2}$ and $2P_{1/2}$ states. If hyperfine states of $2S_{1/2}$ are sorted out, a spin polarized beam would be obtained. Thus, the line shape becomes narrow[6]. Two possible schemes are considered. One is to apply an rf to depopulate $F=1$ states to $2P_{1/2}$. The β_4 state will be selected. Another is a spin selector which applies an rf and a magnetic field to select α_1 or α_2 state. A hyperfine selector is studied for this purpose.

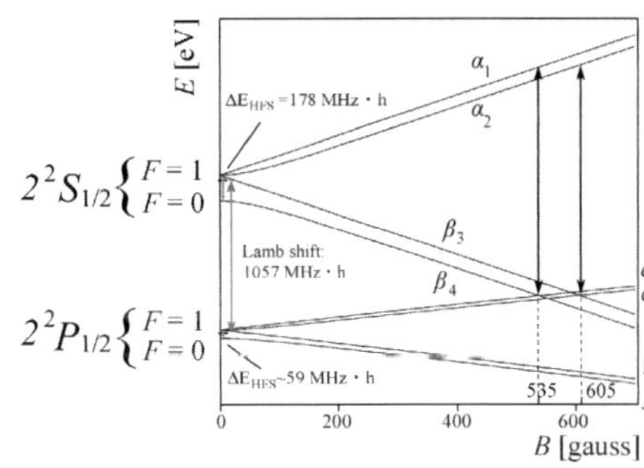

Figure 1. The Breit-Rabi diagram for $2S_{1/2}$ and $2P_{1/2}$ states of antihydrogen atom.

References

[1] S.R. Lundeen and F.M. Pipkin 1981 *Phys. Rev. Lett.* **46** 232

[2] S.G. Karshenboim 1999 *Can. J. Phys.* **77** 241

[3] V.A. Kostelecký and A.J. Vargas 2015 *Phys. Rev. D* **92** 056002

[4] P. Comini and P.-A. Hervieux 2013 *N. J. Phys.* **15** 095002

[5] P. Crivelli *et al.* 2016 *Phys. Rev. D* **94** 05208

[6] G. Newton *et al.* 1979 *Phil. Trans. R. Soc. A* **290** 1373

IOP Conf. Series: Journal of Physics: Conf. Series **875** (2017) 032001 doi:10.1088/1742-6596/875/4/032001

Inelastic squared form factors of the valence-shell excitations of hydrogen studied by high-resolution inelastic x-ray scattering

Long-Quan Xu *, Ya-Wei Liu * and Lin-Fan Zhu * [1]

* Hefei National Laboratory for Physical Sciences at Microscale, Department of Modern Physics, University of Science and Technology of China, Hefei, Anhui 230026, People's Republic of China

Synopsis Utilizing the high-resolution inelastic x-ray scattering technique, absolute inelastic squared form factors of four vibronic series belonging to the $B^1\Sigma_u^+$, $C^1\Pi_u$, $EF^1\Sigma_g^+$ and $B'^1\Sigma_u^+$ electronic states of molecular hydrogen are determined experimentally. In addition, the same inelastic squared form factors are calculated by the multi-reference single and double-excitation configuration-interaction method for the $B^1\Sigma_u^+$, $C^1\Pi_u$, $EF^1\Sigma_g^+$. Apparent differences are observed between the present experimental results and all available calculated ones.

The electronic structures of the excited states of atoms and molecules can be described by the so-called inelastic squared form factors (ISFFs). The ISFFs are directly related to the wave functions of the ground and excited states in the momentum space, and can be used to test the theoretical calculations strictly. The ISFFs of H_2 have been studied by high energy electron energy loss spectroscopy (EELS) experimentally, and only electronic states resolved spectra can be found in the literatures. Moreover, some recent investigations found that the higher Born term can not be ignored in the high-energy electron impact process, especially at large momentum transfer. To avoid the effect of higher order terms, the inelastic x-ray scattering (IXS) method was applied to study the excitation of atoms and molecules lately [1]. The results of IXS have been proven to be accurate and thus can serve as the benchmark data. So we carried out a vibronically resolved IXS measurement of valence shell excitations of H_2 at BL12XU of Spring-8.

As an example, the present ISFFs of $B^1\Sigma_u^+$ (v'=9) and $C^1\Pi_u$ (v'=6) of H_2 are shown in Fig.1 along with the calculated results by Kolos et al[2] and Borges et al[3]. These two theoretical calculations studied vibronic excitations without using of Franck-Condon principle. It is obvious in the figure that the calculated results and the present IXS ones have the similar shape, and the theoretical ISFFs fit the experimental result better in the high momentum transfer region. However, in the low momentum transfer region, both theoretical results are smaller than the experiment results. Similar phenomena have been observed for other excitations. To elucidate them, more elaborate theoretically investigations are greatly recommended.

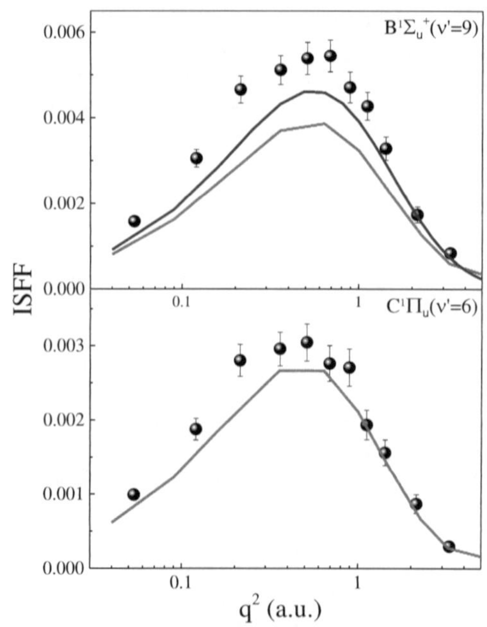

Figure 1. The ISFFs of the vibronic excitation of hydrogen. Solid black points: the present IXS results; solid red lines: the results of Borges et al; solid blue lines: the results of Kolos et al.

References

[1] B. P. Xie, L. F. Zhu, K. Yang et al., 2010 *Phys. Rev. A* **82** 032501.

[2] W. Kolos, H. J. Monkhorst, and K. Szalewwicz, 1983 *At. Data Nucl. Data Tables* **28** 239

[3] I. Jr. Broges and C. E. Bielschowsky, 1999 *Phys. Rev. A.* **60** 1226

[1] E-mail: lfzhu@ustc.edu.cn

The observation of the pair of Lyman-α and Lyman-β photons produced in the photodissociation of H_2

K. Hosaka[*1], Y. Torizuka[*], P. Schmidt[†], T. Odagiri[‡], A. Knie[†], K. Jänkälä[§], A. Ehresmann[†], M. Kitajima[*], and N. Kouchi[*]

[*] Department of Chemistry, Tokyo Institute of Technology, Meguro-ku, Tokyo 152-8551, Japan
[†] Institute of Physics, University of Kassel, D-34132 Kassel, Germany
[‡] Department of Materials and Life Sciences, Sophia University, Chiyoda-ku, Tokyo 102-8554, Japan
[§] Department of Physical Sciences, University of Oulu, PO Box 3000 90014 Oulu, Finland

Synopsis We have measured the coincidence time spectra for the pair of the Lyman-α and Lyman-β photons produced in the photodissociation of H_2 in the range of the incident photon energy 32.5 - 42 eV. The result shows that the pair of H(2p) and H(3p) fragments is produced in the photodossociation of H_2 molecules.

Coincidence detection of the pair of fluorescence photons emitted by the pair of excited photofragments, the $(\gamma, 2\gamma)$ method, is a powerful technique for investigating the dynamics of the doubly excited molecules produced by the photon interaction. The $(\gamma, 2\gamma)$ study on H_2 molecules enabled us to study the dynamics of the doubly excited $Q_2\ ^1\Pi_u(1)$ state[1, 2] and the entanglement of the pair of H($2p$) atoms[3], where the pair of the Lyman-α photons ($2p \rightarrow 1s$, 121.6nm wavelength[4]) was detected in coincidence. In the present study, the $(\gamma, 2\gamma)$ method has been extended to the coincidence detection of the pair of the Lyman-α and Lyman-β ($3p \rightarrow 1s$, 102.6nm wavelength[4]) photons produced in the photoexcitation of H_2 molecules. Our Lyman-α detector is comprised of a 1 mm-thickness MgF_2 window($\sim$60% transmittance for the Lyman-α photon) and microchannel plate coated with CsI, which provides a filter range of approximately 115 - 200 nm[2]. For the Lyman-β detector, the MgF_2 window is replaced with the 100 nm-thickness Indium Oxide(InO) foil[5] ($\sim$8% transmittance for the Lyman-β photon).

The experiments were performed at BL20A of the Photon Factory, KEK, Japan. Linearly polarized light was introduced into a gas cell filled with H_2 gas. The two-photon coincidence time spectra were measured in the range of the incident photon energy 32.5 - 42 eV, and they were summed up to get better statistics. The result is shown in Fig.1, where the accidental coincidence counts were subtracted. The horizontal axis of Fig. 1 expresses the detection time of the Lyman-α photon(T_α) in reference to the detection time of the Lyman-β photon(T_β).

Two decay components are seen in addition to the sharp peak, which originates from the cosmic muons [3]. The slow decay component, which ranges from late time to early time, has the time constant close to the lifetime of $3p$ atom, 5.4 ns[6]. It is difficult to know the time constant of the fast decay component, which ranges from early time to late time, since the sharp peak prevent it from being separated. However, the time constant of the fast decay component seems consistent with the lifetime of the $2p$ atom, 1.6 ns [6]. In conclusion, the coincidence time spectra in Fig.1 shows that the pair of H($2p$) and H($3p$) atoms is produced in the photodissociation of H_2 molecules.

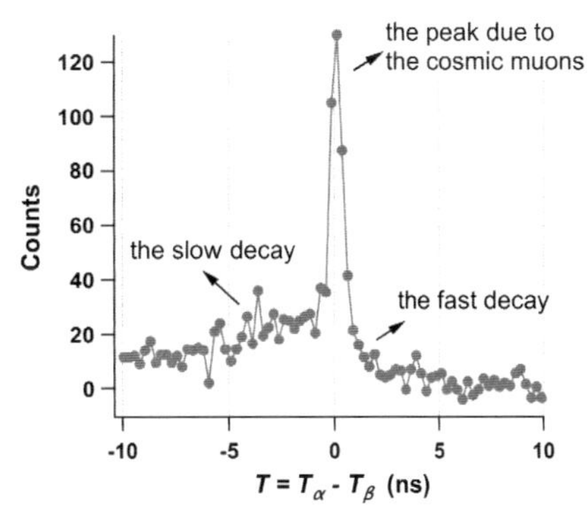

Figure 1. The coincidence time spectrum between the Lyman-α and Lyman-β photons produced in the photodissociation of H_2.

[1] E-mail: hosakak@chem.titech.ac.jp

References

[1] T. Odagiri *et al.* 2004 *J. Phys. B* **37** 3909

[2] K. Hosaka *et al.* 2016 *Phys. Rev. A* **93** 063423

[3] Y. Nakanishi *et al.* 2014 *Phys. Rev. A* **90** 043405

[4] Atomic spectra database, Physical Reference Data, National Institute of Standards and Technology

[5] The center for X ray optics, Materials Sciences Division, Lawrence Berkeley National Laboratory

[6] H. A. Bethe and E. E. Salpeter, Quantum Mechanics of One-and Two-Electron Atoms (Plenum, New York, 1977), p. 266

Dynamics of ion-molecule reactions of $SO_2^{\cdot+}$ with H_2O and CH_4

Antonella Cartoni,[1]* § # Daniele Catone,[†] Paola Bolognesi,[2#] Mauro Satta,[1*°] Pal Markus [#] and Lorenzo Avaldi [#]

[§] Dipartimento di Chimica, Sapienza Università di Roma, Pl.e Aldo Moro 5, 00185, Roma, Italy
CNR-Istituto di Struttura della Materia, Area della Ricerca di Roma 1, 00016 Monterotondo Scalo (RM), Italy
† CNR-Istituto di Struttura della Materia (ISM), Area della Ricerca di Roma 2, , 00133, Roma, Italy
° CNR- Istituto per lo Studio dei Materiali Nanostrutturati (ISMN), Dipartimento di Chimica, Sapienza Università di Roma,P.le Aldo Moro 5, 00185, Roma, Italy

Synopsis. The reaction dynamics and the temperature-dependent kinetic trend of the $SO_2^{\cdot+}$ ion-molecule reactions with water and methane have been studied using the tunable synchrotron radiation to produce excited $SO_2^{\cdot+}$ ions and ab-initio methods. The experimental results show that only one product, HSO_2^+, is formed in both reactions and its yield displays different trends with the photon energy. DFT and VTST calculations have been used to explore the dynamics of the reactions and to calculate the rate constants at different temperatures.

The different reactivity of O-H and C-H bonds in the reaction of water and methane with $SO_2^{\cdot+}$ is of fundamental interest due the dynamical processes occurring along the reaction paths on the potential energy surfaces. Moreover these reactions in the gas phase are considered as a laboratory model of the chemistry of terrestrial and planetary atmosphere as well as of the molecular synthesis in the interstellar space. [1] The experiments have been performed at the "Circular Polarization" beamline (CiPo) of ELETTRA (Trieste, Italy) [2] where monochromatic VUV photons of tunable energy (12-14 eV) were used to ionize the SO_2 molecules from an effusive beam. In this energy range considering the lifetime of the $SO_2^{\cdot+}$ ionic excited states ($^2B_2$ and $^2A_2$), the ion can be considered to be in excited ro-vibrational levels of its ionic ground state X^2A_1 before reacting with the neutrals. In Figure 1 the ratio HSO_2^+/SO_2^+ for the two reactions is reported versus photon energy (i.e. increasing internal energy of the $SO_2^{\cdot+}$ ion). The trend for the two reacting systems is different: the reaction is decelerated in the water case, while it is firstly decelerated and then accelerated in the case of methane. Theoretical calculations allowed to obtain the minimum energy paths (MEPs) for the two reactions and to localize the variational transition state VST, the bottleneck of the reaction, to calculate the rate constants k of the two reactions. The trend of k in the range of 298-500 K (12.4-12.8 eV) reproduces quite well the experimental data [5] and the dynamics of the two reactions explain in exhaustive way the higher reaction efficiency ($\phi=k/k_{coll}$) with water than with methane measured and reported in previous works at 298.15 K [3,4]. Further experimental work has been undertaken to investigate the $SO_2^{\cdot+}$ reactions with H_2/D_2, which are relevant for astrochemistry.

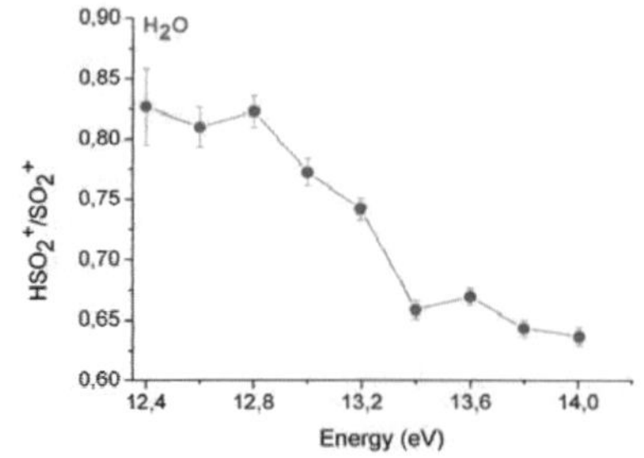

Figure 1. HSO_2^+/SO_2^+ ratio vs photon energy in the reaction of SO_2^+ with methane (top) and water (bottom), at the fixed pressure of 1.0×10^{-5} mbar and nominal collision energy of 0 eV.

References

[1] M. Larsson, W. D. Geppert., G. Nyman 2012, *Rep. Prog. Phys.*, 75, 066901; S. Petrie, D. K. Bohme 2007, *Mass Spectrum. Rev. 26*, 258-280.
[2] A. Derossi *et al.* 1995 *Rev. Sci. Instrum.*, 66, 1718
[3] G. de Petris et al. 2009, *Chem. Eur. J.*, *15*, 4248; G. de Petris et al. 2009, *Phys. Chem. Chem. Phys. 11*, 9976-9978.
[5] A. Cartoni *et al.* 2017 *Chemistry a European Journal*. DOI: 10.1002/chem.201700028.

[1]E-mail: antonella.cartoni@uniroma1.it/mauro.satta@cnr.it

[2]E-mail: paola.bolognesi@cnr.it

IOP Conf. Series: Journal of Physics: Conf. Series **875** (2017) 032004 doi:10.1088/1742-6596/875/4/032004

Evidence for Efficient Pathway to Produce Slow Electrons by Ground-state Dication in Clusters

Daehyun You[*†1], Hironobu Fukuzawa[*†], Yuta Sakakibara[*†], Tsukasa Takanashi[*†], Yuta Ito[*†], Gianluigi G. Maliyar[*†], Koji Motomura[*†], Kiyonobu Nagaya[†‡], Toshiyuki Nishiyama[†‡], Kazuki Asa[†‡], Yuhiro Sato[†‡], Norio Saito[†§], Masaki Oura[†], Markus Schöffler[†¶], Gregor Kastirke[¶], Uwe Hergenhahn[††], Vasili Stumpf[‡‡], Kirill Gohkberg[‡‡], Alexander I. Kuleff[‡‡], Lorenz S. Cederbaum[‡‡], and Kiyoshi Ueda[*†2]**

[*] Institute of Multidisciplinary Research for Advanced Materials, Tohoku University, Sendai 980-8577, Japan
[†] RIKEN SPring-8 Center, Kouto 1-1-1, Sayo, Hyogo 679-5148, Japan
[‡] Department of Physics, Kyoto University, Kyoto 606-8502, Japan
[§] National Metrology Institute of Japan, AIST, Tsukuba 305-8568, Japan
[¶] Institute for Nuclear Physics, Johann Wolfgang Goethe University Frankfurt, Frankfurt 60438, Germany
[**] Leibniz Institute of Surface Modification, Leipzig 04318, Germany
[††] Max-Planck-Institute for Plasma Physics, Greifswald 17491, Germany
[‡‡] Theoretische Chemie, Physikalisch-Chemisches Institut, Universität Heidelberg, Heidelberg 69120, Germany

Synopsis We present an experimental evidence for a so-far unobserved, but potentially very important step relaxation cascades following inner-shell ionization of a composite system: Multiply charged ionic states created after Auger decay may be neutralized by electron transfer from a neighboring species, producing at the same time a low-energy free electron. This electron transfer-mediated decay (ETMD) called process is effective even after Auger decay into the dicationic ground state. Here, we report the ETMD of Ne^{2+} produced after Ne 1s photoionization in Ne–Kr mixed clusters.

When atoms or molecules are irradiated by X-rays, an inner-shell electron is ejected, followed by an emission of an Auger electron producing a dication [1]. If the dication is surrounded by other atoms or molecules, one of the neighbors may donate an electron to the dication and releases energy, which can be transferred to ionize at the same time another neighbor. This process is called electron transfer-mediated decay (ETMD) [2]. It was considered as an inefficient decay channel since it could not compete with other decay processes [3]. In 2013 however, Stumpf *et al.* predicted that the ETMD provides an efficient neutralization pathway for the majority of ions produced by Auger decay in NeKrKr trimer [4]. In this study, we report the experimental identification for that process using Ne-Kr mixed clusters [5].

The experiment was carried out at beamline BL17SU in SPring-8 synchrotron facility. The photon energies were set to 878 eV and 888 eV, corresponding 8 eV and 18 eV above Ne 1s ionization threshold. A COLTRIMS Reaction Microscope was used to image the electron and ion momentum vector in coincidence originating from same cluster [6].

The relaxation cascade triggered by Ne 1s ionization and terminating by ETMD can be written as:

$$\mathrm{NeKrKr} + h\nu \rightarrow \mathrm{Ne^+}(1s^{-1})\mathrm{KrKr} + e_{\mathrm{photo}}$$
$$\mathrm{Ne^+}(1s^{-1})\mathrm{KrKr} \rightarrow \mathrm{Ne^{2+}KrKr} + e_{\mathrm{Auger}}$$
$$\mathrm{Ne^{2+}KrKr} \rightarrow \mathrm{Ne^+Kr^+Kr^+} + e_{\mathrm{ETMD}}$$

One Ne^+ and two Kr^+ ions are produced by this process. We have selected only the events in which we detected these 3 ions, identified the photo- and the ETMD electrons from electron spectra in coincidence with the events. The relative intensity of the ETMD electrons is ~70 % of that of the photoelectrons. This implies that when a Ne 1s photoionization event occurs, almost always a slow electron is emitted from the cluster.

This work was supported by MEXT, JSPS, IMRAM project and Tohoku Univ. DIARE.

References

[1] W. Bambynek *et al.* 1972 *Rev. Mod. Phys.* **44** 716
[2] J. Zobeley *et al.* 2001 *J. Chem. Phys.* **115** 5076
[3] K. Sakai *et al.* 2011 *Phys. Rev. Lett.* **106** 033401
[4] V. Stumpf *et al.* 2013 *Phys. Rev. Lett.* **110** 258302
[5] D. You *et al.* 2017 *Nat. Commun.* **8** 14277
[6] R. Dörner *et al.* 2000 *Phys. Rep.* **330** 95

[1] E-mail: daehyun@mail.tagen.tohoku.ac.jp
[2] E-mail: ueda@tagen.tohoku.ac.jp

IOP Conf. Series: Journal of Physics: Conf. Series **875** (2017) 032005 doi:10.1088/1742-6596/875/4/032005

Direct monitoring of photon induced isomerization, dissociation and electron detachment of the green fluorescent protein chromophore anion

Eduardo Carrascosa [1], Michael S. Scholz, James N. Bull, Evan J. Bieske

School of Chemistry, The University of Melbourne, Parkville VIC 3010, Australia

Synopsis We present the first experimental demonstration of Z↔E photoisomerization the GFP chromophore anion, $HBDI^-$, in the gas phase. In the single photon absorption regime, the photoisomerization action spectra show two maxima at 480 nm and 455 nm. In the multiphoton absorption regime, photodissociation and photodetachment channels modify the appearance of the photoisomerization band. This work provides a new approach to characterize photoisomerization pathways in biomolecular ions.

The green fluorescent protein (GFP) is one of the most widely used species for imaging biological processes *in vivo* [1]. Its optical absorption properties rely on a chromophore based on p-hydroxybenzylidene-2,3-dimethylimidazolinone (HBDI), which is found inside the folded structure of the protein. Many groups have performed gas phase experiments on $HBDI^-$ in order to characterize the main deactivation pathways in GFP following photoexcitation [2, 3, 4]. While many insights into the nature and dynamics of photoinduced fragmentation and electron emission have been gained, the possibility of Z↔E photoisomerization in $HBDI^-$ (see Figure 1, up), a deactivation pathway competing with photodissociation and photodetachment, has received less attention, mainly due to the challenges associated with its experimental detection.

Combining ion mobility mass spectrometry with laser spectroscopy, we have been able to separate the two isomers of the HBDI anion and demonstrate for the first time Z→E and E→Z photoisomerization in the gas phase. The Z→E photoisomerization action spectrum (Figure 1(a)) presents two maxima at 480 nm and 452 nm at low laser intensity conditions (0.4 mJ/pulse) and shows that photoisomerization dominates over detachment at energies above the electron detachment continuum (454 nm). Nearly equivalent features are observed for the photoexcitation of the E isomer, however a slight blue-shift in the E→Z photoisomerization action spectrum is observed. At increasing laser intensities (1.5 mJ/pulse) photoisomerization competes with multiphoton induced dissociation and detachment. The photodissociation maximum lies at 485 nm, coinciding with a relative decrease of photoisomerization intensity if compared with the low laser power conditions (Figure 1(b)). While this feature is ascribed to a multiphoton induced process involving internal conversion to the ground state, as suggested in a recent work [5], the spectra at low laser intensity conditions show that photoinduced action at the origin of the $S_0 \rightarrow S_1$ band does not only correspond to multiphoton induced dissociation but also to photoisomerization. We will present branching ratios for the three main photoinduced channels at different wavelegths and discuss the effect of photoisomerization on the deactivation efficiency of the GFP protein. These findings provide direct insight into the molecular rearrangement of the GFP chromophore anion upon photoexcitation that could only be indirectly inferred to date.

[1] E-mail: eduardo.carrascosa@unimelb.edu.au

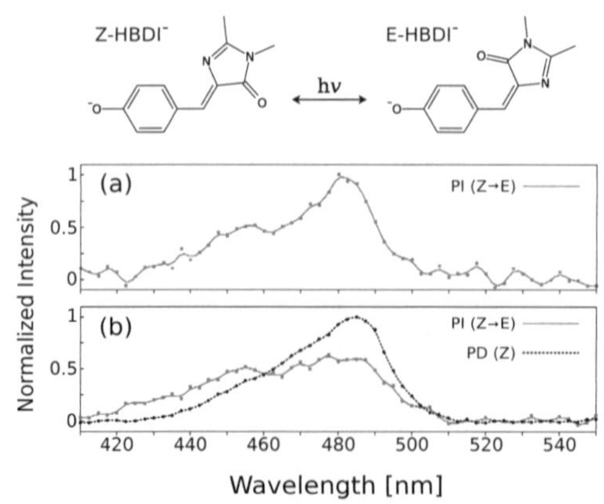

Figure 1. **(a)** Z→E photoisomerization action spectrum (PI) at low light intensity permitting only single photon processes. **(b)** Photoisomerization and photodissociation (PD) action spectra of the Z-HBDI anion at high light intensity.

References

[1] S. R. Meech 2009 *Chem. Soc. Rev.* **38** 2922

[2] D. A. Horke *et al.* 2012 *Phys. Chem. Chem. Phys.* **14** 8511

[3] A. V. Bochenkova *et al.* 2014 *Angew. Chem. Int. Ed.* **53** 9797

[4] C. McLaughlin *et al.* 2017 *Chem. Sci.* **8** 1621

[5] H. V. Kiefer *et al.* 2016 *Phys. Rev. Lett.* **117** 243004

IOP Conf. Series: Journal of Physics: Conf. Series **875** (2017) 032006 doi:10.1088/1742-6596/875/4/032006

X-ray absorption spectra of excited triplet states of organic molecules

Atsunari Hraya[1], **Haruka Inui, Sho Yamahira, and Osamu Takahashi**[†]

Department of Physical Science, Hiroshima University, Higashi-Hiroshima 739-8526, Japan
[†]Institute for Sustainable Science and Development, Hiroshima University, Higashi-Hiroshima 739-8526, Japan

Synopsis X-ray absorption spectra (XAS) have been measured for several organic molecules in the lowest excited triplet states (T_1) prepared by UV irradiation. T_1-XAS deduced from the UV-ON/OFF difference spectra show good agreement, at least in the lowest peak region, with calculated T_1-XAS.

Resonant core-excitation from valence excited states of organic molecules in the soft x-ray region has attracted attentions to elucidate details of the core-hole interaction with valence orbitals [1] and of bond dynamics especially bond dissociation in the valence- and core-doubly excited states. However, experimental difficulties in preparation of valence excited states, especially of excited singlet states having very short lifetime, with enough concentration to detect core excitations have prevented realization of measurements on core excitation from valence excited states. We have been trying to measure XAS from valence excited states by using as the initial state of core excitations the excited triplet states that have much longer lifetime than singlet states. Also theoretical investigation have been carried out to search for molecular system exhibiting experimentally favorable x-ray absorption spectra (XAS) [2].

Measurements on X-ray absorption spectra of excited triplet states of organic molecules were carried out at BL-13 of HiSOR (Hiroshima Synchrotron Radiation Center). X-ray absorption spectra are measured as total electron yields (TEY) method. Thin film samples of organic molecule are prepared on platinum coted titan substrate by drop-and-dry method. Phosphorescence life time of samples were measured prior to XAS measurement to insure experimentally detectable triplet formation yields and time averaged concentration. Organic molecules investigated are 4-mercaptobenzoic acid (MBA), 4-cyanobenzaldehyde (CBA), and benzil (BZL). All these molecules are known to produce the lowest excited triplet state (T_1) with high yield and to exhibit long lifetime in ms order.

Figure 1 shows presently obtained TEY spectra of 4MBA and calculated S_0 and T_1 XAS of benzoic acid for comparison. T_1 component in the TEY spectra is deduced from UV On-Off difference spectrum and S_0 (UV OFF) TEY with assuming T_1 concentration of 0.05.

Experimentally deduced T_1-XAS is in good agreement with calculated XAS especially for the lowest energy peak. Difference between experiment and calculation in the higher energy region is mainly caused by difference of phase, solid for TEY and gas for calculation.

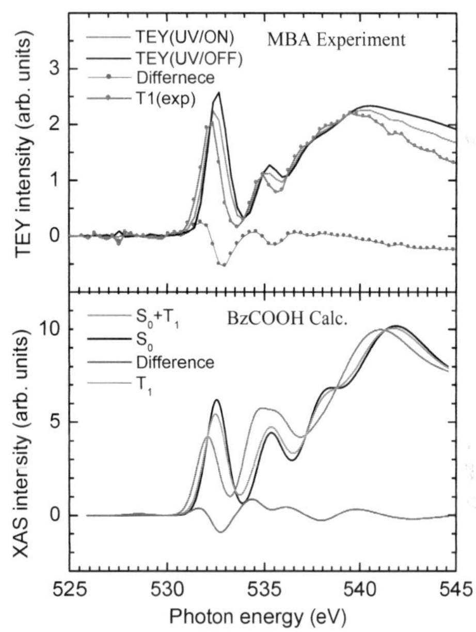

Figure 1. Top: X-ray absorption spectra (top) of 4-mercaptbenzoic acid in low temperature film with (magenta) and without (black) UV irradiation, and UV ON-OFF difference spectrum and T_1-XAS component deduced from difference spectra. Bottom: Calculated S_0 and T_1 XAS of benzoic acid in gas phase together with S_0 and T_1 mixed XAS and expected difference spectrum (green).

References

[1] T. Privalov, F. Gel'mukhanov, H. Ågren, 2003 *J. Electron Spectrosc. Relat. Phenom.* **129** 43.
[2] S. Nagashima, K. Yamamoto, O. Takahashi, A. Hiraya, 2015 *J. Phys. Conf. Ser.* **635** 112058.

[1] E-mail: hiraya@hiroshima-u.ac.jp

IOP Conf. Series: Journal of Physics: Conf. Series **875** (2017) 032007 doi:10.1088/1742-6596/875/4/032007

"Position" does matter : the photofragmentation of the nitroimidazole isomers

J. Chiarinelli[*§1], P. Bolognesi[*2], A. Casavola[*], A. Cartoni[°*], M.C. Castrovilli[*], D. Catone[*], R. Richter[†], S. Borocci[‡], S. Tosic[#], H. Sa'adeh[&], M. Masic[§], B.P. Marinkovic[#], K.C. Prince[†] and L. Avaldi[*]

[*] CNR-Istituto di Struttura della Materia, Area della Ricerca di Roma 1, 00016 Monterotondo Scalo, Italy
[§] Dipartimento di Scienze, Università di Roma Tre, Via della Vasca Navale 84, 00146 Roma
[‡] Dipartimento per l'Innovazione nei Sistemi Biologici, Agroalimentari e Forestali (DIBAF), Università della Tuscia, Viterbo, Italy
[†] Elettra Sincrotrone Trieste, Area Science Park, Basovizza, Trieste, Italy
[#] Institute of Physics Belgrade, University of Belgrade, Belgrade, Serbia
[&] Department of Physics, The University of Jordan, Amman, Jordan
[°] Dipartimento di Chimica, Sapienza Università di Roma, Roma, Italy
[§] School of Chemistry, Cardiff University, Cardiff, United Kingdom

Synopsis Experimental and theoretical spectroscopic methods have been combined to disentangle the fundamental mechanism of VUV induced fragmentation of the three isomers of nitroimidazole radiosensitisers.

Radiotherapy is one of main techniques used in cancer treatment and radiosensitizers are drugs used to selectively improve its effectiveness against tumour cells. Misonidazole and Nimorazole are two radiosensitizers with similar chemical structure. However clinical trials have shown that they have different efficiency in the tumor treatment. In order to understand why this happens we have studied the photofragmentation of 2-nitroimidazole and 4(5)-nitroimidazole, which are the "building blocks" of Misonidazole and Nimorazole, respectively.
To the purpose time of flight mass spectrometry has been used to investigate the fragmentation of these molecules. These measurements have been then extended via the determination of the Appearance Energies (AE) of the different fragments and by Photelectrons-Photoions Coincidences (PEPICO) experiments using synchrotron radiation [1]. To interpret the results DFT calculations [1] have been also performed.
The mas spectra of the two molecules display many similar features and relative intensities, but also a few intriguing peculiarities. The most striking differences are the fragments at m/z 55^+ and 56^+, present exclusively in 4(5)NI and 2NI, respectively, and the fragment at m/z 83^+, which is one of the leading fragmentation channels in 2NI, but is almost absent in the 4(5)NI sample.The results of mass spectrometry are confirmed by the photoelectron–photoion coincidence measurements.

Based on DFT calculations, a model is proposed which fully explains such differences, and reveals the subtle fragmentation mechanisms leading to the release of neutral species like NO, CO and HCN. The present results

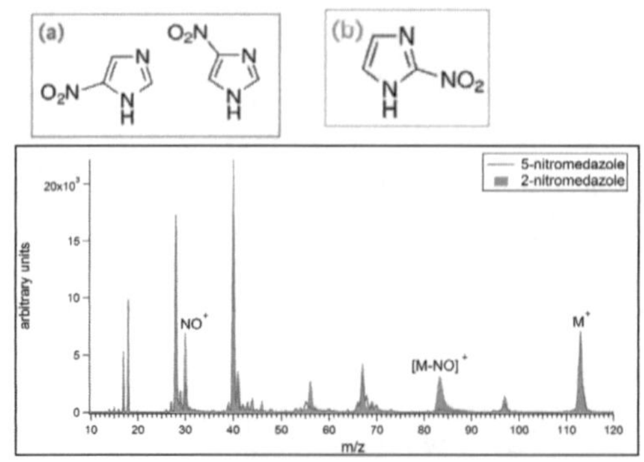

Figure 1. Mass Spectra of 4(5)-nitroimidazole (red line and scheme a) and 2-nitroimidazole (grey area and scheme b) molecules measured with a He lamp.

suggest that the products of decomposition of the different nitroimidazole isomers migth play a role in determining their distinctive degrees of effectiveness in radiotherapy.

Acknowledgments: Work partially supported by the Serbia–Italy Joint Research Project "Nanoscale Insight in the Radiation Damage".

References

[1] P. Bolognesi et al. J Chem Phys Comm. 145, 191102 (2016)

[1] E-mail: jacopochiarinelli@gmail.com

[2] E-mail: paola.bolognesi@cnr.it

Localized or delocalized K-holes in N2: Photoelectron-Auger electron coincidence experiments with high energy resolution

G. Nalin*[1], S. Grundmann*, I. Vela-Perez*, H. Kang*, F. Trinter*, M. Schöffler*, T. Jahnke* and R. Dörner*[2].

* IKF, Johann Wolfgang Goethe Universität, Max-von-Laue-Strasse, 1, 60438, Frankfurt am Main, Germany

Synopsis : For K-shell ionization of N_2 photoelectron auger electron coincident angular distributions show that pathways with *gerade* and *ungerade* K-hole state can interfere. On the other hand high resolution photoelectron spectroscopy seems to indicate that the *gerade* and *ungerade* hole state can be separated in energy. The present experiment aims to combine the high energy resolution with full angular detection of Auger and Photoelectron to resolve this seeming contradiction.

The debate around the localization or delocalization of a K-shell hole induced by x-ray photon ionization in N_2 or O_2 puzzled scientist for a long time. In 2008 M. Schöffler *et al.* [1] gave a well-detailed picture resolving the apparent contradiction of the two views. Using an Auger electron detection in coincidence with the photoelectron, the symmetry breaking (localization) or preservation (delocalization) of the molecule seems to depend on how the entangled Auger electron is detected. This interpretation however assumed that the pathway including a *gerade* or *ungerade* K-hole state can perfectly interfere. This is in seeming contradiction to the known energy splitting of about 100meV between the 1σ *ungerade* and 1σ *gerade* state and raises the question how two pathways can interfere, which can at least partly be separated in the photoelectron energy.

The reaction mechanism consists in the absorption of one photon to ionize the molecule, followed by the emission of an Auger electron and the consequent fragmentation in two N+ ions. The orbital from where the photoelectron is ejected from, could be either a coherent or an incoherent superimposition of the 1σ *ungerade* and 1σ *gerade* molecular orbitals. Due to the entangle state between the Auger probe and the induced photoelectron, angular distribution patterns look different in the two cases.

The present study tries to fully and precisely resolve the intermediate quantum mechanical state of K-shell holes in energy in addition to measuring the molecular frame angular distribution of both electrons in coincidence. The detection setup is a custom modified and optimized COLTRIM system aiming for sub 100meV electron energy resolution. The experiment is performed at beamline P04 PETRA III facility providing a sub 100meV energy resolution at unprecedented photon flux. First results will be presented at the meeting.

References

[1] M. Schöffler *et al*, Science, Vol. 320. no. 5878, pp. 920 - 923 (2008).

[1]E-mail: nalin@atom.uni-frankfurt.de

[2]E-mail: doerner@atom.uni-frankfurt.de

IOP Conf. Series: Journal of Physics: Conf. Series **875** (2017) 032010 doi:10.1088/1742-6596/875/4/032010

Modelling laser-matter interactions for simple molecules

A. Galstyan*[1], Yu.V. Popov†§, F. Mota-Furtado‡, P. F. O'Mahony‡, and B. Piraux *

* Université catholique de Louvain, Louvain-la-Neuve, Belgium
† Lomonosov Moscow State University, Moscow, Russia
‡ Department of Mathematics, Royal Holloway, University of London, United Kingdom
§ Joint Institute for Nuclear Research, Dubna, Russia

Synopsis We consider the interaction of a molecule with an ultrashort laser pulse within the mean field approximation.a model has been generalized to treat the interaction of atomic hydrogen with an electromagnetic pulse. The kernel of the Coulomb potential in the momentum space is replaced by a sum of products of separable potentials supporting a bound state of the atom and built from the corresponding bound state wave function. The molecular orbitals are expressed in terms of gaussians and generated by means of a quantum chemistry software. We analyze the ionization yield as a function of the size of the gaussian basis for H_2 and H_2O molecules and the results are compared to those already existing in the literature.

Treating the interaction of a multi-electron system with an oscillating field is never a simple task. Either one solves numerically a time-dependent Schrödinger equation (TDSE) or one uses the time-dependent density functional theory. Using either of these methods is always time consuming and computationally very expensive. Here, we develop an alternative approach that requires reasonable computer resources. It uses modern quantum chemistry codes to generate molecular orbitals expressed in terms of gaussian type orbitals (GTO). This simplifies the calculations but the basis size becomes crucial for spatially extended system since GTO's are strongly localized near the nuclei.

Our approach is based on a simple model, first introduced for atomic hydrogen [1, 2] and generalized, within the mean field approximation to the treatment of the interaction of a simple molecule with an attosecond laser pulse.

The complicated molecular potential is approximated with a separable potential that describes only one molecular orbital:

$$V(\vec{p},\vec{u}) = v(\vec{p})v^*(\vec{u}), \qquad (1)$$

with

$$v(\vec{p}) = -\frac{1}{a}\left[\varepsilon_0 - \frac{p^2}{2}\right]\Phi_0^{var}(\vec{p}), \qquad (2)$$

where a is a constant. The molecular orbital $\Phi_0^{var}(\vec{p})$ is precalculated variationally in a GTO basis set using the well established quantum chemistry software package GAMESS-US. Assuming the external field is linearly polarized and described in terms of the vector potential $A(t)\vec{e}$ where $\vec{e}$ is a unit vector along the polarization axis, we solve the following one-electron TDSE in momentum space:

$$\left[i\frac{\partial}{\partial t} - \frac{p^2}{2} + \frac{1}{c}A(t)(\vec{e}\cdot\vec{p})\right]\Phi(\vec{p},t)$$
$$+v_n(\vec{p})\int\frac{\mathrm{d}\vec{u}}{(2\pi)^3}v_n^*(\vec{u})\Phi(\vec{u},t) = 0, \qquad (3)$$

the solution of which can be written analytically in terms of a function $F(t)$ which in turn is the solution of a one dimensional Volterra integral equation of the second kind, depending only on time:

$$F(t) = F_0(t) + \int_0^t K(t,\xi)F(\xi)\mathrm{d}\xi \qquad (4)$$

(see [1, 2] for the details of the calculations).

As a first step, this model is applied to atomic hydrogen with the ground state expressed in tems of GTO's. The results are compared to those obtained (i) by using the exact ground state wave function [2] and (ii) by solving the TDSE in the case of the exact Coulomb potential. This allows one to study the adequacy of different GTO basis sets (namely, 6-31G and 6-311++G**) and the applicability of the model in general.

A similar study is performed for molecular hydrogen H_2 and water H_2O. For the latter one, we can consider different molecular orbitals, but it is only for the HOMO orbital that we have an ionization process that keeps the same geometry of the molecule without exciting any vibrational states. Results are compared to those given in the literature.

[1]E-mail: alexander.galstyan@uclouvain.be

References

[1] Tetchou Nganso, H. M. *et al.* 2011 *Physical Review A* **83** 013401

[2] A. Galstyan *et al.* 2017 *To be published in EPJD*

IOP Conf. Series: Journal of Physics: Conf. Series **875** (2017) 032011 doi:10.1088/1742-6596/875/4/032011

Perturbative expansions for laser-atom interactions

Yu. V. Popov[†§1], **A. Galstyan**[*], **F. Mota-Furtado**[‡], **P. F. O'Mahony**[‡], **and B.Piraux** [*]

[†] Lomonosov Moscow State University, Moscow, Russia
[§] Joint Institute for Nuclear Research, Dubna, Russia
[*] Université catholique de Louvain, Louvain-la-Neuve, Belgium
[‡] Department of Mathematics, Royal Holloway, University of London, United Kingdom

Synopsis We analyze the interaction of a one-active electron system with a strong laser field. In order to extract information about the interaction mechanisms, it is important to consider perturbative expansions of the ionization probability amplitude in terms of the Coulomb potential. Based on some of Faddeev ideas, we introduce a new perturbative series that involves both the Coulomb and the Volkov propagators. We show that for photon energies higher than the ionization potential, the series converges rapidly towards the exact result as calculated by solving numerically the time-dependent Schrödinger equation. In the same conditions, the well known strong field approximation series is shown to diverge.

The strong field approximation (SFA) has been the principal tool to describe and explain processes in the domain of laser-matter interactions. It provides valuable insight into the mechanism underlying processes such as the ionization of atoms or molecules exposed to strong infrared fields, high-order harmonic generation and even processes involving more than one electron like non-sequential double photoionization. However, SFA has several flaws which have been discussed in detail in [1]. In particular, the convergence of the SFA series is still an unsolved question. It has been shown in [1] that in a very simple case, the first order SFA can describe the problem reasonably well whereas the series is clearly diverging.

In order to calculate the high-order terms of the SFA series, we first write the full electron wave packet as $|\Phi(t)\rangle = |\varphi_0(t)\rangle + |F(t)\rangle$, where $|\varphi_0(t)\rangle$ is the hydrogen atom ground state. By treating the Coulomb potential V_c as a perturbation, we solve iteratively the following inhomogeneous time-dependent Schrödinger equation (TDSE) for $|F(t)\rangle$ [1]:

$$\left[\mathrm{i}\frac{\partial}{\partial t} - H_0 - V_c - V_d(t)\right]|F(t)\rangle = V_d(t)|\varphi_0(t)\rangle, \quad (1)$$

where $V_d(t)$ is the dipole interaction potential and H_0, the kinetic energy operator.

In our new Faddeev like approach [2] , we write the full wave packet as $|\Phi(t)\rangle = |\Phi_1(t)\rangle + |\Phi_2(t)\rangle$ where $|\Phi_1(t)\rangle$ and $|\Phi_2(t)\rangle$ obey

$$(i\frac{\partial}{\partial t} - H_0)|\Psi_1(t)\rangle = V_c|\Psi(t)\rangle, \quad |\Psi_1(0)\rangle = |\varphi_0(0)\rangle,$$

$$(i\frac{\partial}{\partial t} - H_0)|\Psi_2(t)\rangle = V_d(t)|\Psi(t)\rangle, \quad |\Psi_2(0)\rangle = 0.$$

[1]E-mail: popov@srd.sinp.msu.ru

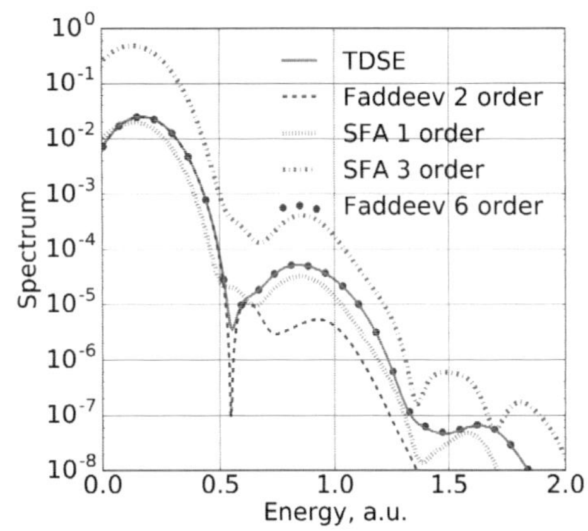

Figure 1. Electron energy spectrum in the forward direction for atomic hydrogen exposed to a 4 optical cycle sine squared pulse of 10^{14} W/cm^2 peak intensity and 0.7 a.u. photon energy. TDSE results in velocity gauge (full line) are compared with SFA results of first order (dotted line) and third order (dot-dashed line) and with our new Faddeev like series of second order (dashed line) and sixth order (dots). The first order of our new expansion coincides with the SFA first order term.

In Fig. 1 and in a simple case, we compare TDSE results for the electron energy spectrum to those obtained within the SFA and the new expansion. While SFA series is diverging as shown in [1], our new Faddeev like approach converges rapidly. Already at the 6th order, it fully coincides with the exact TDSE result. However, for lower frequencies we expect slower convergence.

References

[1] A. Galstyan *et al.* 2016 *Physical Review A* **93(2)**, 023422

[2] Yu. V. Popov *et al.* 2017 *To be published in EPJD*

IOP Conf. Series: Journal of Physics: Conf. Series **875** (2017) 032012 doi:10.1088/1742-6596/875/4/032012

Enhanced ionization of acetylene in intense laser pulses is due to energy upshift and field coupling of multiple orbitals

S. Erattupuzha*, C. L. Covington†, A. Russakoff†, E. Lötstedt‡, S. Larimian*, V. Hanus*, M. Schöffler*, S. Bubin§, M. Koch‖, S. Gräfe♣, A. Baltuska*, X. Xie*, K. Yamanouchi‡, K. Varga†, M. Kitzler*,[1]

* Photonics Institute, Technische Universität Wien, Guβhausstraβe 27-29, A-1040 Vienna, Austria
† Department of Physics and Astronomy, Vanderbilt University, Nashville, Tennessee 37235, USA
‡ Department of Chemistry, The University of Tokyo, 7-3-1 Hongo, Bunkyo-ku, Tokyo 113-0033, Japan
§ Department of Physics, School of Science and Technology, Nazarbayev University, Astana, 010000, Kazakhstan
‖ Institute of Experimental Physics, Graz University of Technology, Petersgasse 16, 8010 Graz, Austria
♣ Institute for Physical Chemistry, Friedrich-Schiller-University Jena, Helmholtzweg 4, 07743 Jena, Germany

Synopsis We describe a new enhanced ionization mechanism for polyatomic molecules. It works via a significant energy up-shift of valence orbitals for stretched bonds and a strong concomitant increase in the coupling between multiple molecular orbitals.

Laser-ionization of molecules is one of the most important processes in the strong-field and attosecond sciences. A key mechanism governing molecular ionization is enhanced ionization, where for diatomic molecules the ionization probability becomes strongly enhanced at a critical internuclear distance [1,2].

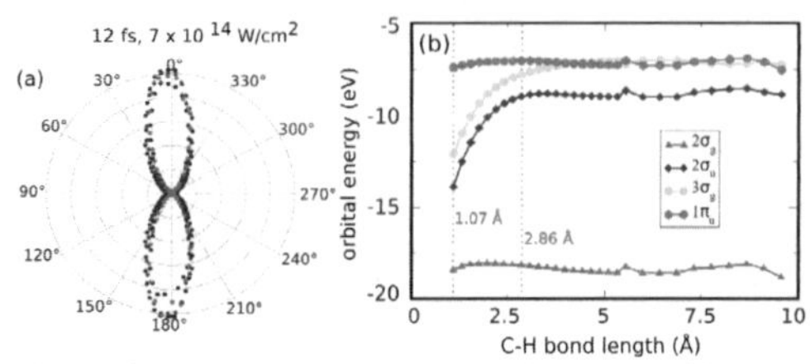

Figure 1. (a) Polar plots of measured proton yields ejected from $C_2H_2^{z+}$ during four-body fragmentation into $C^m/C^n/H^+/H^+$, z=m+n+2. Laser parameters are indicated, polarization direction vertical. Colours brown, green, blue and orange indicate charge states z={+4,+5,+6,+7}. (b) Binding energy of the four most loosely bound valence orbitals as a function of the C-H internuclear distance (symmetric stretch mode) calculated by DFT.

Experiments on polytatomic molecules have revealed remarkably high charge states and indicated the existence of a different ionization-enhancement mechanism [3-5]. However, until now this mechanism has not been fully understood and has been a subject of intense debate.

We present the results of a combined experimental and numerical study on strong-field ionization of acetylene performed with the aim of identifying the mechanism behind the reported surprisingly large multi-electron ionization probabilities of polyatomic molecules [3-5]. Using coincidence momentum imaging techniques and TDDFT simulations, we show that the strong ionization enhancement measured for molecules aligned parallel to the laser polarization [Fig. 1(a)] is due to the combined action of (i) a significant energy up-shift of the most relevant valence orbitals as the C-H distance increases [Fig. 1(b)], and (ii) a concomitant strong increase in the coupling between multiple molecular orbitals. Both processes lead to a strong effective reduction of the binding energy and an enhancement of the emission probability for lower-valence orbitals, making their ionization contributions even the dominant ones. For molecules aligned perpendicularly to the laser polarization direction this mechanism is inhibited because of a dominant emission of π electrons at low charge states, which in turn suppresses the C-H stretch motion and therewith prohibits entering the regime of the enhanced ionization.

To express these two key-ingredients in this mechanism we refer to it as EIC-MOUSE (**E**nhanced **I**onization from laser-**C**oupled **M**ultiple **O**rbitals that are **U**p-**S**hifted in **E**nergy). We argue that the EIC-MOUSE mechanism, which – as we will show in our presentation – is different from the usually considered CREI mechanism [1,2] in many ways, is relevant for a large class of polyatomic molecules.

References

[1] T. Zuo, A. Bandrauk 1995 *Phys. Rev. A* 52, R2511
[2] T. Seideman *et al.*, 1995 *Phys. Rev. Lett.* 75, 2819
[3] S. Roither *et al.* 2011 *Phys. Rev. Lett.* 106, 163001
[4] X. Xie et al. 2014 *Phys. Rev. A* 89, 23429
[5] C. Cornaggia *et al.* 1995 *Phys. Rev. A* 51, 1431

[1] E-mail: markus.kitzler@tuwien.ac.at

IOP Conf. Series: Journal of Physics: Conf. Series **875** (2017) 032013 doi:10.1088/1742-6596/875/4/032013

Above-threshold ionization processes in diatomic molecules driven by strong laser fields

Noslen Suárez[*1], **Alexis Chacón**[*], **Marcelo F. Ciappina**[†§], **Jens Biegert**[*‡] **and Maciej Lewenstein** [*‡]

[*]ICFO - Institut de Ciències Fotòniques, The Barcelona Institute of Science and Technology, Av. Carl Friedrich Gauss 3, 08860 Castelldefels (Barcelona), Spain
[†] Max-Planck-Institut für Quantenoptik, Hans-Kopfermann-Str. 1, 85748 Garching, Germany
[§]Institute of Physics of the ASCR, ELI-Beamlines, Na Slovance 2, 182 21 Prague, Czech Republic
‡ICREA - Pg. Lluís Companys 23, 08010 Barcelona, Spain

Synopsis In this work we develop a general theory to study above-threshold ionization (ATI) in symmetric diatomic molecules. It is based on the Strong Field Approximation (SFA) and allows us to underpin different physical processes corresponding to (i) direct tunneling ionization; (ii) electron rescattering on the center of origin; and, finally, (iii) electron rescattering on a different center. Very good agreement is observed in comparison of the results obtained with the time-dependent Schrödinger equation (TDSE). Our model also captures the dependence of the ATI spectra on the internuclear distance making possible to retrieve molecular structural information.

Strong field photoemission and electron recollision provide a viable route to extract electronic and nuclear dynamics from molecular targets with attosecond temporal resolution. However, since an *ab-initio* treatment of even the simplest diatomic systems is beyond today's computational capabilities, approximate qualitative descriptions are warranted.

In this contribution, we develop a theoretical approach to model the photoelectrons resulting from the interaction of an intense laser with molecular targets. We present a general theory for symmetric diatomic molecules in the single active electron approximation that, amongst other capabilities, allows adjusting both the internuclear separation and molecular potential in a direct and simple way. More importantly, we derive an analytic approximate solution of the TDSE, based on a generalized SFA [1, 2].

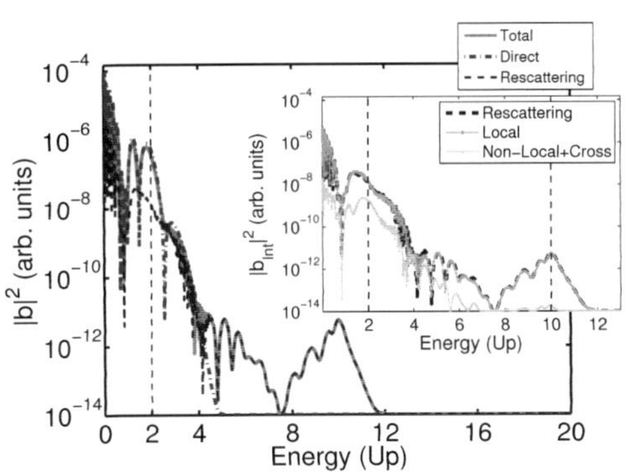

Figure 1. Different contributions to the H_2^+ photoelectron spectra (in logarithmic scale) calculated by using our model. The internuclear distance is set to $R = 2.6$ a.u. and the laser peak intensity to $I_0 = 4 \times 10^{14}\ \mathrm{W \cdot cm^{-2}}$.

Using our approach we obtain expressions for the laser-induced electron emission transition amplitudes from each molecular centres, and accelerated then in the strong laser field. In addition, our model directly underpins different underlying physical processes that correspond to i) direct tunnelling ionization; ii) electron rescattering on the centre of origin; iii) and, finally, electron rescattering on a different centre. In Fig. 1 we show results for H_2^+ molecules, obtained using our the modified SFA model.

One innovative aspect of our theory is the fact that the dipole matrix elements are free from non-physical gauge and coordinate system dependent terms – this is achieved by adapting the coordinate system, in which the SFA is performed, to the centre from which the corresponding part of the time dependent wave function originates.

Our analytic results agree very well with results extracted from the TDSE for H_2^+ molecules. Moreover, the theoretical model was applied to describe laser-induced electron diffraction (LIED) measurements of O_2^+ molecules, obtained at ICFO, and reproduces the main features of the experiment very well. Our approach can be extended in a natural way to more complex molecules and multi-electron systems [3].

References

[1] M. Lewenstein *et al.* 1995 *Phys. Rev. A.* **51** 1507

[2] N. Suárez *et al.* 2015 *Phys. Rev. A.* **92** 063421

[3] N. Suárez *et al.* 2016 *Phys. Rev. A.* **94** 043423

[1]E-mail: noslen.suarez@icfo.eu

High-order harmonic generation in polyatomic systems

Noslen Suárez*[1], Alexis Chacón*, Jose A. Pérez-Hernández §, Jens Biegert*‡, Maciej Lewenstein *‡ and Marcelo F. Ciappina†

*ICFO - Institut de Ciències Fotòniques, The Barcelona Institute of Science and Technology, Av. Carl Friedrich Gauss 3, 08860 Castelldefels (Barcelona), Spain
§Centro de Láseres Pulsados (CLPU), Parque Científico, 37185 Villamayor, Salamanca, Spain
‡ICREA - Pg. Lluís Companys 23, 08010 Barcelona, Spain
†Institute of Physics of the ASCR, ELI-Beamlines project, Na Slovance 2, 182 21 Prague, Czech Republic

Synopsis We present a theoretical approach of high-order harmonic generation (HHG) in polyatomic molecules based on the strong-field approximation (SFA) and within the frame of the single electron approximation. Our model allow us to turn on and off contributions having distinct physical origins or corresponding to different physical mechanisms. Results in di- and tri-atomic systems are shown. In addition a time-frequency analysis is presented to investigate the relative weight of the different electron trajectories contributing to the HHG spectra.

High-order harmonic generation (HHG) results from the interaction of a strong field with atomic and molecular systems. It configures one of the main processes used to extract electron structural and dynamical information about the atomic or molecular targets with sub-femtosecond temporal resolution. Moreover, it is the workhorse for the generation of attosecond pulses.

Here we develop an analytical description of HHG, which extends the well established strong-field approximation (SFA) [1]. Our approach involves two innovative aspects: i) first, using a non-local short-range model, but separable potential, we analytically calculate the bound-free and rescattering transition matrix elements for both atomic and molecular multicenter systems. In comparison with the standard approaches for the HHG process, these analytic derivations of the different matrix elements allow us to study directly how the HHG spectra depend on the atomic target and laser-pulse features. We can turn on and off contributions having distinct physical origins or corresponding to different physical mechanisms. In this way we quantify their weights in the various regions of the HHG spectra; ii) second, as Ref. [2, 3] reports, in our theory the multicenter matrix elements are free from non-physical gauge and coordinate system dependent terms – this is achieved by adapting the coordinate system, in which SFA is formulated, to the centre from which the corresponding part of the time dependent wave function originates.

We present results for di- (H_2) and tri-atomic (CO_2, H_2O) molecular systems exposed to a linearly polarized strong laser field. Additionally, our model captures the existence of an interference minima pattern in every molecular harmonic spectra. These minima can be considered as a multiple-slit type interference phenomenon, ubiquitously present in every strong field process involving a multicenter target [4].

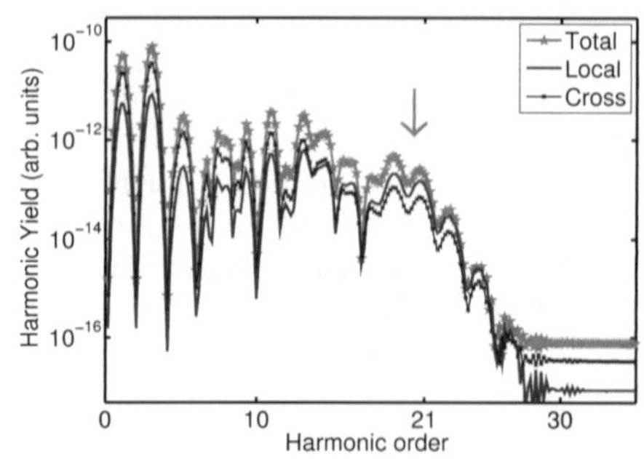

Figure 1. Total, Local and Cross contributions to an H_2O HHG spectra (in logarithmic scale) computed using our quasi-classical SFA model. The HHG spectra are computed averaging 8 different molecular orientations.

References

[1] M. Lewenstein *et al.* 1995 *Phys. Rev. A.* **51** 1507
[2] N. Suárez *et al.* 2015 *Phys. Rev. A.* **92** 063421
[3] N. Suárez *et al.* 2016 *Phys. Rev. A.* **94** 043423
[4] N. Suárez *et al.* 2017 *Phys. Rev. A.* (in press)

[1]E-mail: noslen.suarez@icfo.eu

High-Order Harmonic Generation of hydrogen molecule ions in a large internuclear distance*

Ling-Ling Du, Peng-Cheng Li[1], Hong-Shan Chen, Xiao-Xin Zhou[1]

College of Physics and Electronic Engineering, Northwest Normal University, Lanzhou, 730070, China

Synopsis We study high-order harmonic generation (HHG) by solving a two-dimensional time-dependent Schrödinger equation (TDSE) for a hydrogen molecular ion H_2^+. A double plateau structure is detected in harmonic spectra, the wavelet time-frequency analysis and an extended semiclassical analysis show that the width of the first plateau can be used to estimate the internuclear distance of H_2^+.

The HHG of H_2^+ with internuclear distance R=14 a.u. was obtained by solving 2D TDSE using the time-dependent Crank-Nicolson method [1]. The time-frequency analysis by means of the Morlet transform [2] and a generalized 2D semiclassical approach with the inclusion of the molecular potential [3] are performed to probe the features of the HHG.

Figure 1 shows the HHG spectra of H_2^+ driven by a 1600-nm laser pulse with the peak intensity $I = 2.14\times10^{14}\,\mathrm{W/cm^2}$. It is clear that a double platform structure can be found in the HHG spectra. The cutoff of the first plateau is equal to 37, which corresponds with the energy of classical calculation eE_0R (E_0 is the intensity of laser field) [4]. That is to say the first plateau of the HHG is due to the electron motion by migrating from one nucleus to the other.

Figure 2 shows time-frequency analysis spectra and the semiclassical energy of H_2^+ as a function of return time of the electron. First, the semiclassical energy is in good agreement with those of the time-frequency spectra of the HHG. Second, in Fig.2(a) and Fig.2(b), the trajectories that lie below the harmonic 40 suggest the migratory motion of electron between one nucleus and the neighboring one. Thus, we can estimate the internuclear distance of the diatomic molecule by the width of the first plateau.

Project supported by the National Natural Science Foundation of China (Nos. 11465016, 11364038, 11674268)

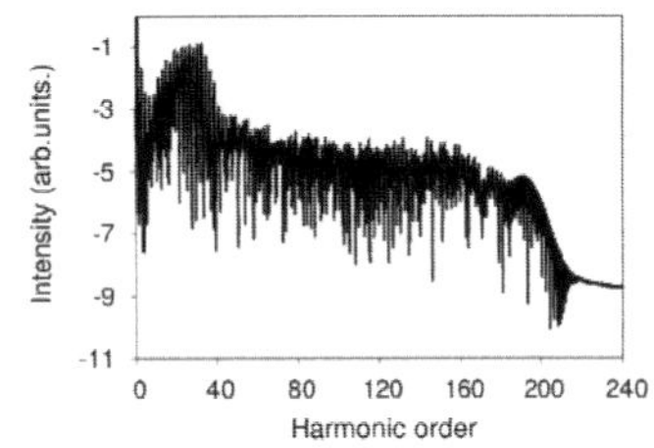

Figure 1. HHG spectra of H_2^+ driven by a 1600-nm laser pulse.

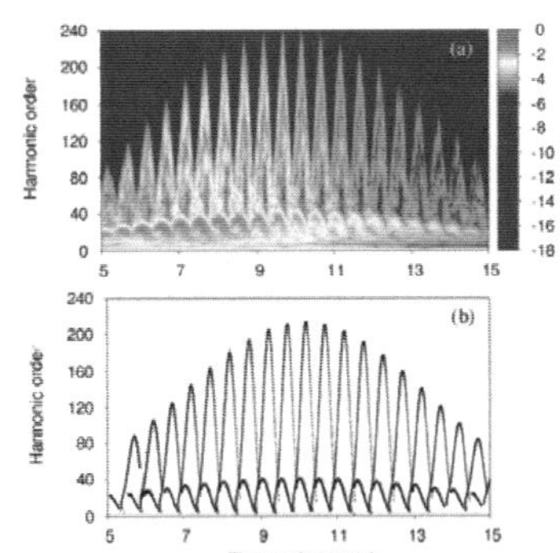

Figure 2. (a) Wavelet time-frequency analysis of HHG spectra; (b) the semiclassical return energy map.

References

[1] W. Li *et al* 2016 *Chin. Sci. Bull.*, 61, 855
[2] X. M. Tong *et al* 2000 *Phys. Rev. A.* **61** **021802(R)**
[3] M. B. Gaarde *et al* 2002 *Phys. Rev. A.* **65** **031406(R)**
[4] L. Liu *et al* 2017 *Sci. Sin-Phys. Mech. Astron.*, **47 033006**

*E-mail: lipc@nwnu.edu.cn, zhouxx@nwnu.edu.cn

IOP Conf. Series: Journal of Physics: Conf. Series **875** (2017) 032016 doi:10.1088/1742-6596/875/4/032016

A dressed harmonic study in laser-assisted photoionization of water molecules by attopulses

L. Martini[*1], **D.I.R Boll**[*2], **O.A. Fojón**[*3]

[*] Laboratorio de Colisiones Atómicas, Instituto de Física Rosario (CONICET-UNR), Rosario, Santa Fe 2000, Argentina.

Synopsis We study theoretically the laser-assisted photoionization of water molecules by attopulses by using a separable Coulomb-Volkov model [1, 2, 3]. We analyze the effect of varying the delay between the attopulse and the assistant laser field on the dressed harmonics lines.

Since organic matter is composed mostly of water, the reactions involving this molecule are essential to understand the interaction between the radiation and the biological tissue. In this work, we study theoretically the photoionization of water molecules by extreme ultraviolet (XUV) attopulses assisted by lasers in the near-infrared (NIR) range.

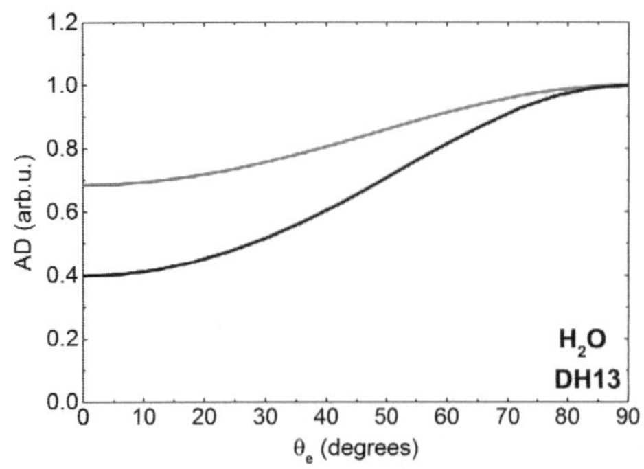

Figure 1. Angular distribution (AD) for the outermost water orbital $1b_1$ as a function of the emission angle θ_e for the 13th dressed harmonic (DH13) line computed using the Moccia's wavefunction for water, for the delays $\phi_L = 0.188\pi$ (——) and $\phi_L = 0.688\pi$ (——).

Firstly, we obtain monochromatic total cross sections in the framework of the dipole approximation within the velocity gauge. We use a simple model where the final state wavefunction is given by a Coulomb continuum wavefunction with optimized effective charges and we employ Moccia's monocentric wavefunctions to representate the H_2O molecule orbitals. We compute total cross sections for randomly oriented molecules and we compare our predictions with more elaborated theoretical results [4, 5] and experimental data (see [4, 5] and references there in). A very good agreement is obtained with the available experiments especially at high enough energies where there is a lack of elaborated results due to their high computational cost.

Secondly, the main objective of our work, we consider the photoionization reaction by XUV attopulses assisted by NIR lasers. We analyze photoelectron angular distributions (AD) for different delays between the attopulse and the assistant laser field. To this end, we use a separable Coulomb-Volkov model in which the temporal evolution of the system can be divided into three stages that allows spatial and temporal separation for the Coulomb and Volkov final state wavefunction. We compare our results for water and Ne atoms as they belong to the same isoelectronic series (results for water are shown in Figure 1 as an example). Furthermore, we contrast our calculations with previous theoretical and experimental work for Ar atoms due to the similarities of the orbitals involved in the reaction [2]. It is expected that these studies promote progress on the control of the chemical reactivity of water molecules.

References

[1] D. I. R. Boll and O. Fojón. 2014 *Phys. Rev. A* **90** 053414

[2] D. I. R. Boll and O. Fojón. 2016 *J. Phys. B: At. Mol. Opt. Phys.* **49** 185601

[3] D. I. R. Boll and O. Fojón. 2017 *J. Phys. B: At. Mol. Opt. Phys.* **50** 055604

[4] S. Gozem *et al.* 2015 *J. Phys. Chem. Lett.* **6** 4532

[5] M. Stener *et al.* 2000 *J. Phys. B: At. Mol. Opt. Phys.* **33** 1081

[1]E-mail: martini@ifir-conicet.gov.ar
[2]E-mail: boll@ifir-conicet.gov.ar
[3]E-mail: fojon@ifir-conicet.gov.ar

The Role of Super-Atom Molecular Orbitals in Doped Fullerenes in a Femtosecond Intense Laser Field

N. Berrah [#1], H. Xiong [#], B. Mignolet [^], L. Fang [&], T. Osipov [*], T. J. A. Wolf [&], E. Sistrunk [&], M. Gühr [&%], and F. Remacle [^]

[#] University of Connecticut, Physics Department, Storrs, CT 06269, USA
[^] Departement de Chimie, B6c, Université de Liege, B4000 Liege, Belgium
[&] Center for High Energy Density Science, University of Texas, Austin, TX 78712, USA
[*] LCLS, National Accelerator Laboratory, Menlo Park, CA 94025, USA
[&] Stanford PULSE Institute, SLAC National Accelerator Laboratory, Menlo Park, CA 94025, USA.
[%] Institut für Physik und Astronomie, Universität Potsdam, 14476 Potsdam, Germany

Synopsis The photoionization of complex targets, such as endohedral fullerenes, in strong laser fields has been studied experimentally and theoretically. We used the prototype $Ho_3N@C_{80}$ interacting with intense (0.1-5×10^{14} W/cm^2), short (30 fs), 800 nm laser pulses giving rise to multiply charged parent ions. The power law dependence of $Ho_3N@C_{80}{}^{q+}$, q=1-2, was found to be different from that of C_{60}. Time-dependent density functional theory computations suggest that the fast ionization of the SAMO states in $Ho_3N@C_{80}$ is responsible for the different power law behavior.

The investigations of the behavior of matter in intense laser fields were extended to larger molecules, such as C_{60} [1]. We increased further the complexity by investigating the photoionization of complex targets, such as endohedral fullerenes in strong laser fields. We used the prototype, $Ho_3N@C_{80}$, interacting with intense (0.1-5×10^{14} W/cm^2), short (30 fs), 800 nm laser pulses giving rise to multiply charged parent ions as well as fragment ions.

We will report on the measurement and theoretical explanation of the power law for singly ionized $Ho_3N@C_{80}$ yields, and we will compare our findings to C_{60} results carried out under similar conditions [1]. In our work, we distinguish two regions in the ion yield spectra with respect to field strength corresponding to different ionization mechanisms and power laws. For low field strengths, multiphoton ionization dominates, while for higher field strengths, tunneling and ionization over the barrier are the main ionization processes. The power law dependence of $Ho_3N@C_{80}{}^{q+}$, q=1-2, was found to be different from that of C_{60}. Our time-dependent density functional theory computations revealed different light-induced ionization mechanisms. Unlike in C_{60}, in doped fullerenes, the breaking of the cage spherical symmetry makes super atomic molecular orbital (SAMO) states optically active. Theoretical calculations suggest that the fast ionization of the SAMO states in $Ho_3N@C_{80}$ is responsible for the measured n=3 power law for $Ho_3N@C_{80}{}^{+}$ at intensities lower than 1.2×10^{14} W/cm^2. The measured $C_{60}{}^{+}$ power law was found to be n=5 [1]. For C_{60}, both super atomic molecular orbitals (SAMO) and Rydberg states played important roles when exposed to a strong laser field [2]. Both kinds of states can be indirectly populated by vibronic coupling, but because the density of Rydberg states is higher than that of SAMO states, Rydberg state ionization in C_{60} dominates the ionization at laser intensities >10^{13} W/cm^2 [2]. Our theory demonstrated that the SAMO states play a more important role in doped fullerenes than in C_{60}. Endohedral fullerenes respond differently to intense near-infrared femtosecond laser fields compared to C_{60}, and this is reflected by the measured and calculated different slope of the power law for low field strengths.

This work was funded by the DOE-BES grants Nos. DE-SC0012376 and DE-SC0012628.

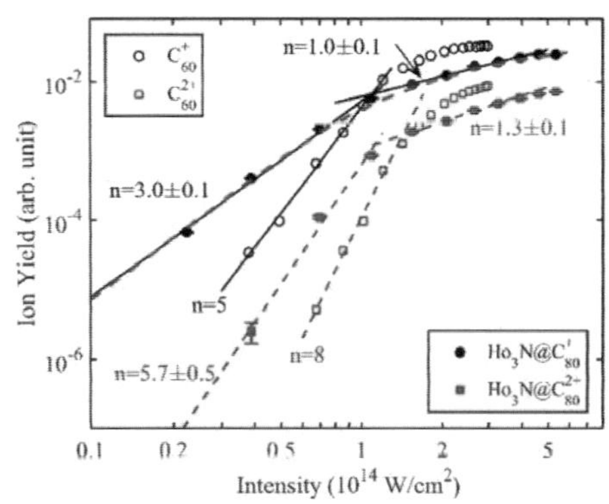

Figure 1. Comparison of the ion yield dependence on peak intensity for the ionization of $Ho_3N@C_{80}{}^{q+}$ (q=1, 2) (filled symbols) and $C_{60}{}^{q+}$ (q=1, 2) (open symbols) [1].

References

[1] I. Shchatsinin *et al.*, *J. Chem. Phys.* 125, 194320, (2006).
[2] H. Li et al. The Journal of Physical Chemistry Letters 7, 4677-4682, (2016).

[1] E-mail: nora.berrah@uconn.edu

Attosecond time delay in harmonic emissions of H_2 and D_2

M. H. Mustary*, D.E. Laban*, J.B.O. Wood*, I. Litvinyuk* and R.T. Sang*[1],

* Center for Quantum Dynamics, School of Natural Sciences, Griffith University, Brisbane 4111, Australia

Synopsis We have measured the relative phase of high harmonic generated radiation from hydrogen (H_2) and deuterium (D_2) using Gouy phase interferometer. This measurement may uncover the underlying information about the nuclear vibrational dynamics of these isotopes.

When a high intense laser field interacts with an atom or molecule, it deforms their potential barrier and electron is ionized. The ionized electron then accelerated away from their parent ion. Finally, as soon as the electric field reverses its direction, the electron accelerates back to its parent ion, recombines and emits high-energy extreme ultraviolet (XUV) photons at odd multiples of fundamental laser frequency. This process is known as high harmonic generation (HHG) and it produces attosecond pulses. In the last 12 years imaging of molecule by HHG attracted a lot of interest.

The amplitude and phase of HHG emission provides a snapshot of the structure and dynamics of the recombining system. A recently developed technique analyse nuclear dynamics with a temporal resolution of roughly 100 attoseconds by measuring HHG intensity ratio of hydrogen molecular isotopes [1]. However, to fully characterize a wavefunction we need both the intensity and phase information. Because of the very short wavelength of HHG radiation it is quite difficult to measure its phase.

Gouy phase interferometer produces XUV pulses in two adjacent gas jets with a variable time delay depending on the separation between the jets in a focus of the Gaussian laser beam. It has <100 zeptoseconds (10^{-21}s) temporal resolution [2]. Thus it can detect up to micro-radian phase shift. The Interferometer consists of 2 individual gas jets; bottom one is fixed in position and top one is movable along 3 translational axes. First we pass H_2 in the top jet and D_2 in bottom and then extract the normalized intensity I_{N1}. Then, by passing D_2 in top jet and H_2 in bottom find I_{N2}. The HHG phase difference between H_2 and D_2 is then given by

$$\phi_{H_2D_2} = \frac{\Delta I_N}{\sin(q\phi_{Gouy})} \quad (1)$$

Where, $\Delta I_N = I_{N2} - I_{N1}$ and $\sin(q\phi_{Gouy})$ is the derivative of normalized intensity for specific separation between the jets when only Gouy phase dominates. Fig.1 shows the phase shift in radian between H_2 and D_2 at two different gas jet separations. That phase shift is nearly constant and equals ~250 mrad for harmonic orders 23 to 33. Reversing the gas order means gas lines are physically altered and this measurement is taken to account for any pressure difference or gas dynamics which could affect our results. It gives a negative phase shift of similar magnitude.

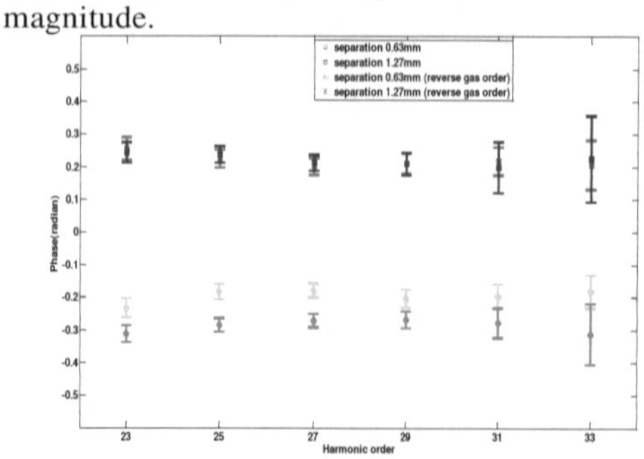

Figure 1. Phase shift between H_2 and D_2

This phase shift corresponds to a time delay of approximately 4 attoseconds between emissions from the two molecular isotopes. We are working on theoretical modeling to explain the nature of this delay.

References

[1] S. Baker *et al.* 2006, *Science,* Vol. **312**, pp 424-427.
[2] D. Laban *et al.* 2012, Phys. Rev. Lett., Vol **109**, pp 263902.

[1]E-mail: r.sang@griffith.edu.au

IOP Conf. Series: Journal of Physics: Conf. Series **875** (2017) 032019 doi:10.1088/1742-6596/875/4/032019

Observation quasi-periodic structures of laser-air plasma ‡

Zhi-Hong Jiao*, Yan-Lei Zuo†, Xiao-Xin Zhou* [1]

* College of Physics and Electronic Engineering, Northwest Normal University, Lanzhou, 730070, China
† Laser Fusion Research Center, China Academy of Engineering Physics, Mianyang, 621900, China

Synopsis We observe the quasi-periodic longitudinal structures of plasma generated by the nanosecond laser breakdown in air. Experimental results indicate that the long foci lens utilized to focus main laser beam can cause periodic structures of plasma.

The laser-gas plasma channel has been of much interest because of many important applications, such as stimulated Raman backscattering [1-4], plasma waveguide [5-7], high harmonic generation [8, 9], plasma accelerator [10, 11], and etc. However, in most cases of laser breakdown in gas the plasma channel has a quasi-periodic longitudinal structure which may limit its potential applications. This experimental work provides quasi-periodic structures of plasma generated by the nanosecond laser breakdown in air through varying three variables of laser energy, plasma evolution, and foci of lens.

In experimental setup, the plasma channels are detected by optical interferometry technique. A short wavelength beam is separated into reference beam and probing one. The probing beam passes through the plasma channel, generated by an intense main laser beam breakdown in air, and is combined with the reference beam to form linear interference fringes. An optical imaging system is utilized to zoom in the interference fringes. And the interference fringes are imaged on charge-coupled-device camera.

The plasma channels, generated through varying laser energy, plasma evolution, and foci of lens, are presented in Fig.1. When short foci lens is utilized to focus main laser beam, under the different laser energies and different plasma evolutions the quasi-periodic structure is absent from plasma channels (Fig.1, two upper fragments). But when a long foci lens is utilized, some clear quasi-periodic structures appear in plasma channels (Fig.1, lower fragment, the periodic structures are marked by the red arrows in). The longer the lens foci is, the more serious periodic structures plasma channels have.

Project supported by the National Natural Science Foundation of China (No. 11465016)

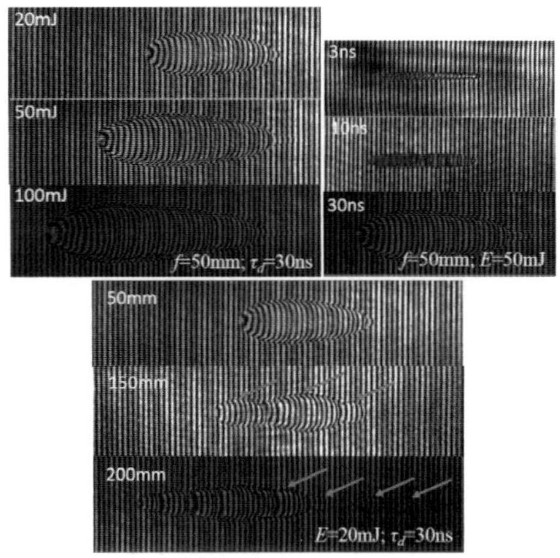

Figure 1. Interferograms of plasma channels: Absence of periodic structures under lens foci f =50mm, different laser energies E=20mJ, 50mJ, 100mJ (upper left) and different expansion times τ_d =3ns, 10ns, 30ns (upper right); Appearance of periodic structures under different lens foci f =50mm, 150mm, 200mm, laser energy E =20mJ and expansion time τ_d =30ns (lower). The red arrows are marks of periodic structures.

References

[1] J. Ren *et al* 2007 *Nat. Phys.* **3 732**
[2] J. Ren *et al* 2008 *Phys. Plasmas* **15 056702**
[3] W. Cheng *et al* 2005 *Phys. Rev. Lett.* **94 045003**
[4] Y. Ping *et al* 2004 *Phys. Rev. Lett.* **92 175007**
[5] C. H. Pai *et al* 2008 *Phys. Rev. Lett.* **101 065005**
[6] D. J. Spence *et al* 2000 *Phys. Rev. E* **63 015401**
[7] Y. F. Xiao *et al* 2004 *Phys. Plasmas* **11 21**
[8] P. L. Shkolnikov *et al* 1994 *Phys. Rev. A* **50 R4461**
[9] U. Teubner *et al* 2009 *Rev. Mod. Phys.* **81 445**
[10] C. G. R. Geddes *et al* 2004 *Nature* **431 538**
[11] P. Sprangle *et al* 2000 *Phys. Rev. Lett.* **85 5110**

[1] E-mail: zhouxx@nwnu.edu.cn

Intense-Field Photoionization of Molecules using Ultrashort Radiation Pulses: REMPI in Toluene, Aniline, Phenol, and Fluorobenzene

Joshua Beck, Collin J. McAcy, Timothy Scarborough*, and Cornelis J. Uiterwaal[1]

Dept. Physics and Astronomy, Jorgensen Hall, 855 N. 16th St., U. Nebraska-Lincoln, Lincoln, NE 68588, USA
*now at Ohio State University, Columbus, OH 43210, USA

Synopsis We investigate the intense-field photodynamics of molecules using 50-fs pulses with intensities of 10^{12} to 10^{14} W/cm^2 at 800-nm from a Ti:sapphire laser. We use a mass spectrometer that avoids focal intensity averaging. We have investigated the isoelectronic target molecules toluene (C_6H_5-CH_3), aniline (C_6H_5-NH_2), phenol (C_6H_5-OH), and fluorobenzene (C_6H_5-F). We discuss a resonance-enhanced multiphoton ionization (REMPI) mechanism that includes the AC Stark shifts of the molecular states. We have also investigated CO_2 and CS_2. For these molecules, no resonance effects are observed.

We report on the ultrafast (50 fs), intense-field (10^{12} to 10^{14} W/cm^2) photoionization and fragmentation of molecules. We investigated a series of four isoelectronic monosubstituted aromatics: toluene (C_6H_5-CH_3, *IP* = 8.83 eV), aniline (C_6H_5-NH_2, *IP* = 7.72 eV), phenol (C_6H_5-OH, *IP* = 8.51 eV), and fluorobenzene (C_6H_5-F, *IP* = 9.20 eV). Data is recorded under intense-field, single-molecule conditions. We focus 50-fs, 800-nm radiation pulses into a dilute target vapor, and record ion mass spectra for intensities of 10^{12} to 10^{14} W/cm^2. Our ion mass spectrometer has an interaction volume of micrometer size in all three dimensions, which allows us to record ion yields without focal intensity averaging, i.e., for sharply determined intensities [1]. This also allows us to estimate absolute ionization probabilities.

For the wavelength and intensities we use, multiphoton ionization is expected to dominate over tunneling ionization (Keldysh parameter > 1). If the smallest number of photons needed to reach the ionization continuum is N, the ion yield is expected to increase with intensity as I^N, at least for intensities low enough to avoid channel closing due to ponderomotive shifts. Remarkably, the ion yields of toluene, aniline, phenol, and fluorobenzene do not follow this simple integer-power dependence. For increasing intensities, but before saturation sets in, the ion yield's intensity dependence changes distinctly from I^N to a lower power of the intensity. We believe this change is evidence of a resonance-enhanced multiphoton ionization (REMPI) pathway. By fitting our data we find that the lower power is typically an integer, which we interpret as the multi-photon order of the rate-determining step. We had previously seen signatures of molecular REMPI in the monosubstituted halobenzenes C_6H_5-X (X = F, Cl, Br, I).[2] Guided by correlations between the excited states of isoelectronic molecules we propose candidates for the molecular resonant states.

To further investigate the role of resonances in molecular intense-field ionization we have started experiments at longer wavelengths, using an optical parametric amplifier source (TOPAS). Work in progress will be discussed.

This material is based upon work supported by the National Science Foundation under Grant No. PHY-0855675 and by the National Science Foundation EPSCoR RII Track-2 CA Award No. IIA-1430519 (Cooperative Nebraska-Kansas grant).

References

[1] J. Strohaber *et al.* 2008 *Phys. Rev. Lett.* **100,** 023002
[2] T. D. Scarborough *et al.* 2011 *Phys. Chem. Chem. Phys.* **13,** 13783

[1]E-mail: cuiterwaal2@unl.edu

IOP Conf. Series: Journal of Physics: Conf. Series **875** (2017) 032022 doi:10.1088/1742-6596/875/4/032022

Femtosecond dynamics of correlated many-body states in C_{60} fullerenes

Sergey Usenko[1], Michael Schüler[2]1, Armin Azima[3,4,5], Markus Jakob[1,5], Leslie L. Lazzarino[3], Yaroslav Pavlyukh[2,6], Andreas Przystawik[1], Markus Drescher[3,4,5], Tim Laarmann[1,5], Jamal Berakdar[2]

[1] Deutsches Elektronen-Synchrotron DESY, D-22607 Hamburg, Germany
[2] Institut für Physik, Martin-Luther-Universität Halle-Wittenberg, D-06099 Halle, Germany
[3] Department of Physics, University of Hamburg, D-22761 Hamburg, Germany
[4] Center for Free-Electron Laser Science CFEL, DESY, D-22607 Hamburg, Germany
[5] The Hamburg Centre for Ultrafast Imaging CUI, D-22761 Hamburg, Germany
[6] Department of Physics and Research Center OPTIMAS, University of Kaiserslautern, D-67653 Kaiserslautern, Germany

Synopsis In this joint theoretical and experimental work we investigate the population and decay dynamics of excited states of the C_{60} molecule by time-resolved two-photon photoemission. We map out how the thermally excited vibrational degrees of freedom lead to a transient redistribution of the photo-excited states. This includes the super-atom molecular orbitals (SAMOs), which are of great interest currently. The measured lifetimes are in line with full-fledged first-principle calculations.

Fullerene complexes hold great promises for molecular junctions or organic solar cells. In particular, extended, nearly atom-like orbitals, the super-atom molecular orbitals (SAMOs) represent – in view of their long lifetimes – potentially very robust transport channels for nano-sized devices. Insights in the control and dynamics of charge currents and their dissipation can be gained on the femtosecond time scale. To this end, ultrafast optical spectroscopy has become a major asset.

In this work [1], we investigate the excitation and dissipation dynamics of correlated states of the C_{60} molecule in the gas phase. Based on first-principle time-dependent density-functional theory (TDDFT) for the excited states of the molecule, a time-resolved two-photon photoemission (2PPE) setup is employed to map out the dynamics of the excited states. The experiments are further supported by time-dependent simulations of the laser-induced dynamics based on *ab initio* calculations of the coupling of the electronic degrees of freedom to the thermally excited vibrations.

Fig. 1(a) shows the simulated dynamics induced by the pump pulse as in the experiment. After initially excited an optically active bright state (BS), the vibrational relaxation and thermalization leads to a transient population of several dark states (DSs), including the SAMOs. The theoretical calculations further allow for inferring the lifetime of the laser-excited states and their manifestation on the pump-probe experiment. The result is compared to the experiment in Fig. 1(b). The decay behavior of the transient 2PPE signal matches the theory very well and thus identifies the vibrational degrees of freedom as major source of the dissipation.

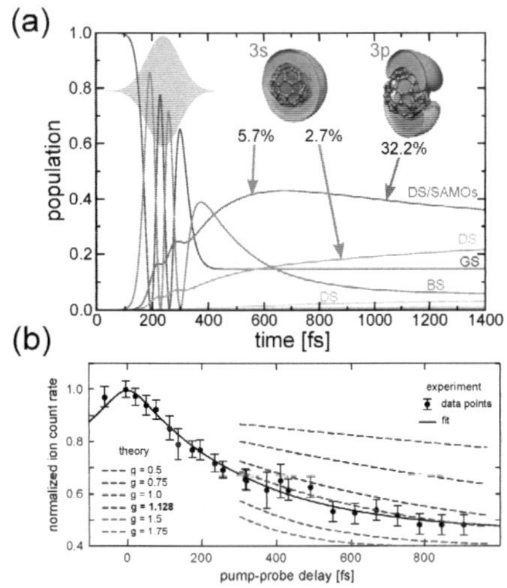

Figure 1. (a) Population dynamics induced by the pump pulse (sketched in the background). The insets show the weight of the SAMOs in the dominantly populated states. (b) Normalized C_{60}^{+} counts as a function of pump-probe delay. The signal derived from theory are compared to the experimental results for different scaling factors g of the electron-vibron interaction strength in the limit of long pump-probe delays. The bold value indicates the best agreement between theory and experiment.

References

[1] S. Usenko, M. Schüler *et al.* 2015 *New. J. Phys.* **18** 113055

1E-mail: michael.schueler@physik.uni-halle.de

Stereochemical configuration and selective excitation of the chiral molecule halothane

Martin Pitzer[*,†1], **Gregor Kastirke**[*], **Phillip Burzinsky**[*], **Miriam Weller**[*], **Daniel Metz**[*], **Jonathan Neff**[*], **Markus Waitz**[*], **Florian Trinter**[*], **Lothar Ph H Schmidt**[*], **Joshua B Williams**[‡], **Till Jahnke**[*], **Horst Schmidt-Böcking**[*], **Robert Berger**[§], **Reinhard Dörner**[*], **Markus Schöffler**[*2]

[*] Institut für Kernphysik, Goethe-Universität, Max-von-Laue-Str. 1, 60438 Frankfurt am Main, Germany
[†] Experimentalphysik IV, Universität Kassel, Heinrich-Plett-Str. 40, 34132 Kassel, Germany
[‡] Department of Physics, University of Nevada, Reno, 1664 N. Virginia Street, Reno, NV 89557, USA
[§] Fachbereich Chemie, Philipps-Universität, Hans-Meerwein-Straße, D-35032 Marburg, Germany

Synopsis The molecular configuration of the chiral ethane derivative halothane ($CHBrClCF_3$) was investigated via Coulomb Explosion Imaging using a COLTRIMS (Cold Target Recoil Ion Momentum Spectroscopy) setup. Several four-particle fragmentation pathways were found that yield information on the absolute stereoconfiguration. In addition, the effect of selective excitation on two-body break-ups was studied.

Coulomb Explosion Imaging can provide structural information of molecules or van-der-Waals-bound systems. In this approach, the system under investigation is multiply ionized within a timescale that is short compared to structural changes. The positively charged ion cores subsequently repel each other due to Coulombic forces so that their momentum vectors carry information on the initial structure. These momentum vectors can be measured for single molecules in coincidence, e.g. with a COLTRIMS setup (Cold Target Recoil Ion Momentum Spectroscopy)[1].

In recent years, Coulomb explosion imaging has been more and more extended to study structural properties of polyatomic molecules by collecting four or more charged fragments. In particular, it has been demonstrated as a way to determine the absolute stereochemical configuration of chiral molecules [2, 3].

This contribution presents Coulomb Explosion results of the chiral ethane derivative halothane ($CF_3CHClBr$), a substance that can be used as an anaesthetic drug. Single X-ray photons with energies corresponding to the carbon K-edge (286 - 305 eV) were used to induce multiple ionization and fragmentation.

A clear separation of enantiomers has been found using the fragmentation into CF_3^+, Cl^+, Br^+ and CH^+ (Figure 1). Additional fragmentation pathways with incomplete detection of fragments show a less pronounced enantiomer signal. For some of these break-ups, a slight increase in the yield was observed when the photon energy was high enough to ionize from either of the carbon 1s shells.

Additional information about the selective excitation can be drawn from the photoelectron spectra measured in coincidence. In this case, a clear correlation between the primary excitation site and the fragmentation pathway could be established.

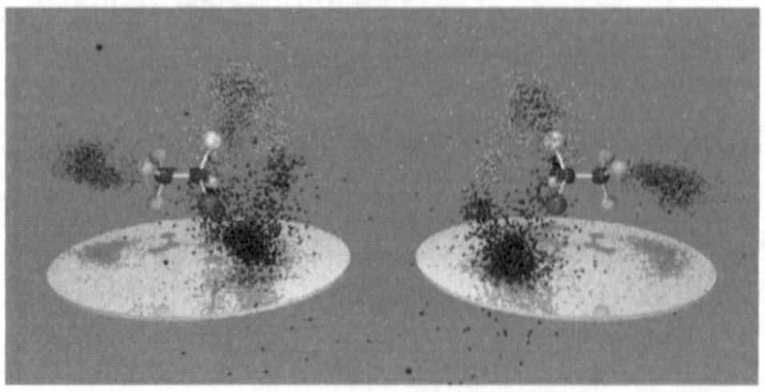

Figure 1. Momenta of the fragments CF_3^+ (green), Cl^+ (yellow), Br^+ (red) and CH^+ (black) in the molecular frame, overlaid with a structural model of the parent molecule halothane. Modified from [4]

References

[1] Dörner R *et al* 2000 *Phys. Rep.* **330** 95

[2] Pitzer M *et al* 2013 *Science* **341** 1096

[3] Herwig Ph *et al* 2013 *Science* **342** 1084

[4] Pitzer M *et al* 2016 *J. Phys. B: At. Mol. Opt. Phys.* **49** 234001

[1]E-mail: martin.pitzer@uni-kassel.de
[2]E-mail: schoeffler@atom.uni-frankfurt.de

IOP Conf. Series: Journal of Physics: Conf. Series **875** (2017) 032024 doi:10.1088/1742-6596/875/4/032024

Laser-Induced Oxygen Formation from Carbon Dioxide

Seyedreza Larimian*, Sonia Erattupuzha*, Markus Schöffler*, Sebastian Mai†, Philipp Marquetand†, Leticia González†, Andrius Baltuška*, Markus Kitzler*, Xinhua Xie*,†1

*Photonics Institute, Technische Universität Wien, A-1040 Vienna, Austria
†Institute of Theoretical Chemistry, University of Vienna, A-1010 Vienna, Austria

Synopsis We report experiments on the direct observation of O_2 formation from CO_2 in strong laser fields. Our simulations and pump-probe measurements suggest that CO_2 undergos bending motion during strong-field ionization which supports the O2 formation.

O_2 production is one of the most important processes for life on the Earth, where oxygen molecules are mainly generated via the photosynthesis by green plants and algae from carbon dioxide and water [1]. Theoretical studies suggested the possibility of producing O_2 through the dissociation of a CO_2 molecule [2], and a recent experiment showed indirect evidences of O_2 formation straightly from CO_2 molecules after ultraviolet excitation through the detection of C^+ (referring the creation of O2 on the other side) [3]. However, so far, O_2 formation from CO_2 has not been directly observed.

In this submission, we report experiments on the direct observation of molecular oxygen formation from CO_2 in strong laser fields with a reaction microscope. In our experiments, strong laser pulses, with peak intensity on the order of 10^{14} W/cm^2 and pulse duration of 25 fs, were employed to doubly ionize CO_2 molecules and further induced molecular fragmentation. Except for the dominate C-O bond breakage process, the O_2 formation process ($CO_2 \rightarrow C^+ + O_2^+ + 2e$) was identified in the photo-ion-photo-ion coincidence distributions [Fig.1(a)]. Our accompanying quantum chemical simulations suggest that the bending motion during the strong-field interaction may trigger the O_2^+ formation. To confirm the mechanism of bending motion-induced O_2 formation, we further performed pump-probe measurements using a Mach-Zehnder interferometer. Fourier analysis on the yields of O_2 formation process in pump-probe measurements contains clear frequency components of bending modes, as shown in Fig.1(b), which support the bending motion mechanism for the O_2 formation.

Our results may trigger experimental and theoretical investigations on further understanding of the O_2 production mechanism from CO_2 and optimizing the efficiency of the O_2 production with laser parameters. Moreover, our observation can also provide useful hints and new concepts for the studies on planetary atmospheres.

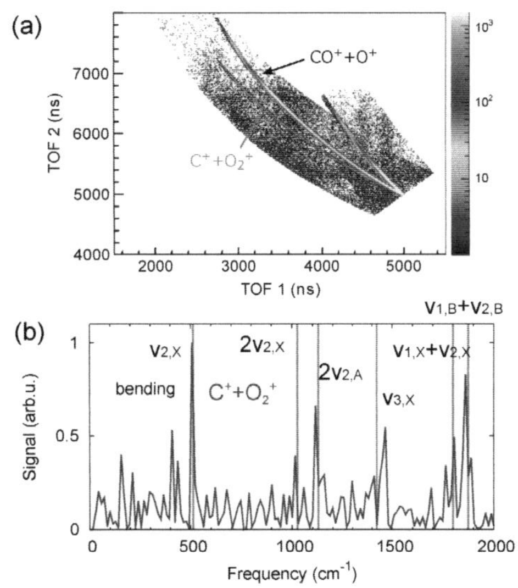

Figure 1. (a) Measured photo-ion-photo-ion coincidence distributions with selection conditions for the $CO^+ + O^+$ channel and the $C^+ + O_2^+$ channel. (b) Fourier transforms of the measured signal for the O_2 formation channel over the pump-probe delay, respectively. Vertical lines indicate the transition frequencies from NIST database [4], where ν_1, ν_2 and ν_3 denote the frequencies of the symmetric stretching mode, the bending mode and the asymmetric stretching mode, respectively. The indexes X, A and B represent for the X, A and B states of CO_2^+, respectively.

References

[1] G. Singhal, *et al.* 2012 *Concepts in photobiology: photosynthesis and photomorphogenesis* (Springer Science & Business Media)

[2] D.-Y. Hwang and A. M. Mebel 2000 *Chemical Physics* **256**, 169

[3] Z. Lu, *et al.* 2014 *Science* **346**, 61

[4] http://http://webbook.nist.gov/chemistry/

[1] E-mail: xinhua.xie@univie.ac.at

IOP Conf. Series: Journal of Physics: Conf. Series **875** (2017) 032025 doi:10.1088/1742-6596/875/4/032025

Observation of molecular rotation during femtosecond laser filamentation in air by a pump-probe longitudinal diffraction method

Li Qingyi*†, Li Suyu*†, Chen Anmin*†, Sui Laizhi*†, Jiang Yuanfei*†, Jin Mingxing*†

*Institute of Atomic and Molecular Physics, Jilin University, Changchun 130012, China

†Jilin Provincial Key Laboratory of Applied Atomic and Molecular Spectroscopy (Jilin University), Changchun, China

Synopsis: Utilizing a pump-probe longitudinal diffraction method, we have experimentally observed dynamics of air molecular rotation during femtosecond filamentation. By adjusting only the pump pulse energy, a recurrence of phase shift variation of the probe throughout a range of delay of at least 100 ps was achieved.

Femtosecond laser filamentation in air is a nonlinear effect and it occurs as a dynamic balance of two main physical phenomena [1]: the optical Kerr self-focusing and the plasma defocusing. Since the observation from Braun et al. in 1995 [2], femtosecond filamentation in air has stimulated tremendous research interest, not only because of the importance of its physical principle behind, but also its potential applications [1].

During femtosecond laser filamentation, lots of phenomena, such as intensity clamping, conical emission, ect, have been observed and extensively studied [1]. However, the microenvironment together with its physical and chemical features around the plasma channel created through filamentation has not deserved the same research enthusiasm. One important feature of the plasma wake of a laser filament is the occurrence of air molecular rotation dynamics which can maintain itself as periodic revivals of rotational periods of air molecules, especially of N_2 and O_2 [3]. By using a pump-probe longitudinal diffraction method [4], we have experimentally observed dynamics of air molecular rotation during femtosecond filamentation.

Here the laser adopted is a 1-KHz, 50-fs Ti:sapphire laser system (Coherent Libra) with an output pulse energy up to 3.6 mJ with a central wavelength of 800 nm. After a beam splitter, the transmitted high-energy pump pulse is focused by a thin lens (f = 400 mm) and transformed into a steady filament in air. The reflected low-energy probe pulse takes on a central wavelength of 400 nm after a BBO crystal and goes through spatial filtering to achieve a uniform wavefront of nearly a Gaussian form. The phase modulation from the pump pulse filamentation felt by the probe is recorded as diffraction patterns using a 12-bit industrial CCD camera.

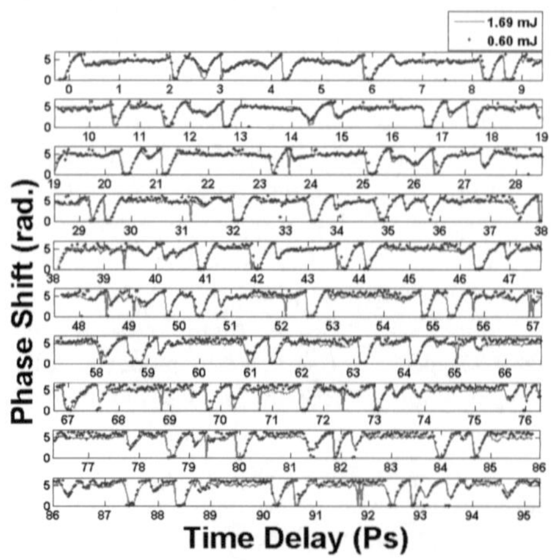

Figure 1. Phase shift of the probe as a function of time delay between pump(800nm) and probe(400nm).

In Fig. 1, the phase shift of the probe extracted from diffraction patterns as a function of time delay is shown. It's easy to figure out periodic revivals of phase variation occur at moments of multiple of the rotational periods of N_2 (T_r=8.3 ps) and O_2 (T_r=11.6 ps).

E-mail: mxjin@jlu.edu.cn

References

[1] A. Couairon et al. 2007 *Physics Reports* **441**, 47.
[2] Braun et al. 1995 *Opt. Lett.* **20** (1), 73 - 75.
[3] F. Calegari et al. 2008 *PRL* **100**, 123006.
[4] JS Liu et al. 2005 *PRE* **72**, 026412.

Studying Molecular Structure and Dynamics via Coulomb Explosion Imaging with X-rays and Ultrafast Laser Pulses

U. Ablikim[*† 1], **C. Bomme**[‡], **E. Savelyev**[‡], **H. Xiong**[§], **R. Obaid**[§], **B. Kaderiya**[*], **S. Augustin**[*], **K. Schnorr**[††], **I. Dumitriu**[¶], **R. Bilodeau**[†], **D. Kilcoyne**[†], **V. Kumarappan**[*], **R. K. Kushawaha**[*], **F. Ziaee**[*], **T. Osipov**[&], **A. Rudenko**[*], **N. Berrah**[§], **and D. Rolles**[* 2]

[*]James R. Macdonald Laboratory, Department of Physics, Kansas State University, Manhattan, KS 66506, USA
[†]Advanced Light Source, Lawrence Berkeley National Laboratory, Berkeley, CA 94704, USA
[‡]Deutsches Elektronen-Synchrotron (DESY), Hamburg, 22607, Germany
[§]Department of Physics, University of Connecticut, Storrs, CT 06269, USA
[††]Department of Chemistry, University of California, Berkeley, CA 94720, USA
[¶]Hobart and William Smith Colleges, Geneva, NY 14456, USA
[&]SLAC National Accelerator Laboratory, Menlo Park, CA 94025, USA,

Synopsis Coincidence momentum imaging is a powerful method to study molecular structure and dynamics. Here, we used the Coulomb explosion imaging method together with synchrotron radiation and ultrafast laser pulses in order to distinguish between geometric isomers of $C_2H_2Br_2$, $C_2H_2Cl_2$, and difluoroiodobenzene and to investigate their fragmentation dynamics following both strong-field and inner-shell photoionization.

Isomers, i.e. molecules with the same chemical formula but different geometric structures, play an important role in many biological processes [1]. Despite containing the same atomic constituents, isomers can have very different physical, chemical, and biological properties. Therefore, it is of particular interest to experimentally distinguish isomers in order to investigate isomer-specific reactions and to study the interconversion between different isomers in time-resolved experiments.

We have conducted coincidence momentum imaging measurements on gas-phase molecular isomers, such as dibromoethane ($C_2H_2Br_2$) and dichloroethene ($C_2H_2Cl_2$), using a double-sided velocity map imaging (VMI) spectrometer at the Advanced Light Source as well as using a COLTRIMS setup and an ultrafast near-infrared laser at the J.R. Macdonald Laboratory at Kansas State University. The goal of the studies is to experimentally identify and separate *cis* and *trans* isomers using the Coulomb Explosion Imaging (CEI) method. Our results show that the geometric structure of the isomers can be distinguished by triply ionizing the molecule into the $C_2H_2^+$+Br^+/Cl^++Br^+/Cl^+ fragmentation channels via inner-shell photoionization using X-ray synchrotron beams [2] or via strong-field ionization with ultrafast femtosecond laser pulses. Furthermore, numerical simulations using a classical Coulomb explosion (CE) model for both isomers match closely with the kinetic energies of the ionic fragments and the momentum correlation between their momentum vectors.

In order to further extend our CEI studies to other types of isomers and more complex molecules, we performed similar coincidence momentum imaging experiments on 2,6- and 3,5-difluoroiodobenzene (DFIB) [3]. Our experimental results and CE simulation indicate a fast charge migration along the phenyl ring after inner-shell photoionization of the iodine *4d* shell. Furthermore, we find that the majority of three-body fragmentation channels result from a sequential two-step fragmentation, where the iodine ion rapidly departs from the system after inner-shell ionization, leaving the remaining DFB dication in a rotationally excited metastable state, which subsequently fragments after a delay that is longer than its rotational period.

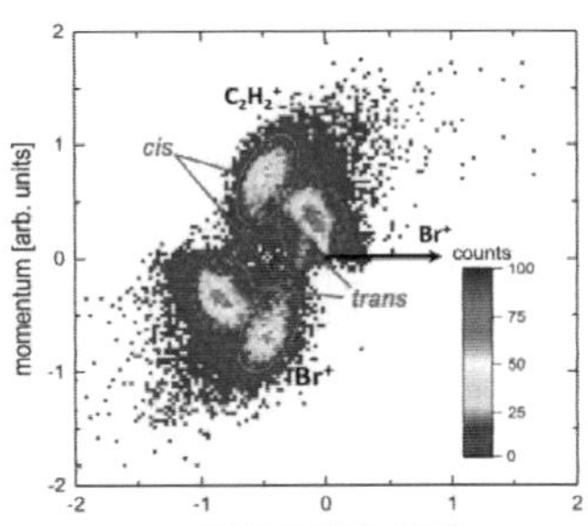

Figure 1. Newton plot of the three-body $C_2H_2^+$+Br^++Br^+ coincidence channel created by inner-shell ionization of $C_2H_2Br_2$ (*cis/trans* mixture) at 140 eV photon energy.

References

[1] U. K. Genick *et al.* 1998 *Nature* **392** 206-209
[2] U. Ablikim *et al.* 2016 *Sci. Reports* **6** 38202
[3] U. Ablikim *et al.* 2017 *submitted*

[1] E-mail: utuq@lbl.gov
[2] E-mail: rolles@phys.ksu.edu

Laser polarization effect on nitrogen fluorescence emission induced by femtosecond filament in air

Suyu Li[*,†,‡1], **Anmin Chen**[*,†] **Yuanfei Jiang**[*,†], **Hang Liu** [†,†], **Mingxing Jin**[*,†2]

[*] Institute of Atomic and Molecular Physics, Jilin University, Changchun 130012, China
[†] Jilin Provincial Key Laboratory of Applied Atomic and Molecular Spectroscopy (Jilin University), Changchun 130012, China
[‡] State Key Laboratory on Integrated Optoelectronics, College of Electronic Science and Engineering, Jilin University, Changchun 130012, China

Synopsis The laser polarization effect on nitrogen fluorescence emission induced by femtosecond filament in air is studied. By comparing the femtosecond filament induced fluorescence emission under different pump laser polarizations in different focal length cases, the emission mechanism of nitrogen fluorescence induced by femtosecond filament is discussed.

During the femtosecond laser filamentation in air, the intensity inside the filament stays around $10^{13} \sim 10^{14}\mathrm{W/cm^2}$ due to the dynamic balance between Kerr self-focusing and plasma defocusing effect [1, 2], which is intense enough to excite, ionize or dissociate the molecules in air, generating a large number of excited molecules and molecular ions. These molecules and molecular ions undergo complicated transition process, emitting characteristic fluorescence [3, 4, 5]. Femtosecond filament induced fluorescence has been has been used to characterize the filamentation process [6], extract the electron temperature and laser intensity inside filaments, and explore the interaction processes between femtosecond laser and atoms and molecules[3, 4, 7]. In addition, the high sensitivity, non-intrusiveness, and real-time analysis of femtosecond laser-induced fluorescence in the field of remote sensing of atmospheric trace species arouses our enthusiasm in the study of it.

In air, the femtosecond filament induced fluorescence mainly comes from nitrogen. However,the emission mechanism of nitrogen fluorescence is still controversial, especially that of the fluorescence from N_2 (e.g., 337-nm line). Since the transition between the singlet state and triplet state of N_2 is spin forbidden, $N_2(C^3\Pi_u^+)$ can not be formed due to the direct high-field photonic excitation.

In this work, the influence of laser polarization state on nitrogen fluorescence emission induced by femtosecond filament in air is investigated as the focal length of focusing lens is different. Fig. 2 shows the backward fluorescence spectra induced by the linearly and circularly polarized pulses. It can be clearly seen from the figure that there exists difference between the emission behaviors of N_2 and N_2^+ fluorescence under different laser polarizations: for fluorescence from N_2 (e.g. 337, 357, 380 and 406 nm lines *etc.*), it is more intense in the circular polarization case, while for that from N_2^+ (391 and 428 nm lines), its intensity is higher in the linear polarization case. Based on the observation, the emission mechanism of nitrogen fluorescence induced by femtosecond filament is discussed.

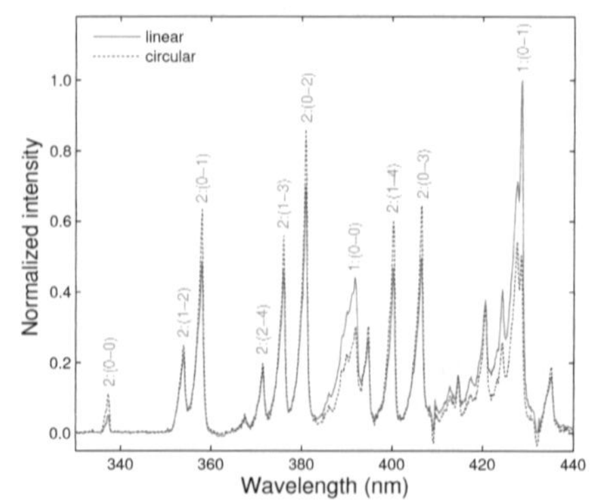

Figure 1. Backward fluorescence spectra induced by the linearly and circularly polarized pulses as the input pulse energy is 1.6 mJ. The focal length is $f = 25$ cm. In the transitions $v - v'$, v and v' denote the vibrational levels of upper and lower electronic states, respectively.

References

[1] S. Y. Li *et al.* 2014 *Phys. Rev. A* **89** 023809
[2] S. Y. Li *et al.* 2015 *Chin. Phys. B* **24** 114207
[3] S. Y. Li *et al.* 2016 *Phys. Rev. A* **93** 013405
[4] Y. Shi *et al.* 2016 *Opt. Commun.* **367** 174
[5] H. Li *et al.* 2016 *Chem. Phys. Lett.* **662** 188
[6] S. Li *et al.* 2015 *Phys. Plasmas* **22** 093113
[7] S. Li *et al.* 2016 *Phys. Plasmas* **23** 023102

[1]E-mail: sylee@jlu.edu.cn
[2]E-mail: mxjin@jlu.edu.cn

IOP Conf. Series: Journal of Physics: Conf. Series **875** (2017) 032028 doi:10.1088/1742-6596/875/4/032028

Strong Field dissociative ionization of the D_2^+

Sándor Borbély[*,1], **Attila Tóth**[†], **G. J. Halász**[‡], **and Ágnes Vibók** [†§]

[*] Faculty of Physics, Babeş-Bolyai University, Kogălniceanu Street 1, 400084 Cluj-Napoca, Romania, EU
[†] ELI-ALPS, ELI-HU Non-Profit Ltd, H-6720 Szeged, Dugonics Square 13, Hungary, EU
[‡] Department of Information Technology, University of Debrecen, H-4002 Debrecen, PO Box 400, Hungary, EU
[§] Department of Theoretical Physics, University of Debrecen, H-4002 Debrecen, PO Box 400, Hungary, EU

Synopsis Strong field dissociative ionization of the D_2^+ molecule in the multiphoton regime is investigated employing *ab initio* calculations for fixed molecular axis orientations. The resulting joint (photoelectron and nuclear kinetic) energy spectra are explored. The number of photons absorbed in each multiphoton peak of these spectra is determined analyzing the angular distribution of the photoelectrons, while the exact path followed by D_2^+ is identified studying the time evolution of the nuclear wave packets.

Despite being the simplest molecular system, the complete dynamical description of the D_2^+ considering all electronic and nuclear degrees of freedom is not a trivial task. Present-day theoretical approaches employed for the investigation of photoionization usually solve the problem in reduced dimensions, and commonly neglect the rotation of the molecule. Our previous studies on the photodissociation of D_2^+ [1, 2] showed the importance of the molecular rotation as the resulting light induced conical intersections (LICIs) influenced strongly the dynamical properties of the molecule.

In this work we incorporate the ionization in our two-state (ground and first excited) model [1, 2] previously developed for the description of photodissociation by considering a third ionized state corresponding to a well defined asymptotic momentum $\vec{k}$. For this three-level system the 1D nuclear Schrödinger equation is solved numerically and the KER spectra is extracted from the final population of the ionized state. The combined ionization and dissociation spectrum is obtained by performing individual calculations for several ionization channels (i.e. continuum electron momenta $\vec{k}$).

Here we report our results obtained for the dissociative ionization of D_2^+ in the multiphoton regime. Although our long term goal is to describe the dynamics in the LICI picture, the present results were obtained for fixed molecular axis orientations. In accordance with previous investigations, the dominant features of the joint energy spectra (JES) were multiphoton peaks (see Figure 1.) [3]. Our model allowed us to analyze the angular distribution of the photoelectrons for each point of the JES. Employing a partial wave analysis we were able to determine the number of absorbed photons in each multiphoton peak as indicated on the transverse energy conservation lines of Figure 1.

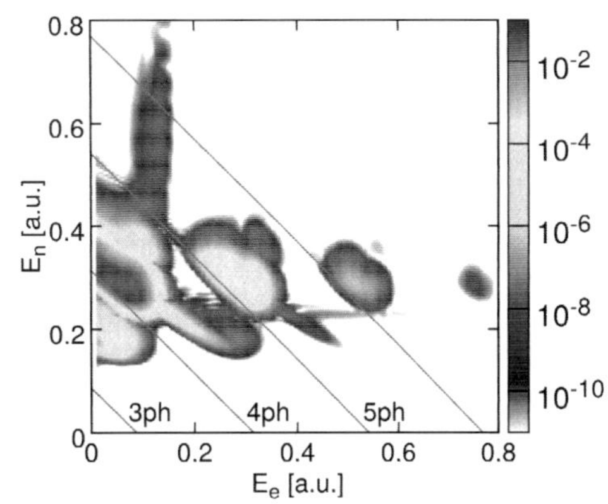

Figure 1. Joint energy spectrum (JES) for D_2^+ as a function of electron (E_e) and nuclear (E_n) kinetic energy. The D_2^+ was irradiated with a 10 fs long laser pulse with $\lambda = 200$ nm and $I_0 = 3 \times 10^{13}\text{W/cm}^2$.

In order to better understand the undergoing processes we also investigated the time-dependent nuclear wave packet density. Taking into account also the topology of the electronic potential energy surfaces we were able to identify the time moment and location (i.e. internuclear separation) where the population transfer between the considered potential energy surfaces occurred [4].

References

[1] G. J. Halász *et al.* 2013 *J. Phys. Chem. A* **117** 8528

[2] G. J. Halász *et al.* 2013 *Phys. Rev. A* **88** 043413

[3] A. Tóth *et al.* 2016 *J. Phys. Chem. A* **120** 9411

[4] A. Tóth *et al.* 2017 *Chem. Phys. Lett.* Accepted for publication

[1]E-mail: sandor.borbely@phys.ubbcluj.ro

IOP Conf. Series: Journal of Physics: Conf. Series **875** (2017) 032029 doi:10.1088/1742-6596/875/4/032029

Ultrafast dynamics in CO_2 studied by XUV-pump – NIR-probe experiments

S. J. Robatjazi, S. Pathak, W. L. Pearson, Kanaka Raju P., J. Powell, X. Li, B. Kaderiya, I. Ben-Itzhak, D. Rolles[1], A. Rudenko[2]

James R. Macdonald Laboratory, Department of Physics, Kansas State University, Manhattan, KS 66506, USA

Synopsis We present the results of an XUV-pump – NIR-probe experiment studying ultrafast dynamics of photo-ionized CO_2. The multi-harmonics XUV pulse, which contains the 11th to 39th harmonics of a 790 nm near-infrared (NIR) laser, is used to excite molecular wave packets in CO_2^+, which are then probed by a near-infrared pulse that induces dissociation. The yields and angle-resolved kinetic energy distributions of all charged fragments are measured as a function of the XUV-NIR delay using a velocity map imaging setup.

We have developed a versatile experimental setup for XUV-pump – NIR-probe experiments using a 10 kHz high-harmonic generation (HHG) source and two different charged-particle momentum imaging spectrometers. The HHG source, based on a commercial KMLabs eXtreme Ultraviolet Ultrafast Source (XUUS), is capable of delivering XUV radiation of less than 30 fs pulse duration in the photon energy range of ~17 eV to 100 eV. It can be coupled either to a conventional velocity map imaging (VMI) setup with an atomic, molecular, or nanoparticle target; or to a double–sided VMI spectrometer equipped with two delay-line detectors for electron-ion and ion-ion coincidence measurements.

We present an overview of the setup and results of first pump-probe experiments, including studies of the ultrafast dynamics of photo-ionized CO_2 molecules. The pump-probe studies are performed using a combination of broadband XUV pulses and near-infrared (NIR) pulses, both of ~25 fs duration. The XUV pulse containing the 11th to 39th harmonics of a 790 nm NIR laser is used to excite molecular wave packets in CO_2^+, which are then probed by near-infrared-induced dissociation.

We use a velocity map imaging setup to measure the yields and angle-resolved kinetic energy distributions of all charged fragments as a function of the XUV-NIR delay. The results, in particular, the delay dependence of O^+ and CO^+ ion production for parallel and perpendicular NIR and XUV polarizations, are compared to the data reported by Timmers *et al.* [1] for a shorter harmonics train, and to our earlier data obtained using narrow-band isolated harmonics (11th or 13th) as a pump.

This project is supported in part by the National Science Foundation (NSF-EPSCOR) Award No. IIA-1430493 and in part by the Chemical Science, Geosciences, and Bio-Science Division, Office of Basic Energy Science, Office of Science, U.S. Department of Energy. K.R.P. and W.L.P. thank NSF-EPSCOR for their support.

(a)

(b)

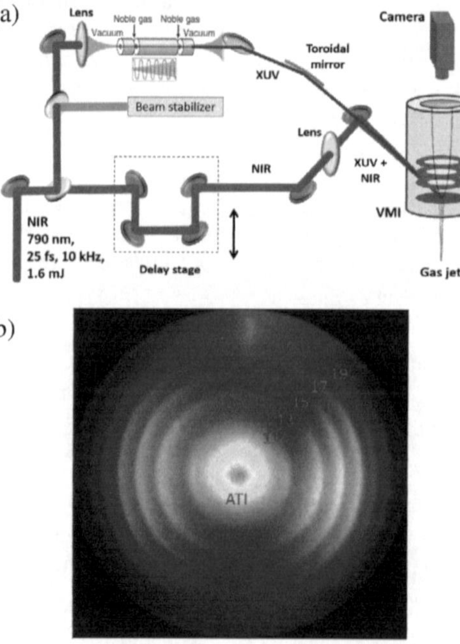

Figure 1. (a) Schematic of the XUV-pump – NIR-probe setup with the VMI spectrometer. (b) VMI image of photoelectrons produced by the ionization of argon atoms with a combination of XUV and NIR fields. The low-energy electrons at the center of the image are from above-threshold ionization (ATI), while the circular structures are the photoelectrons produced by the XUV, with the corresponding harmonic order indicated in red.

References

[1] H. Timmers *et al.* 2014 *Phys. Rev. Lett.* **113** 113003

[1] E-mail: Rolles@phys.ksu.edu

[2] E-mail: Rudenko@phys.ksu.edu

Contribution of resonance excitation on ionization of OCS molecules in strong laser field

Jiaqi Yu*†, Wenhui Hu*†, Lanhai He*†, Chuncheng Wang*†, Sizuo Luo[1]*† and Dajun Ding[2]*†

* Institute of Atomic and Molecular Physics, Jilin University, Changchun, P R China
†Jilin Provincial Key Laboratory of Applied Atomic and Molecular Spectroscopy, Jilin University, Changchun, P R China,

Synopsis The resonant enhanced multiphoton ionization (REMPI) of OCS molecules has been studied by tracing the electron energy shift with the laser intensity dependent photoelectron energy spectra. We find that two excited states ($^1\Delta$ and $^1\Pi$) have been involved in the ionization of OCS at 800 nm and the REMPI through a certain excited state can be selectively controlled by varying the laser intensity. Furthermore, the REMPI of OCS through 4sσ Rydberg states at 400 nm laser fields has also been identified by angular distribution of electrons.

Strong field laser interaction with molecules has attracted extensive attention in recent years. Abundant information on electronic and nuclear dynamics of molecules is encoded in the photoelectron moment distribution, including the contribution of excited states, multi-orbital effect and electron rescattering in photoelectron holograph. Several experimental observations on ionization of OCS molecules show obvious different angular distribution from theoretical calculations. The observed maximum ionization occurs at 90 degrees (i.e., the polarization of ionization laser perpendicular to the molecular axis) [1] that cannot be reproduced by the calculations such as MO-ADK, adiabatic SFA, TD-DFT, and WAFT. In general, the influence of excited states in the process has been proposed for interpreting this discrepancy.

We experimentally investigate ionization of OCS molecules in strong laser fields using velocity-map imaging. A 50 *fs*, 800 nm linearly polarized laser is used to ionize OCS molecules from a pulsed beam. The laser pulses centered at 400 nm are generated by passing an 800 nm linear polarized pulse through a BBO crystal. The obtained electron energy spectra under different laser intensities (10~100 TW/cm^2) are shown in Figure 1. Several peaks generated from above-threshold ionization (ATI), is signed by n=9-13, which energy decreasing as the laser intensity increasing. Three REMPI peaks with the energies of 0.7, 1.2 and 2.2 eV are observed, which energy is not change with the laser intensity. The electrons from Peak1 and Peak2 appear at different laser intensity are generated from resonant ionization from different electronic excited states (Freeman resonance [2]). The Peak1 appeared at laser intensity between 55 and 82 TW/cm^2 is generated from ground state $X^1\Sigma^+$ absorbing 6 photons to excited state $^1\Delta$, and then ionized by absorbing another 4 photons. While the peak 2, appeared above 80 TW/cm^2, can be assigned to transit from the ground state $X^1\Sigma^+$ absorbing 8 photons to excited state $^1\Pi$, and then ionized by absorbing another 3 photons. Peak 3 is the five-photon ionization through the $^1\Delta$ state. Furthermore, we also find Rydberg state 4sσ is involved in REMPI of OCS molecules at 400 nm at laser intensity between 3.5 and 34 TW/cm^2. Therefore, the present results confirm the resonance excitation contribution on strong-field ionization of OCS molecules.

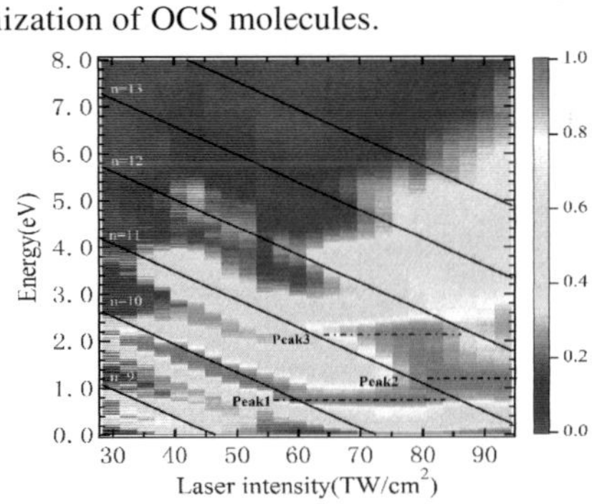

Figure 1. Laser intensity dependent photoelectron energy spectra of OCS molecules obtained from 800 nm.

References

[1] J.L.Hansen *et al.*, J. Phys. B: At. Mol. Opt. Phys. 45, 015101 (2012).
[2] R.R.Freeman *et al.*, Phys. Rev. Lett. 59, 1092 (1987).

E-mail: luosz@jlu.edu.cn
E-mail: dajund@jlu.edu.cn.

THz Streaking of the Autoionization Dynamics of O_2 at the Free-Electron-Laser FLASH

Y.F. Liu[*1], **K. Schnorr**[*], **G. Schmid**[*], **S. Augustin**[*], **S. Meister**[*], **H. Lindenblatt**[*], **A. Rudenko**[†], **M. Kübel**[‡], **C. Burger**[‡], **N. Stojanovic**[§], **R. Treusch**[§], **S. Düsterer**[§], **T. Jahnke**[¶], **M.F. Kling**[‡], **C.D. Schröter**[*], **T. Pfeifer**[*], **R. Moshammer**[*2]

[*] Max-Planck-Institut für Kernphysik, Saupfercheckweg 1, 69117 Heidelberg, Germany
[†] J.R. Macdonald Laboratory, Kansas State University, Manhattan, 66506 Kansas, USA
[‡] Physics Department, Ludwig-Maximilians-Uniersität München, 85748 Garching, Germany
[§] Deutsches Elektronen-Synchrotron (DESY), Notkestrasse 85, 22607 Hamburg, Germany
[¶] Institut für Kernphysik, Goethe Universität, Max-von-Laue-Strasse 1, 60438 Frankfurt, Germany

Synopsis With a XUV-pump/THz-probe scheme the decay of excited O_2 cations that were produced by irradiation with XUV photons has been investigated using a Reaction Microscope (ReMi) at the free-electron-laser FLASH in Hamburg. The temporal profile of electrons emitted due to autoionization of Rydberg states $C^4\Sigma_u^-(nl\sigma_g)$ has been traced using THz streaking. This way the relaxation dynamics was followed. The relative shift in phase between photoelectrons and autoionization electrons was analyzed with respect to emission time delays or autoionization lifetimes.

A proper understanding of the dynamical response of atomic and molecular systems to intense pulses of XUV radiation is of vital importance for the description of the relaxation dynamics and multiple ionization process in the X-ray regime. As a consequence, the interaction of XUV radiation with small molecules leading to the population of highly excited states that evolve on ultrafast time-scales has attracted considerable theoretical and experimental interest. Technologically, attosecond streaking has become a well-established technique to obtain time-resolved information about the electronic processes as well as for the characterization of XUV pulses [1, 2]. Here, we use the FEL pulses (60 eV photons, intensity about 10^{13} W/cm^2) together with synchronized THz pulses (wavelength 152 μm, oscillation period 507 fs).

Within a kinematically complete experiment the emitted electrons are streaked by the simultaneously acting THz-field and they are detected in coincidence with the emerging ion(s). The highly excited states in the molecular ion are created by absorption of one (or more) XUV photons. Both, the photoelectron as well as the secondary Auger or autoionization electron are exposed to the streaking field. This leads to a momentum or energy shift that allows determining the moment of ionization. Thus, the phase shift between the photoelectron and the secondary electron carries information about the time difference between both emission steps or the autoionization lifetime. For autoionization of singly charged O_2 molecules we expect lifetimes between a few 10 and up to 300 fs [3, 4]. Using an XUV-pump/THz-probe scheme, we performed a direct measurement of the autoionization process of singly charged excited oxygen ions. As shown in Fig.1, the autoionizing Rydberg states $C^4\Sigma_u^-(nl\sigma_g)$ have lifetime of 96 ± 21 fs.

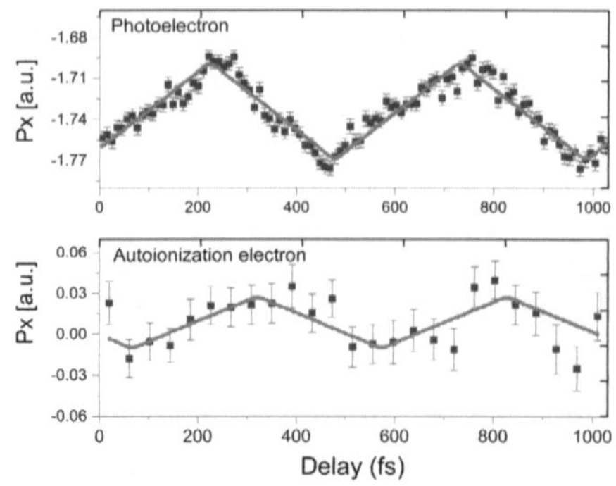

Figure 1. The temporal profiles of the momenta of the photoelectrons and autoionization electrons.

Theoretical work on the autoionization mechanisms of Rydberg series of O_2 cations is needed for further understanding of the relaxation dynamics of the class of excited states.

References

[1] R. Kienberger et al. 2004 *Nature* **427** 817
[2] N. Stojanovie et al. 2013 *J. Phys. B* **46** 192001
[3] A.S. Sandh et al. 2008 *Science* **322** 1081
[4] H. Timmers et al. 2012 *Phys. Rev. Lett.* **109** 173001

[1]E-mail: liuyifan@mpi-hd.mpg.de
[2]E-mail: moshammer@mpi-hd.mpg.de

Spin-polarized electrons upon nondipole photodetachment of fullerene anions

A. C. Edwards*[1], C. G. Lane*[2], V. K. Dolmatov*[3]

* Department of Physics and Earth Science, University of North Alabama, Florence, Alabama, 35632, U.S.A.

Synopsis Spin-polarization of electrons ejected from fullerene anions C_N^- in the geometry where only nondipole photodetachment effects matter is studied at photon energies of only few tens of eV. The degree of electron spin-polarization is found to be on the order of 5 to 10%, at certain energies. This is remarkable, given that this happens at low photon energies.

To the best of the authors' knowledge, little, if any, is known about spin-polarized photoelectron fluxes from fullerene anions C_N^-, in general, not to mention the impact of a nondipole part of the electron-photon interaction on spin-polarization of outgoing photoelectrons, in particular. C_N^-s possess remarkable qualities relating to their huge polarizability, large sizes (they are, in essence, giant atoms with a loosely bound outer electron) and large empty inner spaces in their structures. This makes it a matter of significant interest to gain insight into the impact of these features on spin-polarization of photoelectrons from C_N^-s. In the present work, we focus on the study of spin-polarization of electrons ejected from C_N^-s in the geometry where only nondipole photodetachment effects matter.

To meet the goal, we utilize Cherepkov's theory [1]. Following [1], we consider two photoionization geometries where only electric dipole($E1$)-quadrupole($E2$) interference [2] matters. One geometry, resulting in *longitudinal* spin-polarization of photoelectrons, is when the photon momentum **k**, photoelectron momentum **p** and spin **s** are such that $\mathbf{p} \perp \mathbf{k}$ and $\mathbf{s} \parallel \mathbf{p}$. The degree of longitudinal spin polarization $P_{\rm lon}^{\pm}$ from a single-electron np subshell is given by $P_{\rm lon}^{\pm} = \mp\frac{2\delta}{4+\beta}$ [1]. Here, the signs $\pm$ stand for the left/right circularly polarized light, β is the dipole photoelectron angular-asymmetry parameter and δ is a $E1$-$E2$ nondipole angular-asymmetry parameter [2]. In the other geometry, resulting in *transverse* spin-polarization $P_{\rm tr}$, $\mathbf{p} \perp \mathbf{k}$, $\mathbf{s} \perp \mathbf{k}$ and $\mathbf{p} \perp \mathbf{s}$ [1].

To address photodetachment of a fullerene anion C_N^-, we combine Cherepkov's theory with modelling of a C_N cage by a spherical attractive potential $U_0(r)$ of certain depth U_0, inner radius r_0 and thickness Δ [3] (and references therein). A fullerene anion C_N^- is, then, due to the binding of an external electron by this potential into a $n\ell$-state. To account for the impact of the polarization potential $V_{\rm pol}$ of a fullerene-remainder on the motion of a released electron, $V_{\rm pol}$ is approximated by a static dipole polarization potential: $V_{\rm pol}(r) \approx -\alpha/[2(r^2+b^2)^2]$ [4]. Here, α is the static polarizability of C_N and b is a parameter of the order of r_0. Thus, a released from C_N^- electron moves in the field of an effective potential $U_{\rm eff}(r) = U_{\rm c}(r) + V_{\rm pol}(r)$. Finally, we consider C_N^-s with progressively increasing sizes ($N = 60$, 240 and 540), assume that the attached electron is a 2p-electron and calculate both $P_{\rm lon}$ and $P_{\rm tr}$.

For illustration, calculated degree of longitudinal spin-polarization $P_{\rm lon}$ of the photoelectron flux from $C_{60}^-(2p)$ is depicted in Fig.1. The details and strength of the impact of both nondipole effects and fullerene polarizability on $P_{\rm lon}$ are evident.

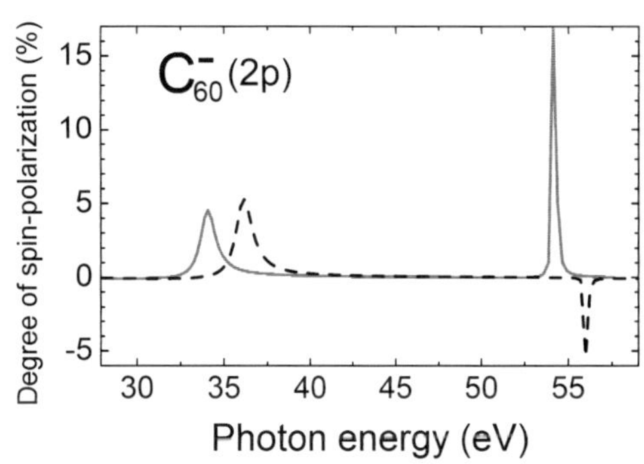

Figure 1. The degree of longitudinal spin-polarization $P_{\rm lon}$ of the photoelectron flux from $C_{60}^-(2p)$, due to $E1$-$E2$ interference upon photodetachment by light of the left circular polarization, calculated with (solid line) and without (dashed line) account for fullerene polarizability α. Parameters: $\alpha \approx 850$, $U_0 \approx -0.26$, $r_0 \approx 5.26$ and $\Delta \approx 2.91$ a.u. [4].

V.K.D. acknowledges the travel grant support from COAS, UNA.

References

[1] N. A. Cherepkov 1983 *Adv. At. Mol. Phys.* **19** 395

[2] O. Hemmers *et al* 2004 *Radiat. Phys. Chem* **70** 123

[3] V. K. Dolmatov 2009 *Adv. Quant. Chem.* **58** 13

[4] V. K. Dolmatov *et al.* 2017 *Phys. Rev. A* **95** 012709

[1]E-mail: aedwards@una.edu
[2]E-mail: cglane@una.edu
[3]E-mail: vkdolmatov@una.edu

Photoionization and photofragmentation of $Sc_3N@C_{80}^+$ at energies from the carbon K edge to the scandium L and nitrogen K edges

A. Müller[*1], **S. Bari**[§], **T. Buhr**[†], **J. Hellhund**[*], **K. Holste**[†], **A. L. D. Kilcoyne**[‡], **S. Klumpp**[ℜ], **M. Martins**[¶], **S. Ricz**[£], **K. Schubert**[§], **S. Schippers**[†]

[*] Institut für Atom- und Molekülphysik, Justus-Liebig-Universität Gießen, 35392 Giessen, Germany
[§] DESY Photon Science, FS-FCS, 22607 Hamburg, Germany
[†] I. Physikalisches Institut, Justus-Liebig-Universität Gießen, 35392 Giessen, Germany
[‡] Advanced Light Source, Lawrence Berkeley National Laboratory, Berkeley, CA 94720, USA
[ℜ] DESY Photon Science, FS-FLASH-D, 22607 Hamburg, Germany
[¶] Institut für Experimentalphysik, Universität Hamburg, 22761 Hamburg, Germany
[£] Institute of Nuclear Research of the Hungarian Academy of Sciences, Debrecen, 4026 Hungary

Synopsis Photoionization of endohedral $Sc_3N@C_{80}^+$ fullerene ions is investigated using synchrotron radiation in the energy range 280 to 420 eV. In the product channels $Sc_3N@C_{80}^{2+}$ and $Sc_3N@C_{78}^{2+}$ clear signatures of the cross section contributions arising from L-shell photoabsorption by the encapsulated Sc atoms have been detected.

Photon-induced reactions of $Sc_3N@C_{80}^+$ ions have been investigated in the energy range 280 – 420 eV employing the **P**hoton-**I**on Spectrometer at **PE**TRA III (PIPE) [1]. The measurements cover photoabsorption near the K edge of carbon, the L edge of scandium and the K edge of nitrogen. Several reaction channels were studied. As an example, Fig. 1 shows the preliminary cross section for single ionization of $Sc_3N@C_{80}^+$ ions. Very clear signatures mainly arising from photoabsorption by Sc atoms near the L edge are visible on top of the contribution of the C_{80} cage. Similar results were obtained for the $Sc_3N@C_{78}^{2+}$ product channel which shows a slightly different spectral shape.

Only the single-ionization channels with and without fragmentation showed significant contributions arising from photoabsorption by one of the encapsulated Sc atoms. This is quite surprising considering that about 400 eV are deposited in the $Sc_3N@C_{80}^+$ molecule when a photon is absorbed by an encapsulated scandium atom with promotion of an L-shell electron. Previous studies on Xe 4d photoabsorption by $Xe@C_{60}^+$ at much lower photon energies of 60 to 150 eV showed strong contributions of double ionization associated with multiple fragmentation [2]. A recent experiment with $Lu_3N@C_{80}^{q+}$, $q = 1, 2, 3$, exploring photoabsorption near the Lu 3d edge [3] suggested that the whole fullerene molecule explodes when a 1500 to 1700 eV photon is absorbed by one of the Lu atoms. In contrast to that, the present findings for $Sc_3N@C_{80}^+$ indicate that an initial vacancy in the L shell of one of the Sc atoms predominantly results in net single ionization of the atom and subsequently also the whole molecule. Redistribution of 2p and 2s absorption oscillator strengths [4] may then be the reason for different spectral shapes of the measured excess cross sections.

We thank G. Hartmann, F. Scholz, J. Seltmann, and J. Viefhaus for their assistance in using beamline P04. SK acknowledges support from the European Cluster of Advanced Laser Light Sources (EUCALL) project which has received funding from the *European Union's Horizon 2020 research and innovation programme* under grant agreement No 654220. Funding by BMBF (contracts 05K10RG1, 05K10GUB, 05K16RG1, and 05K16GUC) and DFG (grant Mu-1068/22) is gratefully acknowledged.

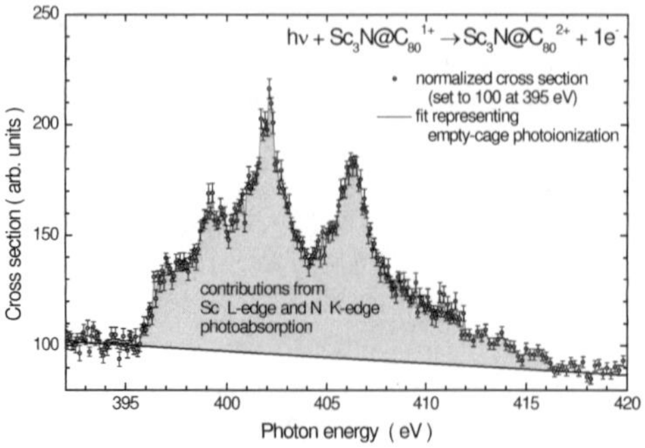

Figure 1. Cross section for single photoionization of $Sc_3N@C_{80}^+$ in the vicinity of the scandium L edge. The data are normalized to 100 Mb at 392 eV.

References

[1] S. Schippers *et al.* 2014 *J. Phys. B* **47** 115602

[2] R. A. Phaneuf *et al.* 2013 *Phys. Rev. A.* **88** 053402

[3] J. Hellhund *et al.* 2015 *Phys. Rev. A.* **92** 013413

[4] A. Müller *et al.* 2008 *Phys. Rev. Lett.* **101** 133001

[1]E-mail: Alfred.Mueller@iamp.physik.uni-giessen.de

IOP Conf. Series: Journal of Physics: Conf. Series **875** (2017) 032034 doi:10.1088/1742-6596/875/4/032034

Photoionization and photofragmentation of $Lu_3N@C_{80}^{q+}$ ions ($q = 1, 2, 3$)

J. Hellhund[*], **A Borovik Jr.**[†], **K. Holste**[†], **S. Klumpp**[¶‡], **M. Martins**[‡], **S. Ricz**[§], **S. Schippers**[†], **A. Müller**[*1]

[*] Institut für Atom- und Molekülphysik, Justus-Liebig-Universität Gießen, 35392 Giessen, Germany
[†] I. Physikalisches Institut, Justus-Liebig-Universität Gießen, 35392 Giessen, Germany
[¶] DESY Photon Science, FS-FLASH-D, 22607 Hamburg, Germany
[‡] Institut für Experimentalphysik, Universität Hamburg, 22761 Hamburg, Germany
[§] Institute of Nuclear Research of the Hungarian Academy of Sciences, 4026 Debrecen, Hungary

Synopsis Cross sections for photoionization and photofragmentation of singly, doubly and triply charged $Lu_3N@C_{80}$ ions have been measured using the photon-ion merged-beams technique at the PETRA III synchrotron light source. The measured spectra exhibit prominent resonance features at energies around the carbon K edge. In contrast to that, no signatures of the nitogen K and lutetium M edges have been found in the experimentally investigated reaction channels.

We have measured cross sections for photoionization and photofragmentation of endohedral fullerene ions $Lu_3N@C_{80}^{q+}$ (q=1,2,3) [1] employing the photon-ion merged-beams technique [2] at the PIPE end-station [3] of beamline P04 [4] of the PETRA III synchrotron at DESY in Hamburg, Germany. Photo-reaction channels $Lu_3N@C_{80}^{q+} \rightarrow Lu_3N@C_{80-2r}^{p+}$ [q=1,2,3; p=2,3,4,5,6; r=0,1,3,4] were investigated in the photon energy ranges 280–330 eV around the carbon K-shell threshold, 380–435 eV around the nitrogen K-Shell threshold, and 1500–1700 eV around the lutetium M-shell threshold. The present work substantially extends previous studies on (endohedral) fullerenes [5, 6, 7, 8].

In the energy range 280 – 330 eV we could identify a group of resonances present in all investigated reaction channels. Since this group of resonances seems to make up for the most prominent structures in all spectra, we attempted to model each spectrum (see Fig. 1) as a sum of seven Fano resonances (full lines), a threshold feature (dashed line), and a constant background (dash-dotted line). In the multiple ionization channels, a threshold can be observed at about 294 eV. We observe a shift of the ionization threshold towards higher photon energies when comparing double ionization of $Lu_3N@C_{80}^{2+}$ with double ionization of $Lu_3N@C_{80}^{+}$. On the basis of this threshold shift the radius of the endohedral fullerene molecule has been determined to be 0.50 ± 0.04 nm. The cross sections measured at higher photon energies of 380–435 eV and of 1500–1700 eV are monotonically decreasing with increasing photon energy and do not exhibit any structures associated with the nitrogen K edge or the lutetium M edge, respectively. Most probably, the absorption of an energetic 1600-eV photon by one of the Lu atoms leads to a much more violent fragmentation event, such that large fragments cannot be observed. This is interesting, for example, from a radiobiological point of view, and will be more closely investigated in future follow-up experiments.

We thank G. Hartmann, F. Scholz, J. Seltmann, and J. Viefhaus for assistance in using beamline P04. SK acknowledges support from the European Cluster of Advanced Laser Light Sources (EUCALL) project which has received funding from the *European Union's Horizon 2020 research and innovation programme* under grant agreement No 654220. Funding by BMBF (contracts 05K10RG1, 05K10GUB, 05K16RG1, and 05K16GUC) and DFG (grant Mu-1068/22) is gratefully acknowledged.

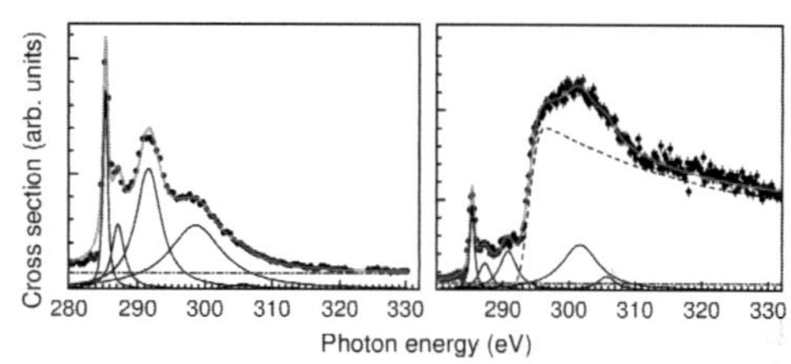

Figure 1. Measured (symbols) and fitted (lines) cross sections for single (left) and double (right) photoionization of $Lu_3N@C_{80}^{+}$ at energies around the C K edge [1].

References

[1] J. Hellhund *et al.* 2015 *Phys. Rev. A* **92** 013413
[2] S. Schippers *et al.* 2016 *Contemp. Phys.* **57** 215
[3] S. Schippers *et al.* 2014 *J. Phys. B* **47** 115602
[4] J. Viefhaus *et al.* 2013 *Nucl. Instrum. Meth. A* **710** 151
[5] K. K. Baral *et al.* 2016 *Phys. Rev. A* **93** 033401
[6] A. L. D. Kilcoyne *et al.* 2010 *Phys. Rev. Lett.* **105** 213001
[7] R. A. Phaneuf *et al.* 2013 *Phys. Rev. A* **88** 053402
[8] A. Müller *et al.* 2008 *Phys. Rev. Lett.* **101**, 133001

[1]E-mail: alfred.mueller@iamp.physik.uni-giessen.de

IOP Conf. Series: Journal of Physics: Conf. Series **875** (2017) 032035 doi:10.1088/1742-6596/875/4/032035

Photodissociation of orange I monoanion studied using an electrostatic storage ring

M. Saito[*1], T. Tanabe[†], M. Lintuluoto[‡], E. B. Starikov[§], K. Noda[¶], T. Majima[*], S. Tomita[||], and K. Takahashi[III]

[*]Kyoto University, Kyoto 615-8540, Japan
[†]High Energy Accelerator Research Organization (KEK), Tsukuba 305-0801, Japan
[‡]Kyoto Prefectural University, Kyoto 606-8522, Japan
[§]Chalmers University of Technology, 412 96 Gothenburg, Sweden
[¶]National Institute of Radiological Sciences, Chiba 263-8555, Japan
[||]University of Tsukuba, Tsukuba 305-8573, Japan
[III]National Institute of Advanced Industrial Science and Technology, Tsukuba 305-8566, Japan

Synopsis Photodissociation of gas-phase orange I monoanions was studied by using an electrostatic storage ring. By comparing their wavelength spectra with those in the liquid phase and with theoretical predictions, it is deduced that the spectra originate from different tautomers of orange I monoanions.

Orange I is one of the most popular dyes and has been studied both experimentally and theoretically. There have been a number of studies on the photoabsorption of orange I molecules in the liquid phase, but not in the gas phase. Recently, we studied photodissociation of gas-phase fluorescein monoanions using an electrostatic storage ring and deduced that the photodissociation spectra originated from various tautomers of fluorescein monoanions [1]. Meanwhile, desodiated orange I monoanions ([orange I-Na]$^-$) can exist in two tautomeric forms: azo and hydrazone forms (Fig. 1). Here, we present the photodissociation of desodiated orange I monoanions in the gas phase studied using an electrostatic storage ring.

Figure 2 shows the experimental setup. The ions were produced in an electrospray ion source. The ions were then stored in an ion trap and accelerated to 20 keV. After mass-analyzed, the ions were injected into the storage ring. The stored ions were irradiated with an OPO laser pulse (410–660 nm) and neutral products were detected using an MCP detector. The storage time in the ring before the laser irradiation was variable up to the order of seconds.

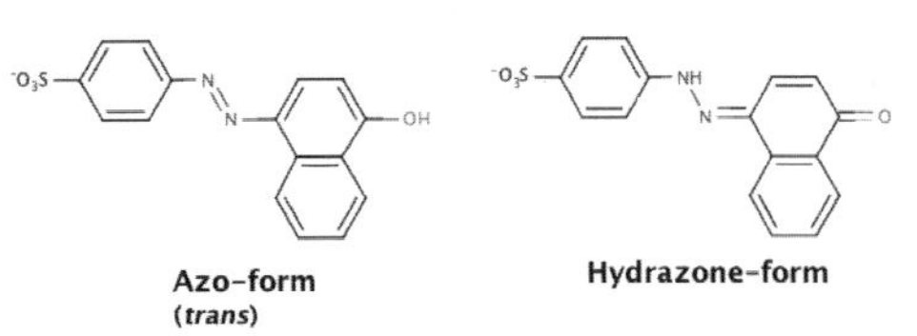

Fig. 1 Chemical structures of orange I monoanions.

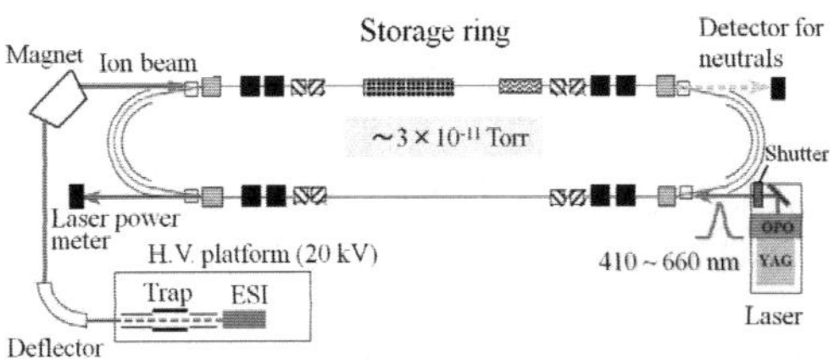

Fig. 2 Experimental setup.

Figure 3 shows the yield of neutral particles produced by the photodissociation as a function of time for different storage times in the ring. As shown in the figure, the time spectra varied depending on the storage time before the laser irradiation. By comparing the yield of neutral particles as a function of the wavelength with the absorption spectra in the solution phase and with theoretical predictions, we deduced that the wavelength spectra originated from two tautomeric forms of the orange I monoanions: azo and hydrazone forms. In addition, we observed that the wavelength spectra depended on the storage time before the laser irradiation.

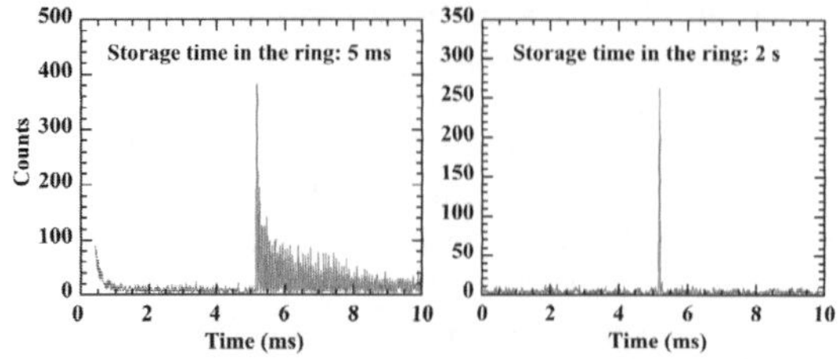

Fig. 3 Counts of neutral particles produced by the photodissociation of orange I monoanion as a function of time, for different storage times in the ring.

Reference

[1] T. Tanabe, M. Saito, K. Noda, E. B. Starikov, Eur. Phys. J. D **66**, 163 (2012)

[1] E-mail: saito@nucleng.kyoto-u.ac.jp

IOP Conf. Series: Journal of Physics: Conf. Series **875** (2017) 032036 doi:10.1088/1742-6596/875/4/032036

Investigation of environmental effects in prototypical noble gas clusters using fluorescence spectroscopy

Andreas Hans[*1], **Xaver Holzapfel**[*], **Philipp Schmidt**[*], **Christian Ozga**[*], **Uwe Hergenhahn**[†], **Till Jahnke**[‡], **Reinhard Dörner**[‡], **Arno Ehresmann**[*], **and André Knie**[*]

[*] Institut für Physik und CINSaT, Universität Kassel, Heinrich-Plett-Str. 40, 34132 Kassel, Germany
[†] Leibniz-Institut für Oberfächenmodifzierung e. V., Permoserstr. 15, 04318 Leipzig, Germany
[‡] Institut für Kernphysik, J. W. Goethe-Universität, Max-von-Laue-Str. 1, 60438 Frankfurt am Main, Germany

Synopsis The investigation of environmental effects in weakly bound clusters has evolved as a comprehensive field in atomic physics during the last two decades. Recently it was shown that fluorescence spectroscopy can be utilized as complementary tool to track such processes. Here, we report on progress in fluorescence spectroscopy experiments on prototypical noble gas clusters.

Recently, it was demonstrated, that non-local relaxation processes in noble gas clusters can be identified and quantitatively characterized by fluorescence measurements [1, 2]. The detection of photons therefore provides an advantageous method for the investigation of environmental effects in weakly bound systems. Using electron and ion spectroscopy, often in coincidence experiments, numerous variants of interatomic Coulombic decay (ICD), electron transfer mediated decay (ETMD) and radiative charge transfer (RCT) were discovered and extensively investigated [3]. Noble gas clusters are regarded as prototype systems for the investigation of these process, because they can easily be produced in a controlled way.

Here, we report on the progress and perspectives on fluorescence measurements concerning these environmental effects. In a proof-of-principle experiment, we show that a resonant variant of ICD triggered by resonant 2s-*np* photoexcitation of neon clusters leads to characteristic fluorescence emission in the vacuum-ultraviolet (VUV) spectral range, which is absent in isolated atoms [1]. This is the first unambiguous observation of an ICD process using photon detection. The complete decay cascade of this path is revealed by dispersed VUV photon detection and simultaneous measurement in the ultraviolet/visible spectral range [4].

Using atomic benchmarks in a partially condensed jet, the cross section of the VUV emission is calibrated on an absolute scale [2]. No absolute cross section of any ICD process following photon absorption was reported before. Complementary to the population of radiative states after resonant excitation, ICD can quench atomic radiative decay, e.g. after neon 2s ionization. This effect is observed by comparing ion yield and photon yield in the vicinity of the neon 2s edge [4].

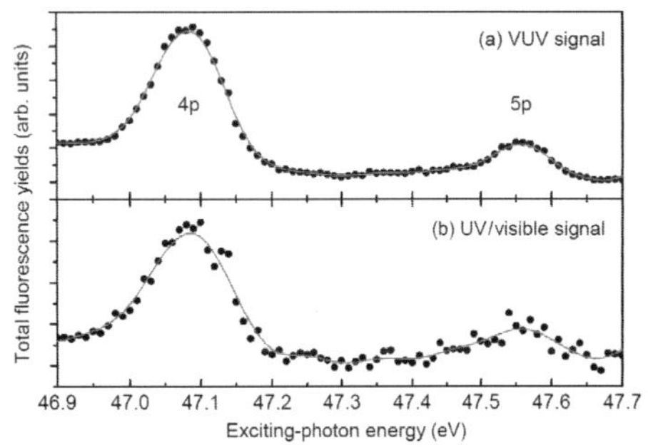

Figure 1. VUV and UV/visible fluorescence yield after 2s-*np* photoexcitation of neon clusters. [4]

Time resolved photon detection, enabled by the bunched nature of the exciting synchrotron radiation, reveales significant differences in the lifetime of radiative states in clusters compared with monomers [1]. For various species it is shown that signals from both condensed and non-condensed parts of a typical experimental cluster jet can be discriminated very efficiently by proper time gating. This is particularly useful, if excitation energy and/or emission wavelength overlap and cannot be separated by other means.

References

[1] A. Knie *et al.* 2014 *New J. Phys.* **16** 102002

[2] A. Hans *et al.* 2016 *J. Phys. B* **49** 105101

[3] U. Hergenhahn 2011 *J. Phys. B* **184** 78

[4] A. Hans *et al.* 2017 *Chem. Phys.* **482** 165

[1]E-mail: hans@physik.uni-kassel.de

Cross sections for the formation of H(2p) atom via doubly excited states in photoexcitation of rotationally cold H_2

Y. Abe*[1], T. Odagiri*[2], S. Ohrui*, T. Taniguchi*, T. Shiratori*, M. Kaida*, K. Yachi†, Y. Kumagai†, K. Hosaka†, M. Kitajima†, and N. Kouchi†

*Dept. of Materials and Life Sciences, Sophia University, Chiyoda-ku, Tokyo 102-8554, Japan

†Dept. of Chemistry, Tokyo Institute of Technology, Meguro-ku, Tokyo 152-8551, Japan

Synopsis Cross sections for the formation of H(2p) atom in photoexcitation were measured for H_2 in the lowest rotational level in the energy range of the doubly excited states for examining contribution of the non-adiabatic transition between the ${}^1\Sigma_u - {}^1\Pi_u$ doubly excited states.

Competing decay of doubly excited states of H_2 has been theoretically and experimentally investigated [1]. Recently, it was found that the non-adiabatic transition would play a role in the dissociation of the $Q_2{}^1\Pi_u$ doubly excited states of H_2 and D_2 [2,3]. In the present study, we measured cross sections for the formation of H(2p) fragment in photoexcitation of para-H_2 in the lowest rotational level, $J'' = 0$, for a detailed piece of information against the non-adiabatic transition. Only the ${}^1\Pi_u^+$ states as well as the ${}^1\Sigma_u^+$ states are populated in photoexcitation from the lowest rotational level of the ${}^1\Sigma_g^+$ ground state [4]. On the other hand, many rotational level ($J'' = 0, 1, 2, 3$) are involved for ordinary-H_2 at room temperature and all the dipole allowed states (${}^1\Pi_u^\pm$ and ${}^1\Sigma_u^+$) can be formed in photoexcitation. It is thus expected that the cross section for the rotationally cold H_2 could be different from those for ordinary-H_2 since the ${}^1\Pi_u^\pm$ states interact with the ${}^1\Sigma_u^+$ states differently with each other due to the Kronig's selection rule [4].

The experiments were carried out at BL20A of the photon factory, KEK. A gas of H_2 in the lowest rotational level was obtained by a cryogenic ortho-para hydrogen converter. The gas cell was kept at approximately -186°C by using liquid-N_2 during the measurement. The rotational distribution in the sample was checked through measuring high-resolution photo-ion yield spectra.

Figure 1 shows the cross sections for the formation of H(2p) fragment for H_2 in the lowest rotational level, $J'' = 0$. The shape of the cross section curve agrees with that for ordinary-H_2 at room temperature within the statistical uncertainty.

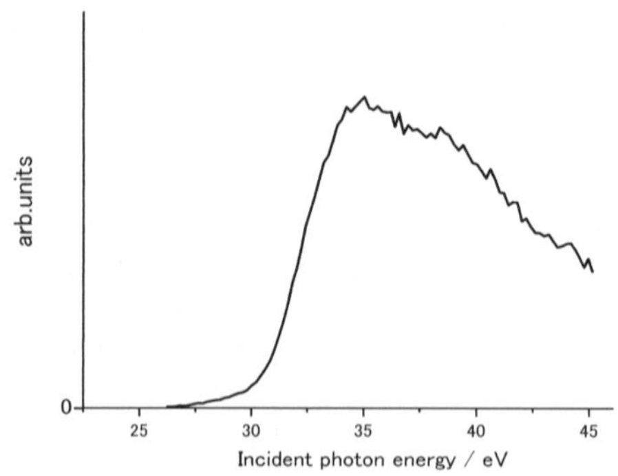

Figure 1. Cross sections for the formation of H(2p) fragment in photoexcitation of H_2 in the lowest rotational level

References

[1] N. Kouchi *et al.*, J. Phys. B **30**, 2319(1997); M.Glass-aujean and H. Schmoranzer, J. Phys. B **38**, 1093(2005)
[2] T. Odagiri *et al.*, Phys. Rev. A **84**, 053401 (2011)
[3] K. Hosaka *et al.*, Phys. Rev. A **93**, 063423 (2016)
[4] G. Herzberg, Molecular spectra and molecular structure: I. Spectra of Diatomic molecules (New York: Van Nostrand)

[1] E-mail: yuta-abe.64@eagle.sophia.ac.jp
[2] E-mail: odagiri.t@sophia.ac.jp

Terahertz generation in crystal driven by two-color laser pulses‡

Lei Zhang*†, Zhong Guan*, Guo-Li Wang*[1], Xiao-Xin Zhou*[1]

* College of Physics and Electronic Engineering, Northwest Normal University, Lanzhou,730070, China
† School of Mathematics and Physics, Lanzhou Jiaotong University, Lanzhou,730070, China

Synopsis We study the generation of terahertz (THz) radiation in 1D crystal driven by two-color laser pulses by solving the time-dependent Schrödinger equations. Our results indicate that stronger THz radiation can be obtained from the interaction of short laser pulse with crystals.

The interaction of intense short laser pulses with solids has attracted considerable attention in the last few years, mainly for the high-order harmonic generation [1-4]. In this work we focus on the other important frequency-conversion process—terahertz (THz) radiation from crystals. We find that stronger THz radiation can be obtained from 1D crystal by using two-color laser pulses with certain laser parameters.

We adopt the method given in Ref. [5] to calculate the total laser-induced current $J(t)$, which comes from both intraband and interband dynamical processes. We then multiply $J(t)$ by a Hanning window [6] to improve the signal-to-noise ratio since the laser intensities used in the present work are relatively low. The THz radiation spectrum is then obtained from a Fourier transform of $\mathrm{d}J(t)/\mathrm{d}t$. In the calculation, the Mathieu-type potential of 1D crystal is chosen as [6]

$$V(x) = -V_0[1+\cos(2\pi x/a_0)], \quad (1)$$

where $V_0=0.37$ and lattice constant $a_0=8$, both in atomic units.

The electric field of the two-color laser pulse can be written as

$$E(t) = f(t)[E_1 \cos \omega t + E_2 \cos(2\omega t + \varphi)], \quad (2)$$

where E_1 and E_2 are the peak amplitudes of the fundamental (ω) and the second harmonic (2ω) fields, φ is the relative phase between them. $f(t)$ is a $\cos^2$ pulse envelope. In our calculations, we use the total laser intensity $I=8.1 \times 10^{11}$ W/cm^2 and the fundamental laser wavelength $\lambda=3.2$ μm. The total pulse duration is 8 cycles.

Email: wanggl@nwnu.edu.cn, zhouxx@nwnu.edu.cn

The corresponding THz radiation spectra are illustrated in Fig. 1. It is obvious that stronger THz radiation (0-30THz) is obtained through the two-color laser pulses. The intensity of the THz radiation is enhanced more than 3 orders of magnitude at 14THz compared to that in the single-color field. We also find that the THz spectrum is sensitive to the intensity ratio and the relative phase between two fields, similar to the case of THz radiation from gas samples [7,8].

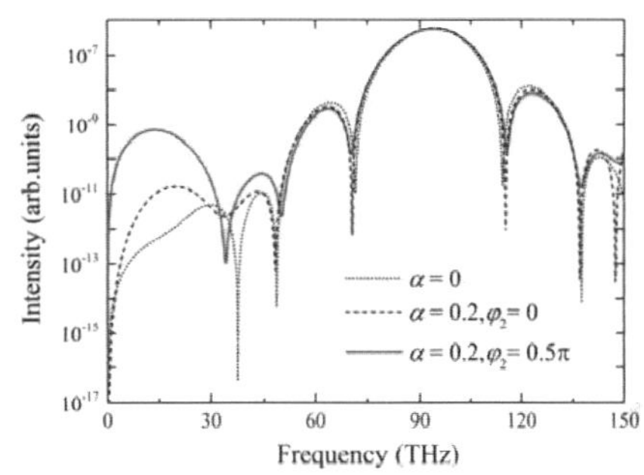

Figure 1. The THz emission spectrum from 1D crystal. Here, $\alpha=I_2/I_1=E_2^2/E_1^2$.

Project supported by the National Natural Science Foundation of China (Nos. 11465016, 11364038)

References

[1] S. Ghimire *et al* 2011 *Nat. Phys.* 7 138
[2] O. Schubert *et al* 2014 *Nat. Photonics* 8 119
[3] M. Hohenleutner *et al* 2015 *Nature* 523 572
[4] G. Vampa *et al* 2014 *Phys. Rev. Lett.* 113 073901
[5] Z. Guan *et al* 2016 *Phys. Rev. A* 93 033852
[6] M. X. Wu *et al* 2015 *Phys. Rev. A* 91 043839
[7] K.Y. Kim *et al* 2009 *Phys. Plasmas* 16 056706
[8] L. Zhang *et al* 2016 *J. Mod. Opt.* 63 2159

Effects due to the induced potential in ultrashort laser interactions with Al(100) and Al(111) surfaces

C. A. Rios Rubiano[§1], R. Della Picca[§], D. M. Mitnik[*], V. M. Silkin[‡¶] and M. S. Gravielle[*]

[§]CONICET and Centro Atómico Bariloche (CNEA) Bariloche, Argentina.
[*] Instituto de Astronomía y Física del Espacio (IAFE, CONICET-UBA)
[‡] Donostia International Physics Center (DIPC) and Fac. de C. Químicas, Univ. del País Vasco, San Sebastián, Spain
[¶] IKERBASQUE, Basque Foundation for Science, Bilbao, Spain

Synopsis Photoelectron emission (PE) from the valence band of a metal surface due to the grazing incidence of ultrashort laser pulses on two faces of aluminum- Al(100) and Al(111) - is investigated in the framework of a distorted-wave formulation. We use the band-structure-based-Volkov (BSB-V) approach [1], including also induced-potential effects originated from the electron response of the surface to the external field. For both crystallographic orientations we found that the induced surface (IS) potential contributes to the emergence of band-structure signatures in the near-threshold region of photoelectron spectra.

When an ultrashort laser pulse interacts with a metal surface, the external electromagnetic field induces not only direct PE from the metal but also collective oscillations of valence electrons, i.e., an IS potential.

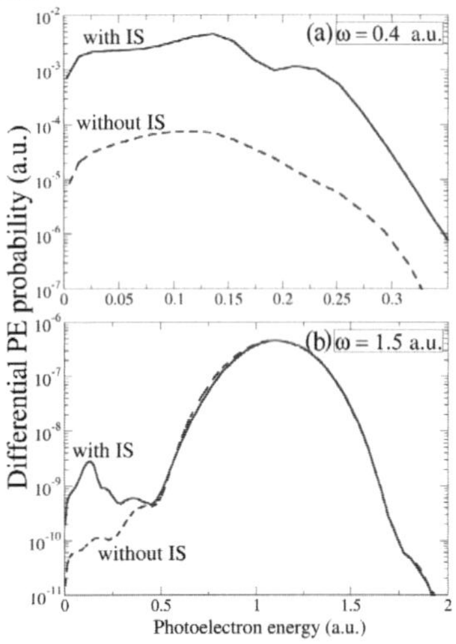

Figure 1. PE distribution from Al(100) for normal emission. The pulse carrier frequency is (a) $\omega = 0.4$ a.u. (b) $\omega = 1.5$ a.u. Solid (dashed) blue lines, BSB-V results with (without) the inclusion of the IS contribution.

In this work we study the role played by the IS field to reveal signatures of the surface-band structure in PE distributions from Al(100) and Al(111), produced by ultrashort laser pulses linear polarized perpendicular to the surface. Calculations were made within an improved version of the BSB-V approach, which includes an accurate description of the electron-surface interaction [2] as well as the contribution of the IS field [3].

Since the IS field depends strongly on the carrier frequency of the laser pulse ω, its contribution is expected to be essential for frequencies close to the surface plasmon frequency ω_s (= 0.4 a.u. for aluminum surfaces). However, we find that also for low and high ω values the IS potential produces an enhancement of band-structure effects in the low-energy region of photoelectron spectra (see Fig. 1).

Partial contributions from surface electronic states (SESs) and states at the top of the occupied states (TOSs) are analyzed, finding that they are more noticeable for the Al(100) face, in contrast to the Al(111) case (see Fig. 2).

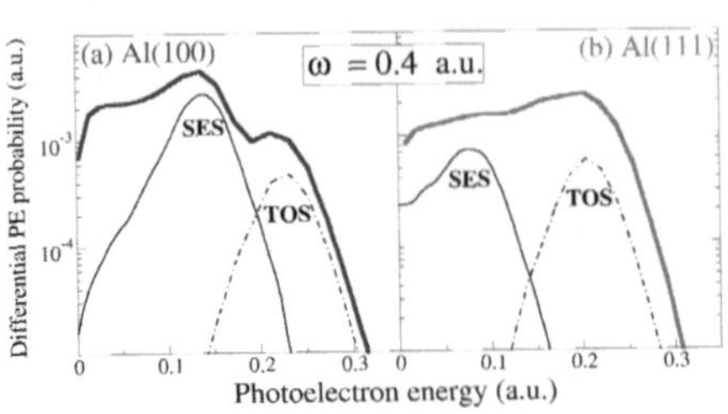

Figure 2. Similar to Fig. 1. In both panels, BSB-V results, including the IS contribution, are displayed with thick solid lines; partial contributions from SESs and TOSs are plotted with thin solid and dot-dashed black lines, respectively.

References

[1] C. Rios Rubiano *et al.* 2012 *Phys. Rev. A* **85** 043422
[2] E.V. Chulkov *et al.* 1999 *Surface Science* **437** 330
[3] C. Rios Rubiano *et al.* 2017 *Phys. Rev. A*. In press.

[1]E-mail: cariosr@iafe.uba.ar

Photoionization of open-shell halogen atoms endohedrally confined in C_{60}

Dakota Shields*, Ruma De* [1], Mohamed Madjet†, Steven T. Manson‡ [2], and Himadri Chakraborty*

* Department of Natural Sciences, D.L. Hubbard Center for Innovation and Entrepreneurship, Northwest Missouri State University, Maryville, Missouri 64468, USA
† Qatar Environment and Energy Research Institute, Hamad Bin Khalifa University, Qatar Foundation, Doha, Qatar
‡ Department of Physics and Astronomy, Georgia State University, Atlanta, Georgia 30303, USA

Synopsis Dipole photoionization of atomically open-shell halogen-endofullerenes Cl,Br,I@C_{60} are calculated using time-dependent local density functional theory and a spherical potential model. Hybridization of the valence *np* orbital of the atom with a C_{60} *p* orbital is found to strongly modify the ionization parameters.

Endofullerene molecules, formed by encapsulating atoms in fullerenes, are extremely stable form of nanoscopic matters that can exist even at room temperature. Since these entrapments have lower-cost sustenance than the laser cooling or magneto-optical trapping processes, endofullerenes are unique natural laboratories to study not only the greatly modified properties of the confined atomic system but also the confining fullerene. Notable effects include the electronic hybridization between guest-host systems [1] and strong spectral processes, such as, the plasmon excitations and quantum diffraction intrinsic to response of these molecules [2].

A number of studies have been conducted for endofullerenes involving closed shell atoms, while there have been little research on systems that confine open shell atoms. In this study, using a jellium-based local density approximation (LDA) method [1], augmented by the LB94 exchange-correlation functional [3], we model the ground state of endofullerenes encaging three halogen atoms, Cl, Br and I, which are one *np* electron shy of the shell closure. Strong atom-C_{60} hybridization in the *np* levels is found which, however, shows a systematic weakening trend from Cl to I.

The dipole photoionization cross sections of these hybrid levels are calculated in a time-dependent LDA which includes linear response electron correlations [3]. Results for Cl@C_{60} are presented in Figure 1. The narrow spikes in the cross sections are routine resonances from the Auger-type decay of inner-shell excitations. Comparing with the free atom 3*p* result, the cross sections for the bonding and antibonding hybrid levels broadly display two energy regions: (i) plasmon-enhanced low-energy domain and (ii) broad oscillations from the coherence of cavity and confinement effects at higher energies. However, even within this generic shape, results reveal significant differences in the details of the results between the hybrid levels. While the bonding level features stronger effects of the plasmons and a strong atomic character at higher energies, the antibonding ionization dramatically diminishes at higher energies suggesting dominant C_{60} emission.

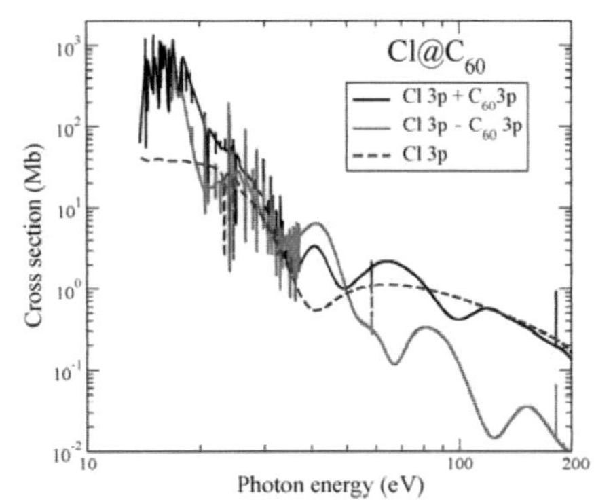

Figure 1. TDLDA cross sections of the two hybrid levels of Cl@C_{60} and the 3*p* level of free Cl.

Furthermore, detailed comparison of these results with those from Ar,Kr,Xe@C_{60}, which involve the nearby close-shell atoms in the periodic table, provides deeper insights into the role of a single shell-closing electron to noticeably influence the ionization dynamics.

The work is supported by the US National Science foundation and the US Department of Energy.

References

[1] H.S. Chakraborty *et al.* 2009 *Phys. Rev. A* **79** 061201(R)
[2] A. Potter *et al.* 2010 *Phys. Rev. A* **82** 033201
[3] J. Choi *et al.* 2017 *Phys. Rev. A* **95** 023404

[1] E-mail: ruma@nwmissouri.edu

[2] E-mail: smanson@gsu.edu

IOP Conf. Series: Journal of Physics: Conf. Series **875** (2017) 042005 doi:10.1088/1742-6596/875/5/042005

Electron-positron pair production in space-time-dependent colliding laser pulses

I. A. Aleksandrov[*1], **G. Plunien**[†], **and V. M. Shabaev**[*]

[*] Department of Physics, St. Petersburg State University, 7/9 Universitetskaya nab., Saint Petersburg 199034, Russia
[†] Institut für Theoretische Physik, TU Dresden, Mommsenstrasse 13, Dresden, D-01062, Germany

Synopsis The Schwinger pair-production process in the presence of two counter-propagating linearly polarised laser pulses is studied by means of a non-perturbative numerical technique. The spatial finiteness of the pulses is taken into account, i.e. the problem is examined beyond the standing-wave approximation.

Electron-positron pair production (PP) from the vacuum in strong external backgrounds is a fundamental non-linear phenomenon predicted by quantum electrodynamics. It is well-known that in a purely electric static field the PP probability is negligible unless the field strength is close to the Schwinger critical value $E_c = m^2c^3/(|e|\hbar) = 1.3 \times 10^{16}$ V/cm. This limit can be effectively lowered if the external field oscillates with time. Let E and ω be the characteristic strength of the external field and its frequency. With the aid of the dimensionless adiabaticity parameter $\xi = |eE|/(mc\omega)$, the non-perturbative (tunneling) and the multiphoton regimes can be distinguished by $\xi \gg 1$ and $\xi \ll 1$, respectively. We focus on the intermediate case $\xi \sim 1$.

Due to the recent rapid development of laser technologies (ELI, XFEL and other facilities), a substantial interest in non-perturbative PP has been revived. In this context a collision of two counter-propagating laser pulses appears to be a very promising scenario. Most theoretical studies carried out so far approximated the resulting field by a spatially homogeneous background depending solely on time (see [1, 2, 3] and references therein). This approach is called the dipole approximation (DA). However, the influence of the spatial variations of the external field as well as its magnetic component may play a very important role in the PP process [4]. In order to analyse these effects, one can take into consideration the spatial variations of the carrier assuming that the envelope is still homogeneous in space. The resulting field becomes a standing wave oscillating with time.

Within the present study we go beyond the standing-wave approximation (SWA) taking into account the spatial dependence of both carrier and envelope. The patterns established within SWA are strongly modified once the pulses contain a small number of cycles. We provide a numerical analysis based on our non-perturbative technique developed previously [5]. By solving the Dirac equation in the momentum representation, we obtain the PP probabilities and study the spectrum of particles created. In Figure 1 we present the mean number of electrons produced as a function of the x-component of their momentum ($\vec{E} \parallel \vec{e}_x$, $\vec{B} \parallel \vec{e}_y$ and z is the propagation direction). In these computations $\xi = 0.5$ for an individual pulse (in the presence of two pulses ξ effectively equals unity), $\omega = 0.5\ mc^2/\hbar$ and the number of cycles $N = 2$. The results were obtained within DA, SWA and beyond these approximations ("exact calculations"). We observe that for the case of a small number of cycles ($N \sim 1$) neither DA nor SWA provides accurate predictions. Within our study this issue is investigated in detail.

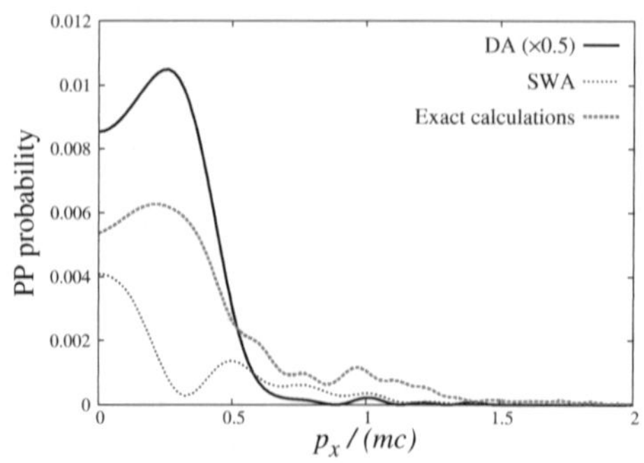

Figure 1. The PP probability as a function of p_x (the field parameters are given in the text).

This work was supported by RFBR (Grant No. 16-02-00334) and SPbU (Grants No. 11.65.41.2017 and 11.38.237.2015). I. A. A. acknowledges the support from G-RISC funded by the German Federal Foreign Office via DAAD.

References

[1] H. K. Avetissian *et al.* 2002 *Phys. Rev. E* **66** 016502
[2] G. R. Mocken *et al.* 2010 *Phys. Rev. A* **81** 022122
[3] I. A. Aleksandrov *et al.* 2017 *ArXiV:* 1701.01058
[4] M. Ruf *et al.* 2009 *Phys. Rev. Lett.* **102** 080402
[5] I. A. Aleksandrov *et al.* 2016 *Phys. Rev. D* **94** 065024

[1]E-mail: i.aleksandrov@spbu.ru

IOP Conf. Series: Journal of Physics: Conf. Series **875** (2017) 042006 doi:10.1088/1742-6596/875/5/042006

Coherent control of two-color above-threshold photoemission from tungsten nanotips

T. Paschen*, M. Förster*†, M. Krüger*†, C. Lemell‡1, G. Wachter‡, F. Libisch‡, T. Madlener‡, J. Burgdörfer‡, and P. Hommelhoff*†§

* Department of Physics, Friedrich-Alexander-Universität Erlangen-Nürnberg, Erlangen, Germany, EU
† Max Planck Institute of Quantum Optics, Garching, Germany, EU
‡ Institute for Theoretical Physics, Vienna University of Technology, Vienna, Austria, EU
§ Max Planck Institute for the Science of Light, Erlangen, Germany, EU

Synopsis We demonstrate coherently controlled two-color above-threshold photoemission from a single crystal tungsten nanotip. With optimized fundamental and second harmonic intensities near perfect phase contrast can be obtained.

The field enhancement at nanotips promises realization of strong-field physics already at moderate laser intensities. Moreover, as the transverse coherence known from dc field emission persists in photoemission from sharp nanotips, they hold the promise to serve as laser-triggered ultrafast electron source with exceptional beam quality.

While most coherent control studies are performed with gaseous systems or macroscopic surfaces we take advantage of the localized emission from nanotips to achieve near perfect control of the electron emission yield. Since the nanotip is much smaller than the focal spot sizes, we can exclude the influence of the inhomogeneous intensity and phase distribution in the laser focus.

Experimental parameters to control electron emission are the mixing ratio and the relative phase of two-color laser fields. Additionally, the dc field present at the tip to extract emitted electrons from the target region can be used to optimize the photoemission.

Without overlap, adding the second harmonic (780 nm, pulse duration 64 fs; 2ω admixture $< 5\%$) to the fundamental (1560 nm, pulse duration 74 fs; both pulses linearly polarized along the tip axis) does not increase the photocurrent noticeably as compared to electron emission in single-color mode. If the pulses overlap, however, a strong increase of electron emission is observed together with a delay-dependent modulation of the measured photocurrent at the frequency of the second harmonic [1, 2]. Optimizing dc field and mixing ratio a modulation depth of 97.5% could be reached.

Based on density-functional theory calculations we have analyzed the density of states (DOS) at the surface of W(310) surfaces, the facet at the tip apex and with the smallest workfunction for tungsten. Close to an energy corresponding to 4ω above the Fermi energy we find a pronounced maximum in the DOS which serves as a doorway state for photoelectron emission. The modulation of the photocurrent can be related to this maximum in the DOS as it can be reached by two equivalent interfering pathways ($4 \times \omega$ or $2 \times \omega + 1 \times 2\omega$).

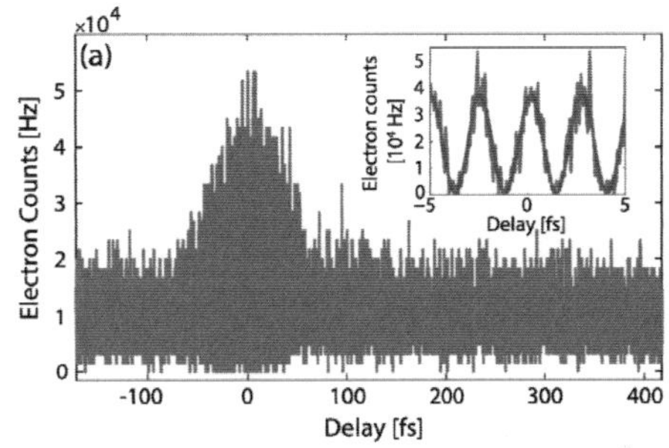

Figure 1. Two-color electron emission from a tungsten nanotip as a function of the delay of fundamental and second harmonic pulses. Depending on the optical phase a dramatic increase or decrease of emission is noticeable in the overlap region (inset).

Coherent control of photoemission from nanotips may in the future allow to use such structures as nanometric probes for light phases. In turn, control of the light phase will pave the way to the generation of femtosecond user-defined electron bunches emitted from nanometric sources.

References

[1] M. Förster *et al.* 2016 *Phys. Rev. Lett.* **117** 217601

[2] T. Paschen *et al.* 2017 *J. Mod Opt.* (published online)

[1] E-mail: lemell@concord.itp.tuwien.ac.at

IOP Conf. Series: Journal of Physics: Conf. Series **875** (2017) 042007 doi:10.1088/1742-6596/875/5/042007

Simulation of High Harmonic Generation in Solids

I. Floss*, **G. Wachter***, **C. Lemell***[1], **S. Sato**†, **X.-M. Tong**‡, **K. Yabana**‡, **and J. Burgdörfer***

* Institute for Theoretical Physics, Vienna University of Technology, A-1040 Vienna, Austria, EU
† MPI for the Structure and Dynamics of Matter, D-22761 Hamburg, Germany, EU
‡ Graduate School of Pure and Applied Sciences and Center for Computational Sciences, University of Tsukuba, Tsukuba 305-8577, Japan

Synopsis We simulate high harmonic generation in dielectrics by ultrashort laser pulses using TDDFT. We show that inclusion of decoherence by dephasing into TDDFT is crucial for a realistic description.

To overcome fundamental limitations in high harmonic generation (HHG) from single atoms, solid targets have moved into the focus of interest. For single atoms the increase of the high-energy cut-off with increasing wavelength, $E_{\rm cut} \propto \lambda^2$, goes hand in hand with the dramatic decrease of the harmonic intensity, $I \propto \lambda^{-5.5}$. In principle, solids promise to overcome this intensity limitation due to the large density of target atoms. Indeed, HHG from dielectrics has already been observed with well-pronounced harmonics up to very high orders. In parallel, theoretical methods are sought for to identify the source of the emitted radiation and to help optimizing the emission process.

We use a real-space real-time implementation of time-dependent density functional theory (TDDFT) employing periodic boundary conditions to model the interaction of ultrashort laser pulses with solids [1]. However, when applied to a realistic 3D solid for one fixed intensity I of the IR pulse (in this case for diamond) only very few harmonics on top of a broad background above the band-gap energy are recognizable (top panel). This is in contrast to the clean high-harmonic spectrum observed in experiment (starting with [2]).

We show that a realistic description of solid-state harmonics requires the treatment of the average over the inhomogeneous intensity distribution as well as decoherent processes. Forming a coherent average over the ensemble of harmonic spectra generated by the inhomogeneous intensity distribution in the laser spot clearly "cleans" the spectrum by amplifying the harmonics while damping other frequencies (bottom panel). Moreover, electron-phonon and electron-electron scattering not accounted for in standard TDDFT further purifies the spectrum and dampens the post-pulse ringing. We have developed an open-system TDDFT which accounts for decoherence without changing the time-depending density of the system [3]. We find an almost noiseless HHG spectrum (decoherence time $\tau = 2$ fs, center panel).

Open-system TDDFT in combination with intensity averaging promises to allow for a realistic simulation of HHG in dielectrics.

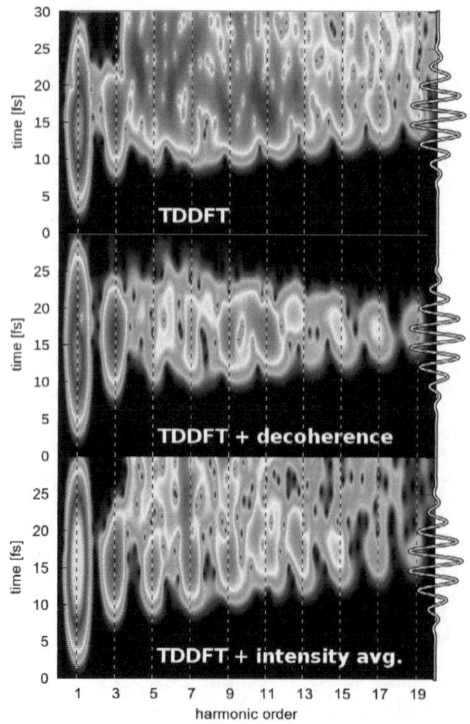

Figure 1. Wavelet transforms of radiation emitted from diamond due to irradiation by an ultrashort laser pulse. TDDFT (top panel), TDDFT with decoherence (middle panel), and intensity average of TDDFT results (bottom panel).

References

[1] T. Otobe *et al.* 2009, *J. Phys.: Condens. Matter* **21** 064224

[2] S. Ghimire *et al.* 2011 *Nature Physics* **7**, 138

[3] I. Floss *et al.* 2017, in preparation

[1]E-mail: lemell@concord.itp.tuwien.ac.at

IOP Conf. Series: Journal of Physics: Conf. Series **875** (2017) 042008 doi:10.1088/1742-6596/875/5/042008

Soft X-ray induced ultraviolet fluorescence emission from bulk and interface of a liquid water microjet

Andreas Hans[*1], **Christian Ozga**[*], **Robert Seidel**[†], **Philipp Schmidt**[*], **Timo Ueltzhöffer**[*], **Xaver Holzapfel**[*], **Marvin N. Pohl**[‡], **Philip Wenzel**[*], **Philipp Reiß**[*], **Isaak Unger**[§], **Emad F. Aziz**[†], **Arno Ehresmann**[*2], **Petr Slavíček**[¶], **Bernd Winter**[‡], **and André Knie**[*]

[*] Institut für Physik und CINSaT, Universität Kassel, Heinrich-Plett-Str. 40, 34132 Kassel, Germany
[†] Helmholtz-Zentrum Berlin für Materialien und Energie, Methods for Material Development, Albert-Einstein-Straße 15, 12489 Berlin, Germany
[‡]Department of Molecular Physics, Fritz-Haber-Institut, 14195 Berlin, Germany
[§]Department of Physics and Astronomy, Uppsala University, Box 516, 75120 Uppsala, Sweden
[¶]Department of Physical Chemistry, University of Chemistry and Technology, Technická 5, 16628 Prague, Czech Republic

Synopsis Tremendous progress has been made in the research on the structure and dynamics of liquids due to the development of advanced experimental techniques such as liquid microjets, enabling investigations on volatile samples in ultrahigh vacuum environments. The spectroscopy of charged particles, e.g. photoelectron or Auger electron spectroscopy on liquids, is an established field by now. Here, we report on the successful application of a fluorescence spectrometer to measure optical emission spectra from liquids irradiated with soft X-ray synchrotron radiation.

The application of various experimental techniques to liquid samples is an ongoing challenge, especially if high vacuum conditions are required for either exciting the sample or detecting reaction products (or both). Some common tools, such as electron spectroscopy, for the investigation of molecular dynamics thus only became applicable to volatile samples with the introduction of liquid microjets [1]. With high-pressure injection of a liquid jet of a typical diameter of some tens of micrometers, sufficient vacuum conditions can be retained. Recent experiments revealed numerous processes of electronic and nuclear dynamics in liquids and solutions [1, 2]. A major remaining constraint is that the detection of electrons is a rather surface sensitive method.

Here, we report on the detection of ultraviolet fluorescence emission from the bulk of a liquid water microjet after irradiation with soft X-ray synchrotron radiation [3]. Different features appear in the emission spectrum, which can be assigned to characteristic processes. The mechanisms observed in the present work are illustrated in Figure 1. Well-known emission features from gas phase water in the jet vicinity can be found. From the liquid phase, two hitherto not reported mechanisms were observed. After inner-shell excitation of water, the Auger electron yield at the liquid–vacuum interface is that high that gas phase molecules can be excited by electron impact. This process (labelled II in Figure 1) is identified by scanning the exciting-photon energy and determining the excitation function of certain gas phase emission features [3]. For inner-shell excitation at the oxygen 1s edge, the absorption cross section is very large, leading to an immense enhancement of this surface-specific process.

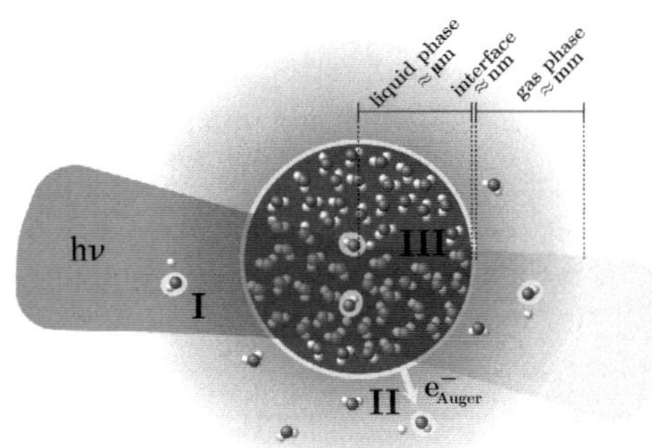

Figure 1. Illustration of radiative processes observed after soft X-ray irradiation of a liquid microjet [3].

Off-resonance, the X-rays can penetrated somewhat deeper into the jet (feature III). Here, we observe a broad genuine emission band in the ultraviolet range with a steep cutoff at the absorption edge of liquid water. Both interpretations of features II and III are supported by lifetime measurements of the respective radiative states.

References

[1] B. Winter 2009 *Nucl. Instrum. Methods Phys. Res. A* **601** 139

[2] S. Thürmer *et al.* 2013 *Nat. Chem.* **5** 590

[3] A. Hans *et al.* 2017 *J. Phys. Chem. B in press*

[1]E-mail: hans@physik.uni-kassel.de
[2]E-mail: ehresmann@physik.uni-kassel.de

IOP Conf. Series: Journal of Physics: Conf. Series **875** (2017) 042009 doi:10.1088/1742-6596/875/5/042009

Measurement of low cross section transition using versatile absorption technique

Gaurav Sharma* [1], **T. Nandi**[†] **and Nitin K. Puri**[† 2]

*Dept. of Applied Physics, Delhi Technological University, New Delhi-110042, India

[†]Inter University Accelerator centre, New Delhi 110067,India

Synopsis The various possibilities of the X-ray absorption spectroscopy is explored in a beam foil experiment. The technique is used for the successful detection of a low cross section process and enhancing the detector resolution which gets degraded due to high intensity ratio. The lifetime measured with the technique matches with the value when absorber is not used.

The versatility of x-ray absorption technique is experimentally explored for enhancing the detector resolution and rejuvenating the very low probable transitions, lost in the pile up region during a beam-foil experiment. The dependence of Beer Lambert's law on the energy plays the key role in measuring the veiled peaks in the spectrum. In a pileup affected x-ray spectrum of collision between 130MeV Ti with 80μg/cm^2 thick carbon target; 4 peaks are successfully rejuvenated. Dependence of absorption coefficient on E^{-3}[1], where E is the energy of incident photons can provide a good tool to handle the intensities of the different energies differently. A transition 10^4 times weaker than one-electron-one-photon (OEOP), possibly originating by filling two-core-vacancies (two-electron-one-photon (TEOP[2]) $1s^{-2} \rightarrow 2s^{-1}2p^{-1}$) in He-like Ti, is detected.

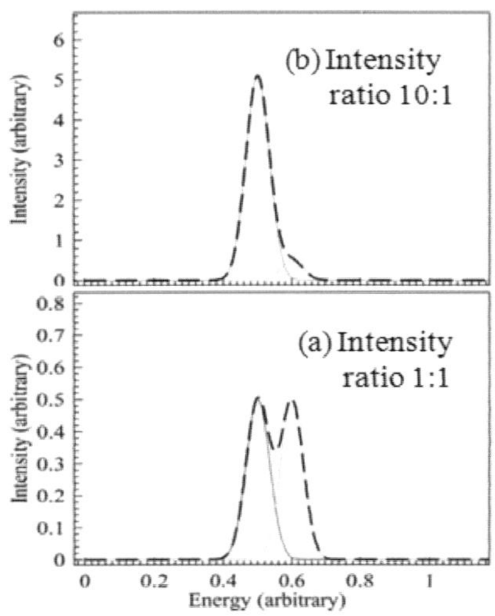

Figure 1. Resolution degradation: Effect of Intensity of the neighbourhood peaks on the resolution of the detector.

Multiple Aluminum absorber layers (10μm each) is used to suppress the pile up and high count rate photons with low energy to rejuvenate the TEOP transition and resolve the K_β and K_α which is obscured due to high intensity ratio and pile-up contamination respectively. We have experimentally confirmed that the lifetime value measured by using this technique (67.11ps) and without absorber (67.34ps) is same within the error limits. The measured branching ratio of this weak transition with OEOP is 1.38E-04. The energy of the spectral lines and the branching ratios of the TEOP/OEOP transitions find potential implications in understanding the multi-electron transitions in ion-solid collisions.

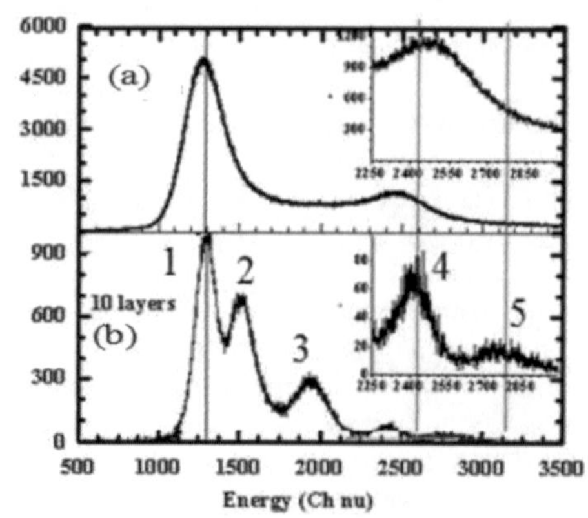

Figure 2. Experimental results: (a) and (b) are experimental x-ray spectrum without and with x-ray absorbers. Part (d) showing new peaks evolved from the spectrum after employing the technique.

References

[1] Newville M 2004 Fundamentals of XAFS *Fundamentals of XAFS* p 1–2 chapter 1

[2] Hoszowska J et.al 2011 *Phys. Rev. Lett.* 107 053001

[1]E-mail: gaurav.kmc891101@gmail.com
[2]E-mail: nitin.phy@dce.edu

IOP Conf. Series: Journal of Physics: Conf. Series **875** (2017) 042010 doi:10.1088/1742-6596/875/5/042010

Ortho-to-para ratio of water desorbed from ice and its implications for astronomy and planetary science

Tetsuya Hama* [1], **Akira Kouchi,*** **Naoki Watanabe**[† 2],

[*] Institute of Low Temperature Science, Hokkaido University, Sapporo 060–0819, Japan

Synopsis The ortho-to-para ratio of water photodesorbed from ice at 10 K shows a statistical value of 3, and does not reflect the formation temperature of the ice.

Astronomical observations have recently revealed the ubiquity of H_2O in interstellar regions. Therefore, chemical history of interstellar H_2O provides better understanding the formation of stars and planetary systems, including our solar system.

A key observable is the abundance ratio of nuclear spin isomers (the ortho-to-para ratio, OPR). Because a proton is a fermion with a nuclear spin angular momentum of $I = 1/2$, nuclear spin isomers exist for H_2O, that is, ortho ($I = 1$, triplet) and para ($I = 0$, singlet) molecules. Nuclear-spin conversion between ortho- and para-molecules by radiation or nonreactive collisions is very slow in the gas phase. The different nuclear spin isomers are thus treated as almost entirely separate species. Moreover, the OPR can be related to the spin temperature (T_{spin}) defined at thermodynamic equilibrium. Therefore, the OPR has been often considered as a valuable tracer for the physical and chemical history experienced by H_2O molecules, including their formation temperature on cosmic dust.

Cometary comae, and star- and planet-forming regions have of OPRs for gaseous H_2O (0.1–2.4), which are anomalously lower than the statistical value of 3. The OPR of H_2O in a comet has been proposed as an indicator that gives the past formation temperature (30 K) of the ice nucleus in the solar nebula some 4.6 billion years ago. The low OPR of interstellar H_2O has also been used to determine the formation temperature (<50 K) of water ice on cold interstellar dust. However, the use of the OPR as a temperature probe requires the assumption that the OPR of H_2O desorbed from ice is related to the ice temperature, which has not been validated, indeed. Thus, the significance of the OPR of water is still unknown and remains one of the most problematic issues in astronomy and planetary science.

The present study performed direct measurements of the OPR forwater molecules desorbed after 157 nm photoexcitation of water ice at 10 K, using resonance-enhanced multiphoton ionization (REMPI) detection methods. We found that the OPR of photodesorbed water showed the statistical value of 3, even when the ice is produced in situ by hydrogenation of O_2, a known formation process of interstellar water. Our results indicate that the OPR does not reflect the formation temperature of the ice (10 K, OPR = 0.3), which invalidates the assumed relation between OPR and temperature. Our results indicate that reinterpretation of previous observations is necessary for improving our understanding of interstellar chemistry, as well as the formation of the solar system and comets [1].

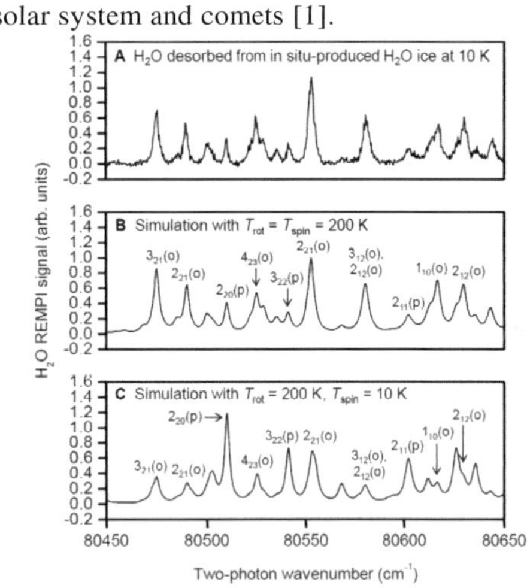

Figure 1. REMPI spectra of photodesorbed H_2O after 157 nm photoirradiation of water ice at 10 K. (A) H_2O ice produced in situ by means of hydrogenation of O_2. (B and C) Simulated spectra with T_{rot} =200 K and (B) T_{spin} =200 K and (C) T_{spin} = 10 K. Indications ($J_{Ka,Kc}$) are rotational assignments, where "o" and "p" denote ortho and para, respectively.

References

[1] T. Hama *et al.* 2016 *Science 351*, 65

[1] E-mail: hama@lowtem.hokudai.ac.jp

[2] E-mail: watanabe@lowtem.hokudai.ac.jp

IOP Conf. Series: Journal of Physics: Conf. Series **875** (2017) 042011 doi:10.1088/1742-6596/875/5/042011

Temperature effects on the isomer's stability of van der Waals clusters

Orlando Carrillo-Bohórquez†, Álvaro Valdés†1, Rita Prosmiti*2

† Departamento de Física, Universidad Nacional de Colombia, Calle 26, Cra 39, Edificio 404, Bogotá, Colombia
* Institute of Fundamental Physics (IFF-CSIC), Serrano 123, 28006 Madrid, Spain

Synopsis Small van der Waals(vdW) complexes have serve as prototypes for investigating energy transfer mechanisms and miscrosolvation processes. High-resolution spectroscopy techniques have provide valuable information, new aspects of these systems emerged from the interplay between theory and experiment, thus such systems are ideal for developing and testing theoretical methods and computational approaches.

High-level *ab initio* electronic structure calculations have shown that triatomic vdW clusters, formed by a dihalogen molecule and a noble gas atom, present double minima at both linear (L) and T-shaped (T) geometries in their ground electronic states [1]. In the case of the He-dihalogen (homonuclear) systems, such like $HeBr_2$ and HeI_2, the energy differences between them counts only to few cm^{-1}, while when nuclear quantum zero-point-energy effects are taking into account the relative stability of the corresponding isomers is less than 1 cm^{-1}. The existence and preferential stability of such isomers have been also supported by recent experimental studies from LIF and action spectra techniques [2]. However, discrepancies between theoretical predictions and even with experimental data of different sets of experiments have been reported, and this arises the question of whether there is a simple explanation of the theoretical point of view on the stability order of such isomers. Although, a number of experimental studies have detected different isomers of such vdW complexes in the supersonic expansions, the mechanisms for their formation, and their relative populations within the expansion under conditions of low temperature (below 3 K) are still uncertain and probably depend on the system. Thus, investigations on how thermodynamic factors (temperature) influence the isomers' formation, and thus their population in the supersonic beam expansion, could serve to improve deficiencies and shed light on our understanding of such processes.

As such a thermodynamic model was introduced [3] to compute the isomers' population as a function of temperature T, $Z^{L,T}(T) = Z^{L,T}_{vib}(T)Z^{L,T}_{rot}(T)$, where Z_{vib} and Z_{rot} the vibrational and rotational partition functions, respectively. Rovibrational quantum calculations for angular momentum values J up to 15 were carried out, all bound states were assigned and then vibrational and rotational partition functions were computed. On the basis of the present results, [3] the relative populations of the linear and T-shaped conformers show a strong dependence on the temperature for both $HeBr_2$ and HeI_2 vdW complexes (see Figure 1). We found that the linear $HeBr_2$ isomer is energetically more stable than the T-shaped ones by 1.14 cm^{-1} at T=0 K, whereas conversion from linear to T-shaped complexes was observed at temperatures above 2.87 K. In turn, for the HeI_2 system the population inversion occurs at 1.04 K, with the energy difference between them being just 0.21 cm^{-1}. Interestingly, for the $HeBr_2$ a nice accord between experiment and theory has been achieved, while the HeI_2 case is still under consideration, pushing for further improvements in the both theoretical and experimental directions.

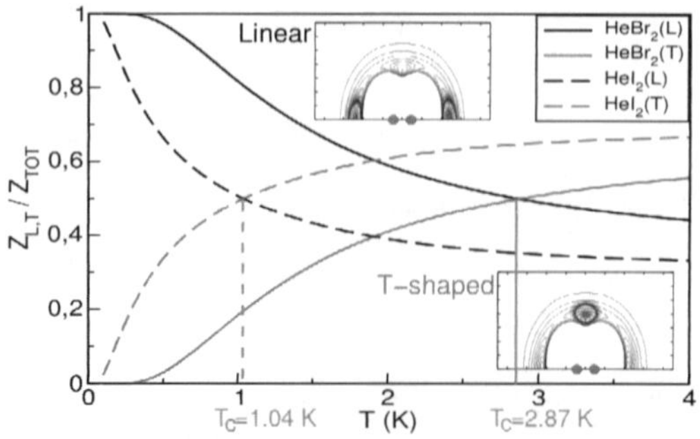

Figure 1. Population ratio of linear and T-shaped isomers as a function of temperature. The two populations become equal at T_C=1.04 and 2.87 K for the $HeBr_2$ (solid lines) and HeI_2 (dashed lines), respectively. In the inset panels the density distribution of each, linear and T-shaped, vibrational state are displayed in the ZX-plane together with the underlying potential energy surface.

References

[1] L. García-Gutierrez *et al.* 2009 *J. Phys. Chem. A* **113** 5754

[2] S.E. Ray *et al.* 2006 *J. Chem. Phys.* **125** 164314

[3] O. Carrillo-Bohórquez *et al.* 2016 *J. Phys. Chem. A* **120** 9458; 2017 *RSC Adv.* submitted

1 E-mail: avaldesl@unal.edu.co
2 E-mail: rita@iff.csic.es

Design of a charge sensitive spectroscopy amplifier for Compton camera

Wei Wang* [1]**, Junliang Liu*, Deyang Yu*, Xin Li**† **and Xiaohong Cai*** [2]

* Institute of Modern Physics, Chinese Academy of Sciences, Lanzhou 730000, China
† School of Nuclear Science and Technology, Lanzhou University, Lanzhou 730000, China

Synopsis A charge sensitive spectroscopy amplifier has been designed to couple with the silicon strip detector of a Compton camera. The amplifier was optimized by using a wave form generator and the output gain of 10^{12}V/C was achieved when the shaping time is 1 μs.

Compton camera is widely used in the radiology and radiation therapy for online dose control and high resolved 3D image reconstruction [1]. A typical Compton imaging system is composed of two detectors where a double-sided silicon strip detector closest to the source of radiation is denoted as the scatterer and a scintillator is absorber. In this system, the incident γ-ray with energy of hv will ideally Compton scatter from an orbital electron in the scatterer, transferring a fraction of its kinetic energy E, and then the scatted γ-ray interact via photoelectric absorption in the absorber, depositing its remaining energy. The scattering angle θ of the incident ray depends on E and hv, and

$$\cos\theta = 1 - \frac{Em_ec^2}{hv(hv-E)}$$

where m_ec^2 is the rest mass energy of the electron, hv is assumed to be known and E is measured value [1].

Usually, the performance of a Compton camera is evaluated based on spatial, energy and time resolution of the SSD and a previous work showed that integrating the electronics into a detector is an optimal method for noise reduction and resolution improvement [2]. In this abstract, we present a low noise spectroscopy amplifier which could be integrated into the SSD.

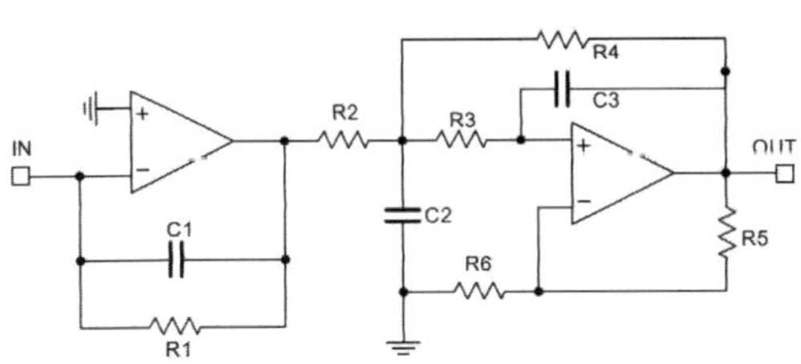

Figure 1. Logic diagram of the spectroscopy.

As shown in figure 1, the amplifier includes a charge sensitive pre-amplifier followed by a two stage Bessel low pass active filter in a multi-feedback topology as the pulse shaper.

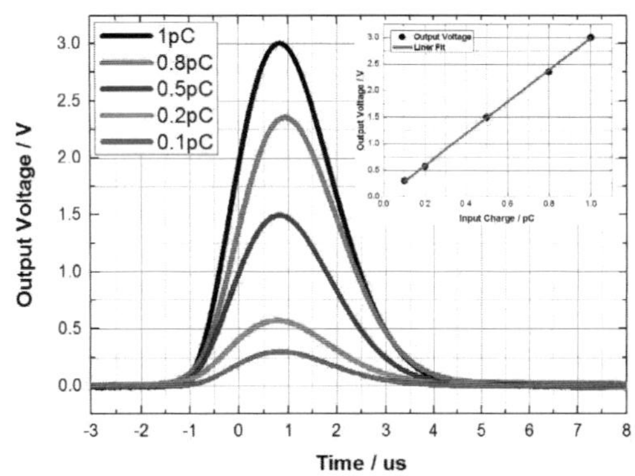

Figure 2. Output signal of the amplifier stimulated by an Agilent 81150A waveform generator.

A wave form generator is employed to simulate the signal from the SSD which enables us to optimize the amplifier. As can be seen from figure 2, the output amplitude is about 3 V when the input charge is 1 pC, which means the gain of the amplifier is about 3×10^{12} V/C. The shaping time is about 1 μs and the full width of the output is about 5 μs, which allows a count rate as high as 25 kHz with a tolerance of 10% pile up. As shown in the inset in figure 2, the output integral nonlinearity of the amplifier is evaluated within 0.3% when the range of input charge is from 0.1 pC to 1 pC, respectively.

The present design will be utilized in a Compton camera which is under construction in Institute of Modern Physics, Chinese Academy of Sciences.

References

[1] F. Roellinghoff *et al.* 2011 *Nucl. Instrum. Methods Phys. Res., Sect.* A **648** 520-523
[2] W. Wang *et al.* 2014 *Rev. Sci. Instrum.* **85** 036103

[1] E-mail: wangwei@impcas.ac.cn
[2] E-mail: caixh@impcas.ac.cn

Gold nanoparticles and their coatings, their effect on cells and their interaction with radiation

Sophie Grellet* [1], Małgorzata A. Śmiałek[§,†], Nigel J. Mason[†] and Jon Golding* [2¶]

[*] School of Life, Health and Chemical Sciences, The Open University, Milton Keynes, UK, MK7 6AA
[§] Dept. of Control and Energy Engineering, Gdansk University of Technology Gabriela Narutowicza 11/12, Gdansk, Poland, 80-233
[†] School of Physical Sciences, The Open University, Milton Keynes, UK, MK7 6AA

Synopsis Two nanometre gold nanoparticles (AuNPs), bearing sugar moieties and/or polyethylene glycolamine (PEG-amine), were synthesised and evaluated for their in vitro toxicity and ability to radiosensitise cells with 220 kV and 6 MV X-rays, using four cell lines representing normal and cancerous skin and breast tissues. In addition to the observed intrinsic cancer-selective chemotoxicity, these AuNPs acted as radiosensitisers in combination with 220 kV or 6 MV X-rays. The ability of AuNPs bearing simple ligands to act as cancer-selective chemoradiosensitisers is a novel discovery that holds great promise in developing low-cost cancer nanotherapeutics

Radiotherapy is currently used in around half of all cancer treatments. Although generally effective, it is damaging to surrounding healthy tissues and needs to be improved by better targeting of cancer cells. One promising approach is to use nanoparticles composed of high atomic number elements, such as gold, hafnium, gadolinium, platinum or iron, which have large X-ray photon capture cross-sections, and can therefore locally increase the energy deposition near the nanoparticle through secondary electron emission from the nanoparticles [1]. Because of their biocompatibility and amenability to surface modification for tumour targeting, gold nanoparticles (AuNPs) have predominantly been used for tumour radiosensitisation studies [2]. AuNP radiosensitisation with external beam sources is more effective when using kilovoltage X-ray photons than with megavoltage X-ray photons, although megavoltage is preferable due to its deeper tissue penetration.

AuNPs are generally non-toxic, except at high concentrations where they generate appreciable levels of reactive oxygen species (ROS) [3]. AuNPs smaller than 6 nm are preferable for therapeutic applications, since these can be excreted from the body by renal clearance, reducing long-term exposure to other organs [4].

Chemoradiosensitisers are dual-action drugs that are directly toxic to cells and also render the DNA more susceptible to radiation-induced damage. They include inhibitors of topoisomerase I, poly ADP-ribose polymerase (PARP), histone deacetylase (HDAC) and heat-shock protein 90 (Hsp90). However, current chemoradiosensitisers lack the ability to locally increase the deposited dose of radiation within cells. With that goal in mind, we have designed novel 2 nm gold nanoparticles coated with sugar ligands to improve aqueous solubility [5], and PEG-amine to improve biocompatibility [6] and cellular uptake [7]. Although originally envisaged as radiosensitising platforms to co-deliver anti-cancer drugs, these novel AuNPs were found to be selectively toxic for cancer cells at nanomolar concentrations and also act as radiosensitisers.

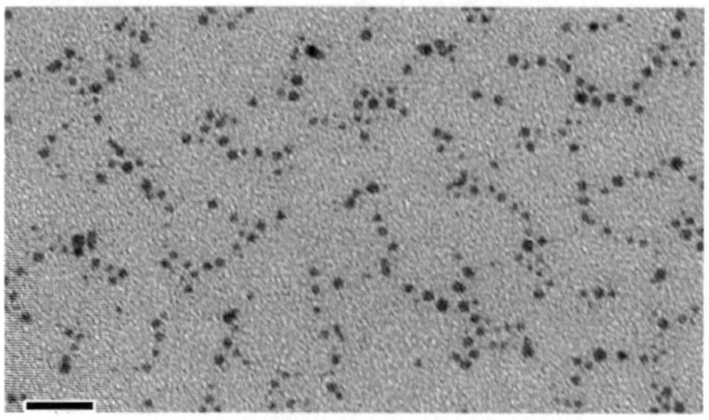

Figure 1. Representative TEM image, showing 50:50 βGlc:PEG-amine. Scale bar is 20 nm.

References

[1] Kwatra D *et al.* 2013 *Transl Cancer Res.* **2** 330
[2] Hainfeld JF *et al.* 2010 *Phys Med Biol.* **55** 3045
[3] Fratoddi I *et al.* 2015 *Nano Res.* **8** 1771
[4] Alric C *et al.* 2013 Nanoscale **5** 5930
[5] Irure A *et al.* 2013 *Biomater Sci.* **1** 658
[6] Jokerst J V *et al.* 2011 Nanomedicine **6** 715
[7] Chhour P *et al.* 2017 Bioconjug Chem. **28** 260

[1] E-mail: Sophie.Grellet@open.ac.uk
[2] E-mail: Jon.Golding@open.ac.uk

Electron impact L and M-subshell ionization cross sections for atoms (14$\leq$ Z $\leq$92) including the relativistic effects

B. C. Saha[*1], **A. K. Basak**[†], **M. A. Uddin**[†], **A. K. F. Haque**[† 2], **M. A. R. Patoary**[†], **M. M. Haque**[†], **M. Ismail Hossain**[†], **and M. Maaza** ‡

[*]Department of Physics, Florida A & M University, Tallahassee, FL-32307, USA
[†] Department of Physics, University of Rajshahi, Rajshahi-6205, Bangladesh
‡University of South Africa (UNISA), Muckleneuk Ridge, PO Box392, Pretoria, South Africa

Synopsis Electron impact inner-subshell (L and M) ionization cross sections for various atoms (14$\leq$ Z $\leq$92) are calculated and compared with experimental and other theoretical results.

The electron impact ionization cross sections (EI-ICS) are needed in various fields ranging from biological to chemical, plasma to astrophysics, and laser to medical physics. At energies above the breakup threshold the scattering in a three-body system remains a most challenging quantum mechanical problem. For simpler targets, like H and He, there are few accurate calculations available but for many electron targets it is rather disappointing. Experimental data for the K-shell ionization are rife; they are, however, limited for both the L-shell and M-shell but their subshell data [1] are practically zero. On the other hand the huge demands for EIICS in applications cannot be fulfilled easily by generating a database either by experimental or by quantal studies. So resorts are made to fulfill this void by using *simple* − *to* − *use* models that must yield accurate results. In this study we report few such models that are capable of describing reasonably the experimental data for a wide range of atomic targets over a wider energy domain. Two easy-to-implement models, namely XMCN [1, 2] and XMUIBED [1], are presented here; the former modifies the DM relativistic factor of the MCN model with a Z-dependent factor, while the latter replaces the Bethe part of the BED model by a simple two-parameter Born term. Both the XMCN and XMUIBED models, however, generate reliable EIICS quickly. Our results in Fig. 1 are compared with the experimental data and found good agreements for L- and M-subshell.

B. C. Saha thankfully acknowledges the partial support by National Nuclear Security Administration (NNSA), Award Number(s) DE- NA0002630.

[1]E-mail: bidhan.saha@famu.edu
[2]E-mail: fhaque2001@yahoo.com

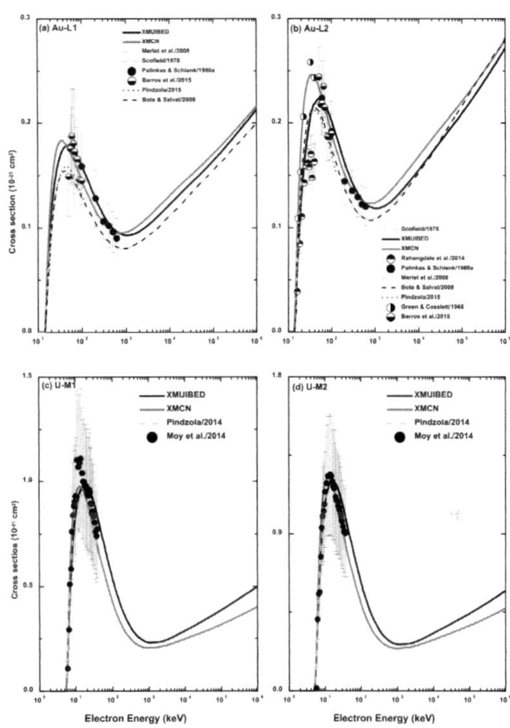

Figure 1. EIICS for Au and U atoms. Upper panel represents L1 and L2 subshells of Au and the lower panel for M1 and M2 subshells of U. Symbols denote experimental data and the lines predicted cross section from various sources including XMUIBED and XMCN.

References

[1] A. K. F. Haque et al. 2017 AQC, in press, https://doi.org/10.1016/bs.aiq.2017.02.002

[2] A. K. F. Haque et al. 2017 J. Phys. B: At. Mol. Opt. Phys. 50, 055005

IOP Conf. Series: Journal of Physics: Conf. Series **875** (2017) 052002 doi:10.1088/1742-6596/875/6/052002

Electron impact excitation of O III

S. S. Tayal*[1] and O. Zatsarinny†[2]

* Department of Physics, Clark Atlanta University, Atlanta, GA 30314, USA
† Department of Physics and Astronomy, Drake University, Des Moines, Iowa 50311, USA

Synopsis The B-spline Breit-Pauli R-matrix method has been used to calculate electron excitation collision strengths for a wide range of transitions between the first 220 fine-structure levels of O III. The calculation includes levels of the $2s^22p^2$, $2s2p^3$, $2s^22p3l$, $2s^22p4l$ and $2s2p^23l$ configurations. The present work considerably improves the existing calculations for oscillator strengths and collision strengths of O III.

Accurate transition rates and electron collision excitation rates of O III are important for the analysis and diagnostics of a wide range of astrophysical spectra. Electron scattering from O III has been performed by using highly accurate target wave functions and by including fine-structure effects in the close-coupling expansions directly. The present calculations have been carried out using the B-spline Breit-Pauli R-matrix (BSR) method [1]. The multi-configuration Hartree-Fock method in combination with B-spline expansions is employed for accurate representation of the target wave functions. This allows us to optimize the atomic wave functions for different states independently, resulting in a more accurate target description than those used in previous collision calculations.

Figure 1 shows collision strengths of the fine-structure $2s2^2p^2$ ($^3P_0$-$^3P_1$, $^3P_1$-$^3P_2$) transitions in O III. High resolution at near-threshold energies is necessary for accuracy in rate coefficients at low energies. We will present detail comparison with available calculations. Generally, all recent R-matrix calculations [2-4] agree to within 10% for the thermally averaged collision strengths for the forbidden transitions among the five lowest levels of the ground $2s^22p^2$ configuration. An exception is for the transitions from the lowest three ^3P_J levels to the $^1S_0$ level. Here our results support the most recent calculation by Story *et al* [4]. Note also that the calculations Palay *et al* [3] show intensive resonance structure at electron energies above 3 Ry not supported by the present calculations.

For higher lying states the agreement is more diverse. The comparison of effective collision strengths for a set of temperatures is presented in figure 2 where the ratios of our results with other calculations [2] are shown. About 80% of transitions to excited states are within 30%, though for some weak transitions the rate coefficients differ by several factors. The present calculations are the direct Breit-Pauli calculations whereas the results in the works of Aggarwal and Keenan [2] and Storey *et al* [4] obtained from nonrelativistic LS calculation using the intermediate coupling frame transformation.

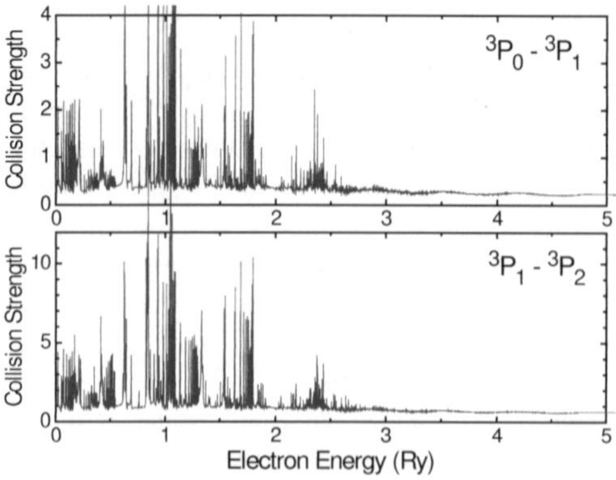

Figure 1. Collision strengths for the $2s^22p^2\ ^3P_0$ - $^3P_1$ and $^3P_1$ - $^3P_2$ fine-structure transitions in O III.

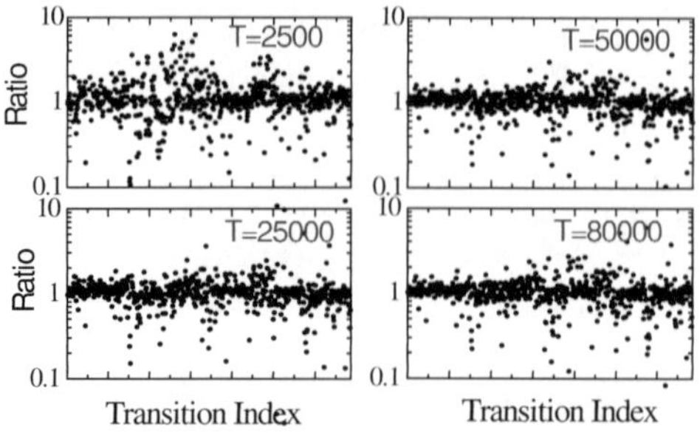

Figure 2. Comparison of effective collision strengths in O III with other calculations. The ratios between the R-matrix calculation [2] and the present results are shown.

This work was supported by NASA.

References

[1] O. Zatsarinny 2006 Comp. Phys. Commun. **174** 273
[2] K. Aggarwal and F. Keenan 1999 ApJS **123** 311
[3] E. Palay *et al* 2012 MNRAS **423** L35
[4] P. Storey *et al* 2014 MNRAS **441** 3028

[1] E-mail: stayal@cau.edu
[2] E-mail: oleg.zatsarinny@drake.edu

Benchmark calculations for electron-impact excitation of Mg^{4+}

K. Wang[*1], **L. Fernández-Menchero**[†2], **O. Zatsarinny**[†3], **K. Bartschat**[†4]

[*] College of Physics and Materials Science, Henan Normal University, Xinxiang, 453007, P.R. China
[†] Department of Physics and Astronomy, Drake University, Des Moines, Iowa, 50311, USA

Synopsis In order to resolve major discrepancies between a number of previous calculations for electron impact excitation of Mg^{4+}, we carried out extensive B-Spline R-matrix calculations for this problem. In particular, we investigate the effect of the target structure description and the number of states included in the close-coupling expansion on the final results. Finally, we provide an uncertainty estimate the current predictions.

There are major discrepancies between recent B-spline R-matrix (BSR) [1] and Dirac Atomic R-matrix Code (DARC) [2] calculations regarding electron-impact excitation rates for transitions in Mg^{4+}. To identify possible reasons for these discrepancies and to estimate the accuracy of the various results, we carried out independent BSR calculations with the same 86 target states as in the previous calculations, but with a different and more accurate representation of the target structure.

We find close agreement with the previous BSR results for the majority of transitions, thereby confirming their accuracy. At the same time the differences with the DARC results are much more pronounced. The discrepancies in the final results for the collision strengths are mainly due to differences in the structure description, specifically the inclusion of correlation effects, and due to the likely occurrence of pseudoresonances in the DARC model.

To further check the convergence of the predicted collision rates, we carried out even more extensive calculations involving 316 states of Mg^{4+}. Figure 1 presents a comparison of effective collision strengths from the present BSR-86 and BSR-316 models. As expected, the corrections for the strong transitions are small and the corresponding rates are expected to be converged. The rate coefficients for all transitions between the lowest 10 levels with configurations $2s^2 2p^4$, $2s2p^5$, and $2p^6$ are stable against changes in the size of the CC expansion, for both strong dipole and weak intercombination transitions.

To quantify the likely uncertainties, Table 1 shows the average deviations of the BSR-86 predictions from the BSR-316 results for three electron temperatures, with relatively large ($\Upsilon \geq 1$), medium ($10^{-2} < \Upsilon < 1$), and small ($\Upsilon < 10^{-2}$) effective collision strengths. As expected, the results for the strong transitions agree well in both models, while those for the weaker transitions deviate most from each other.

[1]E-mail: 021141@htu.edu.cn
[2]E-mail: luis.fernandez-menchero@drake.edu
[3]E-mail: oleg.zatsarinny@@drake.edu
[4]E-mail: klaus.bartschat@drake.edu

Table 1. Average difference (in %) between the effective collision strengths obtained in the BSR-86 model from the results generated with BSR-316 at three electron temperatures for strong, intermediate, and weak transitions.

$T_e =$	10^4 K	10^5 K	10^6 K
$\Upsilon \geq 1$	8.4	9.3	6.0
$10^{-2} < \Upsilon < 1$	33.2	40.0	20.4
$\Upsilon < 10^{-2}$	61.8	67.6	56.3

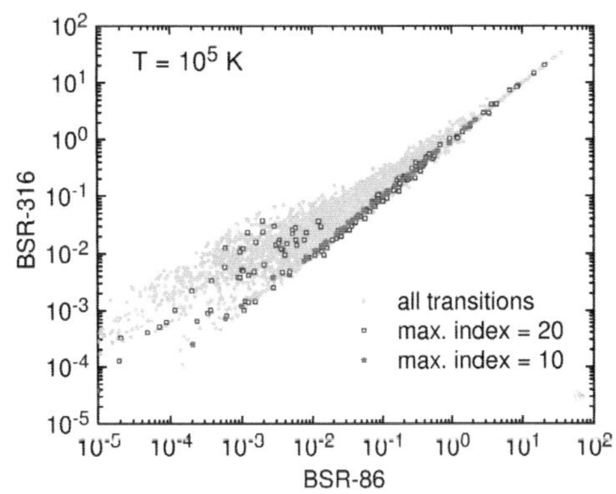

Figure 1. Comparison of effective collision strengths obtained in the present BSR-86 and BSR-316 models for an electron temperature of 10^5 K. In addition to all transitions, we illustrate the situation for the lowest 10 and 20 states, respectively.

This work was supported by the China Scholarship Council (K.W.) and the US National Science Foundation under grants PHY-1403245 and PHY-1520970, and the XSEDE allocation PHY-090031.

References

[1] S. S. Tayal and A. M. Sossah 2015 *Astron. Astroph.* **574** A87
[2] K. M. Aggarwal and F. P. Keenan 2017 *Can. J. Phys.* **95** 9

B-spline R-matrix with pseudo-states calculations for electron-impact excitation and ionization of magnesium

Oleg Zatsarinny [1] and Klaus Bartschat [2]

Department of Physics and Astronomy, Drake University, Des Moines, Iowa 50311, USA

Synopsis The B-spline R-matrix with pseudo-states method was employed to treat electron collisions with magnesium atoms. Predictions for elastic scattering, excitation, and ionization (including ionization plus excitation) were obtained for all transitions between the lowest 25 states of magnesium. The accuracy of the predictions was checked by comparing with available experimental data and with results obtained in different approximations with increasing number of coupled states.

Electron scattering from magnesium has been intensively studied both experimentally and theoretically. Most previous calculations, however, were devoted to selected individual transitions and a few energies (see [1] for references). The purpose of the present work, therefore, is to provide a comprehensive set of results that covers essentially all recent experimental data. This systematic comparison of available experimental and theoretical results makes it possible to assess the accuracy of the existing data and to search for possible sources of the main discrepancies between experiment and theory.

In the present work, the B-spline R-matrix method [2] was used to study electron collisions with neutral magnesium over an energy range from threshold to 100 eV. The largest scattering model included 712 coupled states, most of which were pseudo-states that simulate the effect of the high-lying Rydberg continuum and, most importantly, the ionization continuum on the results for transitions between the discrete states of interest. This effect is particularly strong at "intermediate" incident energies of a few times the ionization threshold.

As an example, Fig. 1 shows the $(3s3p)^1P^o$ excitation. The differences between the BSR-37 and BSR-712 results illustrate the influence of the target continuum. Our most extended BSR-712 cross sections are in close agreement with recent CCC results [3]. This indicates the likely convergence of both calculations. At the same time there is a noticeable difference with the experimental measurements [4] at lower energies. This calls for further checking the absolute normalization in [4], in particular since these data are used for the normalization of cross sections for other transitions [5] and hence may severely affect their accuracy.

Figure 2 shows ionization cross sections, including ionization plus excitations. We see relatively large contributions from the latter processes. The good agreement with existing experimental data over a wide range of incident electron energies indicates the completeness of the pseudo-state basis and can be considered as an additional confirmation of the accuracy of the entire data set.

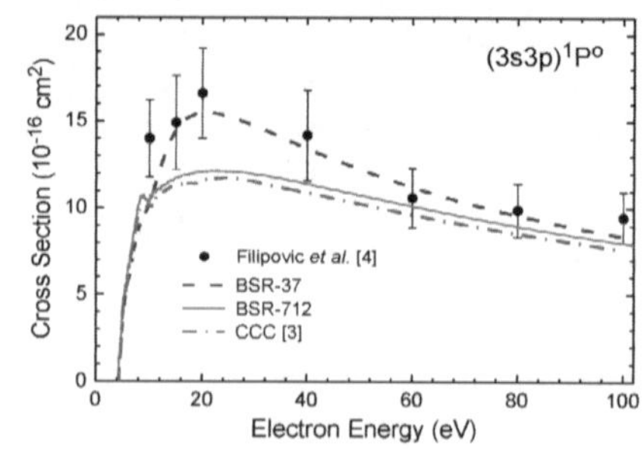

Figure 1. Electron-impact excitation cross sections of the $(3s3p)^1P^o$ resonance transition in Mg.

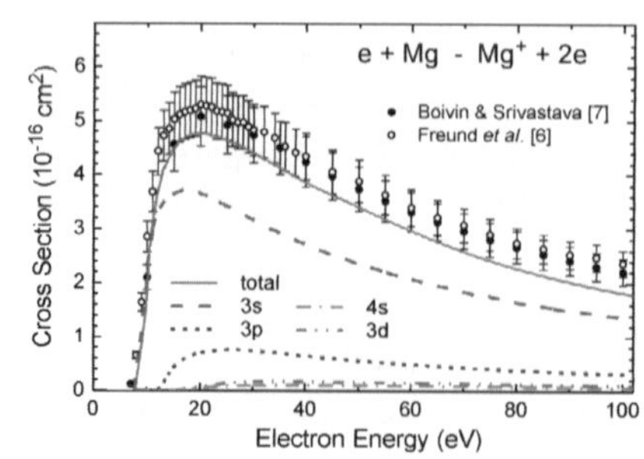

Figure 2. Electron-impact ionization cross sections for neutral Mg in different final ionic states.

This work was supported by the US National Science Foundation under grants PHY-1403245 and PHY-1520970, and the XSEDE allocation PHY-090031.

References

[1] O. Zatsarinny and K. Bartschat 2009 *Phys. Rev. A* 79 052709
[2] O. Zatsarinny 2006 *Comp. Phys. Commun.* **174** 273
[3] I. Bray and D. Fursa 2017 *private communication*
[4] D. M. Filipović *et al.* 2006 *J. Phys. B* **39** 2583
[5] B. Predojević *et al.* 2008 *J. Phys. B* **41** 015202
[6] R. S. Freund *et al.* 1990 *Phys. Rev. A* **41** 3575
[7] R. Boivin and S. Srivastava 1998 *J. Phys. B* **31** 2381

[1] E-mail: oleg.zatsarinny@drake.edu
[2] E-mail: klaus.bartschat@drake.edu

IOP Conf. Series: Journal of Physics: Conf. Series **875** (2017) 052005 doi:10.1088/1742-6596/875/6/052005

Photorecombination of berylliumlike and boronlike silicon ions

D. Bernhardt*, A. Becker†, C. Brandau‡,§, M. Grieser†, M. Hahn¶, C. Krantz†, M. Lestinsky‡, O. Novotný†,¶, R. Repnow†, D. W. Savin¶, K. Spruck*,†, A. Wolf†, A. Müller*, S. Schippers§1

* Institut für Atom- und Molekülphysik, Justus-Liebig-Universität Gießen, 35392 Giessen, Germany
† Max-Planck-Institut für Kernphysik, 69117 Heidelberg, Germany
‡ GSI Helmholtzzentrum für Schwerionenforschung, 64291 Darmstadt, Germany
§ I. Physikalisches Institut, Justus-Liebig-Universität Gießen, 35392 Giessen, Germany
¶ Columbia Astrophysics Laboratory, Columbia University, New York, NY 10027, USA

Synopsis Absolute rate coefficients for electron-ion recombination of astrophysically relevant Si^{9+} and Si^{10+} ions were measured at a heavy-ion storage ring. Hyperfine quenching of long-lived metastable levels was used for preparing practically pure ground-state beams in the storage ring. From the measured results rate coefficients for recombination of Si^{9+} and Si^{10+} ions in a plasma are derived and represented by simple parameterizations.

We report measured rate coefficients for electron-ion recombination of Si^{10+} forming Si^{9+} and of Si^{9+} forming Si^{8+}, respectively [1]. The measurements were performed using the electron-ion merged-beams technique at the Heidelberg heavy-ion storage ring TSR [2]. Electron-ion collision energies ranged from 0 to 50 eV for Si^{9+} and from 0 to 2000 eV for Si^{10+}, thus extending previous measurements for Si^{10+} [3] to much higher energies (Fig. 1). Experimentally-derived rate coefficients for the recombination of Si^{9+} and Si^{10+} ions in a plasma are presented along with simple parametrizations. These rate coefficients are useful for the modeling of the charge balance of silicon in photoionized plasmas (Si^{9+} and Si^{10+}) and in collisionally ionized plasmas (Si^{10+} only). In the corresponding temperature ranges, the experimentally-derived rate coefficients agree with the latest corresponding theoretical results within the experimental uncertainties.

One of the hallmarks of the storage-ring technique is that it allows for the preparation of ions in well defined energy levels. This has been exploited in particular for reducing the Si^{10+} ion beam contamination by long-lived $2s2p\ ^3P$ metastable levels to almost insignificance. To his end, the isotope ^{29}Si was used in the Si^{10+} experiment. For this isotope, with nonzero nuclear spin, hyperfine quenching shortens the $^3P_0$ lifetime by several orders of magnitude [4, 5] as compared to the more abundant isotope ^{28}Si. In fact, the storage-ring technique can also be used for precisely measuring these hyperfine induced lifetimes [6, 7]. Hyperfine quenching has been exploited already previously for state-preparation in a storage-ring measurement of electron-impact ionization of Be-like sulfur ions [8].

We would like to thank the MPIK accelerator and TSR crews for their excellent support during the experiments. Financial support by the Max-Planck-Gesellschaft and by the Deutsche Forschungsgemeinschaft (DFG, contract no. Schi 378/8-1) is gratefully acknowledged. MH, ON, and DWS were supported in part by the NASA Solar Heliospheric Physics program.

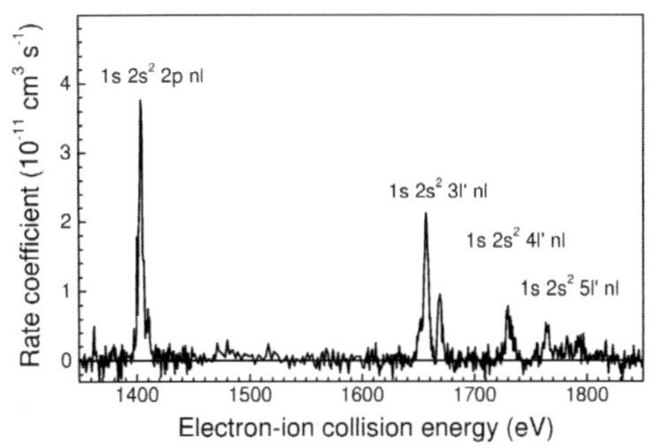

Figure 1. High-energy portion of the measured merged-beams rate coefficient (solid black line) for photorecombination of Si^{10+} in the energy range of dielectronic recombination resonances associated with K-shell excitations [1]. Resonance groups are labelled by the according $1s^2 2s^2 \rightarrow 1s 2s^2 N'l'$ core excitations.

References

[1] D. Bernhardt *et al.* 2016 *J. Phys. B* **49** 074004

[2] M. Grieser *et al.* 2012 *Eur. Phys. J ST* **207** 1

[3] I. Orban *et al.* 2010 *Astrophys. J.* **721** 1603

[4] K. T. Cheng *et al.* 2008 *Phys. Rev. A* **77** 052504

[5] M. Andersson *et al.* 2009 *Phys. Rev. A* **79** 032501

[6] S. Schippers *et al.* 2007 *Phys. Rev. Lett.* **98** 033001

[7] S. Schippers *et al.* 2012 *Phys. Rev. A* **85** 012513

[8] M. Hahn *et al.* 2012 *Phys. Rev. A* **85** 042713

[1]E-mail: stefan.schippers@physik.uni-giessen.de

IOP Conf. Series: Journal of Physics: Conf. Series **875** (2017) 052006 doi:10.1088/1742-6596/875/6/052006

Study on *KLL* dielectronic recombination for highly charged tungsten ions at Shanghai EBIT

B. Tu, J. Xiao, K. Yao, Y. Shen, Y. Yang, D. Lu, Y. Q. Fu, B. Wei, C. Zheng, L. Y. Huang, R. Hutton, and Y. Zou[1]

Shanghai EBIT Laboratory, Institute of Modern Physics, Fudan University, and the Key Laboratory of Applied Ion Beam Physics, Chinese Ministry of Education, China

Synopsis This work reported both experimental and theoretical study on *KLL* dielectronic recombination for He-like up to O-like tungsten ions, including resonance levels, resonance strengths and interference effect.

Dielectronic recombination (DR) is an important process in hot plasma physics as well as in atomic structure and collision theory. It significantly affects the plasma temperature, the charge state distribution, and the ion level population. The radiative processes in DR often cause unresolvable satellites, which may disturb the main line shape, line intensity, and line width, while the resolved satellite lines are often used for electron temperature diagnostics. Furthermore, DR of highly charged ions contributes significantly to radiation energy loss in fusion plasmas, and thus leading to the flameout of fusion.

The electron beam ion trap (EBIT) has an advantage on high energy investigation; researches on DR processes which involve a *K*-shell electron excitation have been performed for many medium heavy elements and for a few high Z elements. In these researches, resonance energy and strength of DR, Breit interaction, polarization and interference effect between DR and RR have been studied [1-6].

In this presentation, we reported the recent study on *KLL* dielectronic recombination processes of He- up to O-like W ions through both experiment and calculation. The measurement was performed on Shanghai electron beam ion trap [7] by employing a fast electron beam-energy scanning technique. During the experiment, $W(CO)_6$ was continuously injected into the EBIT, and the electron beam energy was fast scanned through the resonant energies of *KLL* DR of He-like up to O-like W ions, while the X ray photons were detected by a HPGe detector. A fully relativistic configuration interaction method implemented in the flexible atomic code was employed to calculate DR process and also radiative recombination (RR).

The interference effect between DR and RR were analyzed in experiment and a dual Fano and Lorentzian line profile properties in an autoionizing state was demonstrated via selecting the DR resonances which go through an intermediate state with decay channels to both the final states with no excited electrons and the final states with two excited electrons. In addition, the *KLL* DR resonance strengths for W were determined with uncertainty below 11% in this work. Our experimental results show good agreements with the calculations. The comparison between two experimental results (analysis with or without the consideration of interference effect) indicates that the interference effect between DR and RR is revealed to be necessary to determine the resonance strength.

[1] E-mail: zouym@fudan.edu.cn

References

[1] D. A. Knapp *et al.*, Phys. Rev. Lett. **74**, 54 (1995)
[2] A. J. G. Martínez *et al.*, Phys. Rev. Lett. **94**, 203201 (2005)
[3] Z. Hu *et al.*, Phys. Rev. Lett. **108**, 073002 (2012)
[4] B. Tu, *et al.*, Phys. Rev. A **91**, 060502(R) (2015)
[5] B. Tu, *et al.*, Phys. Rev. A **93**, 032707 (2016)
[6] B. Tu, *et al.*, Phys. Plasmas **23**, 053301 (2016)
[7] D. Lu *et al* *Rev. Sci. Instrum.* 85, 093301 (2014)

IOP Conf. Series: Journal of Physics: Conf. Series **875** (2017) 052007 doi:10.1088/1742-6596/875/6/052007

Spin evolution of two valence electrons in two separated ions

Biplab Goswami† [1], **Andrey V. Volotka**† and **Stephan Fritzsche**†,‡

† Helmholtz-Institut Jena, Fröbelstieg 3, D-07743 Jena, Germany
‡ Theoretisch-Physikalisches Institut, Friedrich-Schiller-Universität Jena, D-07743 Jena, Germany

Synopsis The spin evolution of two spin-1/2 valence electrons, located at two separated ions, is investigated with the help of Breit equation. This evolution that arise solely depends on their dipole-dipole magnetic interaction. Special attention is paid to the effects beyond the leading order of the dipole-dipole magnetic interaction. In contrast to just including the leading order, it is found that the spin-spin interactions flips the spin of both electrons together, while the spin-orbit interactions typically changes the spin of just single electron. As a consequence, different behaviour is observed in the time evolution of two different entangled Bell states.

Understanding the spin evolution of two or more qubits is a key task in physics and quantum engineering. In particular, the transport and relaxation of spin waves in metals and semiconductor are of fundamental interest not only in basic research, but also in the field of spintronics [1], in which the spin of electrons is utilized *instead of* or *in addition* to the charge degrees of freedom. In spin-field effect transistors, for example, the spin-orbit interaction is known to control the motion of electrons inside the semiconductor channel.

The evolution of two coupled spins have been measured recently by Kotler *et al* [2] for two separated ions. In their simplified model, however, these authors describe the associated spin-magnetic moment of the electrons as point-like dipole, an approximation that breaks down at small distances. More generally, therefore, all the spatial and spin degrees of freedom have to be accounted for the two valence electrons. In present work, we aim to work out a theory which goes beyond the leading order in dealing with the spin-dependent interactions of electrons.

In more detail, we have explored probable spin-flip transitions of the two electrons. Since the spin-flip of different spin states can be represented by a parity sign, we have studied the difference in parity for the different spin states. We especially considered the initial spin state of two electrons at $|\uparrow\downarrow\rangle$ and calculated the time evolution of this state as a function of interaction time (t) at different inter-ionic distances (d) and for a typical strength of the external magnetic field ($\vec{B}$).

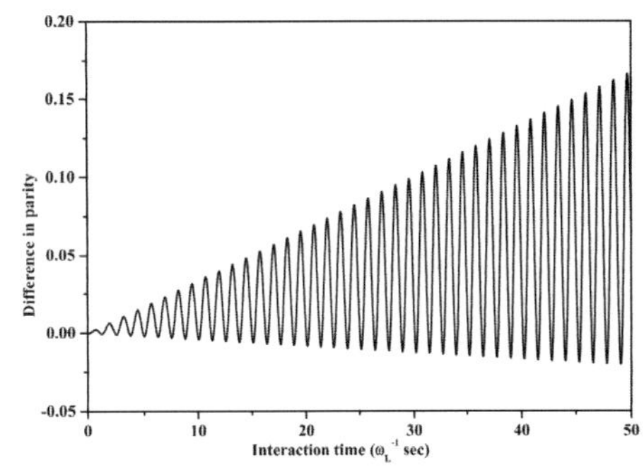

Figure 1. The difference of parity at $d = 10^3 a_0$ and $\omega_L = 1.4$ MHz (where, a_0 is Bohr radius and ω_L is spin Larmor frequency) with $\vec{B} = 0.1$ mT.

Fig. 1 displays how the parity changes exponentially with t, if terms beyond the leading order interactions are taken into account in the electron-electron interaction operator. Our results show that the spin-spin interaction always changes the spin of two electrons together, while the additional terms in the Hamiltonian may flip also the spin of just a single electron, and which results in a different behaviour of the entangled Bell states. Moreover, it is found that the parity difference (of Fig. 1), is highly sensitive to the inter-ionic distances as well as to the strength of the external magnetic field.

References

[1] S.A. Wolf *et al* 2001 *Science* **294** 1488

[2] S. Kotler *et al* 2014 *Nature* **510** 376

[1]E-mail: b.goswami@gsi.de

Electron impact ionization of calcium atoms inside quadrupole trap

Łukasz Kłosowski[*1], **Mariusz Piwiński**[*], **Szymon Wójtewicz**[*], **Daniel Lisak**[*], **Michael Drewsen**[†], **Katarzyna Pleskacz**[*], **Stanisław Chwirot**[*]

[*] Institute of Physics, Faculty of Physics, Astronomy and Informatics, Nicolaus Copernicus University in Toruń, Grudziadzka 5, 87-100 Torun, Poland

[†] Department of Physics and Astronomy, Aarhus University, DK-8000 Aarhus C, Denmark

Synopsis We present experimental method for electron impact ionization cross sections of calcium. The technique involves a Paul trap equipped with low energy electron gun. The general assumptions of the method are discussed.

Loading a Paul trap with ions requires application of ionization process of any type. In the case of calcium ions, two photoionization schemes [1, 2] are commonly used. High-energy (keV) electron impact technique was also used before development of the optical methods.

Our experimental results confirmed that the electronic ionization can be relatively easily achieved for energies below 100 eV. The proposed technique provides ion production rate at the level of hundreds of new ions per second.

An efficiency of ionization strictly depends on experimental conditions such as energy and density of the electron beam, atomic beam density and trapping potential parameters. Thus, analysis of the cloud of trapped ions for given conditions delivers information about integral cross sections for electron impact ionization processes.

The ion production rate is given by equation:

$$\dot{N} = \sigma \cdot l \cdot \frac{I}{e} \cdot n, \quad (1)$$

where σ is integral ionization cross section, l is geometrical parameter (approximately diameter of the trap), I is the effective current of electron beam, n is number density of neutral atoms and e is elementary charge. Thus, by determining number of ions trapped in certain time in well known conditions, ionization cross section can be found. The method can provide experimental data complementing existing data sets obtained with different experimental methods (time of flight [3]), (crossed beam experiment [4, 5]) and theoretical models [6, 7, 8].

In the experiment [9] cross-fired electronic and atomic beams were used. The interaction region was situated in central part of a linear, segmented Paul trap. The ions produced in collisions were captured by the trap's potential and subsequently cooled with Doppler red-shifted optical laser beams of 397 and 866 nm. The presence of the ions was detected by observation of 397 nm fluorescence with a CCD camera. Number of ions in cooled ensemble can be determined by analysis of the image. Geometry of the experiment is presented in Fig. 1.

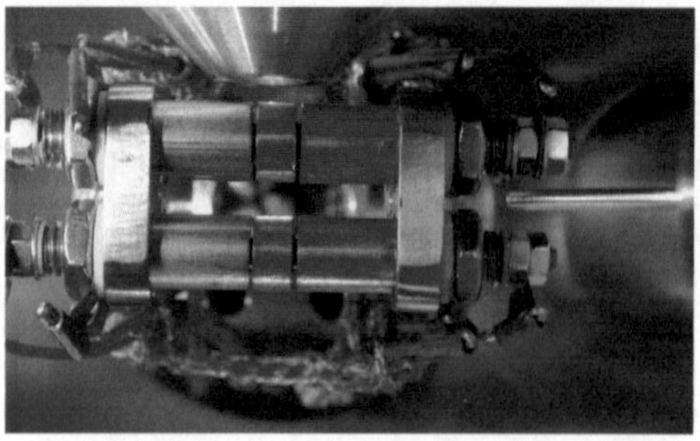

Figure 1. Photograph of the main part of the experimental setup. The linear segmented trap of 30 mm in length and 4 mm gap between electrodes is accompanied by electron gun (top) and the source of atomic beam (on the right, aligned at 45° to the image plane). The laser beams propagate along the trap main axis. The imaging system is aligned perpendicular to the plane of the picture.

This work has been supported by the National Science Centre, Poland, project no. 2014/13/B/ST2/02684.

References

[1] S. Gulde *et al.* 2001 *Appl. Phys. B* **73** 861

[2] N. Kjargaard *et al.* 2000 *Appl. Phys. B* **71** 207

[3] D. Cvejanović, A. J. Murray 2003 *J. Phys. B* **36** 3591

[4] S. Okudaira 1970 *J. Phys. Soc. Japan* **29** 409

[5] Y. Okuno 1971 *J. Phys. Soc. Japan* **31** 1189

[6] F. Fiquet-Fayard, M. Lahmani 1962 *J. Chim. Phys.* 59 1050 **59** 1050

[7] L. A. Vainstein *et al.* 1972 *Sov. Phys. - JETP* **34** 271

[8] B. N. Roy, D. K. Rai 1983 *J. Phys. B* 16 4677 **16** 4667

[9] Ł. Kłosowski *et al.* 2015 *J. Phys. Conf. Ser.* **635** 072003

[1]E-mail: lklos@fizyka.umk.pl

IOP Conf. Series: Journal of Physics: Conf. Series **875** (2017) 052009 doi:10.1088/1742-6596/875/6/052009

Energy and momentum transfer in electron-ion elastic collisions

Łukasz Kłosowski[1], **Mariusz Piwiński**

Institute of Physics, Faculty of Physics, Astronomy and Informatics, Nicolaus Copernicus University in Toruń, Grudziadzka 5, 87-100 Torun, Poland

Synopsis Heating of cold trapped calcium ions by elastic collisions with electrons of various energies was observed. The effect is analyzed theoretically using a multi-center Coulomb potential model.

A new experimental setup using ion trap with electronic beam for ionization process was applied for investigation of trapped ions [1]. There are two interesting phenomena connected with such technique: heating of the trapped ions by Coulomb interactions between ions and the incoming electrons and pressure of electrons acting on the trapped ion cloud. Only the former was observed in the experiment.

The experiment has been performed in a Paul trap with Ca^+ ions as a target. The experimental procedure includes loading the trap by electron-impact ionization of calcium atoms at chosen collision energy. After cutting off the electronic and atomic beams, the trapped ions are Doppler-cooled to the temperature below 1 mK using 397 nm and 866 nm laser beams. They are detected by observation of 397 nm fluorescence using a CCD camera.

After introduction of electron beam to the ion cloud, the image of the ion ensemble becomes less explicit, which can be explained by increase in amplitude of ion motion in the trap's potential well. Example of such effect is presented in Fig. 1. Shifting of ion cloud was not observed.

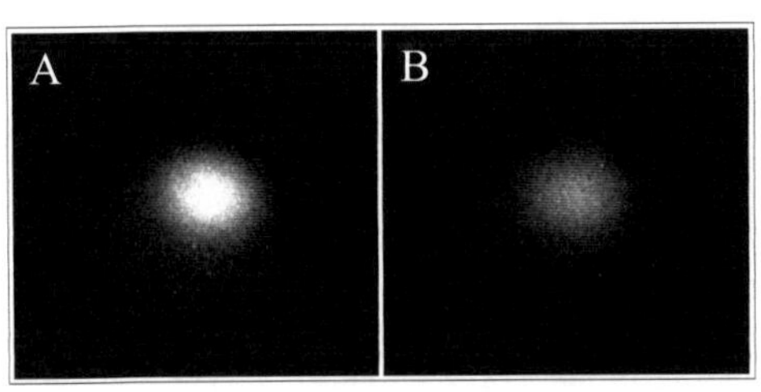

Figure 1. Example of cold ion clouds with the electron beam turned off (A) and on (B). Several thousand of ions interacting with 1 μA, 100 eV electron beam were observed. Size of the cloud was approximately 100 μm. The electrons propagated from the top to the bottom of the image.

Both effects of heating and pushing of the ion ensemble can be described using classical scattering model applying sum of Coulomb potentials from all the ions present in the ion trap. In such model the average energy transfer from a single electron passing the ion cloud is given by equation:

$$\Delta E = \frac{4\eta^2 m_e}{E m_{Ca}} \log\left(1 + (E/\eta)^2\right), \quad (1)$$

where $\eta = \frac{e^2}{2\pi\varepsilon_0 r}$ (r is a geometrical parameter depending on cloud thickness and density). At typical experimental conditions, the parameter has value of $\eta = 0.3$ meV. The average momentum transfer is given by equation:

$$\Delta p = \frac{\sqrt{8}\eta^2 m_e^{1/2}}{E^{3/2}} \log\left(1 + (E/\eta)^2\right). \quad (2)$$

The energy and momentum transfer dependences on electron energy are presented in Fig. 2.

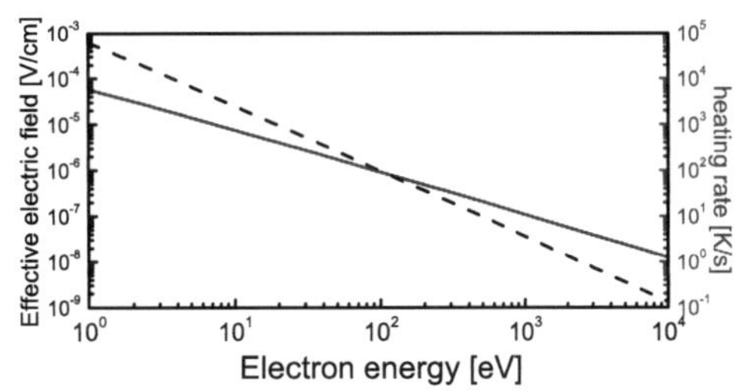

Figure 2. Heating rate (solid) and dislocating force (dash) of electron beam acting on calcium ion cloud. Both graphs are normalized to 1 μA electron beam. The force has been presented in electric field units (normalized to single elementary charge).

According to the proposed model, the pressure of electrons is negligible comparing to the trap's field. Despite of relatively high heating rates, the heated-up ion cloud stays in equilibrium defined by electronic heating and optical cooling.

This work has been supported by the National Science Centre, Poland, project no. 2014/13/B/ST2/02684.

References

[1] Ł. Kłosowski *et al.* 2015 *J. Phys. Conf. Ser.* **635** 072003

[1]E-mail: lklos@fizyka.umk.pl

Electronic excitation of P-state of cadmium and zinc atoms

M. Piwiński [1], Ł. Kłosowski, D. Dziczek, S. Chwirot

Institute of Physics, Faculty of Physics, Astronomy and Informatics, Nicolaus Copernicus University in Toruń, Grudziądzka 5, 87-100 Toruń, Poland

Synopsis The electron-photon coincidence technique in coherence analysis version was used to study the electron impact excitation of first excited $^1P_1$ states of cadmium and zinc atoms. The experimental results of the Stokes and Electron Impact Coherence Parameters (EICPs) for various collision energies are compared and presented alongside CCC and RDWA theoretical predictions.

Atoms with two valence electrons outside an inert electron core are still interesting target for both experimental and theoretical electron-atom collision studies [1-4]. The collisional excitation process can be characterised using a set of the so-called Electron Impact Coherence Parameters [5]. The EICPs determining angular distributions of the electron charge cloud of the excited atoms can be obtained in electron-photon coincidence experiments [6-9]. However, due to experimental problems and long time consuming measurements there is only a limited number of the experimental data enabling verification of proposed theoretical models [10-12].

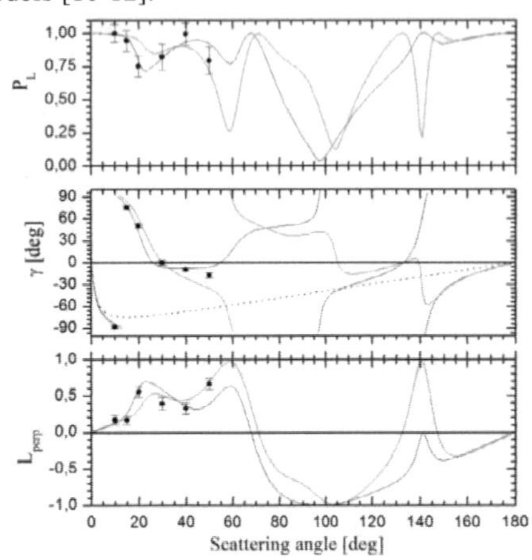

Figure 1. EICP (P_L, γ, L_{perp}) for electronic excitation (80 eV) of 5^1P_1 Cd state. Experimental data (•) are presented together with (—) CCC, (—) RDWA and (···) FBA theoretical predictions [10, 12].

We will present the values of Stokes and EICP parameters for excitation of the lowest singlet P-state of cadmium and zinc atoms [10,13,14]. The experiments were performed using electron-photon coincidence technique in the coherence analysis version. The data are presented with existing CCC [12] and RDWA [4] theoretical predictions.

[1] E-mail: Mariusz.Piwinski@fizyka.umk.pl

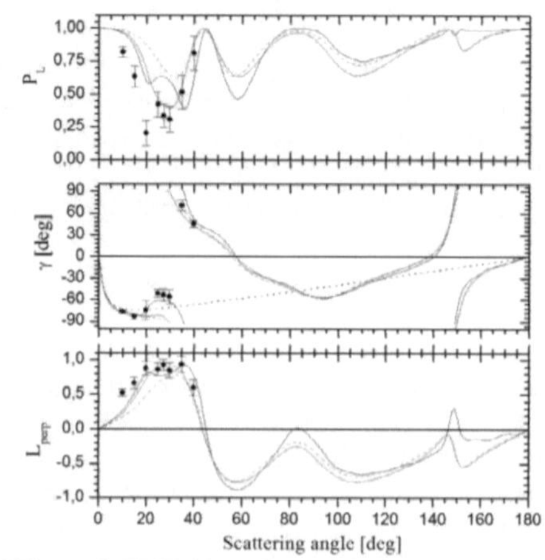

Figure 2. EICP (P_L, γ, L_{perp}) for electronic excitation (80 eV) of 4^1P_1 Zn state. Experimental data (•) are presented together with (—) CCC, (—) RDWA, (– –) SC RDWA and (···) FBA theoretical predictions [4,14].

References

[1] D. Dziczek *et al.* 1998 *Acta Phys. Polon. A* **93**, 717–721
[2] D. Dziczek *et al.* 2003 *Acta Phys. Polon. A* **103**, 3–11
[3] J. F.Williams *et al.* 2012 *Phys. Rev. A* **85**, 022701
[4] T. Das *et al.* 2014 *Phys. Lett. A* **378**, 641–643
[5] N. Andersen *et al.* 1988, *Phys. Rep.* **165**, 1–188
[6] D. Dyl *et al.* 1999 *J. Phys. B: At. Mol. Opt. Phys.* **32**, 837–844
[7] M. Piwiński *et al.* 2002 *J. Phys. B: At. Mol. Opt. Phys.* **35**, 3821–3827
[8] Ł. Kłosowski *et al.* 2007 *Meas. Sci. Technol.* **18**, 3801–3810
[9] Ł. Kłosowski *et al.* 2009 *Phys. Rev. A* **80**, 062709
[10] M. Piwiński *et al.* 2006 *J. Phys. B: At. Mol. Opt. Phys.* **39**, 1945–1953
[11] M. Piwiński *et al.* 2012 *Phys. Rev. A* **86**, 052706
[12] M. Berrington *et al.* 2012 *Phys. Rev. A* **85**, 042708
[13] M. Piwiński *et al.* 2013 *Eur. Phys. J.-Spec. Top.* **222**, 2273–2277
[14] M. Piwiński *et al.* 2015 *Phys. Rev. A* **91**,062704

IOP Conf. Series: Journal of Physics: Conf. Series **875** (2017) 052011 doi:10.1088/1742-6596/875/6/052011

Benchmark calculations of double ionization of Ne in Graphics Processing Units

F. D. Colavecchia[*1], **M. Schulz**[†], **T. Kirchner**[‡] **and M. F. Ciappina**[§]

[*]División Física Atómica, Molecular y Óptica, y Conicet, Centro Atómico Bariloche , 8400 San Carlos de Bariloche, Argentina.

[†]Department of Physics and LAMOR, Missouri University of Science & Technology, Rolla, MO 65409, USA

[‡]Department of Physics and Astronomy, York University, Toronto, Ontario M3J 1P3, Canada

[§]Institute of Physics of the ASCR, ELI-Beamlines, Na Slovance 2, 182 21 Prague, Czech Republic

Synopsis In this work we present the analysis of the numerical performance and accuracy of the Monte Carlo Event Generator method applied to the double ionization of Neon by fast electrons. This implementation of the code runs in Graphics Processing Units, giving a clear advantage over other computational approaches. The code can provide five hundred cross sections per second with the optimal configuration, an order of magnitude faster than CPU counterparts.

Theoretical calculations of multiple ionization processes have become more and more feasible along with the advances in computer power. One of the most recent approaches is the Monte Carlo Event Generator (MCEG). It is a numerical technique that is able to compute differential cross sections for a variety of complex ionization processes and can simulate a given setup of a collision[1, 2]. The simulated events are convolved with experimental resolutions of the detectors, providing an accurate representation of many ionization processes[2]. The main shortcoming of this scheme is that it is usually necessary to compute billions of cross sections in order to get a physically meaningful statistics. Since each cross section can be computed independently, this calculation belongs to the so-called *embarrasingly parallel* numerical problems. Therefore, it perfectly suits the new computational features provided by latest generation of Graphical Processing Units (GPUs).

In this work we report benchmarks of the GPU implementation of the MCEG method, applied to the double ionization of Ne by high energy electron impact[3]. The collision is modeled as a two step process: the first collision induces the ionization of Ne, while the second one is the single ejection of one electron from the residual target Ne^{+}. Each step is computed with the First Born Approximation. The code procedes as follows: first, a large set of cross section is computed, and second, favourable events are selected from this set. These two computational steps can be repeated to obtain a given number of events. The code is written in C language using the Compute Unified Device Architecture (CUDA) extensions[4].

The running times for the calculation of 10^5 cross sections are depicted in Fig. 1, as a function of the number of threads and the total number of events per iteration. The performance exhibits a good scaling with the number of total events 2^J, and optimal times are achieved for 32 threads per calculation block.

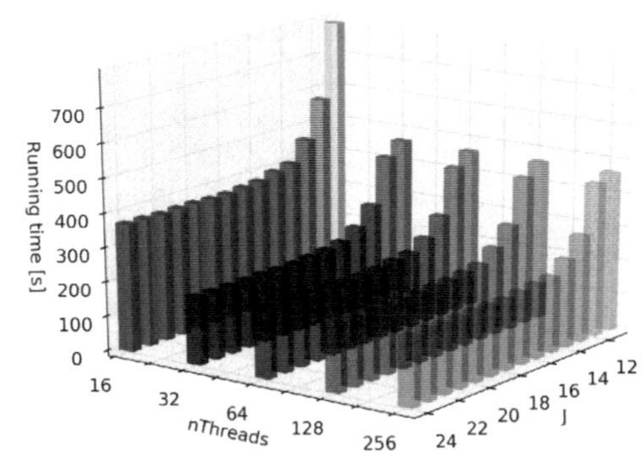

Figure 1. Running times of 2^J (e,3e) cross sections with different number of threads per computational block, to obtain 10^5 double ionization favourable events.

Our preliminary results show that the MCEG is accelerated in the GPU more than ten times, compared to the standard CPU code, providing the same accuracy.

References

[1] M. F. Ciappina, T. Kirchner and M. Schulz, *Comp. Phys. Comm.* **181** 813-820 (2010).

[2] M. F. Ciappina, T. Kirchner and M. Schulz, *Phys. Rev. A* **82** 062701 (2010); ibid., *Phys. Rev. A* **84** 034701 (2011).

[3] E. M. S. Casagrande, C. Li, A. Lahmam-Bennani, C. D. Cappello, M. Schulz and M. F. Ciappina, *J. Phys. B: At., Mol. Op. Phys.*, **44**, 055201-7 (2011).

[4] *NVIDIA CUDA Programming Guide 7.5*, NVIDIA Corp. (2015).

[1]E-mail: flavioc@cab.cnea.gov.ar

IOP Conf. Series: Journal of Physics: Conf. Series **875** (2017) 052012 doi:10.1088/1742-6596/875/6/052012

Multiple scattering of slow muons in an electron gas

Claudio Archubi* [1], **Diego Arbó***, **and Nestor R. Arista**†

* Instituto de Astronomía y Física del Espacio, Universidad de Buenos Aires, Ciudad Universitaria, 1428 Buenos Aires, Argentina

† División Colisiones Atómicas, Centro Atómico Bariloche, 8400 S. C. Bariloche, Argentina,

Synopsis We have performed a study of velocity threshold effects in the angular dispersion of slow muons in an electron gas using different dielectric models and a non-linear model to represent metals and band-gap materials.

With the advent of low energy muon beams capable of stopping in thin layers of crystalline material it becomes important to know the depth distribution of these muons in order to properly interpret muon spin rotation experiments. These experiments are performed to learn more about the surrounding dipole contributions and the nature of local magnetic field distribution of a crystalline sample. Recent computational simulations [1] conclude that channeling conditions may increment the penetration depth, but these simulations do not take into account the importance of the multiple scattering of the projectile with the electrons of the channel.

In a previous work [2] we developed a formalism based on the theory of multiple scattering and the calculus of the elastic scattering cross section for a dielectric model. Using this formalism, we aim here to describe the behaviour of the half width of the angular dispersion of muons with a material characterized in terms of different dielectric approaches which represent the cases of metals (Lindhard free electron gas model), semiconductors and insulators (Levine and Louie model [3] and Brandt and Reinheimer model [4] for systems with an energy band gap Eg). However, at low velocities non-linear effects become important. So we have incorporated the results using the non-linear approach [2], which applies also in the cases of very strong screening and large scattering amplitudes, and reduces to the linear treatment when the magnitude of the perturbation is small. The non-linear approach is based on transport cross section calculations according to quantum scattering theory.

Figure 1 shows the results for the half width $\alpha_{1/2}$ of the angular dispersion as a function of the projectile velocity for two different cases. For velocities lower than 1 a.u. the dispersion becomes important as the half width raises to several degrees, increasing the probability of dechanneling. However, in insulators and semiconductors this effect is attenuated by the sudden decreasing of the dispersion near the velocity threshold.

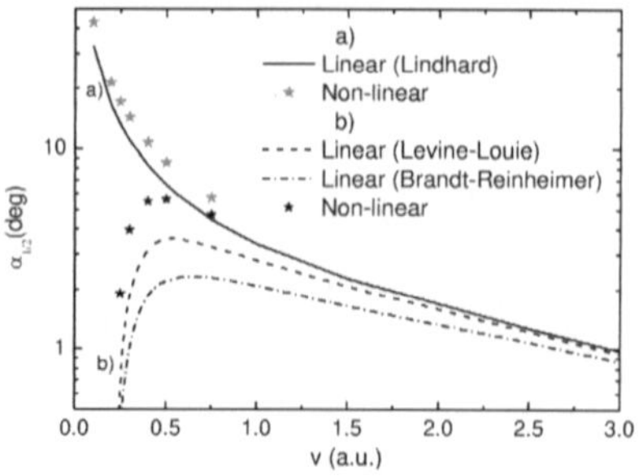

Figure 1. Muons traversing an electron gas target of width 800 a.u. with r_s.=1.5 a.u. for two cases: a) Eg=0 eV (Lindhard dielectric model and non-linear model) , b) Eg=14 eV (Levine and Louie and Brandt-Reinheimer dielectric models and non-linear model) The velocity threshold for this case is v=0.21 a.u.

References

[1] N. Saquib, W. J. Kossler and M. Deady 2012 *Physics Procedia.* **30** 42

[2] C. D. Archubi and N. R. Arista 2006 *Phys Rev A* 74 052717

[3] Z. H. Levine, S. G. Louie 1986 *Phys Rev B* 25.6310

[4] W. Brandt and J. Reinheimer 1970 *Phys Rev B* 2 3104

[1] E-mail: archubi@iafe.uba.ar

Spin entanglement in elastic electron scattering from quasi-one electron atoms

K. Bartschat[1], **S. Fonseca dos Santos**[2]

Department of Physics and Astronomy, Drake University, Des Moines, Iowa, 50311, USA

Synopsis We present results for the differential cross section and the spin exchange asymmetry in elastic electron collisions with atomic hydrogen and light alkali atoms. Such systems were recently suggested as candidates for continuously varying the degree of entanglement between the elastically scattered projectile and the valence electron. The feasibility of an experimental realization in light of the expected signals is discussed.

We extended our work on e-Li collisions [1] to investigate low-energy elastic electron collisions with atomic hydrogen and other alkali targets (Na, K, Rb). Measured in Bielefeld [2] and at NIST [3] years ago, these systems were only recently suggested for the possibility of continuously varying the degree of entanglement between the elastically scattered projectile and the valence electron [4, 5]. In order to estimate how well such a scheme may work in practice, we carried out overview calculations for energies between 0 and 10 eV and the full range of scattering angles $0° - 180°$.

The relative exchange asymmetry parameter is defined as

$$A_{\rm ex} = -P = \frac{\sigma^{\uparrow\downarrow} - \sigma^{\uparrow\uparrow}}{\sigma^{\uparrow\downarrow} + \sigma^{\uparrow\uparrow}} = \frac{\sigma^s - \sigma^t}{\sigma^s + 3\sigma^t}, \qquad (1)$$

where $\sigma^{\uparrow\uparrow}(\sigma^{\uparrow\downarrow})$ and $\sigma^s(\sigma^t)$ denote the angle-differential cross sections (DCSs) for parallel (anti-parallel) spin orientations of the projectile and target spins or singlet (triplet) scattering. The DCS for unpolarized projectile and target beams is given by

$$\sigma_u = \frac{1}{4}\sigma^s + \frac{3}{4}\sigma^t, \qquad (2)$$

and hence the limiting values for P are $+1/3$ for pure triplet and -1 for pure singlet scattering. The latter extreme case corresponds to the well-known situation of two spins forming a combined spin-0 system.

In addition to P, which characterizes the entanglement, we present the differential cross section in order to estimate whether the count rates in the most interesting energy-angle regimes are sufficient to make such experiments feasible in practice.

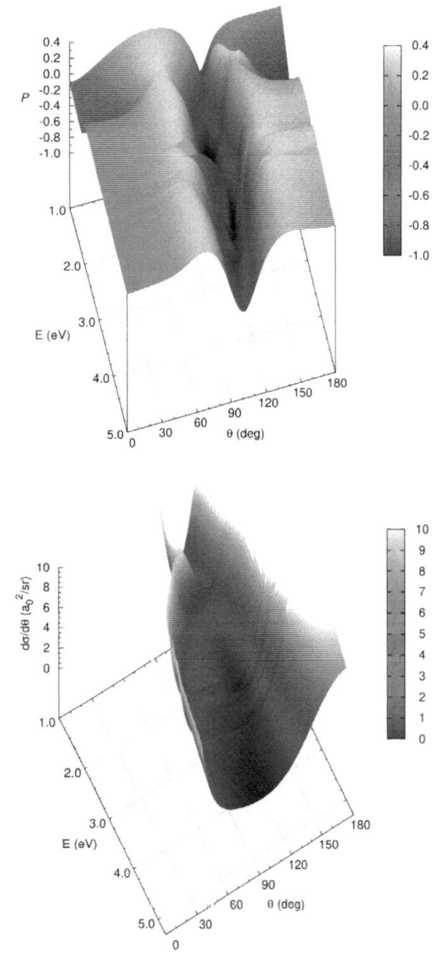

Figure 1. Spin asymmetry parameter $P = -A_{\rm ex}$ (top) and differential cross section (bottom) for elastic electron scattering from lithium atoms.

References

[1] K. Bartschat and S. Fonseca dos Santos, arXiv:1611.06180 (2016)
[2] G. Baum *et al.*, *Phys. Rev. Lett.* **57** (1986) 1855
[3] J. J. McClelland *et al.*, *Phys. Rev. A* **46** (1992) 6079
[4] K. Blum and B. Lohmann, *Phys. Rev. Lett.* **116** (2016) 033201
[5] B. Lohmann, K. Blum, and B. Langer, *Phys. Rev. A* **94** (2016) 032331

This work was supported by United States National Science Foundation under grant PHY-1403245.

[1]E-mail: klaus.bartschat@drake.edu
[2]E-mail: samantha.fonseca@drake.edu

Novel mechanism for creating long-lived metastable atomic negative ions

Zineb Felfli[†][1] **and Alfred Msezane**[†][2]

[†] Department of Physics and CTSPS, Clark Atlanta University, Atlanta, Georgia 30314, USA

Synopsis A novel mechanism is proposed for creating long-lived metastable atomic negative ions in complex atoms. It exploits the orbital collapse of the 5d orbital in Gd (Z=64) into the 4f orbital of Tb (Z=65). In the region of collapse the properties of the 5d and 4f orbitals are quite sensitive to the changes in the effective potential. Consequently the collapse phenomenon impacts significantly the core-polarization interaction in the relevant atoms. The mechanism is demonstrated in Tb and Dy through the appearance of long-lived Tb^- and Dy^- anions in the Regge-pole calculated electron elastic cross sections.

The impact of orbital collapse on the inner and outer wells and the potential barrier was discussed in [1, 2], while the importance of the d orbital collapse phenomenon enhancing the influence of the core-polarization has also been demonstrated in the isoelectronic sequence of the singly ionized Yb [1].

In the region of collapse the properties of the 5d and 4f orbitals are quite sensitive to the changes in the effective potential. Consequently the collapse phenomenon impacts the core-polarization interaction significantly in the relevant atom, namely Tb. As a result, a new excited state is induced in the Tb atom. Its TCS resembles that of the ground state TCS with a dramatically sharp resonance appearing in its second R-T minimum at 1.20eV, corresponding to the Tb^- metastable negative ion formation, in excellent agreement with the measured EA [3].

The collapsed 5d orbital in Gd leads to maximum polarization in the Dy atom. Consequently, the proposed mechanism of creating long-lived metastable negative ions should be observable in both the Z=65 and Z=66 atoms through the appearance of dramatically sharp resonances in the newly generated TCSs. The characteristic TCSs for Dy in the presence of the induced long-lived metastable state (blue curve) appearing at the energy of 1.37 eV are displayed in Fig. 1. We expect the mechanism to be general and realizable in other atomic systems where orbital collapse occurs. The isoelectronic sequences of Tb and Dy are also expected to exhibit long-lived metastable anionic TCSs.

Notable from the Fig. is that the metastable could easily be mistaken for the ground state resonance. Indeed, identifying and delineating the resonance structures is essential for understanding the mechanism of electron attachment.

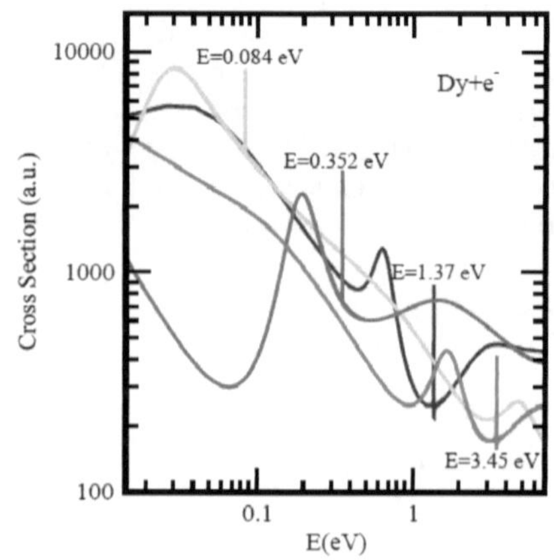

Figure 1: Total cross sections (a.u.) for electron elastic scattering from atomic Dy versus E (eV). The pink, blue, orange and green curves represent TCSs for the ground, induced metastable and first excited and second excited states, respectively. The dramatically sharp resonances correspond to Dy^- negative ions formation during the collisions.

We conclude by noting that the measured EA of Tb [3] actually corresponds to the metastable negative ion resonance Tb^- at 1.20 eV.

Research is supported by U.S. DOE, OBES, OS

References

[1] J. Migdałek and W. Siegel 2000 *Phys. Rev. A* **61** 062502
[2] J. P. Connerade and V. K. Dolmatov 1998 *J. Phys. B* **31** 3557
[3] V. T. Davis *et al* 2005 *NIMB* **241** 118
[4] Z. Felfli and A. Z. Msezane 2017 *J. Phys. B* (Submitted)

[1] E-mail: zfelfli@cau.edu

[2] E-mail: amsezane@cau.edu

Low-Cost Computer-Controlled Power Supplies for Optimization and Control of Electron Spectrometers

Ahmad Sakaamini*[2], Matthew Harvey*, Andrew James Murray*[1]

* Photon Science Institute, School of Physics & Astronomy, University of Manchester, Manchester M13 9PL, UK.

Synopsis New Arduino-based power supply designs are presented that have direct applicability to electron spectrometers such as the (e,2e) spectrometer that is used in Manchester. Voltages from these supplies can be manually selected using digital potentiometers, or can be remotely controlled using LabVIEW. These new control routines allow the apparatus to explore both ground state and laser-excited targets.

New computer controlled power supplies have been designed using an Arduino microcomputer as the interface between the power supplies and personal computer system, as shown in figure 1. The supply voltages to the experiment can either be adjusted manually using a set of digital potentiometers, or can be directly controlled by a personal computer using LabVIEW. All power supplies are optically isolated from ground, so that they can be used for either electron detection or ion detection by changing their relative common point.

The Arduino also monitors analogue signals from the apparatus to determine different experimental parameters. These range from the vacuum pressure through to the angles of the electron analyzers and electron gun. All signals to and from the Arduino to the control computer are via a USB interface.

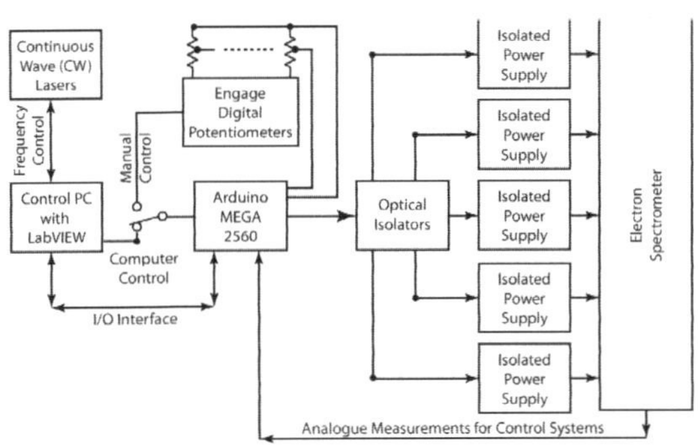

Figure 1. Block diagram of the new interface and power supplies that are currently being installed on the (e2e) spectrometer in Manchester.

The control computer can hence be used to optimize the spectrometer for different experiments, using the same type of Simplex routine that has been adopted in the (e,2e) spectrometer in Manchester since 1992 [1]. In this way the control systems operate and optimize the spectrometer 24 hours/day for long periods of time without user intervention. The new supplies are currently replacing the older supplies that have been operating successfully on this experiment for many years [2].

A further cost advantage of the new design is that the coincidence signal from the experiment is monitored by a software-built Multichannel analyzer (MCA) that operates under LabVIEW. The electron detectors are amplified using home-built preamplifiers [3], further saving on the cost of the instrument, and making any component failures easy to fix.

By modernizing the computer interface to the spectrometer, new experiments become possible, including ionization measurements from laser-excited atoms. These new experiments require the laser systems to be frequency controlled to ~1 part in 10^{10} for long periods of time, and this becomes possible by interfacing the laser control systems through the common LabVIEW hierarchy. Control of the lasers will be achieved by using the same optimization routines as adopted for the electron spectrometer, as shown in figure 1.

Full details of the new supplies and the interface will be presented at the conference, as well as experimental results that have been taken with the spectrometer using these new supplies.

References

[1] A. J. Murray *et al Rev. Sci. Inst.* **63** 3346 (1992)
[2] A. Sakaamini *et al J Phys. B* **49** 195202 (2016)
[3] A. J. Murray *Meas. Sci. Tech.* **23** 107001 (2012).

[1] E-mail: Andrew.Murray@manchester.ac.uk

[2] E-mail: Ahmad.Sakaamini@postgrad.manchester.ac.uk

IOP Conf. Series: Journal of Physics: Conf. Series **875** (2017) 052016 doi:10.1088/1742-6596/875/6/052016

Positronium formation in positron-hydrogen collisions in Debye plasma

Jia Ma[*,1], **Yuancheng Wang**[†2], **Yongjun Cheng**[§], **Yajun Zhou**[‡], **and Heng Wang**[*]

[*] College of Science, Shenyang Aerospace University, Shenyang, 110136, China
[†] College of Physics Science and Technology, Shenyang Normal University, Shenyang, 110034, China
[§] Natural Science Research Center, the Academy of Fundamental and Interdisciplinary Science, Harbin Institute of Technology, Harbin 150080, China
[‡] Institute of Atomic and Molecular Physics, Jilin University, Changchun, 130012, China

Synopsis Positronium (Ps) formation cross sections (n=1, 2) in positron-hydrogen collisions for various Debye lengths from the Ps formation thresholds to 40 eV are calculated using the screening approximation model, taking account of the influence of the plasmas screening on the Coulomb potentials and the structures of target states and Ps states. The influence of considering modified bound-state wave functions and eigenenergies for the Ps by the screened Coulomb potential on the Ps formation process is investigated.

Positronium formation is a special rearrangement process in positron-atom collision. Recently, the effects of Debye plasmas on this Ps formation process have received attention from many atomic physics research groups (see, e.g. Refs. [1, 2, 3] and references therein).

Screening approximation model was applied to investigate positron scattering on the hydrogen atom in Debye plasma, where interesting features were observed with the reduction of Debye length D [4]. In this work, Ps formation process is further investigated with the screening approximation model with the inclusion of the modified structure of Ps. Ps formation cross sections (n=1, 2) are calculated for various Debye lengths from the Ps formation thresholds to 40 eV. Consistent point is obtained with the previous calculation [4] that the effect of screening causes an enhancement in the values of the Ps formation cross sections at low energies. However, the increase is not substantial as the previous results in which the modified structure of the Ps in Debye plasmas was not taken into consideration.

In figure 1, a comparison is displayed between the previous results [4] and the present results. We observe that the inclusion of the modified structure of the Ps causes a reduction in the value of Ps formation cross section. Moreover, the value decreases increasingly as the screening length decreases. That is, it reduces the values of the Ps formation cross section at low energies and this influence is enhanced as the screening effect increases, whereas it can not counteract the enhancement of the Ps formation by the Debye screening. The screened Coulomb potential always increases the values of the Ps formation cross sections at low energies.

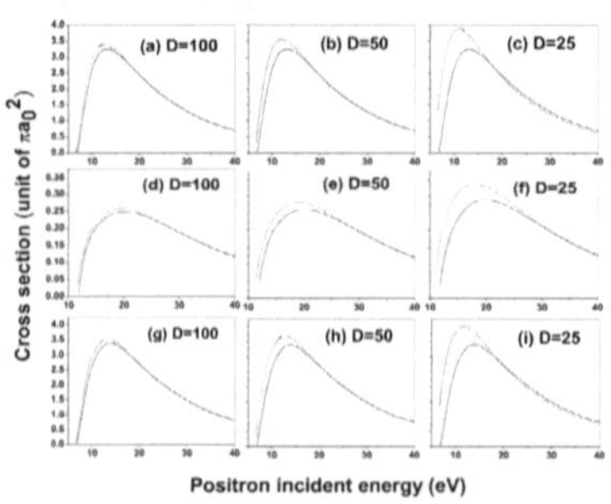

Figure 1. A comparison of positronium formation cross sections under different Debye lengths between the previous calculations [4] (dashed lines) and the present calculations (solid lines) with the inclusion of the modified structure of the Ps. (a)-(c): Ps (n=1) formation cross sections; (d)-(f): Ps (n=2) formation cross sections; (g)-(i): Ps (n=1+2) formation cross sections.

This work was supported by the National Natural Science Foundation of China (Grant Nos. 11404223, 11447158 and 11604223), and the Doctoral Program Foundation of Shenyang Aerospace University under Grant No. 13YB26. Authors are thankful to Dr. Kar for providing the Highly correlated bound-state wave functions and eigenenergies of the positronium.

References

[1] R. K. Janev *et al.* 2016 *Matter and Radiation at Extremes* **1** 237

[2] P. Rej *et al.* 2016 *J. Phys. B: At. Mol. Opt. Phys.* **49** 125203

[3] M. K. Pandey *et al.* 2016 *J. Phys. B: At. Mol. Opt. Phys.* **49** 034007

[4] J. Ma *et al.* 2012 *Phys. Plasmas* **19** 063303

[1]E-mail: majia@sau.edu.cn
[2]E-mail: rickywangyc@aliyun.com

Positronium formation in positron-lithium collisions with Debye potentials

Yuancheng Wang[†1], **Jia Ma**[*2], **Liguang Jiao**[‡], **and Yajun Zhou**[§]

[†] College of Physics Science and Technology, Shenyang Normal University, Shenyang, 110034, China
[*] College of Science, Shenyang Aerospace University, Shenyang, 110136, China
[‡] College of Physics, Jilin University, Changchun, 130012, China
[§] Institute of Atomic and Molecular Physics, Jilin University, Changchun, 130012, China

Synopsis Positronium (Ps) formation processes in positron-lithium collisions in Debye plasma environments are investigated using the screening approximation model. We present the Ps(1s) formation cross sections in the incident energy range 2-40eV for screening parameters μ=0.00-0.25.

Recent years, atomic collision processes in plasma environments have received considerable attention[1]. The interactions among the charged particles in weakly coupled plasmas have been reduced to Debye-Huckel potentials. The ambient plasma temperature and its density are related to the screening parameter μ. The eigen energies and wave functions of the atomic systems are modified depending on the plasma screening parameter. Under the influence of no external environment, a number of studies have been conducted on the problem of positron-lithium collisions. However, in plasma environments such calculations are limited.[2]

We are reporting the Ps formation cross sections in positron-lithium collisions of the screening approximation model, which has been used in studying of positron-hydrogen collisons[3]. A complex polarization potential has been derived in describing the Ps formation in positron-lithium system. We have considered modified bound-state energies and wave functions for lithium due to Coulomb screening determined by the parameter μ.

In Figure 1 Ps(1s) formation cross sections for incident positron energies in the range 2-40eV are presented, for different values of the screening parameter μ=0.00-0.25. The Ps formation cross section decreased at larger values of the screening parameter μ, in this incident energy range. The screening effect on the electron capture probability decreases as the projectile energy increases.

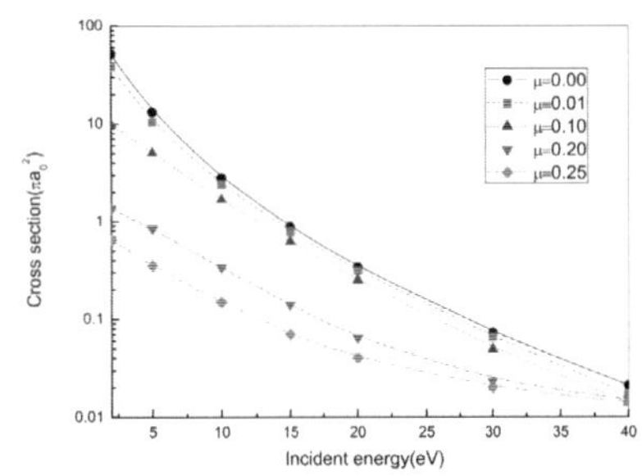

Figure 1. Positronium formation(1s) cross sections under different screening parameters. solid lines, μ=0.00; dash lines, μ=0.01; dot lines, μ=0.10; dash dot lines, μ=0.20; dash dot dot lines, μ=0.25.

So far, the core polarization term has not been include in these calculations. We are planning to include this term to determine the effect on low energy Ps formation.

This work was supported by the National Natural Science Foundation of China (Grant Nos. 11404223, 11447158 and 11504128), and the Doctoral Program Foundation of Shenyang Normal University.

References

[1] R. K. Janev *et al.* 2016 *Matter and Radiation at Extremes* **1** 237

[2] P. Rej *et al.* 2016 *J. Phys. B: At. Mol. Opt. Phys.* **49** 125203

[3] J. Ma *et al.* 2012 *Phys. Plasmas* **19** 063303

[1]E-mail: rickywangyc@aliyun.com
[2]E-mail: majia@sau.edu.cn

IOP Conf. Series: Journal of Physics: Conf. Series **875** (2017) 052018 doi:10.1088/1742-6596/875/6/052018

Cross sections for electrons scattering from silver at low energies

Harsh Mohan

Department of Physics, M. L. N. College, Yamuna Nagar, 135 001, Haryana, India

Synopsis Differential cross sections for scattering of electrons from silver atoms are calculated in the low energy region using the relativistic Dirac equation. The projectile target interaction is represented as sum of three local and real terms i.e. the static, a parameter free correlation polarization potential and the modified semi classical exchange potentials. We compare our results for differential cross sections with the available calculations and experimental measurements.

Scattering of electrons with metallic atoms provides useful knowledge for understanding their structures and reaction mechanism in terms of cross sections [1, 2]. This information is of great value in the field of astrophysics, radiation physics, plasma and biomedical sciences. In the present study, we have considered electron scattering with silver atom in low energy region. The reason for choosing silver as a target atom is that it has the highest thermal and electrical conductivity with the number of applications. It is a precious metal that has been used for many years to make fine pieces of jewellery and coins.

Our goal here is to use the relativistic Dirac equation for calculating the differential cross sections (DCS) at energies 10, 15, 20, 30 and 100 eV. The motion of the projectile electron in a central field $V(r)$ is described by the Dirac equation given as:

$$[c\alpha.p + \beta m_0 c^2 + V(r)]\psi = E\psi$$

where $E = m_0\gamma c^2 = E_i + m_0 c^2$ is the total energy of incident electron, $\gamma = (1 - v^2/c^2)^{-1/2}$ and E_i is the kinetic energy of incident electron. α and β are the 4 x 4 Dirac matrices. The spinor ψ has four components and $\psi = (\psi_1, \psi_2, \psi_3, \psi_4)$, where (ψ_1, ψ_2) are the large components and (ψ_3, ψ_4) are the small components of ψ. The total interaction between an electron and the target atom is approximately represented by an effective potential which can be expanded in terms of symmetry-adapted function of the A_1 irreducible representation (totally symmetric).

$$V_{\text{eff}}(r) = \sum_{l,h} V_{lh}(r)\, X_{lh}^{A_1}(r)$$

In spherical approximation, we need only the first term (l = 0, h = 1) in the expansion of above equation, in order to evaluate all potential terms. The details of the potentials i.e. the static (V_{st}), the exchange (V_{ex}) and the polarization (V_{pol}) potentials are given in our earlier work [3]. We have displayed our DCS results at 100 eV (in Figure 1) along with other calculations and measurements. Present results are in good agreement with the experimental measurements compared to other calculations. Detailed results will be discussed in the conference.

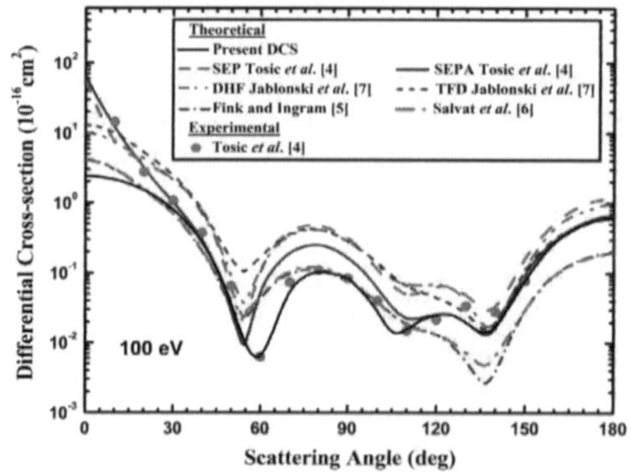

Figure 1. Differential cross section for e – Ag scattering at 100 eV.

E-mail: mohan_harsh@yahoo.com

References

[1] Z. Felfli *et al.* 2011 *J. Phys.* B: *At. Mol. Opt. Phys.* **44** 135204.
[2] R. P. McEachran and A. D. Stauffer 2009 *J. Phys.* B: *At. Mol. Phys.* **42** 075202.
[3] G. Kaur *et al.* 2015 *Phys. Rev.* A **91** 022702.
[4] S. D. Tosic *et al.* 2009 *Nucl. Instr. Meth. B* **267** 283.
[5] M. Fink and J. Ingram 1970 *At. Data Nucl. Data Tables* **4** 129.
[6] F. Salvat *et al.* 2005 *Comp. Phys. Comm.* **165** 157.
[7] A. Jablonski *et al.* 2004 *J. Phys. Chem. Ref. Data* **33** 409

IOP Conf. Series: Journal of Physics: Conf. Series **875** (2017) 052019 doi:10.1088/1742-6596/875/6/052019

Vortex rings in the ionization of atoms by positron impact

F. Navarrete[*,†], **R. O. Barrachina**[*,†1]

[*]Centro Atómico Bariloche and Instituto Balseiro (Comisión Nacional de Energía Atómica and Universidad Nacional de Cuyo), Avda. Bustillo 9500, Bariloche, Argentina.
[†] Consejo Nacional de Investigaciones Científicas y Técnicas (CONICET), Argentina.

Synopsis We uncovered the morphology of quantum vortex structures in positron-atom ionization collisions. By performing an exhaustive calculation of the position of vortices in the final momentum space of the problem with only one kinematic restriction, we could elucidate that quantum vortices form a closed structure, which is one of the two possible scenarios for their emergence in three dimensions. This is the first time that these structures are fully calculated for positron-atom collisions.

Quantum vortices have been subject of both theoretical and experimental interest in the last decade for the ionization of atomic and molecular targets by the impact with proton, electron and positron projectiles [1, 2, 3]. They are submanifolds of co-dimension 2 on the phase space of the T-matrix element of the collision. This means that they appear as points in two dimensions, and as lines in three dimensions. The *velocity field* associated with T [4] exhibits a solenoidal shape similar to that of irrotational vortices in *Fluid Dynamics*.

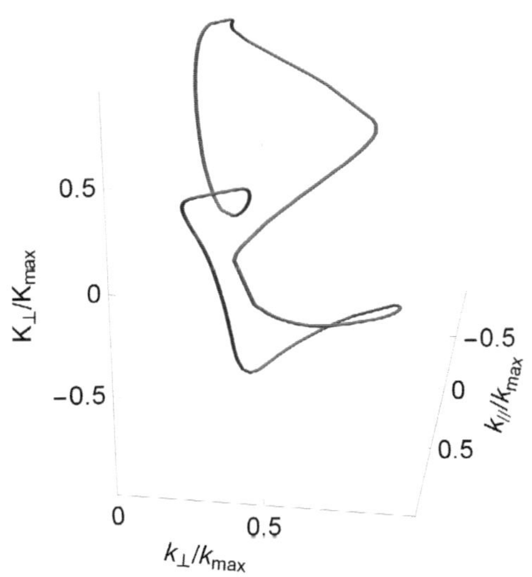

Figure 1. Vortex ring on the T-matrix element for the ionization of hydrogen by positron impact at an energy of 275 eV in coplanar geometry. It is represented in the phase space of the final momenta of the electron $k_{//}$ and $k_\perp$, and the positron $K_\perp$ (see the text), normalized by the maximum momenta that the particles can acquire due to energy-momentum conservation.

A previous systematic study [3] of the ionization of atoms for different positron impact energies, contributed to the understanding of how these structures emerge in a *collinear geometry* as pairs of isolated points with a velocity field of opposite circulation to each other. In a subsequent work [5] we analysed the same problem for a fixed collision energy in a *coplanar geometry* which proved that many structures that seemed isolated when studied in two dimensions where in fact part of a single *vortex line*.

In the current study, we deepen the analysis of the aforementioned vortex line for the whole coplanar phase space. We were able to fully trace the vortex structure on the phase space of the final components of the electron momentum $k_{//}$ and $k_\perp$, parallel and perpendicular (with respect to the initial direction of the projectile) and the perpendicular component of the final positron momentum $K_\perp$. As it can be seen in the figure, which corresponds to a positron impact energy of 275 eV, it forms a closed structure, named *vortex ring*, which is one of the possible scenarios for their emergence [6].

References

[1] J. H. Macek *et al.* 2010 *Phys. Rev. Lett.* **104** 033201

[2] Schmidt *et al.* 2010 *Phys. Rev. Lett.* **112** 083201

[3] F. Navarrete, R.O. Barrachina 2015 *J. Phys. B: At. Mol. Phys.* **48** 055201

[4] F. Navarrete, R. Della Picca, J. Fiol and R. O. Barrachina 2013 *J. Phys. B: At. Mol. Phys.* **46** 115203

[5] F. Navarrete, R.O. Barrachina 2016 *Nucl. Instrum. Methods. B* **369** 72

[6] I. Bialynicki-Birula *et al. Phys. Rev. A.* **61** 032110

[1]E-mail: barra@cab.cnea.gov.ar

IOP Conf. Series: Journal of Physics: Conf. Series **875** (2017) 052020 doi:10.1088/1742-6596/875/6/052020

The AC-MOT Cold Atom Electron Source (CAES)

Michael Jones*[1], Matthew Harvey*, William Bertsche*[†], Guoxing Xia*[†], Swapan Chattopadhyay[‡], Andrew James Murray*[2] and Robert Appleby*[†]

[*] Photon Science Institute, School of Physics & Astronomy, University of Manchester, Manchester M13 9PL, UK.
[†] The Cockcroft Institute, Daresbury Laboratory, Sci-Tech Daresbury, Daresbury, Warrington, WA4 4AD UK.
[‡] Department of Physics, Northern Illinois University, 1425 W. Lincoln Hwy DeKalb, IL, USA 60115.

Synopsis. A new cold atom electron source based on an AC-driven magneto-optical trap (AC-MOT) is described for electron diffraction studies. The advantages of the new source which uses an AC-MOT are detailed.

Progress towards observing ultra-fast dynamics on atomic length scales is limited by the achievable brightness, or current per unit solid angle, of available electron beams. The brightness of an electron beam is related to its characteristic temperature [1], and hence by reducing the temperature of the electron beam, higher brightness beams can be produced.

In a cold atom electron source (CAES), cold atoms held in a magneto-optical trap (MOT) are photo-ionized near threshold by a laser. This process yields electrons with an energy spread on the order of 1 meV, corresponding to a characteristic temperature of ~10 K [2]. Electrostatic fields are then used to extract the electrons into a bunched beam for delivery to experiments.

The magnetic **B**-fields found in a conventional MOT perturb the electron beam trajectory and its quality. The new source described here uses an AC-MOT [3] to ensure the trapping fields are zero at the time of electron extraction. This allows field-free electron extraction whilst retaining a high source density. In the AC-MOT, the trapping laser polarization is switched synchronously with an alternating **B**-field. Once the **B**-fields are zero, the atoms are ionized using pulsed, two colour photo-ionisation, and an electron beam is extracted.

To determine the brightness B of the CAES electron beam, its beam current **J** and characteristic temperature T are measured, since

$$B = \frac{\mathbf{J}}{4\pi}\frac{mc^2}{k_B T}.$$

The charge and temporal profile of the CAES electron bunches are measured using a charge pick-off circuit from a multi-channel plate (MCP). Figure 1 shows the cold atom source and the electrostatic extraction and acceleration optics. These consist of trapping coils, acceleration and focusing electrostatic optics, and an MCP and phosphor screen. The temperature of the beam is measured by the expansion of the electron bunch over the ~500 mm flight path, as determined from the spot size imaged using the MCP relative to the initial ionization volume. This expansion is linearly dependent on the initial divergence of the beam. When combined with the beam energy, the temperature can hence be estimated [1].

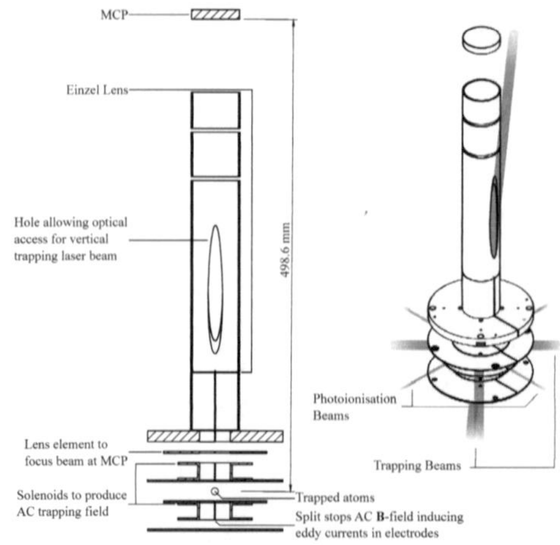

Figure 1. (Main): Cross-section of the CAES apparatus. Electrons are extracted upwards by biasing the MOT plates. (Insert): 3-D render showing the trapping (red) and ionization (blue, orange) laser beams.

A description of the source, temperature measurements, and the progress towards initial electron diffraction studies will be presented at the conference.

References

[1] W Engelen *et al.* 2014 *Ultramicrosc.* **136** 73 80
[2] AJ McCulloch *et al.* 2011 *Nat. Phys.* **7** 7 785
[3] M Harvey *et al.* 2008 *Phys Rev Lett* **101** 173201

[1] E-mail: Michael.Jones@manchester.ac.uk
[2] E-mail: Andrew.Murray@manchester.ac.uk

IOP Conf. Series: Journal of Physics: Conf. Series **875** (2017) 052021 doi:10.1088/1742-6596/875/6/052021

Study of multiply charged Argon-ions formed by decay of 2p-hole state under e^-- Ar collision employing energy selected ion coincidence technique

Sunil Kumar[1], Suman Prajapati, Bhupendra Singh, B.K.Singh, R.Shanker[2]

Atomic Physics Laboratory, Department of Physics, Institute of Science, Banaras Hindu University, Varanasi. India -221005

Synopsis In this presented work we have performed the energy selected electron-ion coincidence technique to investigate the formation of multiply charged argon ions and determined correlation probabilities for individual charge state of formed ions and several other features of the employed technique is presented and discussed.

Inner shell de-excitation of atoms and molecules give rise to rearrangement of internal electron distribution, for light atoms non-radiative and for heavier atoms radiative transition are dominant de-excitation mechanisms. Correlated many-particle dynamics in Columbic systems is one of today's grand challenges in physics. Multi-ionization studies on atoms and molecules provide information on the correlation between the electrons and also on the dynamics of the ionization processes [1]. There are various applications of such study particularly in the field of atomic, molecular and optical (AMO) physics, atmospheric physics, plasma physics and many more fields.

In the present work we have employed energy selected electron-ion coincidence technique to study the decay pathways of 2p hole created under the impact of continuous electron beam having kinetic energy 3.5 keV. Several non-radiative inner shell transitions e.g. Auger process, Auger cascade transition, Coster-Kronig, auto-ionization, shake processes etc. have been investigated which lead to the formation of multiply charged Argon ions from Ar^+ to Ar^{4+} for 18 energy selected electrons (see Fig.1). The relative correlation probability for the formation of various charge state of Argon ions as a function of electron energy have been determined experimentally.

Several other aspects and implications of the formation of multiply charged Argon ions using our recently developed experimental set-up [2] will be presented and discussed.

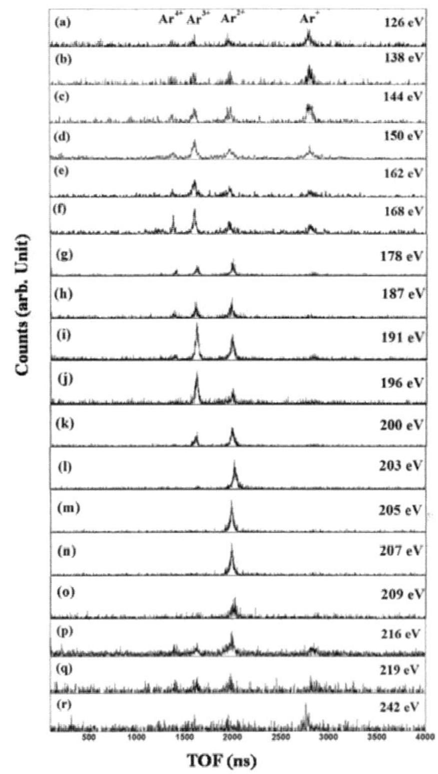

Figure 1 : TOF spectra of multiply charged Argon ions observed in coincidence with 18-energy selected electrons in 3.5 keV electron collisions with free argon atoms [3].

References

[1] Egil Andersson, 2010, Faculty of Science and Technology, Uppsala Dissertations 743.

[2] S. Kumar et al., 2017 (Accepted, Ind. J. Phys)

[3] S. Kumar et al., 2017 (Accepted, EPJD)

[1]E-mail: sunilphy88@gmail.com
[2]E-mail: shankerorama@gmail.com

IOP Conf. Series: Journal of Physics: Conf. Series **875** (2017) 052022 doi:10.1088/1742-6596/875/6/052022

Dielectronic Recombination of Tungsten Ions

Bowen Li*[1], Gerry O'Sullivan†, Chenzhong Dong‡, Yang Li*, Xiaokai Xu*, Ximeng Chen*

* School of Nuclear Science and Technology, Lanzhou University, Lanzhou 730000, China
† School of Physics, University College Dublin, Belfield, Dublin 4, Ireland
‡ Key Laboratory of Atomic and Molecular Physics & Functional Materials of Gansu Province, College of Physics and Electronics Engineering, Northwest Normal University, Lanzhou, 730070, China

Synopsis *Ab-initio* calculations of dielectronic recombination rate coefficients of Ne-, Pd-, Rh- and Ag-like tungsten have been performed. The present calculated rate coefficients are compared with other calculations where available. A large discrepancy was found for Pd-like W, which implies that more *ab initio* calculations and experimental measurements are badly needed. The data obtained are expected to be useful for modelling plasmas for fusion applications, especially for the ITER community.

The DR process plays an important role in high-temperature plasmas, where it affects both the ionization balance and radiative energy losses. Tungsten is being considered as a plasma-facing component (PFC) in magnetically confined fusion devices, such as ITER, because of its low sputtering rate, high temperature characteristics and low tritium absorption, so atomic data for tungsten has become of major importance [1, 2]. A significant number of publications on computational DR rate coefficients of have appeared for tungsten during the past decade. DR rate coefficients from the atomic data and analysis structure (ADAS) databases are currently the most widely used to model various fusion plasmas in the nuclear fusion community. The computations of such DR rate coefficients are based on the semiempirical Burgess-Merts formulas and the generalized collisional-radiative framework. The accurancy of such rates is another problem.

We carried out a detailed calculation on dielectronic recombination rate coefficients of Rh- [3], Ne-, Pd- and Ag-like W [4]. As an example, only status of Pd-like W (W^{28+}) is summarized in Fig. 1. For W^{28+}, which has a ground level [Kr]$4d^{10}$, The doubly excited configurations $4d^9 4fnl$ $(4d-4f)$, $4d^9 5l'nl$ $(4d-5l')$ and $4d^9 6l'nl$ $(4d-6l')$ as well as $4p^5 4d^{10} 4fnl$ $(4p-4f)$ and $4p^5 4d^{10} 5l'nl$ $(4p-5l')$ were included in the calculations. Energy levels, radiative transition probabilities and autoionization rates were calculated up to $n \leq 18$ and $l \leq 6$, and the contributions from higher-n levels were extrapolated up to $n = 1000$. Calculated radiative recombination (RR) rate coefficients obtained by Trzhaskovskaya *et al.* [5] are also presented for comparison. There is thus no doubt that the DR process can have a greater influence than RR on plasma balance. An *ab-initio* calculation by Safronova *et al.* [6] using the Cowan code is also available, but the total rate coefficient predicted from this calculation is about ten times smaller than that of by Foster [7], Pütterich *et al.* [8] and the FLYCHK [9] code, as well as present FAC calculation at energies over 500 eV. This is most likely due to omission of significant resonances [4]. Considering the current status of these discrepancies, further studies, especially experimental measurements of DR cross-sections are urgently needed.

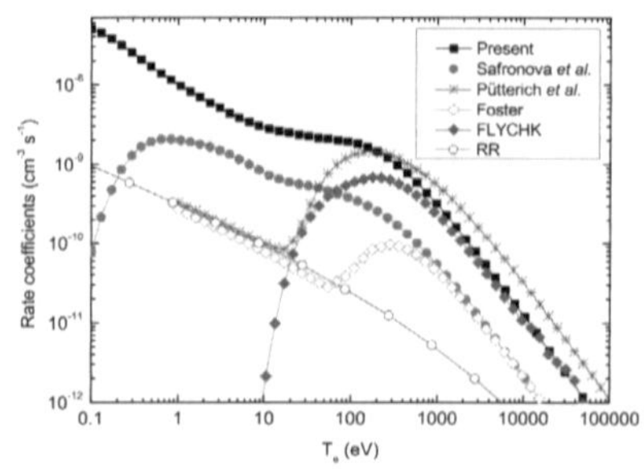

Figure 1. Total DR rate coefficients as a function of Te in Pd-like W.

References

[1] P. Beiersdorfer, *et al.* 2015 *Atoms*. **3** 260

[2] A. Müller, *et al.* 2015 *Atoms*. **3** 120

[3] B. Li, *et al.* 2012 *Phys. Rev. A*. **85** 052706

[4] B. Li, *et al.* 2016 *J. Phys. B*. **49** 155201

[5] M. B. Trzhaskovskaya, *et al.* 2013 *At. Data Nucl. Data Table*. **99** 249

[6] U. I. Safronova, *et al.* 2012 *J. Phys. B*. **45** 085001

[7] A. Foster, PhD thesis, University of Strathclyde, 2008.

[8] T. Pütterich, *et al.* 2008 *Plasma Phys. Control. Fusion*. **50** 085016

[9] *http://www.hydkin.de/plotform/index_tungsten.html*

[1]E-mail: libw@lzu.edu.cn

IOP Conf. Series: Journal of Physics: Conf. Series **875** (2017) 052023 doi:10.1088/1742-6596/875/6/052023

Absolute excitation-emission cross section of electron induced argon excitation

Juraj Országh[1], **Michal Ďurian, Štefan Matejčík**

Dept. of Experimental Physics, Comenius University in Bratislava, Mlynská dolina F2, 842 48 Bratislava, Slovakia.

Synopsis The electron impact excitation-emission cross section of argon $4P^2[1/2] \rightarrow 4S^2[1/2]°$ [1] was experimentally investigated in the electron energy range 5eV – 100eV. The absolute value of the cross section was calibrated by comparison with the known cross section for the helium transition He $3^1S \rightarrow 2^1P^0$ [2].

The argon deexcitation $4P^2[1/2] \rightarrow 4S^2[1/2]°$ produces one of the strongest argon emission lines (750.39 nm) and the absolute value of the cross section for electron-atom reaction is often utilized in optical emission spectroscopy of electrical discharges. The cross section value was re-evaluated experiementaly in this work.

The experiment was conducted on Electron Induced Fluorescence (EIF) apparatus (described in detail in [3]) utilizing crossed beams of monochromatic electrons and atoms/molecules. The electron beam is formed by thermal emission from hairpin tungsten filament and subsequently it is energy filtered by trochoidal electron monochromator. For this experiment the electron energy resolution was set to ~200 meV (FWHM) and the electron current was in the range of 600 – 800 nA. Argon beam was created by effusion through a capillary. The background pressure in vacuum chamber was set to ~10^{-4} mbar to ensure single-collision conditions. Photons emitted from excited atoms were collected by an optical system, consisting of collimating optics, Jobin Yvon THR1500 Czerny-Turner monochromator and thermoelectrically cooled Hamamatsu R3869 photomultiplier working in photon counting mode with background noise averaging at 3.6 cps, primarily due to photomultiplier thermal noise.

The narrow spectral range around the expected argon line was measured at several different electron energies (25 eV, 50 eV and 100 eV) well above the excitation limit for the process in question at high optical resolution to ensure there are no other spectral lines present in the vicinity that might influence the measured cross section. Identical procedure was used to measure photon efficiency curve of the He 728.13 nm line. The pressure in the effusive capillary was kept constant for both measurements ensuring equal sample densities in the reaction region. Electron energy step was 0.5 eV before the excitation threshold, 0.1 eV around the threshold and cross section maximum and 1 eV for the rest of the curve.

Process of calculating the absolute cross section has multiple steps, as there are some properties of the experiment for which we have to make corrections, specifically pressure, electron current and spectral sensitivity corrections.

For the argon 4P2[1/2] → 4S2[1/2]° transition the main peak and broader secondary peak was observed. It is typical for cross sections with contribution of cascade processes from the higher excited states, in addition to direct excitation. This phenomenon can be observed also at the rising slope of the cross section, which exhibits multiple threshold structure. This will be a subject of further study using higher electron beam resolutions and also more complex numerical fits of the rising edge.

The values of the argon cross section were determined at electron energies 25 eV, 50 eV and 100 eV along with the value of excitation threshold. Experimental uncertainties were estimated by taking into account the statistical uncertainties of the measurements, pressure and current uncertainties and finally the uncertainty of helium reference cross section.

Acknowledgments

This work was supported by the Slovak Research and Development Agency, project Nr. APVV-15-580 and received funding from the European Union's Horizon 2020 research and innovation programme under grant agreement No 692335.

References

[1] J. B. Boffard *et al.* 2007 *Atomic Data and Nuclear Tables* **93** 831
[2] B. Van Zyl *et al.* 1980 *Physical Review A* **22** 5.
[3] M. Danko *et al.* 2013 *J. Phys. B: At. Mol. Opt. Phys.* **46**.

[1] E-mail: juraj.orszagh@uniba.sk

IOP Conf. Series: Journal of Physics: Conf. Series **875** (2017) 052024 doi:10.1088/1742-6596/875/6/052024

Measurement of the Integrated Stokes Parameters for Zn 468 nm Florescence Excited by Polarized-Electron Impact

N. B. Clayburn[1] **and T. J. Gay**[2]

Jorgensen Hall, University of Nebraska, Lincoln, NE 68588-0299, USA

Synopsis These measurements resolve a standing inconsistency between experiment and state-of-the-art theory for Zn atom excitation by spin-polarized electrons. Our results are consistent with theory but not previous experiments.

We have measured the integrated Stokes parameters P_1, P_2, and P_3 (shown schematically in Fig. 1) of Zn ($4^3P_0 - 5^3S_1$) 468 nm florescence resulting from transversely-spin-polarized electron impact excitation of the Zn $(4s5s)5^3S_1$ state. This work was motivated by similar studies reported several years ago by Pravica *et al.*[1], in which they measured non-zero values of the integrated Stokes parameter P_2 between the threshold for the $(4s5s)5^3S_1$ excitation and the first cascading $(4s5p)5^3P_J$ threshold.

In our experiment, the electrons scattered in the excitation process are not measured (hence the designation of the Stokes parameters as "integrated"), and the fluorescence is observed in the direction of the initial electron spin polarization. For this geometry (Fig. 1), Bartschat and Blum [2] have shown that P_2 must be identically zero, based on the symmetry properties of the 9j symbol used in the algebra required to describe the dipole fluorescence, and the assumption that the L and S angular momenta of the total wavefunction (atom plus incident electron) are factorizable throughout the scattering process. This assumption is invalid if (a) the excited state producing the fluorescence is not well LS-coupled or (b) the spin of the continuum electron precesses during the collision under the influence of a motional magnetic field, i.e., the electron undergoes Mott scattering. Since both possibilities are ruled out by state-of-the-art theory [3,4], which predicts P_2 values of order 10^{-4}, the results of reference [1], which are as large as 10^{-1}, are remarkable.

We used a standard GaAs polarized electron source to produce beams of electrons with a polarization of 0.25(1) and an energy width of *ca.* 400 meV. The atomic Zn target was produced by a Zn oven and a heated effusive channel that directed an atomic beam at right angles to both the fluorescence observation direction and the electron beam axis.

We observe optical excitation functions in agreement with those of Kontrosh *et al.* [5]. In the cascade-free range of excitation between 6.7 eV and 7.6 eV, we have measured P_2 at 7.0 eV and 7.3 eV and find its values in this range to be consistent with zero and inconsistent with those measured in [1].

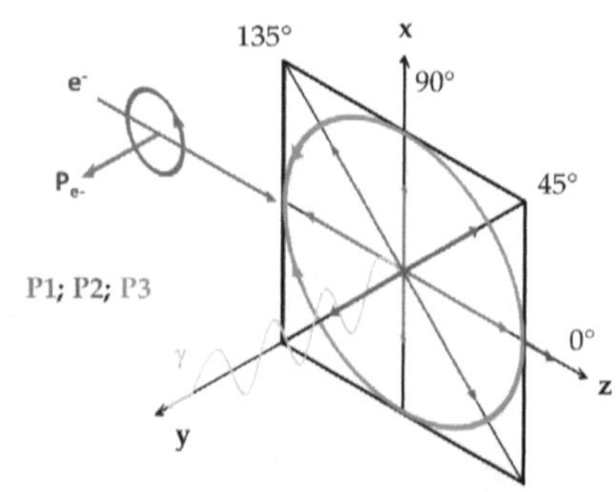

Figure 1. Geometry of integrated Stokes parameter measurements when transversely-polarized electrons, incident along $\hat{\mathbf{z}}$, excite atoms, and the subsequent fluorescence is observed in the $\hat{\mathbf{y}}$ direction.

Work supported by the USA NSF Award PHY-1505794.

References

[1] L. Pravica *et al* 2011 Phys. Rev. A **83** 040701

[2] K. Bartschat and K. Blum 1982 Z. Phys. A **304** 85

[3] C. J. Bostock *et al* 2013 Comments Phys. Rev. A **87** 016701

[4] K. Bartschat and O. Zatsarinny 2013 Comments Phys. Rev. A **87** 016702

[5] E. É. Kontrosh *et al* 2001 Opt. Spectrosc. **90** 339

[1] E-mail: nathan.clayburn@huskers.unl.edu

[2] E-mail: tgay1@unl.edu

IOP Conf. Series: Journal of Physics: Conf. Series **875** (2017) 052025 doi:10.1088/1742-6596/875/6/052025

The Tungsten Project: A complete set of isonuclear dielectronic recombination rate coefficients for use in magnetically confined fusion plasmas

S. P. Preval[*1], **N. R. Badnell**[*2], **M. G. O'Mullane**[*3]

[*] Department of Physics, University of Strathclyde, Glasgow, G4 0NG, United Kingdom

Synopsis A requirement of collisional-radiative models of magnetic fusion plasmas is that a detailed set of atomic data for the constituent ions is included. For the upcoming fusion reactor ITER, this is a difficult requirement to fulfill as the divertor will be made of tungsten. For this metal, partial final-state resolved dielectronic recombination (DR) rate coefficients are required. For this purpose we created *The Tungsten Project*. We give a brief overview of the methods used, and present results for a revised steady-state ionization balance.

The upcoming experimental reactor, ITER, is due to produce its first plasma in 2020. The divertor is a plasma-facing component, and will be composed of Tungsten. Tungsten has been chosen for its ability to withstand large power loads. As a plasma facing component, tungsten impurities will be introduced into the main body plasma through sputtering. The high atomic number of tungsten means it is an excellent radiator, which can result in the plasma being cooled, and potentially quenched. Due to the plasma densities involved, understanding the impact of tungsten impurities in the plasma requires detailed collisional-radiative modelling. This in turn requires partial final-state resolved DR rate coefficients for the complete isonuclear sequence. Multiple efforts resulted, including that of Pütterich et al. [1], who used an average-ion method to calculate recombination rate coefficients for the sequence.

To address the need for partial final-state resolved DR rate coefficients for tungsten, we created *The Tungsten Project* [2, 3]. The project used the distorted-wave code AUTOSTRUCTURE [4] to calculate the data using κ-averaged relativistic wavefunctions in level (IC), term (LS), and configuration (CA) resolution. The data are freely available on the Atomic Data Analysis Structure (Open-ADAS) website (http://open.adas.ac.uk) in the standard adf09/48 formats, the definitions of which can also be found on Open-ADAS.

In Figure 1, we have plotted the ionization fractions calculated using Putterich et al's [1] recombination rate coefficients, and the present data. For our fraction, we have replaced Pütterich's data with ours for ionization stages W^{74+}–W^{38+}. In both cases, we used the ionisation rate coefficients of Loch et al. [5]. It can be seen that by using the present data, there are significant shifts in both the peak fractional abundance, and peak abundance temperature.

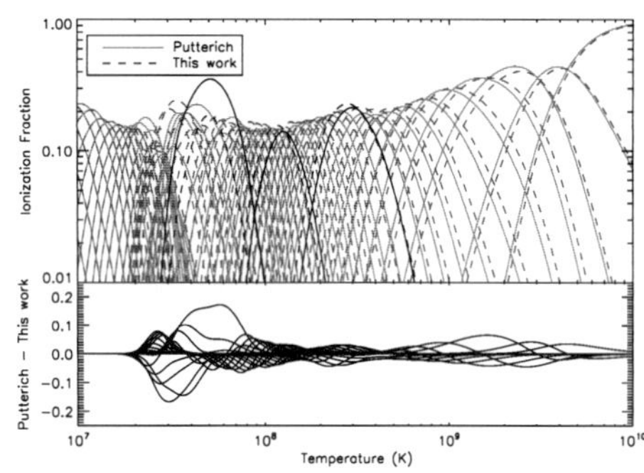

Figure 1. Comparison of ionization fractions calculated using Pütterich et al's [1] data (solid red), and the present data (blue dash) for W^{74+}–W^{38+}. The bottom plot shows the arithmetic difference between the two fractions.

With the completion of the tungsten project attention now turns to modelling efforts. The data described will be used not only in models of the ITER plasma, but will also help inform the construction of the first commercial demonstration fusion plant, DEMO.

References

[1] Pütterich T, Neu R, Dux R, Whiteford A D, O'Mullane M G, ASDEX Upgrade Team, 2008, *Plasma Phys. Control. Fusion*, **50**, 8

[2] Preval S P, Badnell N R, O'Mullane M G, 2016, *Phys. Rev. A*, **93**, 042703

[3] Preval S P, Badnell N R, O'Mullane M G, 2017, *Submitted to J. Phys. B*

[4] Badnell N R, 2011, *Comput. Phys. Commun.*, **182**, 1528

[5] Loch S D, Ludlow J A, Pindzola M S, Whiteford A D, Griffin D C, 2005, *Phys. Rev. A*, **72**, 052716

[1]E-mail: simon.preval@strath.ac.uk
[2]E-mail: badnell@phys.strath.ac.uk
[3]E-mail: martin.omullane@strath.ac.uk

IOP Conf. Series: Journal of Physics: Conf. Series **875** (2017) 052026 doi:10.1088/1742-6596/875/6/052026

Dielectronic resonances in highly-charged heavy ions observed in ion traps

A. Borovik, Jr.[*†‡1], **J. M. Dreiling**[*], **R. Silwal**[*§], **Dipti**[*], **E. Takács**[*§], **J. Gillaspy**[*¶], **R. Lomsadze**[†#], **V. Ovsyannikov**[‡], **K. Huber**[‡], **S. Schippers**[†], **A. Müller**[‡], **and Yu. Ralchenko**[*]

[*] Quantum Measurement Division, National Institute of Standards and Technology, Gaithersburg MD, USA
[†] I. Physikalisches Institut, Justus-Liebig-Universität Gießen, 35392 Giessen, Germany
[‡] Institut für Atom- und Molekülphysik, Justus-Liebig Universität Gießen, 35392 Giessen, Germany
[§] Department of Physics and Astronomy, Clemson University, Clemson SC, USA
[¶] Division of Physics, National Science Foundation, Arlington VA, USA
[#]Department of Physics, Ivane Javakhishvili Tbilisi State University, Tbilisi, Georgia

Synopsis Series of inner-shell dielectronic resonances were measured in the 7-20-keV x-ray spectra of highly charged tungsten and iridium ions produced in ion traps by highly dense electron beams. A detailed analysis of the spectra with a non-Maxwellian collisional-radiative model allowed us to reliably assign the experimentally observed features.

Satisfactory understanding of ionized-matter environments ranging from hot plasmas in laboratory devices to ionized gases in interstellar clouds in outer space requires reliable data on elementary atomic processes. Such data are routinely calculated using advanced methodes and codes that may provide good accuracy for simple atoms and ions. However, for multi-electron atomic systems, the accuracy of general theoretical approaches may not be sufficient to adequately interpret the observations. This calls for benchmark measurements of spectra from various multi-charged ions, in particular, those of heavy elements for which available spectroscopic data is very limited.

We present the x-ray spectra of highly-charged tungsten and iridium ions measured with the Electron-Beam Ion-Trap (EBIT) at the National Institute of Standards and Technology [1] and with the Main Magnetic Focus Ion Trap (MaMFIT) at the Justus-Liebig-Universität Gießen [2], respectively. The data are shown in Fig. 1. The NIST EBIT spectra (*a*), measured with a high-purity Ge detector at the electron beam energies between 6.44 and 11.27 keV, reveal series of resonances corresponding to LMn and LNn inner-shell dielectronic recombination (DR) involving stabilizing transitions $2\ell - 3\ell'$ and $2\ell - 4\ell'$ in Na-like through Ti-like tungsten. The spectra from the MaMFIT (*b*), measured at energies of 7.1 to 8.65 keV, show resonances corresponding to DR processes involving transitions between $2p - 3d$ subshells in K-like through Ni-like iridium. Detailed modeling of the observed spectra was performed employing the non-Maxwellian collisional radiative code NOMAD [3] which accounts for a large number of atomic levels and all important physical processes (radiative and collisional). Comparisons of the experimental and theoretical results show a good agreement for both the W^{q+} and Ir^{q+} spectra.

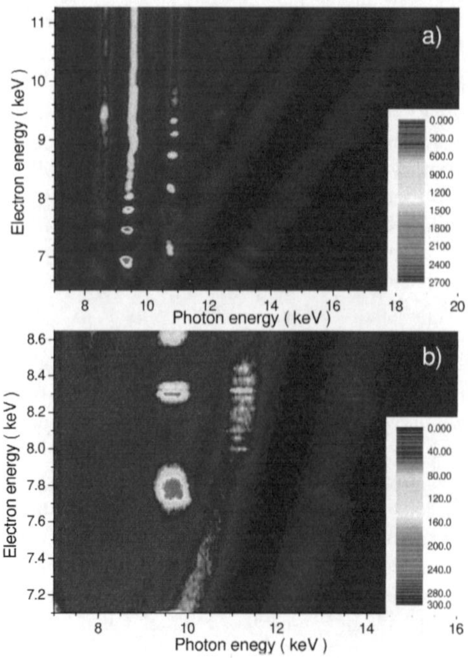

Figure 1. X-ray spectra from a) EBIT at NIST and b) MaMFIT in Giessen.

References

[1] J. D. Gillaspy 2012 *AIP Conf. Proc.* **1438** 97

[2] V. P. Ovsyannikov 2014 *arXiv*: 1403.2168

[3] Yu. V. Ralchenko and Y. Maron 2001 *J. Quant. Spec. and Rad. Trans.* **71** 609

[1]E-mail: Alexander.Borovik@physik.uni-giessen.de

Guiding of high-current electron beam by macro-insulating units

Mingwu Zhang*[1], Yingli Xue*, Xin Li*, Junliang Liu*, Bian Yang*, Wei Wang*, Yipan Guo*, Wei Xi*, Xiaona Zhu*, Liping Yang*, Deyang Yu*, Caojie Shao*, Xiaohong Cai*

* Institute of Modern Physics, Chinese Academy of Science, Nanchang Rd. 509, Lanzhou, China

Synopsis The guiding effect of high-current electron beams transmitting through different macro-insulating materials has been investigated by following our previous study. It is very surprised that the guiding effect by different materials was found to be very similar, which may be due to the combined contributions of the individual atom scattering and the repulsive field on material surface.

Since the unexpected guiding effect of low energy ions by PET nanocapillaries was reported, considerable work [1, and references therein] has been done to investigate slow highly charged ions (SHCIs) interacting with inner surface of various insulating capillaries. Now it is clear that the preceding SHCIs will deposit positive charges on the surface, building up an electric field by forming a charged patch and therefore preventing the succeeding ions from close contact with the surface. The deposited charge patch more than one may be created until stable guiding is achieved. However, when an electron beam passes through an insulating capillary, besides the low energy secondary electrons, a notable portion of the transmitted electrons suffers significant energy loss, and the transmission efficiency is much lower than that of SHCIs. The current studies of electron guiding indicate that it is essentially different from that of SHCI [2, and references therein].

In our previous work [2], using a pair of grooved SiO_2 parallel plates, we had obtained stably guided electron beams without energy loss in the incident energy range of 800-2000V. Different beam fluxes and different ground boundaries were employed to explore the discharge mechanisms, but no obvious influences have been observed. It shows that the self-organizing repulsive electric field played the dominant role to guide the electrons, by comparing with a pair of asymmetric parallel plates that were constructed by replacing one of the plates by a ground metal. This is similar to the case of SHCI beams. However, our results suggest that the complicated mechanisms involved in charging and discharging of insulators bombarded by electrons.

Following the above investigation, we used several insulators which were all made into grooved plates as same as the above grooved SiO_2 plates, to study the dependence relative to the guiding beams on different materials. As shown in Figure 1, it is very surprised that, because the conductivities of these materials may differ several magnitudes from each other, the guiding behaviors among these different insulators are quite same. Moreover, we found that the gap between two symmetric plates might play significant role in guiding process of electron beams. The present results suggest that the guiding effect of high-current electron beams by macro-insulating units may be due to the combined contributions of the individual atom scattering and the repulsive field on material surface.

This work was supported by the National Natural Science Foundation of China (Nos. 11205224, 11275240, and U1332206).

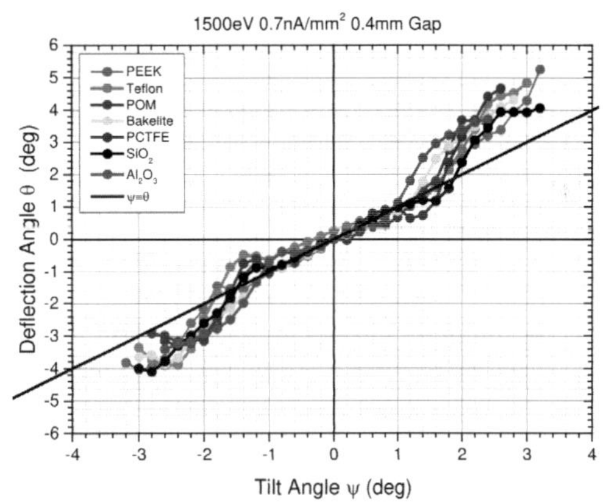

Figure 1. Deflection angle as functions of the tilt angle for 1500 eV, 0.7 nA/mm^2 electrons passing through different parallel plates with 0.4 mm gap.

References

[1] N. Stolterfoht *et al.* 2016 *Phys. rep.* **629 1-107**
[2] Yingli Xue *et al.* 2015 *Appl. Phys. Lett.* **107 254102**

[1] E-mail: zhangmingwu@impcas.ac.cn

IOP Conf. Series: Journal of Physics: Conf. Series **875** (2017) 052028 doi:10.1088/1742-6596/875/6/052028

Emission line spectra of the 2s-2p transitions of S VIII-S XIII in the extreme ultraviolet region 160-300 Å

Safdar Ali[*1], **Hiroyuki Kato**[*], **Nobuyuki Nakamura**[*2]

[*] Institute for Laser Science, The University of Electro-Communications, Tokyo 182 8585, Japan

Synopsis The extreme ultraviolet spectra of S VIII-S XIII have been measured using compact electron beam ion trap in the wavelength range of 160-300 Å. We have identified several lines by observing electron energy dependence between 365 and 680 eV and with the support of collisional radiative modeling data obtained with the FAC code. The measured wavelengths are found in good agreement with the calculated and previous reported results.

Most of our knowledge of high-temperature astrophysical plasma is based on the information collected by means of spectroscopy. One of the most important parameters that we acquire through spectroscopy is wavelength, which is a prerequisite for extracting the information contained in the spectra collected with space based observatories [1]. The accurate wavelengths, is thus required for modeling astrophysical spectra and to get plasma parameters such as densities, temperatures, elemental abundances, emission measures, etc. The extreme ultraviolet (EUV) region (regarded here as 160-300 Å), is in particular rich with the emission lines important for modeling solar atmosphere plasmas.

Sulfur is the 10th most abundant element in the solar atmosphere with a photospheric abundance of 2.14×10^{-5} that of hydrogen [2, 3]. This abundance is sufficiently high to produce spectral lines from different charge states of sulfur and in particularly from S VIII-S XIII in the wavelengths region 160-300 Å. Several of the emission lines from these ions provide density diagnostics for solar flares and active regions [4].

In the present work, we report laboratory observations for the emission of EUV radiation from highly charged S VIII-SXIII in the wavelength range of 160-300 Å. The measurement was performed with a compact electron beam ion trap called CoBIT [5]. Sulfur ions were produced by injecting SF_6 into the center of the trap via a gas injection system. Different charge states of sulfur were produced by changing the electron beam energy between 365 and 680 eV. Spectra were recorded with a flat-field grazing-incidence spectrometer [6] with a groove number of 1200 mm^{-1} (Hitachi 001-0660) in a slitless configuration.

A typical spectrum of S XI acquired at electron beam energy of 530 eV is shown in Fig. 1a, while in Fig. 1b we plotted CR model spectra obtained using FAC code at an electron density of 10^{10} cm^{-3} and beam energy of 530 eV. Tentative identifications are given for several lines corresponding to 2s-2p transitions in S XI ions (indicated with arrows). The identification is based on the comparison with calculated spectra and beam energy dependence, since lines corresponding to different charge states show different energy dependence. The preliminary analysis shows the identification of total 32 lines from S VIII-S XIII. Several identified lines, in particularly from S X, XI and XII are useful for the density diagnostics of solar plasmas.

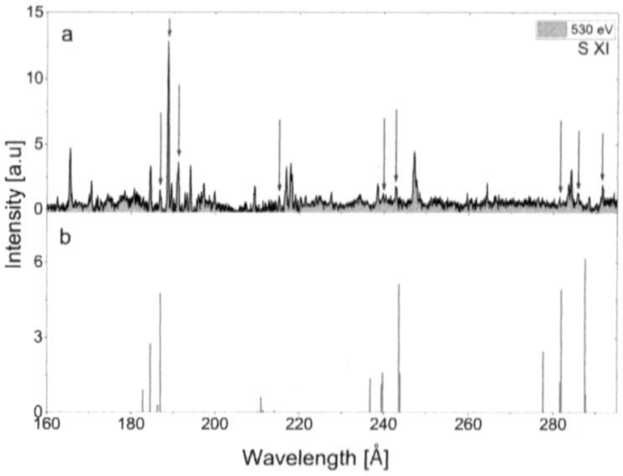

Figure 1. (a) Experimental and (b) CR model spectra from the 2s-2p transitions of S XI

[1]E-mail: safdaruetian@gmail.com
[2]E-mail: n_nakamura@ils.uec.ac.jp

References

[1] P. Beiersdorfer et al. 2012 *APJ* **745** 167

[2] S. Chakraborty et al. 2012 *PNAS* **110** 17655

[3] U. Feldman& J. M. Laming 2002 *Phys. Scr.* **61** 222

[4] S. O. Kastner& H. E. Mason 1978 *A&A* **67** 119

[5] N. Nakamura et al. 2008 *Rev. Sci. Instrum.* **79** 063104

[6] H. Ohashi et al. 2011 *Rev. Sci. Instrum.* **82** 083103

Theoretical study on the soft X-ray spectra of E1 transition of W LV ion

Jiaoxia Yang*, **Xiaobin Ding***[1], **Rui Sun***, **Fumihiro Koike**†, **Izumi Murakami**‡, **Daiji Kato**‡, **Hiroyuki A Sakaue**‡, **Nobuyuki Nakamura** §, **Chenzhong Dong***

* Key Laboratory of Atomic and Molecular Physics and Functional Materials of Gansu Province, College of Physics and Electronic Engineering, Northwest Normal University, Lanzhou 730070, China
† Department of physics, Sophia University, Tokyo 102-8554, Japan
‡ National Institute for Fusion Science, Toki, Gifu 509-5292, Japan
§ Institute for Laser Science, The University of Electro-Communications, Chofu, Tokyo 182-8585, Japan

Synopsis A collisional-radiative model was constructed to investigate the spectra of W LV ion from EBIT. A good agreement with previous experiment was obtained and some new transition lines were predicted.

The research on the energy level structure and radiation properties of various Tungsten(W) ions was largely spurred by the decision of chosen tungsten as the armor material of ITER divertor due to its favourable properties. In order to monitor and control the tungsten ion impurities produced when the plasma interact with the divertor, the throughout knowledge of the tungsten ions especially the radiation properties are necessary.

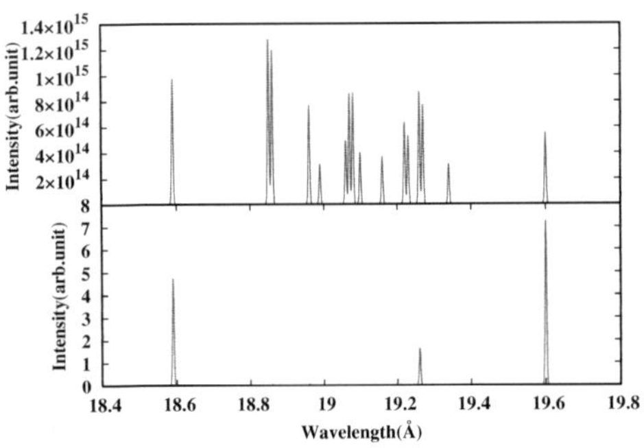

Figure 1. Theoretical spectra of W LV ion from EBIT with $n_e \approx 10^9 cm^{-3}$ and electron beam energy 18.2 keV in 18.4-19.8 Å. Upper: obtained by weighted transition probability. Lower: obtained by CRM.

The energy level structure and radiation of W LV ion had been studied experimental and theoretically [1, 2]. Lennartsson et al. observed the spectrum of this ions in 18.4-19.8 Å [1]. The intensity I_{ij} of an observed transition line can be defined as:

$$I_{ij} \propto N(i)A_{ij}\phi(\lambda) \quad (1)$$

where A_{ij} is the transition probability which can be obtained by relativistic configuration interaction method, $N(i)$ is the population of the ions in the upper level i which can be acquired by constructing and solving the Collisional-Radiative model [3] in which all the necessary atomic data such as energy level, transition energy, probability and electron impact (de)excitation cross section has been calculated with the implementation of FAC code [4] , and $\phi(\lambda)$ is the normalized line profile which was taken as Gaussian profile.

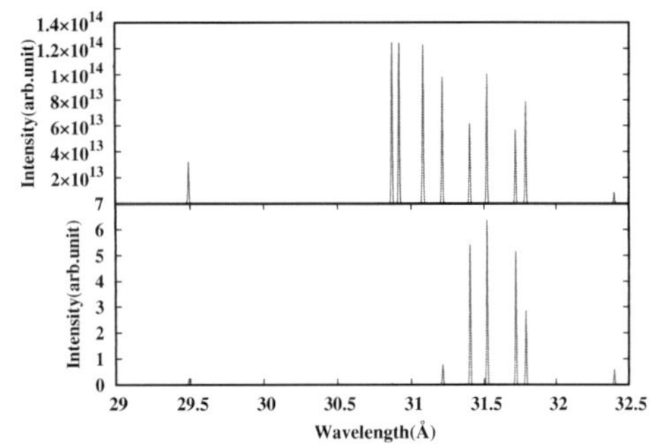

Figure 2. Theoretical spectra of W LV ion from 29-32.5 Å. The same as in Fig. 1.

The present work explained the experiment by Lennartsson *et al.* in which only three strong E1 $3d \rightarrow 3p$ transition were observed although there were 20 transitions with large transition probability. This result means that the electron excitation process plays curial role in the excited states population. Another several observable transition were predicted in 29-32.5 Å in Fig.2.

This work was supported by National Nature Science Foundation of China, Grant No:11264035 and Specialized Research Fund for the Doctoral Program of Higher Education(SRFDP), Grant No: 20126203120004, the Young Teachers Scientific Research Ability Promotion Plan of Northwest Normal University (Grant No:NWNU-LKQN-15-3).

References

[1] Lennartsson *et al.* 2013 *Phys. Rev. A* **87** 062505

[2] X.B. Ding *et al.* 2017 *J. Phys. B: At. Mol. Opt. Phys.* **50** 045004

[3] X.B. Ding *et al.* 2017 *Phys. Lett. A.* **380** 874

[4] M. F. Gu *et al.* 2008 *Cana. J. Phys.* **86** 675

[1] E-mail: dingxb@nwnu.edu.cn

IOP Conf. Series: Journal of Physics: Conf. Series **875** (2017) 052030 doi:10.1088/1742-6596/875/6/052030

Influence of Breit interaction on the linear polarization of radiation following electron impact excitation of B-like ions

C. Ren, F. Jiao, X. Wang, J. Jiang, L. Y. Xie, D. H. Zhang and C. Z. Dong[1]

Key Laboratory of Atomic and Molecular Physics and Functional Materials of Gansu Province, College of Physics and Electronic Engineering, Northwest Normal University, Lanzhou, 730070, China

Synopsis The influences of Breit interaction on the total cross section and linear polarization of highly charged Ca^{15+}, Xe^{49+}, and W^{69+} ions in the electron impact excitation process are calculated by fully relativistic distorted-wave program REIE06. It is found that the influence of Breit interaction on the linear polarization is more obvious with the increase of atomic number Z.

The cross section of electron impact excitation(EIE) and the corresponding linear polarization of light emitted after the EIE have widely applied for testing atomic structure theory, revealing collision kinetics, predicting new phenomena and so on [1]. Recently, Holger Jörg *et al.* measured the linear polarization of x rays produced by dielectronic recombination of highly charged B-like xenon ions [2], discussed the influence of Breit interaction on the linear polarization, and they found that the Breit interaction have no effects on it.

In order to test the effects in the similar transitions but formed from the EIE process, the EIE linear polarization of 1s to 2p and 2s to 2p of highly charged Ca^{15+}, Xe^{49+} and W^{69+} ions is investigated by the fully relativistic distorted-wave program REIE06 [3,4]. It is found that the Breit interaction makes the linear polarization of the transitions $[(1s^2 2s2p_{1/2})_0 2p_{3/2}]_{3/2}$ to $1s^2 2s^2 2p_{1/2}$ and $[(1s^2 2s2p_{3/2})_1 2p_{3/2}]_{5/2}$ to $1s^2 2s^2 2p_{3/2}$ decrease, while the linear polarization of the transitions $[(1s2s^2 2p_{1/2})_0 2p_{3/2}]_{3/2}$ to $1s^2 2s^2 2p_{1/2}$ and $[(1s2s^2 2p_{1/2})_1 2p_{3/2}]_{5/2}$ to $1s^2 2s^2 2p_{3/2}$ increase. And the effects from the latter case are more obvious.

Figure 1 shows the linear polarization of the transition 1s to 2p for the highly charged Ca^{15+}, Xe^{49+} and W^{69+} ions in the EIE process. For Ca^{15+} ions, the results of N+N, B+N and B+B increase monotonically by the increase of incident electron energy. For Xe^{49+} and W^{69+} ions, however, the results of B+N and B+B also increase monotonically by the increase of incident electron energy, while the results of N+N decrease at first then increase. This characteristic becomes more obvious with increase of atomic number Z. It is quite different with the case in the dielectronic recombination process [2].

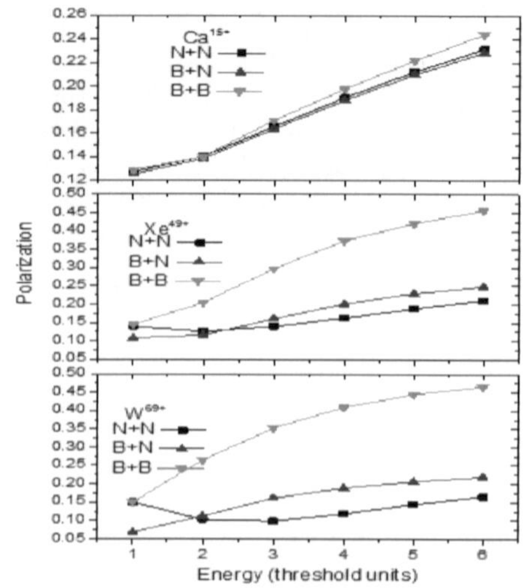

Figure 1. The linear polarization of the transition 1s to 2p of Ca^{15+}, Xe^{49+} and W^{69+} ions. N+N represents without Breit interaction in calculations of the target wave functions and impact matrix element, B+N represents with Breit interaction in calculations of the target wave functions and without Breit interaction in calculations of the impact matrix element, B+B represents with Breit interaction in calculations of the target wave functions and impact matrix element.

This work was supported by National Natural Science Foundation of China (Grant No.U1332206, U1331122,11464046).

References

[1] Z. W. Wu *et al.* 2011 *Phys. Rev. A* **84** 032713
[2] H. Jörg *et al.* 2015 *Phys. Rev. A* **91** 042705
[3] J. Jiang *et al.* 2008 *Phys. Rev. A* **78** 022709
[4] J. Jiang *et al.* 2007 *Chin. Phys. Lett.* **24** 691

[1]E-mail:dongcz@nwnu.edu.cn

IOP Conf. Series: Journal of Physics: Conf. Series **875** (2017) 052031 doi:10.1088/1742-6596/875/6/052031

A coupled rearrangemnet channel analysis of positronium antihydride Ps$\overline{\mathrm{H}}$

Takuma Yamashita[*1], **Yasushi Kino**[*2], **Emiko Hiyama**[†], **Svante Jonsell**[‡], **and Piotr Froelich**[§]

[*] Departmant of Chemistry, Tohoku University, Sendai 980-8578 Japan
[†] RIKEN Nishina Center, RIKEN, Wako 351-0198, Japan
[‡] Department of Physics, Stockholm University, SE-10691 Stockholm, Sweden
[§] Department of Chemistry, Uppsala University, Box 518, 751-20 Uppsala, Sweden

Synopsis Ground and resonance states of Ps$\overline{\mathrm{H}}$ are calculated with a coupled rearrangement-channel, four-body calculation. The structures of these states are examined with help of correlation functions and probabilities of finding two-body states in the total wavefunction. Asymptotic configurations of the ground and resonance states are determined from the tail behaviour of the correlation functions.

We report a theoretical investigation of positronium antihydride Ps$\overline{\mathrm{H}}$ which consists of an antiproton ($\overline{\mathrm{p}}$), an electron (e^-) and two positrons (e^+). The Ps$\overline{\mathrm{H}}$ is a charge-conjugated system of the positronium hydride PsH. The Ps$\overline{\mathrm{H}}$ has only one truly bound state below $\overline{\mathrm{H}}$ (1s) + Ps (1s) threshold.

Resonance states of Ps$\overline{\mathrm{H}}$ are of interest for the ongoing experiments of GBAR project [1] as intermediate states for rearrangement collisions, $\mathrm{Ps} + \overline{\mathrm{H}} \to \mathrm{e}^- + \overline{\mathrm{H}}^+$. Since below the dissociation threshold of $\mathrm{e}^- + \overline{\mathrm{H}}^+$ there exist several other dissociation thresholds, all possible channels for the rearrangement dissociation should be accounted for the branching ratio for the decay of the resonance state to various states.

To obtain a four-body wavefunction of Ps$\overline{\mathrm{H}}$ including all coupled rearrangement channels, a Gaussian expansion method [2] is adapted. The role of the various rearrangement channels in our total wavefunctions provides an insight to the structure of the ground and resonance states.

The structure of Ps$\overline{\mathrm{H}}$ is analyzed with help of a correlation function between two particles ('i' and 'j') defined as $C(r'_{\mathrm{ij}}) = \langle \Psi \mid \delta(\mathbf{r}_{\mathrm{ij}} - \mathbf{r}'_{\mathrm{ij}}) \mid \Psi \rangle$ [3]. As shown in Fig. 1 (a),the correlation functions $C(r_{\overline{\mathrm{p}}\mathrm{e}^+})$ has the same slope as isolated $\overline{\mathrm{H}}$ (1s) around the origin, while the middle-long range tail of $C(r_{\overline{\mathrm{p}}\mathrm{e}^+})$ shows a large deviation from the $\overline{\mathrm{H}}$ (1s). Since the bound state lies only 1 eV below the $\overline{\mathrm{H}}$ + Ps threshold, the expected configuration of the ground state may be a molecule-like structure. The asymptotic form of the correlation function between $\overline{\mathrm{H}}$ and Ps should be proportional to $\exp(-\kappa r)$ where $\kappa = \sqrt{2\mu\varepsilon}$ (μ is a reduced mass and ε is binding energy). This asymptotic slope, however, appears over 20 bohrs and its amplitude is $\sim 10^{-11}$ which indicates that in the ground state coexistence of atomic/molecular structure is remarkable. The Ps formation is manifested by $C(r_{\mathrm{e}^+\mathrm{e}^-})$ and $C(r_{\mathrm{e}^+\mathrm{e}^+})$ in which the $C(r_{\mathrm{e}^+\mathrm{e}^-})$ has the same slope as isolated Ps (1s) and the $C(r_{\mathrm{e}^+\mathrm{e}^+})$ almost overlaps with $C(r_{\mathrm{e}^+\mathrm{e}^-})$, as shown in Fig. 1 (b).

[1]E-mail: t.ymst@dc.tohoku.ac.jp
[2]E-mail: y.k@m.tohoku.ac.jp

The coexistence of molecular and atomic structures in Ps$\overline{\mathrm{H}}$ is an intriguing feature of a quantum few-body system. The analysis of the four-body resonances in terms of the three-body resonant subsystem ($\overline{\mathrm{H}}^+$, $\mathrm{e}^-\overline{\mathrm{H}}$) will also be presented. Since in the energy region of resonances the dissociation thresholds associated with various configurations exist, a hybridization of different configurations and a small asymptotic amplitude of wavefunction can seriously affect the interpretation of the decay.

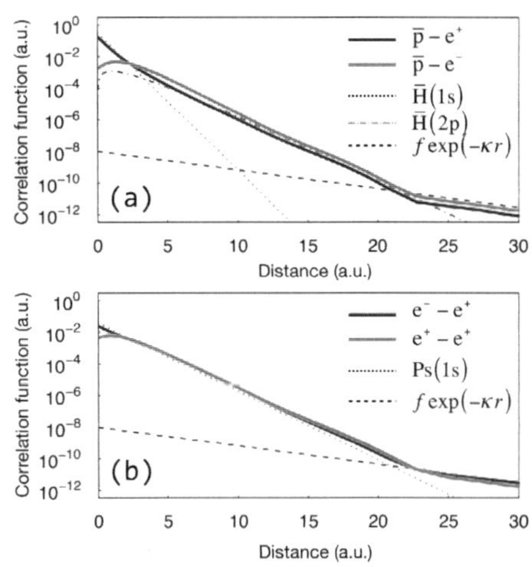

Figure 1. The correlation functions $C(r_{\overline{\mathrm{p}}\mathrm{e}^{+/-}})$ and $C(r_{\mathrm{e}^+\mathrm{e}^{+/-}})$ defined in text is shown together with that of isolated $\overline{\mathrm{H}}$ (1s), $\overline{\mathrm{H}}$ (2p), Ps (1s) and an asymptotic behavior ($f = 10^{-8}$ is a scaling constant).

References

[1] P. Perez and Y. Sacquin, 2012 *Class. Quantum Grav.* **29** 184008

[2] E. Hiyama, Y. Kino and M. Kamimura, 2003 *Prog. Part. Nucl. Phys.* **51** 223

[3] J. Usukura, K. Varga, and Y. Suzuki, 1998 *Phys. Rev. A* **58** 1918

IOP Conf. Series: Journal of Physics: Conf. Series **875** (2017) 052032 doi:10.1088/1742-6596/875/6/052032

KLL-dielectronic recombination and polarization of X-ray emissions of H-like to B-like barium ions

Q. Q. Man *, L. Y. Xie*,[1], H. Y. Cui *, J. Jiang* , D. H. Zhang*, C.Z.Dong*

* Key laboratory of Atomic and Molecular Physics & Functional Materials of Gansu Province, College of Physics and Electronic Engineering, Northwest Normal University, Lanzhou, 730070, China

Synopsis KLL dielectronic recombination processes and linear polarization degree of emitted x-rays are investigated for the H-like to B-like barium ions in the framework of Multi-configuration Dirac-Fock (MCDF) method and the density matrix theory. The present theoretical results are compared with the experimental measurements of electron beam ion trap, good consistency is obtained.

Dielectronic recombination (DR) is an important process for high-temperature laboratory and astrophysical plasmas [1,2], it strongly affects both the charge-state distribution and the x-ray spectrum. The x-ray radiations from the DR process can provide informations on the directionality of the electron currents and the orientation of the magnetic-field lines, and anisotropy of the electron energy distributions in plasmas, and can also be used as a diagnostic tool [3]

In this report, we calculated the resonance energies, resonance width, and corresponding radiative and auger transition rates for all of individual KLL resonance states of the highly charged H-like to B-like barium ions and obtained the DR resonance strengths and linear polarization degree of X-ray emissions[4] for each of the ions mentioned above. In the calculations, the electron correlation effects, QED effects, and Breit interactions are well considered. The influence of Breit interactions on Auger rates, DR resonance strengths and polarization of X-ray are analyzed and discussed. Partly of the present results for the differential cross sections and a comparison with the EBIT experimental measurements [5] are plotted in Fig.1 and 2, good consistency is obtained.

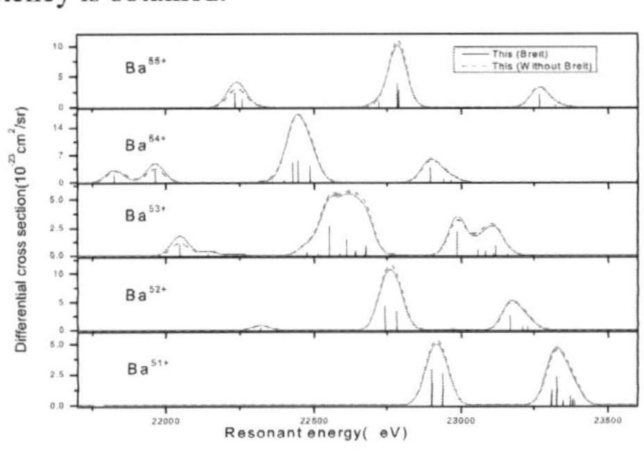

Fig. 1. Resonance strengths and differential cross sections (convolved with a Gaussian of FWHM=70eV) for the H-like to B-like barium ions .

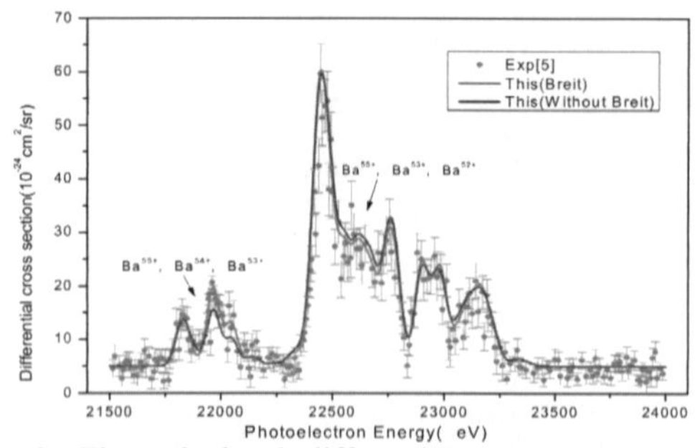

Fig. 2. The calculated differential cross section (convolved with a Gaussian of FWHM=70eV) with considering Breit interactions (red line) and without Breit interactions (blue line) for H-like to B-like Barium ions (the ratio of the relative proportion of Ba^{54+}:Ba^{53+}:Ba^{52+}:B^{51+}=30.11%:42.99%:23.76%:3.14 %), and compared with EBIT results [5].

This work is partly supported by the National Natural Science Foundation of China (Grant Nos. U1331122,U1530142,11464042,11564036,U1332206), the Young Teachers Scientific Research Ability Promotion Plan of Northwest Normal University (NWNU-LKQN-15-3).

References

[1] A. Burgess et al .1964 *Astrophys. J.* 139 776
[2] A.H. Gabriel et al 1972 *Mon. Not. R. Astron. Soc.* 160 99.
[3] H. Jörg, Z.M.Hu, H. Bekker et al. 2015 *Phys. Rev. A* **91** 042705.
[4] Z. Hu, X. Han, Y.Li et al., 2012 *Phys. Rev. Lett.* **108** 073002.
[5] D.A. Knapp, R.E. Marrs, M. B. Schneider, 1993 *Phys Rev. A* 47 2039.

[1] E-mail: xiely@nwnu.edu.cnu

The influence of Breit interaction and E1-M2 quantum interference on polarization following inner-shell electron-impact excitation of Li-like ions

L. Y. Xie*,[1], Y. L. Ma *, Q. Q. Man *, J. Jiang*, D. H. Zhang*, C. Z. Dong*

* Key laboratory of Atomic and Molecular Physics & Functional Materials of Gansu Province, College of Physics and Electronic Engineering, Northwest Normal University, Lanzhou, 730070, China

Synopsis The total and magnetic sublevel cross sections of electron-impact excitation from the ground state $1s^22s\ ^2S_{1/2}$ to the inner-shell excited states 1s2s2p (J=3/2) are calculated based on a relativistic distorted-wave method for highly charged Li-like ions. The important influence of the Breit interactions and quantum interferences between electronic dipole (E1) and magnetic quadrupole (M2) transition on polarization of emitted X-ray are systematically studied.

The magnetic sublevels of excited states, which are produced by electron-impact excitation, may not be populated statistically. The radiation emitted during the decay of these unequally populated sublevels to a lower level is polarized. The degrees of polarization of X-ray, which depend on the degree of anisotropy of the velocity distribution, can provide an important diagnostic tool studying the vacuum spark plasmas, Z-pinches, and laser-produced plasmas [1-3].

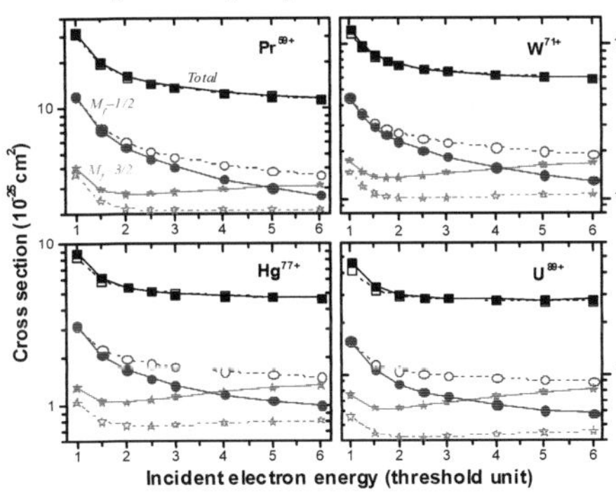

Figure 1. The total and specific magnetic sublevel cross sections for the excited state $1s2s2p(^1P)\ ^2P_{3/2}$ from the ground state $1s^22s\ ^3S_{1/2}$ for the Li-like Pr^{59+}, W^{71+}, Hg^{77+}, and U^{89+} ions as functions of incident electron energy in threshold unit. The solid line with symbols indicate the results with inclusion of Coulomb + Breit interactions, and the dash line with symbols indicate the results with only Coulomb interaction included in the calculations.

In this report, we performed a detailed calculation for the total and magnetic sublevels cross sections of inner-shell electron-impact excitation of the highly charged Li-like Pr^{59+}, W^{71+}, Hg^{77+}, and U^{89+} ions from the ground state to the excited states $1s2s2p(^3P)\ ^2P_{3/2}$, $1s2s2p(^1P)\ ^2P_{3/2}$, and $1s2s2p(^3P)\ ^4P_{3/2}$, based on our developed relativistic distorted-wave code [4]. The polarization of X-ray emissions following the electron-impact excitation are obtained for $1s2s2p(^3P)\ ^2P_{3/2}$-$1s^22s\ ^2S_{1/2}$ (labeled q), $1s2s2p(^1P)\ ^2P_{3/2}$-$1s^22s\ ^2S_{1/2}$ (s), $1s2s2p(^3P)\ ^4P_{3/2}$-$1s^22s\ ^2S_{1/2}$ (u) transition lines. Figure 1 and 2 illustrate the present results of the cross sections and polarizations for s-line of Li-like ions, which show significantly influences of the Breit interaction [5] and E1-M2 interferences.

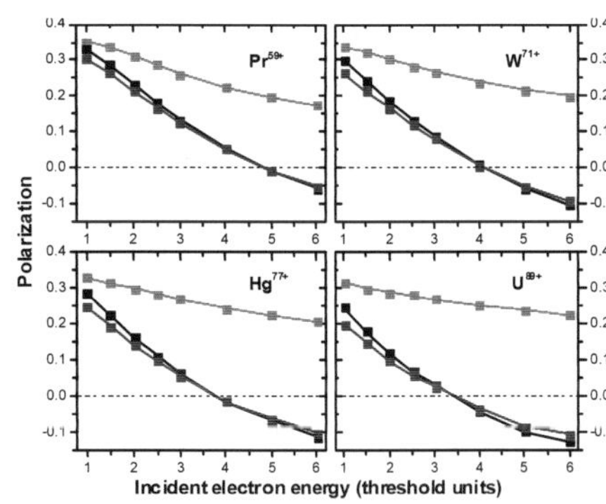

Figure 2. The polarization of s-line in Li-like Pr^{59+}, W^{71+}, Hg^{77+}, and U^{89+} ions. red solid (Coulomb interaction included only), black solid (Coulomb + Breit interactions included), and blue solid (Coulomb + Breit + E1-M2 quantum interference considered).

This work is partly supported by the National Natural Science Foundation of China (Grant Nos. U1331122,U1530142,11464042,11564036,U1332206), the Young Teachers Scientific Research Ability Promotion Plan of Northwest Normal University (NWNU-LKQN-15-3).

References

[1] R. Beier *et al* 1981 *J. Phys. D* 14, 643
[2] A. S. Shlyaptseva *et al* 2003*Proc. SPIE* 5196, 16
[3] J. C. Kieffer *et al* 1993 *Phys. Rev. E* 48, 4648
[4] J. Jiang *et al* 2008 *Phys. Rev. A* 78, 022709
[5] C. J. Fontes *et al* 1993 *Phys.Rev.A* 47, 1009

[1] E-mail: xiely@nwnu.edu.cnu

IOP Conf. Series: Journal of Physics: Conf. Series **875** (2017) 052034 doi:10.1088/1742-6596/875/6/052034

Manipulating quantum interferences in laser dressed-helium by the transferred momentum between the electron projectile and helium target

Hicham Agueny* [1], **Abdelkader Makhoute**†, **and Alain Dubois**‡

*Department of Physics and Technology, Allegt. 55, University of Bergen, N-5007 Bergen, Norway
† UFR de Physique du Rayonnement et des Interactions Laser-Matière, Faculté des Sciences. Université Moulay Ismail, B.P. 11201, Zitoune, Meknès, Morocco
‡ Sorbonne Universités, UPMC Univ Paris 06, UMR 7614, Laboratoire de Chimie Physique-Matière et Rayonnement, F-75231 Paris Cedex 05, France

Synopsis We present theoretical study of energetic electron-impact excitation of "six-level helium atom" embedded in a laser field. During the scattering process, the combined interactions lead to a change in the kinetic energy and the angular distribution of the scattered electron. The latter contains all information about the combined interactions. The analysis of the angular distribution reveals the emergence of a peak profile. This is will be discussed in terms of quantum-path interferences in bound-bound electronic transitions. Furthermore, the sensitivity of the phenomenon to the collision dynamics will also be discussed.

The scattering processes of electrons on atoms in the presence of a laser field, also called laser-assisted electron atom scattering, have long been recognized as a useful tool to make evident the multiphoton signals [1, 2]. The characteristics that distinguishes this class of elementary collision physics is the presence of the laser photon, which plays a role of a third body , in addition to the electron projectile and target atom. The latter enables to control the atomic and molecular dynamics and manipulate the reaction rates by a suitable choice of laser parameters.

In this context, electronic transitions induced by a laser field can lead to new effects, which may emerge in the angular distribution of the scattered electron [3, 4]. During the conference we shall present a study of quantum-virtual path interferences produced in energetic electron-impact excitation of a helium atom embedded in a low-frequency laser field. The process under investigation is dealt with a nonperturbative approach based on Born-Floquet method [5, 4]. Under the framework of this approach, the dressed atomic states are first obtained via Floquet theory. The incoming and scattered dressed-electron projectile are described using non-relativistic Volkov solutions; while their interactions with the target are treated within the first Born approximation. Here, we will show that the observed peak profile in the angular distribution of the scattered electron (cf. Fig. 1) is a signature of quantum interference between different virtual pathways that the excitation may follow in order to end up in a common final channel. These findings are supported by additional calculations based on Kroll-Watson approximation and time-dependent perturbation theory, in which no quantum interferences are accounted for. Furthermore, the possibility of stimulating the peak profile by means of the transferred momentum between the collision partners during the scattering process will also be addressed.

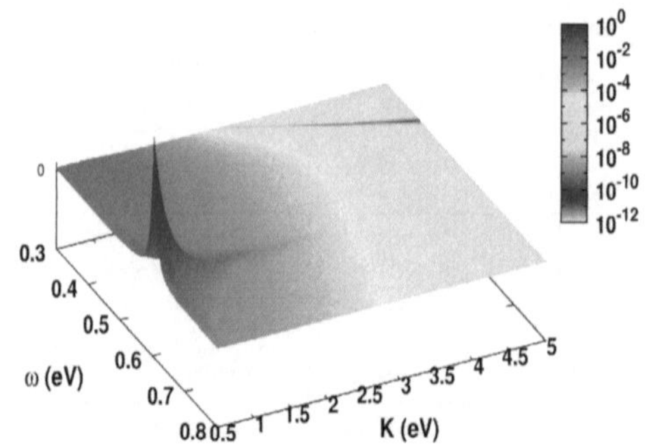

Figure 1. Angular distribution of the excitation of the dressed-state 1s2s of the helium atom for one photon emission (i.e., photon transferred between the colliding system and laser field) as a function of the transferred momentum K and the frequency ω of the laser field. The polarization of the laser field is kept along the momentum transfer. The peak intensity is fixed at 10^9 W/cm^2, the incoming electron energy is 100 eV, and the scattering angle is varied from 0 to 180°.

References

[1] D. Andrick and L. Langhans 1976, *J. Phys. B* **9** L459
[2] Y. Morimoto *et al* 2015, *Phys. Rev. Lett.* **115** 123201
[3] H. Agueny *et al* 2017, *submitted to New J. Phys.*.
[4] H. Agueny *et al* 2015, *Phys. Rev. A* **92** 013423
[5] M. Dorr *et al* 1994, *Phys. Rev. A* **49** 4852

[1]E-mail: hicham.agueny@uib.no

IOP Conf. Series: Journal of Physics: Conf. Series **875** (2017) 052035 doi:10.1088/1742-6596/875/6/052035

Electron transfer processes in potassium collision with nitroimidazoles: the role of methylation at N1 site

M. Mendes* [1], **F. Ferreira da Silva*** [2], **G. García**† [3] **and P. Limão-Vieira*** [4] ¶

* Laboratório de Colisões Atómicas e Moleculares, CEFITEC, Departamento de Física, Faculdade de Ciências e Tecnologia, Universidade NOVA de Lisboa, 2829-516 Caparica, Portugal
† Instituto de Física Fundamental, Consejo Superior de Investigaciones Científicas, Serrano 113-bis, 28006 Madrid, Spain

Synopsis We present site selectivity studies by electron transfer induced dissociation experiments in potassium collisions with nitroimidazoles and methylnitroimidazoles. In this contribution we report on H loss leading to the dehydrogenated parent anion formation in such experiments.

The nitroimidazoles are a family of molecules that have been used as radiosensitizers in radiation treatments to enhance the effect of ionizing radiation in radioresistant hypoxic tumours [1]. These compounds are electroaffinic structures that react with free radicals as oxygen producing DNA lesions. However detailed knowledge of the underlying molecular mechanisms of the radiosensitization process is still unclear. It is well-known that low-energy electrons (<30 eV) are the most abundant secondary species produced by primary radiation along the ionization track. The interaction of these electrons with biological tissues and molecules, like DNA, water represents a key issue for radiation damage studies, but also for the performance of radiosensitizers used in cancer therapy. These secondary electrons interact with the bio-environment creating large quantities of reactive species, like radicals, anions and cations, which will cause a series of chemical damages in biological tissues leading to double and single strand breaks, mutations and cluster lesions [2].

A literature survey reveals some studies on the formation of radical ions from radiosensitizers using different spectroscopic techniques [3][4]. It has been shown that low-energy electrons (0-8 eV) effectively decompose 4-nitroimidazole and two methylated isomers via a dissociative electron attachment (DEA) process. Moreover, these studies also demonstrated that the fragmentation pattern of 4NI is completely inhibited after methylation at the N1 site.

The importance of controlling chemical reactions in electron induced collisions with molecules has been one of the most challenging key aspects of the physical chemistry community over the last decades. This has been achieved using several different experimental techniques where mode-selective excitation, stereodynamic and orbital alignment control in molecular collisions. In atom-molecule collisions, the electron donor projectile interacts with a given molecule transferring an electron to a specific molecular orbital creating instability to the molecule. In the collision energy regime where formation of a transient negative anion (TNI) occurs, leads to several pathways including fragmentation into stable anionic species and access to parent molecular states which are not accessible in free electron attachment experiments [5].

Here we show novel results on time-of-flight (TOF) negative ion formation in potassium (electron donor) and nitroimidazoles collisions in the energy range between 8 and 1000 eV, exploring site and bond selective excision of H leading to the dehydrogenated parent anion formation. Negative ion mass spectra were obtained at several collision energies for different molecular targets as 4-nitromidazole, its methylated analogue in the N1 site and 1-methyl-5-nitromidazole.

References

[1] J. Overgaard, J. 2011 *Radiother. Oncol.* 100 22
[2] B. Boudaïffa *et al.* 2000 *Science* 287 1658
[3] K. Tanzer *et al.* 2014 Angew. Chem. Int. Ed. Engl. 53 12240
[4] K. Tanzer *et al.* 2015 *J. Phys. Chem.* 119 6668
[5] D. Almeida *et al.* 2013 *Phys. Rev. Lett.* 110 023201

[1] Email: mf.mendes@campus.fct.unl.pt
[2] Email: f.ferreiradasilva@fct.unl.pt
[3] Email: g.garcia@iff.csic.es
[4] Email: plimaovieira@fct.unl.pt

IOP Conf. Series: Journal of Physics: Conf. Series **875** (2017) 052036 doi:10.1088/1742-6596/875/6/052036

New numerical techniques to determine ionization cross sections: series representation for r_{12}^{-1} integrals and Bohm's velocity field

Juan M. Randazzo[†1] **and Lorenzo Ugo Ancarani**[*2]

[†] División Física Atómica, Molecular y Óptica, Centro Atómico Bariloche, 8400 S. C. de Bariloche, Río Negro, Argentina and Consejo Nacional de Investigaciones Científicas y Técnicas (CONICET), Argentina

[*] Equipe TMS, SRSMC, UMR CNRS 7565, Université de Lorraine, 57078 Metz, France

Synopsis Given an *ab initio* calculation of a three-body scattering wave function in spherical coordinates, we present a combination of new numerical tools for the evaluation of ionization cross sections.

In the last years we have developed an innovative tool to deal with three body scattering problems, based on a direct solution of the Schrödinger equation with appropriate flux boundary conditions [1]. The method uses a Generalized Sturmian Function (GSF) expansion of the three body wave function, in terms of two of the inter-particle coordinates of the problem. By using spherical coordinates, a partial wave series is performed both for the wave functions and the representation of the Coulomb interaction which depends on the third inter-particle coordinate, usually named r_{12}. The r_{12}^{-1} interaction is the only one not considered by the basis, as generally done in atomic calculations with not explicitly correlated basis functions.

Once all partial wave terms of the scattering wave function are evaluated, one may use integral formulas to extract the scattering amplitude. This is the case, for example, for the triple differential cross sections for double photoionization of Helium or for electron-Hydrogen ionization; from a practical point of view, this approach provides a good overall agreement with experimental results [1, 2]. However, such formulas have a major disadvantage since the integrals involve continuum wave functions and, besides, present some arbitrariness in the definition of the asymptotic final channels (see Ref. [3], for the discussion in the electron-hydrogen case).

In an attempt to overcome these difficulties, we recently proposed a new method [3] to recover the flux formula for ionization amplitudes [4] which was discarded after observation of apparently unphysical behaviors [5]. Our method, based on the use of Bohm's velocity field, has been applied successfully to s-wave electron-hydrogen ionization. In this contribution, we use it to deal with the ionization process in its full dimensionality, i.e., including the partial wave terms which are necessary for convergence.

Furthermore, we will combine this approach with a not so new technique to deal with the r_{12}^{-1} partial wave expansion. The idea, proposed in atomic calculations [6], is based on using a Taylor series in $\cos\theta_{12}$ instead of the usual Legendre polynomial expansion; the technique provides several alternatives as different series can be used for different angular regions. The method introduces a significant improvement when calculating atomic bound state energies; this can be understood as a more efficient representation of the potential for a given number of partial waves included in the expansion. We test this approach, for the first time here, for electron-hydrogen ionization and double photoionization of Helium.

References

[1] J. M. Randazzo *et al.* 2015 *Eur. Phys. J. D* **69** 189
[2] M. Baertschy *et al.* 2001 *Phys. Rev. A* **63** 022712
[3] J. M. Randazzo and L. U. Ancarani 2015 *Phys. Rev. A* **92** 062706
[4] R. Peterkop, Theory of Ionization of Atoms by Electron Impact (Colorado Associated University Press, Boulder, 1977)
[5] P. L. Bartlett *et al.* 2003 *Phys. Rev. A* **68** 020702
[6] C. F. Bunge 1989 *Phys. Rev. A* **40** 477

[1]E-mail: randazzo@cab.cnea.gov.ar
[2]E-mail: ugo.ancarani@univ-lorraine.fr

IOP Conf. Series: Journal of Physics: Conf. Series **875** (2017) 052037 doi:10.1088/1742-6596/875/6/052037

Commissioning of a high-power electron gun for electron-ion crossed-beams experiments

B. Ebinger*[1], A. Borovik Jr.*, B. M. Döhring*, T. Molkentin*, A. Müller†, S. Schippers*

* I. Physikalisches Institut, Justus-Liebig-Universität Gießen, 35392 Giessen, Germany
† Institut für Atom- und Molekülphysik, Justus-Liebig-Universität Gießen, 35392 Giessen, Germany

Synopsis A high-power electron gun delivering a ribbon-shaped electron beam with energies of 10-3500 eV and high currents at all energies is currently being commissioned at the electron-ion crossed-beams setup in Giessen. As compared to the formerly used electron gun [1], the new one will enable us to perform measurements of cross-sections for electron-impact ionization of ions in a wider energy range and, consequently, for ion charge states that were not accessible before.

In an electron-ion crossed-beams experiment, the experimental sensitivity is mainly determined by the densities of both beams in the interaction region. Formerly, an electron gun producing an electron beam with energies of up to 1000 eV has been successfully operated at our institute [2]. Aiming at the extension of the available range of accessible electron energies and densities, a new high-power electron gun has been developed and built. It delivers a ribbon-shaped beam with high currents at all energies variable between 10 and 3500 eV [3, 4]. The predicted high electron currents and good beam transmission have already been demonstrated [5].

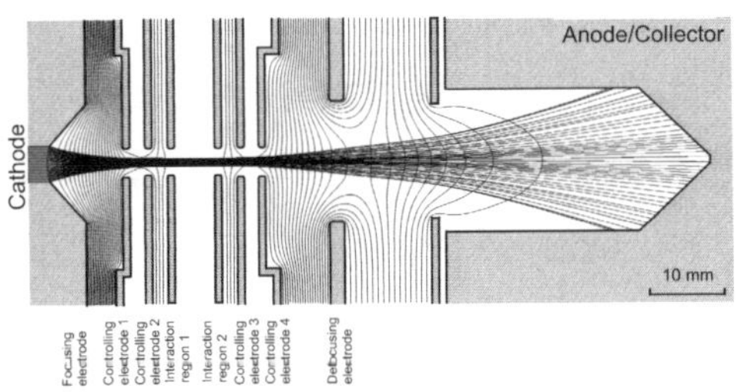

Figure 1. Simulated electron propagation in the new electron gun. Reprinted from [5] with permission from Elsevier.

This type of electron gun consists of nine electrodes (Figure 1): The cathode/focusing electrode, the controlling electrodes 1-4, the interaction region 1 and 2, the defocusing electrode and the anode/electron collector. This configuration allows for a variety of operational modes (Table 1) and, thus, the independent control of the energy and density of the electron beam available in the interaction region by variation of the electrostatic potential between the cathode and the controlling electrode 1 (the extraction electrode) while leaving the potential between the cathode and the interaction region constant. Moreover, a potential trap in the interaction region can be controlled, and also a deceleration of the electrons right in front of the collector is possible.

[1]E-mail: benjamin.ebinger@physik.uni-giessen.de

Table 1. Some possible operation modes of the electron gun. Column 2 provides the ratios of potentials on the cathode/focusing electrode, controlling electrode 1, controlling electrode 2, controlling electrode 3, controlling electrode 4, defocusing electrode and collector, respectively. The interaction region is grounded and, thus, corresponds to the relative zero point.

Operation mode	Potentials on electrodes
High-energy	-1:0.05:0.05:0.05:0.05:-1:0
High-current	-1:4:0.05:0.05:4:-1:0
No potential trap	-1:0.05:-0.15:-0.15:0.05:-1:0
Reduced power	-1:0.05:0.05:0.05:0.05:-1:-0.5

Here, we report on the current status of the commissioning of this electron gun. The electron gun is integrated into the experimental electron-ion crossed-beams setup in Giessen. Employing the animated crossed-beams technique [6], first cross sections for electron-impact ionization of xenon and helium ions were measured [5]. The measurement of more cross sections is intended for the near future. Further investigations concerning, e.g., space-charge effects in the high-density electron beam are currently performed.

The work is supported by the German Federal Ministry of Education and Research (BMBF) within the "Verbundforschung" funding scheme (contract no. 05P15RGFAA) and by the GSI Helmholtz Center for Heavy Ion Research (Darmstadt, Germany).

References

[1] R. Becker *et al.* 1985 *Nucl. Instr. Meth. Phys. Res. B* **9** 385

[2] A. Borovik Jr. *et al.* 2013 *J. Phys B* **46** 175201

[3] W. Shi *et al.* 2003 *Nucl. Instr. Meth. Phys. Res. B* **205** 201

[4] A. Borovik *et al.* 2014 *J. Phys. Conf. Ser.* **488** 142007

[5] B. Ebinger *et al.* 2017 *Nucl. Instr. Meth. Phys. Res. B* In press

[6] A. Müller *et al.* 1985 *J. Phys. B* **18** 2993

IOP Conf. Series: Journal of Physics: Conf. Series **875** (2017) 052038 doi:10.1088/1742-6596/875/6/052038

Strong higher-order resonant contribution to Fe K_α x-ray line polarization in hot anisotropic plasmas

Chintan Shah*,†, Pedro Amaro†, René Steinbrügge*, Sven Bernitt*,‡, Stephan Fritzsche‡, Andrey Surzhykov♮, José R. Crespo López-Urrutia*, and Stanislav Tashenov†

*Max–Planck–Institut für Kernphysik, Saupfercheckweg 1, 69117 Heidelberg, Germany
†Physikalisches Institut, Im Neunheimer feld 226, 69120 Heidelberg, Germany
‡Friedrich-Schiller-Universität Jena, Fürstengraben 1, 07743 Jena, Germany
♮Physikalisch-Technische Bundesanstalt, Braunschweig, Germany

Synopsis We report the first systematic measurement of x-ray angular distribution and polarization following dielectronic, trielectronic and quadroelectronic recombination into He-like through O-like iron and krypton ions. The experimental data revealed that the the degree of polarization of x rays is dominated by relativistic Breit interaction and hitherto neglected higher-order trielectronic and quadroelectronic recombination resonances.

For a wide range of temperatures, resonantly captured electrons with energies below the excitation threshold are the strongest source of x-ray line excitation in hot plasmas containing highly charged Fe ions. The angular distribution and polarization of x rays emitted due to these processes were experimentally studied using an electron beam ion trap. The electron-ion collision energy was scanned over the *KLL* dielectronic, trielectronic and quadruelectronic recombination resonances of $Fe^{18+..24+}$ and $Kr^{28+..34+}$ with an excellent resolution of ~ 6 eV. The angular distribution of x rays was measured along and perpendicular to the electron beam propagation direction [1]. Subsequently, polarization due to dielectronic recombination of $Kr^{28+..34+}$ was measured using a novel Compton polarimetry as well [2, 3].

The experimental data reveal the alignment of the populated excited states and exhibit a high sensitivity of these parameters to the relativistic Breit interaction [2, 4]. We observed that most of the transitions lead to polarization, including hitherto neglected trielectronic and quadruelectronic resonances. Furthermore, these channels dominate the polarization of the prominent K_α x rays emitted by hot anisotropic plasmas in a broad temperature range, see Fig. 1. The present results comprehensively benchmark full-order atomic calculations carried out with the FAC [5] and RATIP [6] codes. We conclude that accurate polarization diagnostics of hot anisotropic plasmas, e. g., of solar flares and active galactic nuclei, and laboratory fusion plasmas of tokamaks can only be obtained under the premise of careful inclusion of relativistic effect and higher-order resonances which were most often neglected in previous works [1]. Furthermore, the present experiments also demonstrate the suitability for accurate directional diagnostics of electron or ion beams in the hot plasmas [7].

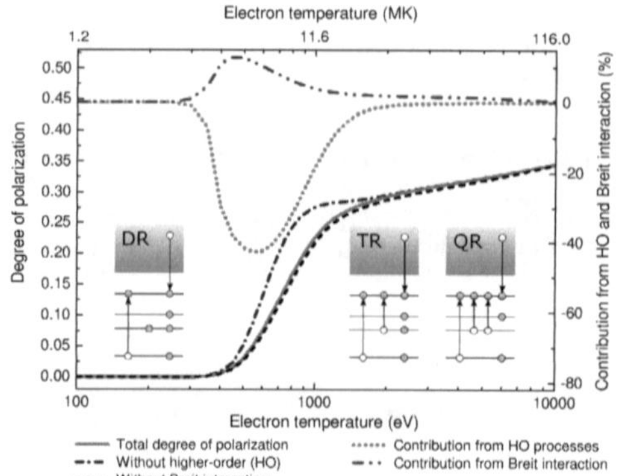

Figure 1. Maximum polarization of iron K_α x rays due to resonant recombination as a function of the plasma temperature.

References

[1] C. Shah *et al.* 2016 *Phys. Rev. E* **93** 061201(R)

[2] C. Shah *et al.* 2015 *Phys. Rev. A* **92** 042702

[3] S. Weber *et al.* 2015 *Rev. Sci. Instr.* **86** 093110

[4] P. Amaro, C. Shah *et al.* 2017 *Phys. Rev. A* **95** 022712

[5] M. F. Gu 2008 *Can. Phys. J* **86** 675

[6] S. Fritzsche 2012 *Comput. Phys. Commu.* **183** 1525-1559

[7] C. Shah *et al.* 2017, to be submitted

IOP Conf. Series: Journal of Physics: Conf. Series **875** (2017) 052039 doi:10.1088/1742-6596/875/6/052039

Laboratory measurements compellingly supports a charge-exchange mechanism for the "Dark matter" $\sim$ 3.5 keV X-ray line

Chintan Shah*, Stepan Dobrodey*, Sven Bernitt*,†, René Steinbrügge*, Liyi Gu‡, Jelle Kaastra‡,♮, and José R. Crespo López-Urrutia*

*Max–Planck–Institut für Kernphysik, Saupfercheckweg 1, 69117 Heidelberg, Germany
†Friedrich-Schiller-Universität Jena, Fürstengraben 1, 07743 Jena, Germany
‡SRON Netherlands Institute for Space Research, Utrecht, The Netherlands
♮ Leiden Observatory, Leiden University, 2300 RA Leiden, The Netherlands

Synopsis The reported observations of an unidentified X-ray line feature at $\sim$ 3.5 keV from galaxy clusters have driven a lively discussion about its possible dark matter origin. Motivated by this, we have investigated the X-ray spectra of highly ionized bare sulfur ions following charge exchange with residual gas in the electron beam ion trap, as a source or a contributor to this X-ray line. The X-ray feature at about 3.5 keV shows up in the experiment, which could explain the the astrophysical observations and confirm the predictions of Gu *et al.*

A mysterious X-ray signal at 3.5 keV from nearby galaxies and galaxy clusters [1] recently sparked an incredible interest in the scientific community and given rise to a tide of publications attempting to explain the possible cause for this line. The origin of this line has been hypothesized as the result of decaying sterile neutrinos–potential dark matter particle candidate, presumably on the fact that this X-ray line is not available in the standard spectral databases and models for thermal plasmas. Cautiously, Gu *et al.* [2] have pointed out to an alternative explanation for this phenomenon: charge exchange between bare ions of sulfur and atomic hydrogen. Their model shows that X-rays should emitted at 3.5 keV by a set of S^{15+} transitions from $n \geq 9$ to the ground states, where n is the principle quantum number.

With this motivation, we tested the hypothesis of Gu in the laboratory by measuring K-shell X-ray spectra of highly ionized bare sulfur ions following charge exchange with gaseous molecules in an electron beam ion trap. We produced bare S^{16+} and H-like S^{15+} ions and let them capture electrons in collision with those molecules with the electron beam turned off while recording X-ray spectra. The 3.5 keV transition clearly shows up in the charge-exchange induced spectrum under broad range of conditions. The inferred X ray energy, 3.47 $\pm$ 0.06 keV, is in full accord agreement with both the astrophysical observations and theoretical calculations, and confirms the novel scenario proposed by Gu [2, 3]. Taking the experimental uncertainties and inaccuracies of the astrophysical measurements into account, we conclude that the charge exchange between bare sulfur and hydrogen atoms can outstandingly explain the mysterious signal at around 3.5 keV [3].

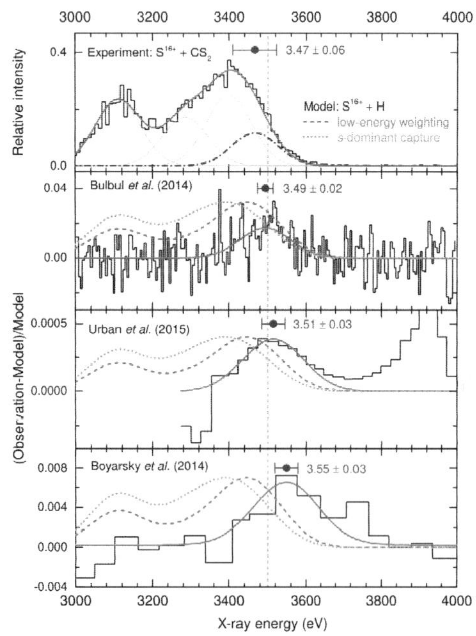

Figure 1. Charge-exchange-induced X-ray spectrum in comparison with recently reported astrophysical observations. The experimental data and observations are compared with the charge exchange model of Gu.

References

[1] E. Bulbul *et al.* 2014 *Astrophys. J* **13** 789
[2] L. Gu *et al.* 2015 *A & A* **L11** 584
[3] C. Shah *et al.* 2016 *Astrophys. J* **833** 52

Developments towards a transverse free-electron target for the storage ring CRYRING@ESR

C. Brandau[*,†1], **A. Borovik, Jr.**[*], **B. M. Döhring**[*], **B. Ebinger**[*], **C. Kozhuharov** [†], **T. Molkentin**[*], **A. Müller**[§], **Th. Stöhlker**[†,‡,¶], **and S. Schippers**[*2], **for the SPARC working group "Electron Targets"**

[*] I. Physikalisches Institut, Justus-Liebig-Universität Gießen, 35392 Giessen, Germany
[†] GSI Helmholtzzentrum für Schwerionenforschung, 64291 Darmstadt, Germany
[§] Institut für Atom- und Molekülphysik, Justus-Liebig-Universität Gießen, 35392 Giessen, Germany
[‡] Institut für Optik und Quantenoptik, Friedrich-Schiller-Universität Jena, 07743 Jena, Germany
[¶] Helmholtz-Institut Jena, 07743 Jena, Germany

Synopsis A transverse free-electron target for crossed-beams collision studies is currently being developed. The target will be installed in the main experimental section of the FAIR storage ring CRYRING@ESR. The setup will allow for electron-ion collision studies with highly charged ions on an unprecedented level of detail that comprises, e.g., the angle-differential observation of photon emission.

The storage ring CRYRING@ESR will be one of the first operational devices at the upcoming anti-proton and heavy-ion accelerator facility FAIR. As a part of the program of the SPARC collaboration it is intended to install a dedicated sheet-beam free-electron target at CRYRING@ESR [1]. Electron-ion collision studies with transverse electron targets were conducted at low-energy single-pass beam lines since the 1960s [2] but a crossed-beams setup has never been realized at a heavy-ion storage ring, yet.

Currently, a new electron gun is being developed (Fig. 1) that advances previous multi-electrode arrangements [3, 4, 5] with adaptations to the special requirements at CRYRING@ESR, namely, free space for the circulating ion beam and stringent vacuum conditions with residual gas pressures of the order of 10^{-11} mbar. The multi-electrode layout enables a detailed control of electron-beam parameters such as beam size, electron density and electron energy. Maximum electron energies up to 12.5 keV are envisaged. The typical beam-size in the interaction region is 8×80 mm^2 at an electron density of up to 10^9 cm^{-3}.

The set-up is optimized for spectroscopy of photons thus enabling many new opportunities for experimental access to fundamental electron-ion collision processes including excitation, recombination, elastic scattering and ionization. In contrast to collision studies at a gas-jet target, no target nucleus is present. Hence, observations of electron-impact excitation or electron-ion recombination are not obscured by competing processes like proton-impact excitation or non-radiative capture. A further advantage is that the electron-ion collision energy can easily be changed over a wide range without changing the ion energy.

This work is supported by German Federal Ministry of Education and Research BMBF (contract 05P15RGFAA), and the Helmholtz International Center for FAIR, a Hessian state funded Center of Excellence.

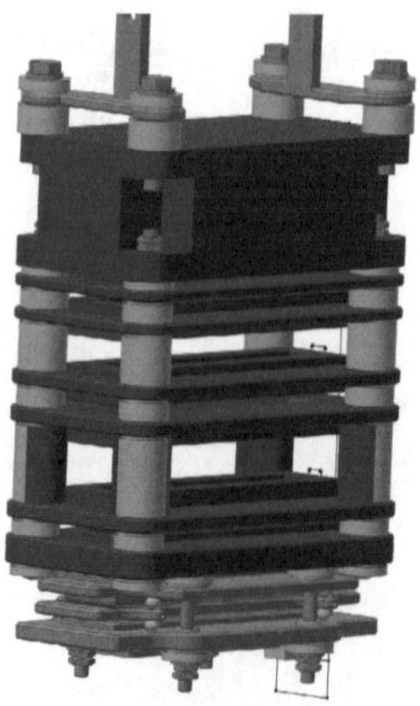

Figure 1. Current design of the new electron gun, which is still being optimized. The electron beam is directed bottom to top.

References

[1] M. Lestinsky *et al.* 2016 *Eur. Phys. J. ST* **225** 797

[2] A. Müller 2008 *Adv. At. Mol. Opt. Phys.* **55** 293

[3] W. Shi *et al.* 2003 *Nucl. Instrum. Meth. B* **205** 201

[4] S. Geyer *et al.* 2013 *Phys. Scr.* **T156** 014093

[5] A. Borovik Jr. *et al.* 2014 *J. Phys.: Conf. Ser.* **488** 142007

[1]E-mail: c.brandau@gsi.de
[2]E-mail: Stefan.Schippers@physik.uni-giessen.de

IOP Conf. Series: Journal of Physics: Conf. Series **875** (2017) 052041 doi:10.1088/1742-6596/875/6/052041

Inner Shell Excitations through Laser Induced Electron Recollision

Gilad Marcus* [1]**, Yunpei Deng**†**, Zhinan Zeng**‡**, Zhengmao Jia**‡**, Pavel, Komm*, Yinhui Zheng**‡**, Xiao-chun Ge**‡ **and Ruxin Li**‡

* Department of Applied Physics, Hebrew University of Jerusalem, Jerusalem 91904, Israel.
† SwissFEL, Paul Scherrer Institut, 5232 Villigen PSI, Switzerland.
‡ State Key Laboratory of High Field Laser Physics, Shanghai Institute of Optics and Fine Mechanics, Chinese Academy of Sciences, Shanghai 201800, China

Synopsis Sub-femtosecond excitation of inner-shell electrons is a prerequisite step toward time domain "pump-probe" study of core-hole dynamics. By using a few cycles 1800nm source, we show that the recolliding electron might provide this necessary step.

Excitations of electron from deep inner shells are usually accompanied by multi-electron dynamics such as double excitation, the Auger decay, Cooper minima, and the giant resonance, which cannot be explained by the single active electron approximation. Such excitations are unstable and usually decay on a timescale ranging from few femtosecond to few attosecond. The decay may take place in a single step, but more often occurs as a cascade of radiative and non-radiative channels. To follow such dynamics, one resorts to a time domain pump-probe spectroscopy Since the relevant timescale for such dynamics spans from attosecond to femtosecond and the relevant energy-scale spans from 10^2 -10^5 eV, x-ray attosecond bursts may be the choice to serve as the pump and the probe events. However, with the low photon flux of current soft x-ray attosecond sources (hν >300 eV) and the low absorption cross-sections in this spectral range, it is currently impossible to both pump and probe these processes with attosecond x-ray pulses. Excitation of inner-shell dynamics by laser-induced electron recollision might be the solution. Such an excitation process occurs in a sub-femtosecond timescale, thus, provides the necessary "pump" step in "pump-probe" studies of inner-shell dynamics.

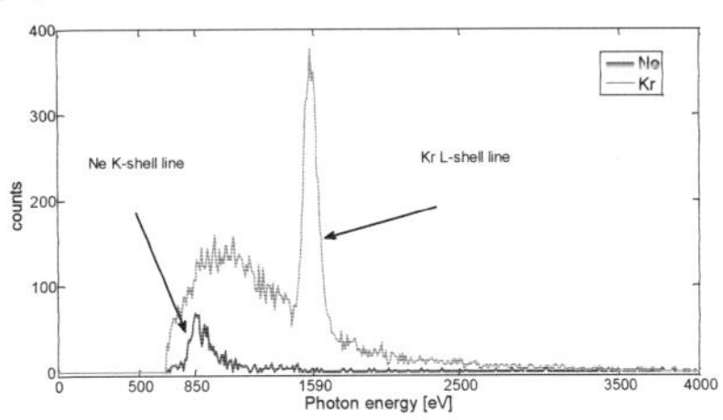

Figure 1. Characteristic fluorescence lines and continuum x-ray induced by electron recollision (Ne K-shell: blue, Kr L-shell: red).

In this work, we show core-hole excitation by the recolliding electron [1], as opposed to previous indirect evidences [2,3]. In our experiment, we focus a 12 fs, 1mJ infrared radiation source (λ_0 =1.8 μm) on a pulsed gas-jet and observed the soft x-ray radiation from the interaction region. The reason for using infrared laser instead of 800 nm laser is the quadratic scaling of the ponderomotive energy with the wavelength of the drive laser. According to the semi-classical three steps model, the maximum energy an electron may come back with, when it collides with the parent ion is equal to $3.17U_p$ and proportional to $I_0\lambda^2$. Here, U_p is the ponderomotive energy, I_0 and λ are the peak intensity and the central wavelength of the drive field respectively. This quadratic scaling with the laser wavelength allow for high kinetic energy of the recolliding electron without ionizing the target atoms too much. We observed x-ray fluorescence radiation from krypton L-shell and neon K-shell along with a continuum x-ray (see figure 1). To prove that this excitation is coming from recollision and not just from mere heating of the plasma, we measured both the directionality of the x-ray emission with respect to laser linear polarization and the x-ray yield against the laser polarization

References

[1] Y. Deng *et.al.2016 Phys. Rev. Lett.* **116**, 073901
[2] G. Marcus *et.al 2012 Phys. Rev. Lett.* ***108**, 023201*
[3] A. D. Shiner *et al.* 2011 *Nature Physics 7, 464–467*

[1] E-mail: Gilad.marcus@mail.huji.ac.il

IOP Conf. Series: Journal of Physics: Conf. Series **875** (2017) 052042 doi:10.1088/1742-6596/875/6/052042

A Perturbative Treatment For the Dielectronic Recombination of the Si-Like Isoelectronic Sequence

J. Kaur,*1 T. W. Gorczyca,*2 and N. R. Badnell†

*Department of Physics, Western Michigan University, Kalamazoo, MI 49008-5252, USA
†Department of Physics, University of Strathclyde, Glasgow G4 0NG, United Kingdom

Synopsis A detailed investigation of Si-like dielectronic recombination is performed for the entire Si-like isoelectronic sequence using a perturbative multi-configurational Breit-Pauli method.

We present total and final-state level-resolved DR rate coefficients for the silicon-like ions from P^+ through Zn^{16+}, relevant to the modelling of astrophysical and laboratory plasmas, and for the ions Kr^{22+}, Mo^{28+} and Xe^{40+} that are important in fusion research. Both $\Delta n_c = 0$ and $\Delta n_c = 1$ core excitations are included in *LS* and *IC* (intermediate coupling) schemes. Our calculations were performed using the atomic structure and collision code AUTOSTRUCTURE [1, 2], a perturbative multi-configurational Breit-Pauli (MCBP) method that relies on the independent proceses, isolated resonance, distorted-wave (IPIRDW) approximation. Energy levels, radiative rates, and autoionization rates are calculated in both *LS* and *IC* approximations, where the latter also include semi-relativistic corrections such as the spin-orbit interaction that gives rise to significant fine-structure effects. The electronic orbitals are obtained using the Thomas-Fermi-Dirac-Amaldi (TFDA) model potential, where the scaling parameters λ_{nl} are optimized so as to reproduce the experimental (NIST) fine-structure splitting for the low-lying levels.

The computed Maxwellian-averaged DR rate coefficients for the entire silicon-like isoelectronic sequence are shown in Fig. 1 and compared with the earlier recommended data of [3, 4], as incorporated in all previous plasma models. Those data were based on crude *LS* calculations, and thus were unable to reproduce the low-temperature DR contributions that are due to fine-structure splitting [5]. This effect becomes more predominant with increase in the effective charge z. At higher temperatures, the more complete IC cross sections instead lie below the non-relativistic LS results, again due to less obvious fine-structure effects [5]. This work is a part of an assembly of a dielectronic recombination database for the modelling of dynamic finite-density plasmas.

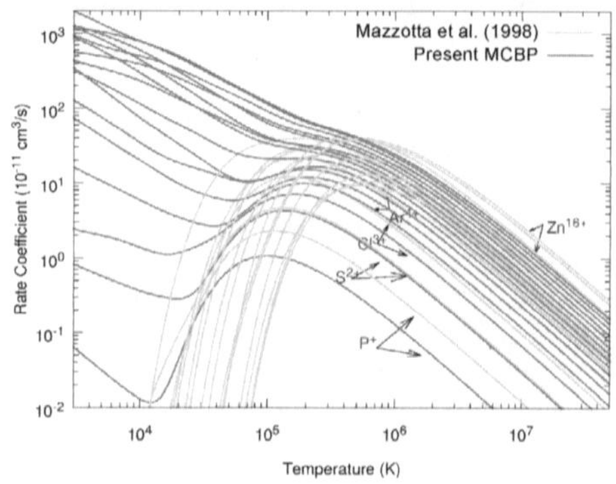

Figure 1. Comparison between present Maxwellian-averaged DR rate coefficients and recommended compilation [3, 4] for the entire silicon-like isoelectronic sequence.

References

[1] N. R. Badnell 1986 *J. Phys. B* **19** 3827

[2] N. R. Badnell 1997 *J. Phys. B* **30** 1

[3] P. Mazzotta, G. Mazzitelli, S. Colafrancesco and N. Vittorio 1998 *Astron. Astrophys. Suppl. Ser.* **133** 403

[4] G. Mazzitelli and M. Mattioli 2002 *at. data nucl. data tables* **82** 313

[5] S. A. Abdel-Naby, D. Nikolić, T. W. Gorczyca, K. T. Korista and N. R. Badnell 2012 *Astron. Astrophys.* **537** A40

1E-mail: fnu.jagjitkaur@wmich.edu
2E-mail: Thomas.Gorczyca@wmich.edu

IOP Conf. Series: Journal of Physics: Conf. Series **875** (2017) 052043 doi:10.1088/1742-6596/875/6/052043

Accurate Electron Spin Optical Polarimetry (AESOP)

T.J.Gay[1]

Jorgensen Hall, University of Nebraska, Lincoln, Nebraska 68588-0111 USA

Synopsis The next generation of high-energy parity violation experiments will require knowledge of the electron beam spin polarization to unprecedented accuracy. I propose to develop high-accuracy (< 0.4%) electron optical polarimetry, based on impact excitation of the noble gases to spin triplet Russell-Saunders states. Such measurements will allow us to calibrate, e.g., the 5 MeV Jefferson Lab injector Mott polarimeter to 0.5%.

The ultra-high accuracy parity-violation experiments being planned at Jefferson Lab (MØLLER, PVDIS) and the Mainzer Mikrotron (P2) will require that the electron polarization of the incident beam be known to < 0.5% to optimally interpret the physics asymmetries of order 10^{-9} that they will measure. Current electron polarimetry at these facilities is limited to accuracies ≥ 1% at both their experimental energies (of order 10 GeV) and at their injectors. Improvement at both the low- and high-energy ends of the experiment will be needed to achieve the 0.5% accuracy goal [1].

Mott polarimeters are used at the injectors of both MAMI and JLab, and operate at energies between 3 and 5 MeV. Mott polarimetry requires theoretical calculation of the analyzing power (the "Sherman function") which can currently be done with an accuracy of about 1% at 5 MeV. We thus propose to calibrate these Mott polarimeters to the requisite 0.5% by using electrons of polarization known to better than 0.4%. This would be accomplished by the use of optical electron polarimetry, which has currently been carried out to a *precision* of ~0.8% [2].

Optical electron polarimetry involves the exchange excitation of a well-*LS*-coupled state, and the measurement of the Stokes parameters of its subsequent fluorescence. This process has an analyzing power that can be determined analytically with no error, in the absence of contamination by cascades and strong scattering resonances. If these effects can be well characterized, the challenge becomes one of measurement accuracy alone. Figure 1 shows a design for a prototype polarimeter. It will use 5-cm-diameter optics, very high extinction polarizers, zeroth-order quartz retarders, and an *in vacuo* calibration light source. In order to eliminate the polarization effects of birefringent vacuum windows, we will operate the analysis optical polarimeter inside the vacuum as well.

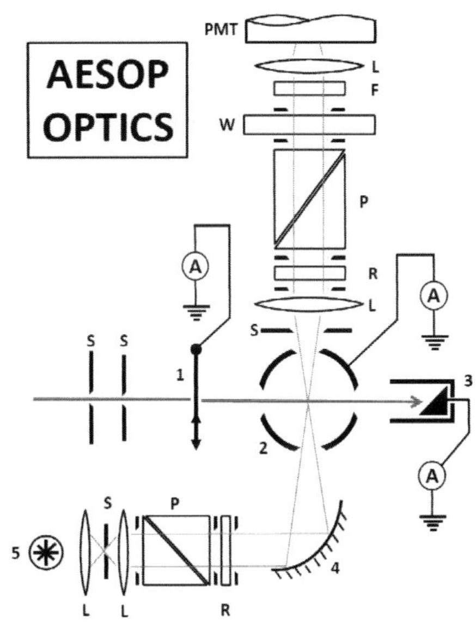

Figure 1. The AESOP setup showing electron or photon beam-defining slits (S), linear polarizers (P), retarders (R), a vacuum window (W), lenses (L), an interference filter (F), ammeters (A), and a photomultiplier detector (PMT). Also indicated are (1) a movable beam-current flag, (2) the effusive target gas containment cylinder, (3) the Faraday cup, (4) a mirror for the calibration source immediately to its left, and (5) a calibration light source. The light source in the vacuum system would use a noble-gas discharge lamp, whose fluorescence is focused at the electron-beam gas intersection point. Its polarization would be varied using the elements shown.

This work is supported by the US NSF (Award PHY – 1632778).

References

[1] J. Grames *et al.* 2004 Phys. Rev. ST-AB 7 042802

[2] T. J. Gay *et al.* 1996 Phys. Rev. A 53, 1623

[1] E-mail: tgay1@unl.edu

On $e^- + \mathrm{Mn}$ elastic scattering at $\varepsilon = 20$ eV impact energy

V. K. Dolmatov[1]

Department of Physics and Earth Science, University of North Alabama, Florence, Alabama, 35632, U.S.A.

Synopsis Current situation relating to 20-eV electron elastic scattering off Mn is discussed and complemented by the author's calculation. Noticeable discrepancies between available calculated and experimental data for large scattering angles are highlighted. Existing understanding of electron scattering off Mn is, thus, inconclusive. Further probes into the matter are urged.

$e^- + \mathrm{Mn}$ elastic scattering is an interesting and important case study in view of both the high spin of $\mathrm{Mn}(3d^5 4s^2, {}^6S)$ and its open $3d^5$ subshell. To date, however, conclusive agreement exists neither between experiment [1] (absolute scale) and [2] (relative scale), nor between experiments and available theoretical data [2, 3, 4, 5, 6] relating to the total $\sigma_{\rm tot}(\varepsilon)$ and differential $d\sigma/d\Omega$ elastic scattering cross sections at the electron energy of $\varepsilon = 20$ eV. Corresponding R-matrix results [2] are inconclusive because, first, they do not agree well with experiment and, second, more complete 83-state R-matrix calculated data are in worse agreement with experiment than a only 5-states calculation. Calculated data [3, 4, 5] (spin-polarized RPAE) are incomplete and, thus, inconclusive because of accounting for too few (for $\varepsilon = 20$ eV) electronic ℓ-partial waves, up to only $\ell_{\rm max} = 5$, for a reason. Some of calculated data of work [6] (spin-polarized LDA) agree somewhat with experiment [1] until about $\theta \approx 50^{\rm o}$, but sharply fall considerably below experiment at larger θs, and calculated $\sigma_{\rm tot} \approx 11 \cdot 10^{-16}$ cm^2 [6] makes it only one-half of experiment [1]. Clearly, another study of $e^- + \mathrm{Mn}$ elastic scattering is in order. This constitutes the aim of the present work.

To meet the goal, the author embeds a semiempirical polarization potential $V_{\rm p}(r)$ into spin-polarized Hartree-Fock equations [3, 5], rather than into spin-polarized LDA equations used by the authors of work [6], to account for the polarization impact on $e^- + \mathrm{Mn}$ scattering. The potential $V_{\rm p}(r)$ is approximated by a static dipole polarization potential in the form of $V_{\rm p} = -\alpha/(r^2 + \bar{r}^2)^2$, for a reason. Here, α and $\bar{r}$ are the static dipole polarizability and mean radius of an atom, respectively. This is a well detailed potential used broadly in atomic physics since 1930th [7]. Excellent agreement between thus calculated $\sigma_{\rm tot} \approx 21 \cdot 10^{-16}$ cm^2 and experimental $\sigma_{\rm tot} = 22 \cdot 10^{-16}$ cm^2 [1] is obtained. Present calculated data for $d\sigma/d\Omega$ are depicted in Fig. 1 along with experiments [1, 2] and theories [2, 6]. Excellent agreement between present theory and experiment [1] persists up to $\theta \approx 65^{\rm o}$. At larger angles, theory falls noticeably below experiment; this is in line with predictions by other theories [2, 6]. Thus, either there exists an important part of the interaction that affects electron scattering at large angles, which either of the reviewed theories fails to underpin, or experiment is incorrect. Hence, knowledge of $e^- + \mathrm{Mn}$ elastic scattering is incomplete. The author urges the renewing of corresponding experimental and theoretical studies of $e^- + \mathrm{Mn}$ elastic scattering, to dot the i's and cross the t's concerning this important case study.

[1]E-mail: vkdolmatov@una.edu

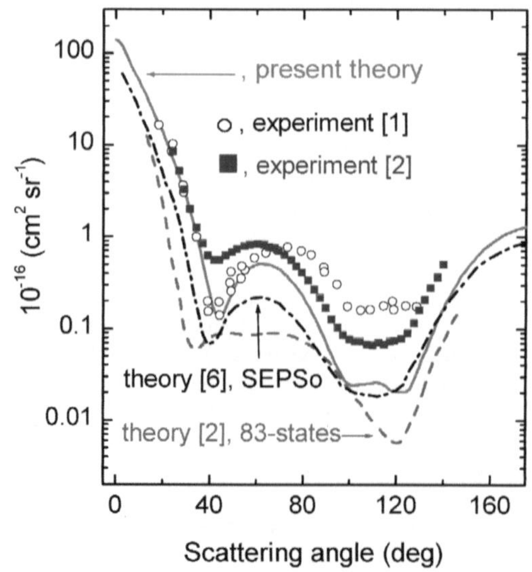

Figure 1. $d\sigma/d\Omega$ for $e^- + \mathrm{Mn}$ elastic scattering at $\varepsilon = 20$ eV. Parameters of theory used in the present calculation: $\alpha \approx 100$, $\bar{r} \approx 3.6$ a.u. and $\ell_{\rm max} = 40$.

References

[1] W. Williams *et al.* 1978 *J. Phys. B* **11** 2031

[2] R. Meintrup *et al.* 2000 *J. Phys. B* **33** L289

[3] M. Ya. Amusia *et al.* 1990 *Sov. Phys. JETP* **70** 632

[4] M. Ya. Amusia *et al.* 2012 *unpublished*

[5] V. K. Dolmatov *et al.* 2013 *Phys. Rev. A* **88** 042706

[6] E. Yu. Remeta *et al.* 2010 *J. Phys. B* **43** 045202

[7] D. R. Bates *et al.* 1933 *Trans. R. Soc. A* **239** 269

 doi:10.1088/1742-6596/875/6/052045

CoBIT Spectroscopy of Mo and Y ions relevant to beyond EUV source development

Emma Sokell* [1], **Safdar Ali**[†2], **John Sheil***, **Hiroyuki Kato**[†], **Nobuyuki Nakamura**[†]¶

*School of Physics, University College Dublin, Belfield, Dublin 4, Ireland,
† Institute for Laser Science, University of Electro-communications, 1-5-1 Chofugaoka, Chofu, Tokyo 182-8585, Japan,

Synopsis A CoBIT has been used to produce Mo and Y ions and EUV spectra have been recorded. The data indicate that the CoBIT spectra are significantly different to laser-produced plasma emission spectra for similar ion stages. This difference is attributed to the different number density of electrons in the two kinds of plasmas.

The relevance of the spectroscopy of highly charged ions for EUV (13.5 nm) and beyond EUV (6.x nm) light source development is well established [1]. In 2009, ASML prompted by the availability of La/B_4C mirrors with a reflectivity of ~40% in a 0.06 nm bandwidth near 6.7 nm, announced that sources would be needed at 6.x nm for future lithography [2]. This would necessitate focus on shorter wavelengths than 13.5 nm which has been extensively studied [1]. Experiments on laser produced plasmas (LPPs) of elements in the second transition row of the periodic table show significant emission in the spectral range 2-8 nm. This is due to 3d-4p and 3d-4f transitions [3]. The present study was conducted using a CoBIT [4] and focuses on two of these elements; Yttrium (Z=39) and Molybdenum (Z=42) in the 5-12 nm range. Spectroscopic studies using electron beam ion traps are complimentary to spectra obtained from LPPs, as the electron beam energy of the CoBIT can be set to optimize the abundance of a particular ion stage in the trap. This study focused on ion stages in the vicinity of Ni-like Y^{11+} and Mo^{14+}, with ionization energies of 374 eV and 544 eV, respectively [5]. For each of the elements studied, the CoBIT spectra are far simpler than those obtained from LPPs. The major difference between the two plasmas is the number density of electrons and this factor strongly influences the lines that appear in the BEUV spectra. In LPPs the number density of electrons is sufficiently high that metastable levels are likely to be collisionally depopulated before they can radiate. In contrast in CoBIT plasmas, such metastable levels live long enough to radiate and may give rise to optically forbidden lines, such as the $3d^{10}$-$3d^{9}4s$ (J=2) line of MoXV [6]. The prominent appearance of these lines in the CoBIT spectra, indicate that the population of the $3d^{9}4s$ state is enhanced within the electron beam ion trap.

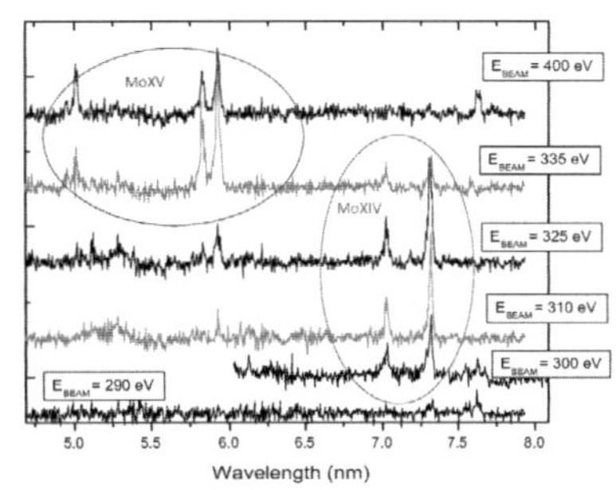

Figure 1. CoBIT spectra of Mo ions obtained at the electron beam energies indicated.

References

[1] G. O'Sullivan *et al* 2015 *J. Phys. B: At. Mol. Opt. Phys.* **48** 144025
[2] E. Louis *et al* 2011 *International Workshop on EUV and Soft X-ray Sources Proceedings,* Dublin, ed. by V. Bakshi, (2011)
[3] R. Lokasini, *et al* 2015 J. Phys. B: At. Mol. Opt. Phys. **48** 245009
[4] N. Nakamura *et al* 2008 Rev. Sci. Inst. **79** 063104
[5] http://physics.nist.gov/PhysRefData/ASD/ionEnergy.html
[6] M. Klapisch *et al* 1978 Phys. Rev. Lett. **41** 403

[1] E-mail: Emma.Sokell@ucd.ie
[2] E-mail: safdaruetian@gmail.com

IOP Conf. Series: Journal of Physics: Conf. Series **875** (2017) 052047 doi:10.1088/1742-6596/875/6/052047

On the effects of Orbital Angular Momentum (OAM) and impact parameters on (e, 2e) process on atoms by twisted electron

Aditi Mandal[1] **and R. Choubisa**[2]

Department of Physics, Birla Institute of Technology and Science (BITS), Pilani, Pilani Campus: 333031 (India)

Synopsis We present our theoretical study of (e, 2e) process on atoms by the twisted electron, with special emphasis given on the effects of OAM and impact parameters on TDCS.

Electron impact scattering/coincidence ionization processes in the atomic physics are important to study the basic few-body Coulomb problem, correlation effects, target structure and the spin dependent effects at microscopic level. Hitherto, most of the reported studies have been confined for the incident electron beam which does not carry any angular momentum and transverse linear momentum (untwisted electron beam). Recently, there is a growing interests to revisit theses processes with the twisted electrons beam [1-2]. Serbo *et al* [2] found that the scattering of the relativistic twisted electron from atoms/solid targets is sensitive to the longitudinal and transverse components of the linear momentum as well as on the projection of angular momentum (OAM) number m of the twisted electron state.

Here, we report our study on the K-shell ionization of the atoms by the twisted electron. To the best of our knowledge, the K-shell (e, 2e) process with the impinging twisted electron is not reported earlier. We compute the triple differential cross sections (TDCS) in different geometrical arrangements of the scattered and ejected electrons and study the TDCS as a function of different projections of the angular momentum and transverse linear momentum of the twisted electron. The calculation has been performed in the first Born approximation in the frozen core approximation wherein the effects of the other bound electrons in the (e, 2e) process are ignored. We use twisted plane wave for the incident electron, plane wave for the scattered electron and hydrogen like wave function with Darwin spinors and Coulomb wave function with Dirac spinors for the bound and the ejected electron respectively. Our preliminary results show that the binary and recoil peak are shifted from the momentum transfer direction for the twisted electron (e, 2e) process. This is highly in contrast with the normal (e, 2e) process by untwisted electron wherein we get symmetric angular profile of TDCS in the first Born approximation with binary and recoil peak in the momentum transfer and opposite to momentum transfer direction respectively. We also observe that the amount of shifting in the peak depends on the projection of angular momentum of the twisted electron state (see figure 1). Our preliminary results in the relativistic electron energy observe more interesting results. The transverse spin asymmetry in the TDCS is found to be dependent on the various parameters which describe the twisted state of the electron beam. We are in process to study the effects of the impact parameters on the TDCS and the transverse spin asymmetry.

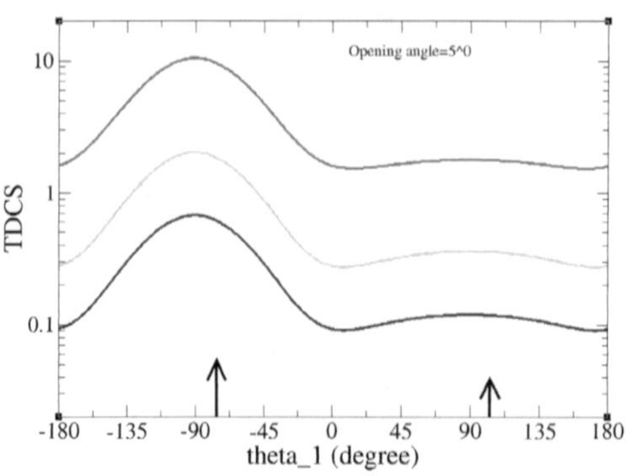

Figure 1. TDCS vs the angle of the ejected electron for the apex angle 5^0 and for m= 0.5 (red curve), 1.5 (green curve) and 2.5 (blue curve) of the incident twisted electron beam carrying energy 5530 eV on He atom. The other kinematics parameters are same as in [3]

References

[1] R. Van Boxem *et al*, 2015, *Physical Review A* 92,032703
[2] V. Serbo *et al*, 2015, *Physical Review A* 92, 012705
[3] Lahmam-Bennani, 1991, *J. Phys. B* 24, 2401

[1] E-mail: p2016009@pilani.bits-pilani.ac.in

[2] E-mai: rchoubisa@pilani.bits-pilani.ac.in

IOP Conf. Series: Journal of Physics: Conf. Series **875** (2017) 052048 doi:10.1088/1742-6596/875/6/052048

Prominent role of indirect processes in electron-impact ionization of Xe^{24+} ions

P. Liu*, **J. Zeng***†, **A. Borovik, Jr.**‡§1, **S. Schippers**§, **and A. Müller**‡

*Department of Physics, College of Science, National University of Defense Technology, Changsha Hunan, PR China

†IFSA Collaborative Innovation Center, Shanghai Jiao Tong University, Shanghai, PR China

§Institut für Atom- und Molekülphysik, Justus-Liebig-Universität Gießen, 35392 Giessen, Germany

‡I. Physikalisches Institut, Justus-Liebig-Universität Gießen, 35392 Giessen, Germany

Synopsis The cross section for electron-impact single ionization of Xe^{24+} has been studied both experimentally and theoretically. The cross section exhibits significant features from resonant ionization processes such as resonant-excitation double autoionization. These are well described by the theoretical calculations if the latter take into account excitations to atomic shells with principal quantum numbers of up to $n = 39$.

Electron-impact ionization of ions is the fundamental atomic collision process. Accurate data on ionization cross sections for atoms and ions are of crucial importance for the correct description and understanding of the processes occurring in ionized gases. With their many-electron nature and rather low binding energies, xenon ions are well suited for studying atomic processes in complex many-electron atomic systems, where electron-impact ionization results from a complicated pattern of ionization mechanisms. Depending on impact energy, the ionization vent might involve the simultaneous rearrangement of several atomic subshells. Previous studies on ionization of xenon ions revealed, that mainly mechanisms of direct ionization (DI) and non-resonant excitation-autoionization (EA) contribute to the single-ionization cross section. For intermediate charge states, resonant processes such as resonant-excitation double autoionization (REDA) yield sizeable contributions, in addition [1, 2].

Here, we report on case where resonant processes are as significant as direct ionization, i.e., single ionization of Xe^{24+} ions [3]. Experimental measurements were performed at the Giessen crossed-beams setup at electron energies ranging from the lowest ionization threshold up to 1000 eV. In addition, fine-structure resolved distorted-wave calculations of the ionization cross section employing the Flexible Atomic Code (FAC) [4] were performed including all relevant partial cross-section contributions. Therefore, DI, EA, as well as REDA processes were considered. Figure 1 shows the comparison of the experimental and the theoretical results. The experimental data exhibit distinct resonant structures which, at certain energies, reach up to 50% of the total measured cross section. According to our calculations, these originate from REDA processes involving, e.g., $3d \rightarrow 4f$ excitations and simultaneous capture of the impacting electron into high-n levels with principal quantum numbers of up to at least $n = 39$. A similarly slow convergence was found for the EA contributions to the total cross section. This strongly suggests, that calculations of accurate cross sections for collision processes involving many-electron ions are, in general, very demanding, and that, conversely, simple scaling formula as often applied in plasmaphysical and astrophysical applications are bound to fail.

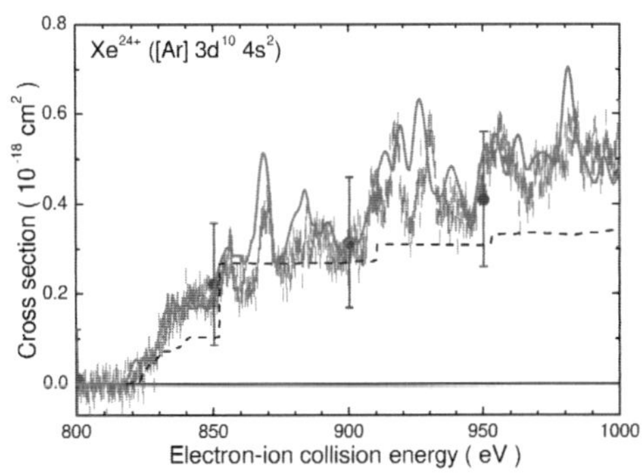

Figure 1. Cross section for electron-impact ionization of Xe^{24+} ions [3]. Symbols represent the experimental data. The solid (red) curve is the result of the theoretical calculation including DI, EA and REDA processes, while the dashed (navy) curve represents the summed contributions of the non-resonant processes DI and EA, only.

References

[1] M. S. Pindzola *et al.* 2013 *J. Phys. B* **46** 215202

[2] A. Borovik, Jr. *et al.* 2015 *J. Phys. B* **48** 035203

[3] P. Liu *et al.* 2015 *Phys. Rev. A* **92** 012701

[4] M. F. Gu 2008 *Can. J. Phys.* **86** 675

1E-mail: Alexander.Borovik@physik.uni-giessen.de

IOP Conf. Series: Journal of Physics: Conf. Series **875** (2017) 052049 doi:10.1088/1742-6596/875/6/052049

Higher-order contribution in the resonance recombination of electron-ion interaction

G.Y. Liang*[1], J. R. Crespo López-Urrutia†, C. Beilmann‡, H.G. Wei*, and G. Zhao*

* Key Laboratory of Optical Astronomy, National Astronomical Observatories, CAS, 100012 Beijing, P.R. China
† Max-Planck-Institut für Kernphysik, Saupfercheckweg 1, 69117 Heidelberg, Germany
‡ Physics Institute, Heidelberg University, Im Neuenheimer Feld 226, 69120 Heidelberg, Germany

Synopsis Due to the correlation between different quantum states, such high-order correlation in the electron-ion resonance recombination can not be resolved in those published state-of-art calculations. Wherever such correlation effect is embodied in the resonant recombination spectroscopy, and were measured on electron beam trap to benchmark theory recently. In this abstract, we theoretically investigate the high-order contribution in the total resonance recombination for carbon-like iron, and it is found to be significant.

Resonance recombination plays a major role on the ionization equilibrium of collisional plasmas [1], and further the temperature diagnostic for the plasmas. Sine about 10 years ago, serial calculations for dielectronic recombination have been done with the state-of-the-art method, that significantly improve the ionization equilibrium data used extensively by astronomical community [2]. Beyond the well-known DR, resonant recombination processes also involve higher-order (HO) correlations, that is, two or even three bound electrons can be simultaneously excited by the resonantly captured electron in trielectronic or even quadruelectronic recombination (TR or QR). However, such HO process are weak and difficult to be decomposed in experiments and theories. By using advanced forced evaporate technology, Beilmann *et al.* [3] disintegrated such HO processes in K-L intershell recombination and found it to be prominent relative to the corresponding DR by comparing their line strength.

In this work, we make a detailed investigation to the HO contributions for carbon-like iron ion including L-L plus L-M, K-L inter- and intra-shell resonant recombinations. We found that such HO contributions are partly included in published Maxwellian-averaged DR rate coefficients. In the state-of-the-art calculation, it is very difficult to disintegrate the TR and/or QR from DR channels. So a decomposition procedure is developed and applied to carbon-like iron ion, and found such K-L intershell HO contribute about ∼35-40% in total resonant recombination rates, and the L-L intrashell plus L-M intershell HO occupies about up to 40% for photoionized plasma zone (at temperatures of 20–70 eV).

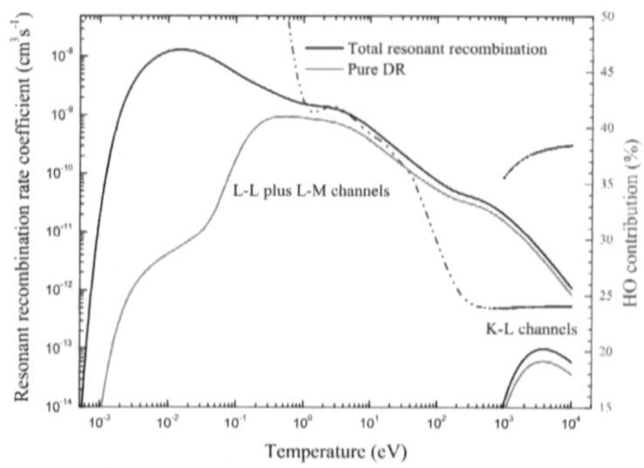

Figure 1. Resonant recombination rate coefficients with and without HO contribution from K-L intershell and L-L intrashell plus L-M intershell cases.

References

[1] N.R. Badnell 2016, *J. Phys. B: At. Mol. Opt. Phys.* **39** 4825

[2] P. Bryans, N.R. Badnell, T.W. Gorczyca, *et al.* 2006 *Astrophys. J. Supp. Ser.* **167** 343

[3] C. Beilmann, P.H. Mokler, S. Bernitt, *et al.* 2011 *Phys. Rev. Lett* **107** 143201

[1]E-mail: gyliang@bao.ac.cn

Low energy elastic cross section of fluoroacetylene by electron impact

Dhanoj Gupta, Heechol Choi and Mi-Young Song

Plasma Technology Research Center, National Fusion Research Institute, 37 Dongjangsan-ro, Gunsan, Jeollabuk-do 54004, South Korea

Synopsis Electron impact low energy cross sections for fluoroacetylene is investigated for which there are no reported results in the literature. We have used various target models to compute accurate target properties and the best target model is used for the final scattering calculation. The fluorination effect in the cross section is studied by comparing the results with difluoroacetylene and acetylene.

Fluoroacetylene (HCCF) is a simple linear structure molecule similar to acetylene (HCCH) and difluoroacetylene (FCCF) and offers a very good opportunity to study the effect of fluorination by replacing the hydrogen atom in HCCH with fluorine atom and by replacing the fluorine atom in the case of FCCF with hydrogen atom in the scattering cross sections. Recently we have investigated FCCF [1] and compared its cross section with HCCH [2] and obtained interesting results in terms of cross section and resonances. Hence it is of great interest to study HCCF with respect to HCCH and FCCF for various cross sections and resonance along with fluorination effect.

The R-matrix method [3] is used for the present study at low energies. The cc-pVTZ basis set is used for the representation of the target wave function with 16 target states included in the calculation. The calculations are done in the C_{2v} point group symmetry. The absolute value of the ground state energy estimated is -175.75 Hartree. The present target model predicts the dipole moment of 0.285 *au* which is in very good agreement with the experimental value of 0.283 *au* [4]. The first excitation energy of 7.09 eV is also in reasonable agreement with the recent calculation of Khiri *et al.* [5] at 6.326 eV. The target properties are listed in Table 1 with available comparison.

Table 1. Target properties for HCCF.

Properties	Present	Others
Absolute Ground state energy (Hartree)	-175.75	-
Dipole moment (au)	0.285	0.283[4]
First excitation energy (eV)	7.09	6.326[5]

The scattering calculations are carried out further with this model and the sample result for elastic cross section (Q_{el}) of HCCF along with comparison of FCCF by us [1] and HCCH by Gauf *et al.* [2] is presented in Figure 1. The π^* resonance is predicted at 5.30 eV for HCCF due to ($^2B_1$, $^2B_2$) states. The π^* resonances for HCCH and FCCF is detected at 2.5 eV by Gauf *et al.* and 5.86 eV by us. It is clearly observed that by replacing the hydrogen atom with fluorine atom shifts the resonance to the higher energy side as seen in the case of HCCH, HCCF and FCCF in Figure 1. In addition to π^* resonance we have also detected another resonance at 3.17 eV due to $^2A_1$ state for HCCF.

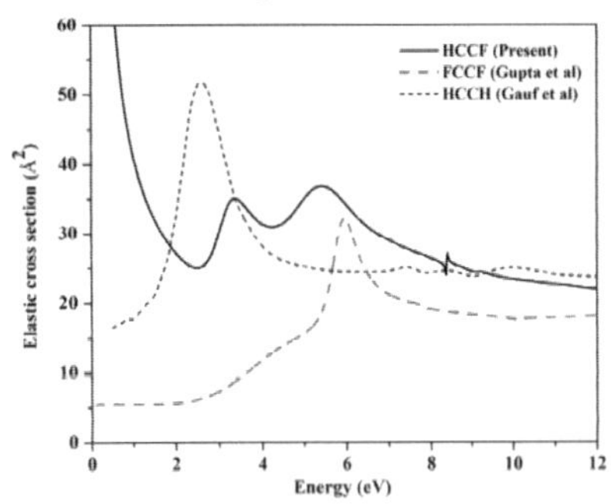

Figure 1. Q_{el} for HCCF, FCCF [1] and HCCH [2].

At low energies below 3 eV, the cross section for HCCF shows sharp rise unlike HCCH and FCCF which is due to dipole moment of the molecule as given in Table 1. The experimental investigations will be useful for further understanding and to verify the present results.

References

[1] D. Gupta *et al.* 2016 *J. Phys B. (under revision)*.
[2] A. Gauf *et al.* 2013 *Phys. Rev. A* **87** 012710.
[3] J. Tennyson 2010 *Phys. Rep.* **491** 29.
[4] T. Tanaka e*t al.* 1976 *J. Mol. Spectrosc.* **63** 142.
[5] D. Khiri *et al.* 2016 *J. Phys. Chem. A* **120** 5985.

E-mail: dhanojsanjay@gmail.com
E-mail: hcchoi@nfri.re.kr
E-mail: mysong@nfri.re.kr

IOP Conf. Series: Journal of Physics: Conf. Series **875** (2017) 062003 doi:10.1088/1742-6596/875/7/062003

Low-energy electron interactions with chromium hexacarbonyl $Cr(CO)_6$

J. Khreis[† 1], J.Ameixa[* 2], M. Neustetter[† 3], J. Reitshammer[† 4], F. Ferreira da Silva[* 5], and S. Denifl[† 6]

[*] Laboratório de Colisões Atómicas e Moleculares, CEFITEC, Departamento de Física, Faculdade de Ciências e Tecnologia, Universidade NOVA de Lisboa, Campus de Caparica, 2829-516 Caparica, Portugal
[†] Institut für Ionenphysik und Angewandte Physik, Leopold-Franzens Universität Innsbruck, Technikerstraße 25, 6020 Innsbruck, Austria,

Synopsis Positive and negative ion formation upon low energy electron interaction with chromium hexacarbonyl precursor molecules used in FEBID applications was studied. The obtained results revealed a noteworthy different fragmentation pattern compared to other precursors. This work contributes to the development and the possible production concerning highly pure nanoscale structures.

The technique of Focused Electron Beam Induced Deposition (FEBID) allows the development of functional three-dimensional nanoscale structures on surfaces using reactions of highly energetic (tens of keV) electrons and metalorganic molecules so called precursors. Major limitations in FEBID occur due to the purity-level of the created deposit, which influences the function and the resistivity of the deposit [1][2]. This limitation results as a consequence of incomplete decomposition of the precursor, i.e., the presence of precursor organic ligands. Another problem is the spatial resolution of the built nanoscale structure which leads to inadvertent surrounding deposit. Therefore the understanding of chemical reactions during FEBID is crucial so that the amount of impurities can be minimized, aiming for a pure deposit. In particular, it is assumed that both the purity and the spatial resolution of the deposit depend on interactions of the metalorganic precursor with secondary low-energy electrons [3]. These low-energy electrons are formed due to the interaction of the primary high-energy electrons with the precursor, the substrate, and the deposit.

Chromium containing compounds are used in FEBID as precursor molecules due to the high purity of the deposit-structure [1][4][5]. In this gas phase study, the interaction of organometallic chromium hexacarbonyl ($Cr(CO)_6$) precursor molecules and low-energy electrons (between ~ 0–70 eV) was investigated. All measurements were carried out using a double focusing two sector field mass spectrometer in reversed Nier-Johnson geometry. This experimental setup allowed the investigation of positive and negative ions as well as the dissociation of metastable ions in the microsecond timescale. The present results can be compared to a previous measurement with the precursor $W(CO)_6$ [6]. Remarkably, a completely different anion yield pattern regarding the loss of CO-ligands is observed.

Like the anionic case, the formation of cations (Figure 1) indicated a strong sequential loss of CO-ligands. Additionally, the bare chromium ion ($Cr^+/Cr(CO)_6$ and $Cr^-/Cr(CO)_6$) was observed. This is in agreement with the high purity deposit measurements of $Cr(CO)_6$ observed in FEBID.

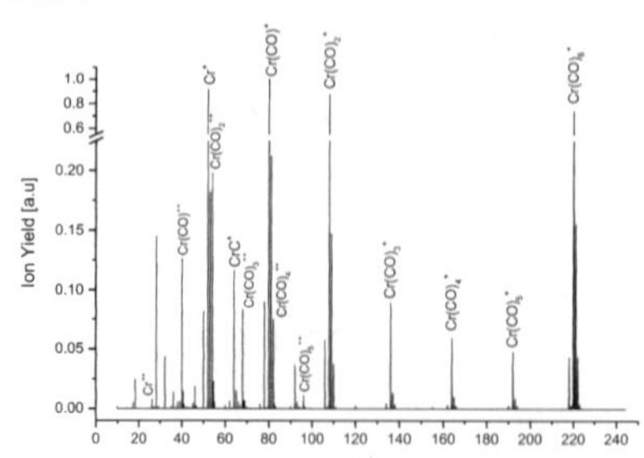

Figure 1. Ionization mass spectrum of chromium hexacarbonyl obtained at the electron energy of ~70 eV

E-mail: fr.silva@fct.unl.pt

E-mail: ste-phan.denifl@uibk.ac.at

References

[1] A. Botman *et al.* 2009 *Nanotechnology* **20** 372001
[2] M. Huth *et al.* 2012 *Beilstein J. Nanotechnol.* **3** 597-619
[3] R. Thorman *et al.* 2015 *Beilstein J. Nanotechnol.* **6** 1904-1926
[4] N. Kislov *et al.* 1996 *Scanning* **18** 114-118
[5] R. Kunz *et al.* 1988 *J. Vac. Sci. Technol. B.* **6** 1557-1564
[6] K. Wnorowski *et al.* 2012 *Rapid Commun. Mass Spectrom.* **26** 2093-2098

IOP Conf. Series: Journal of Physics: Conf. Series **875** (2017) 062004 doi:10.1088/1742-6596/875/7/062004

Shake-up processes in Auger Cascades of Light and Medium Elements

R. Beerwerth[*1], **S. Fritzsche**[†2],

[*] Helmholtz-Institut Jena, Fröbelstieg 3, 07743 Jena, Germany
[†] Theoretisch-Physikalisches Institut, Friedrich-Schiller-Universität Jena, 07743 Jena, Germany

Synopsis Many-electron computations are applied to model the de-excitation of inner-shell excited atoms by a cascade of Auger decays. In particular, the effects of shake-up transitions on Auger cascades of different elements are discussed together with various theoretical models that are needed to account for the underlying electron-electron correlations.

In recent years, coincidence spectroscopy of photo and Auger electrons helped to investigate the de-excitation of atoms, molecules and solids. These techniques are, for example, used at synchrotrons to analyze the emission of multiple electrons due to Auger processes that follow the excitation or ionization of inner-shell electrons. The total kinetic energy of the emitted electrons enables one to obtain information about the spectrum and population of the final states, while the individual electron energies reveal details about the intermediate states, and therefore the decay pathways of an Auger cascade.

To analyze such Auger cascade processes, we have extended the RATIP program [1] to model the multiple emission of electrons after the creation of inner-shell hole states. Our studies reveal that many Auger cascades are strongly affected by shake-up (or shake-down) transitions, in which the two-electron Auger process is accompanied by an additional (de-) excitation of a valence electron. A careful treatment of the inter-electronic correlation is necessary to describe shake-up transitions in the computation of the Auger transition rates. A subsequent analysis of all possible decay paths allows to obtain quantities that are easily accessible in experiments, such as ion yields, the population of final and intermediate fine-structure levels and electron spectra.

Fig. 1 shows the electron spectrum that arises in the Auger decay of 1s $\rightarrow$ 3p resonantly excited neon to the $1s^2 2s^2 2p^4 np$ configurations of Ne^{1+}. In this transition about 30% of the decays result in the additional excitation of one 3p electron. We also show the experimental data from [2] that was obtained by a multi-electron coincidence method and that agrees very well with our computed result.

The formation of Auger cascade processes that leads to the multiple ionization of atoms after the ionization or excitation of a single inner-shell electron is in some cases strongly affected by shake-up transitions. For example, the photo ionization of a 4p electron in atomic cadmium, as well as the Auger decay of this hole state, can be accompanied by an additional excitation of one 5s valence electron [3]. The gain in energy due to this shake-up transition is sufficient to allow the emission of a second Auger electron. Therefore, a two-step cascade can be observed that would otherwise be forbidden energetically.

Another example for the prominent influence of shake-up transitions is the de-excitation of resonantly excited negative oxygen ions [4]. Complex electron correlation effects lead to a strong contribution of shake transitions to the total decay width. Here, the population of higher lying intermediate states even enables the occurrence of three-step Auger cascade decays that are otherwise not possible energetically.

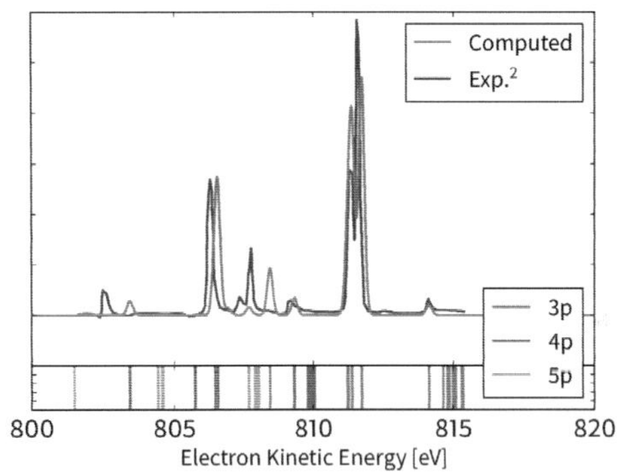

Figure 1. Auger electron spectrum of the first-step decay of 1s $\rightarrow$ 3p resonantly excited neon. This spectrum is particularly influenced by shake-up transitions 3p $\rightarrow$ 4p, and is here compared with experimental data from Ref. [2].

References

[1] S. Fritzsche, Comput. Phys. Commun. **183**, 1525–1559 (2012).

[2] Y. Tamenori and I. H. Suzuki, J. Phys. B: At. Mol. Opt. Phys. **47**, 145001 (2014).

[3] J. Andersson, R. Beerwerth, et al., Phys. Rev. A **92**, 023414 (2015).

[4] S. Schippers, R. Beerwerth, et al., Phys. Rev. A **94**, 041401 (2016).

[1]E-mail: Randolf.Beerwerth@uni-jena.de
[2]E-mail: S.Fritzsche@gsi.de

Electron-Impact Ionization of Laser-Aligned Atoms – Contributions from both Natural and Unnatural-Parity States.

Andrew James Murray*[1], James Colgan[£], Don Madison[†], Matthew Harvey* and Ahmad Sakaamini*

[*] Photon Science Institute, School of Physics & Astronomy, University of Manchester, Manchester M13 9PL, UK.
[£] Los Alamos Labs, Los Alamos, New Mexico, NM 87545, USA.
[†] Department of Physics, Missouri Science & Technology, Missouri, Rolla, MO 65409, USA.

Synopsis. The progress of experimental and theoretical measurements for (e,2e) ionization cross sections from laser-aligned atoms is presented here. It is found that both natural and unnatural parity contributions must be included in the models to emulate the experimental data.

In this work experimental and theoretical results will be presented for (e,2e) ionization measurements from laser-excited and aligned atoms. The experimental data are taken in Manchester, whereas theoretical results are from the groups of Don Madison in Missouri, and James Colgan at Los Alamos Labs in the USA.

The motivation for these studies arises since time-independent distorted wave (DWBA, 3DW) models predict zero flux for atoms aligned orthogonal to the scattering plane (as in fig 1,2), in disagreement with experimental data. By contrast, time-dependent close coupling (TDCC) models predict a *non-zero* cross-section under these conditions, and conclude it is the *unnatural parity* contributions to the cross section that produce this flux [1-4]. An unnatural parity state has parity $(-1)^{L+1}$, compared to a natural parity state that has parity $(-1)^{L}$.

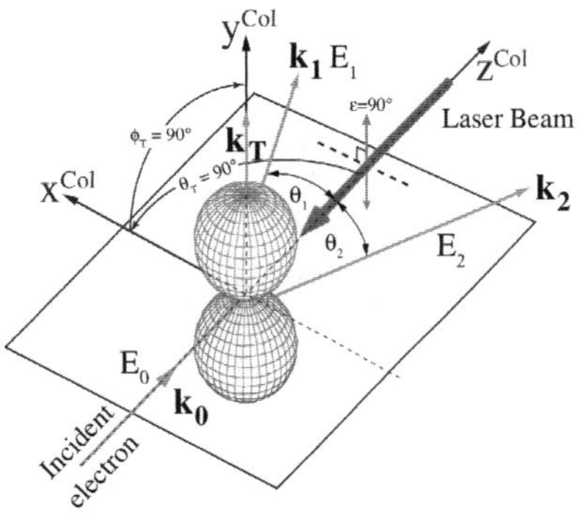

Figure 1. Ionization from atoms aligned orthogonal to the scattering plane using linearly polarized radiation at ε = 90°. DW theories predict the cross section to be identically zero in this configuration, whereas the TDCC model predicts a finite cross section, as found by experiment.

It is important in these studies that the laser-excited P-state is fully aligned (L=1, m_L=0), with minimum contribution from any orientation of the target (L=1, $m_L = \pm1$). For this reason alkali-earth atoms are chosen since they have no hyperfine structure, and so can be aligned to better than 99% accuracy. Mg, Ca and Sr are to be used in these studies, so the effects of mass can also be determined.

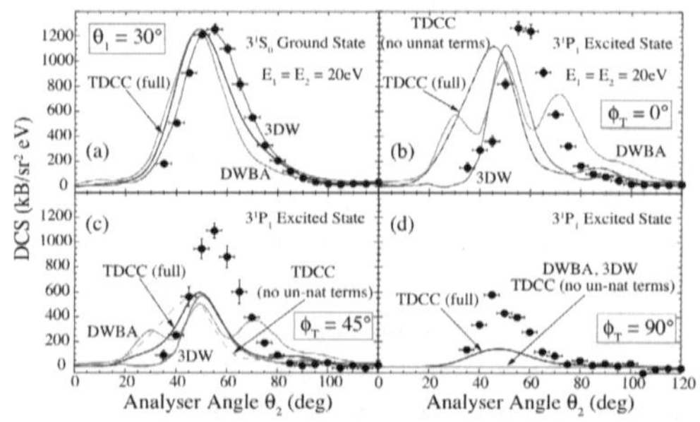

Figure 2. Comparison of experiment and theory for Mg, in the 3^1S_0 ground state and 3^1P_1 excited state at different angles ϕ_T to the plane. The data are normalized to the TDCC calculation for the 3^1S_0 state. The TDCC calculations are shown both with and without unnatural parity contributions.

The atoms are aligned using continuous wave radiation. Cross-sections will be determined for the incident electron in the scattering plane (as in fig. 1), and for out-of-plane geometries. The laser radiation will be injected into the interaction region through angles determined by theory, so that the cross-section sensitivity to different parity contributions can be explored.

The progress of this combined study will be presented here.

References

[1] KL Nixon *et al.* 2011 *Phys Rev Lett* **106** 123201
[2] KL Nixon *et al.* 2014 *Phys Rev Lett* **112** 023202
[3] S Amami *et al.* 2014 *Phys Rev A* **90** 062707
[4] G Armstrong *et al.* 2015 *Phys Rev A* **92** 032706

[1] E-mail: Andrew.Murray@manchester.ac.uk

(e,2e) Ionization Studies of N_2 at Low to Intermediate Energies from a Coplanar Geometry to the Perpendicular Plane.

Ahmad Sakaamini*, Matthew Harvey*, Sadek Amami†, Andrew James Murray*[1], Don Madison† and Chuangang Ning£

* Photon Science Institute, School of Physics & Astronomy, University of Manchester, Manchester M13 9PL, UK.
† Department of Physics, Missouri Science & Technology, Missouri, Rolla, MO 65409, USA.
£ Dept. of Physics, State key lab of low dimensional quantum physics, Tsinghua University, Beijing 100084, China.

Synopsis. The progress of experimental and theoretical measurements for (e,2e) ionization cross sections from Nitrogen molecules is presented. Results are given for energies from ~10 eV above the ionization potential (IP) through to ~100 eV above the IP for the $3\sigma_g$, $1\pi_u$ and $2\sigma_g$ states.

Ionization triple differential cross sections (TDCS) have been determined experimentally and theoretically for neutral N_2 over a range of geometries and energies, from a coplanar geometry through to the perpendicular plane. Data were obtained at incident electron energies from ~10 eV to ~100 eV above the ionization potential (*IP*) of the $3\sigma_g$, $1\pi_u$ and $2\sigma_g$ states, using equal and non-equal outgoing electron energies, and using symmetric and asymmetric geometries. Data were taken with the incident electron beam in the scattering plane ($\psi = 0°$), as well as at angles $\psi = 45°$ and $\psi = 90°$ (see figure 1).

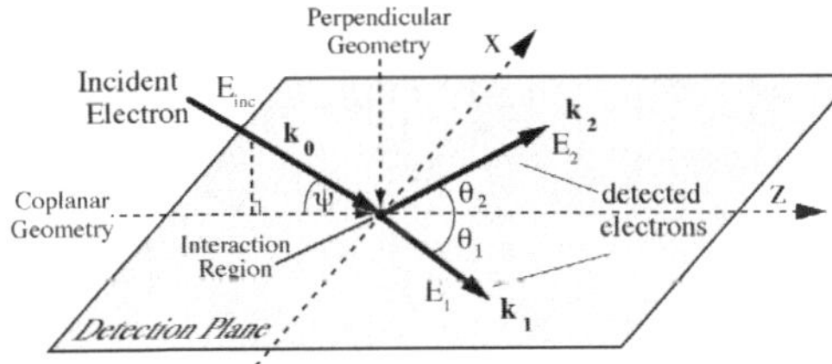

Figure 1. The (e,2e) geometry for these experiments. The incident electron can move from a coplanar geometry ($\psi = 0°$) to the perpendicular plane ($\psi = 90°$) while the analyzers rotate in the detection plane.

The measured differential cross sections at a given energy were inter-normalized to each other by linking the data through a set angle. Binding energy spectra were obtained at each energy, so relative cross sections could be obtained for the different ion states. An example of binding energy spectra is shown in figure 2 at three different incident electron angles, for outgoing electrons having 4.6 eV energy.

The experimental data are compared to new calculations using various distorted wave methods, and differences between theory and experiment are discussed. New results for non-equal angles in a coplanar geometry are also presented, where one of the analyzers is fixed in position (θ_1 = constant) while the second (θ_2) sweeps around the detection plane. Results from these studies are linked to the symmetric results through their common angle when ($\theta_1 = \theta_2$), so that all data are then normalized to each other at a given energy.

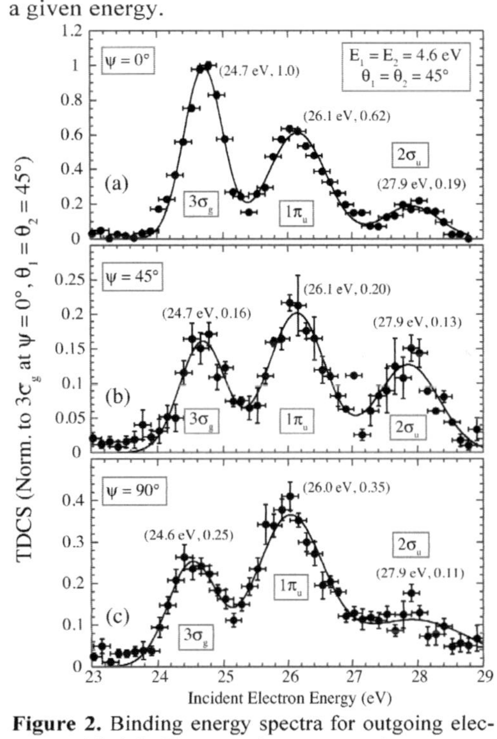

Figure 2. Binding energy spectra for outgoing electrons with 4.6 eV energy in a symmetric geometry, for the electron gun at different angles [1].

The progress of these combined experimental and theoretical studies will be presented.

References

[1] A Sakaamini *et al.* 2016 *J Phys B* **49** 195202

[1] E-mail: Andrew.Murray@manchester.ac.uk

Dissociative electron attachment to CO molecule probed by velocity slice imaging technique

Pamir Nag [1] **and Dhananjay Nandi** [2]

Indian Institute of Science Education and Research Kolkata, Mohanpur 741246, India

Synopsis We have studied dissociative electron attachment to CO molecule using the well-established velocity slice imaging spectrometer. We have conclusively determined the symmetries of the TNI states involved in both the channels producing O^- ions. Also in contrast to a recent report we observed that there is no need to invoke coherent interference between different states. Recent calculations and experimental study by other groups also strongly support our claims.

Low-energy electron-molecule collision leading to dissociative electron attachment (DEA) is an important process from the fundamental as well as the application point of view. DEA is a two-step resonant process resulting into a final anionic and neutral fragments from a parent neutral molecule via intermediate temporary negative ion (TNI) state. DEA study of molecules are important starting from electrical discharges, atmospheric chemistry, installer medium chemistry to radiation induced damage of living cell and biologically important molecules.

Carbon monoxide (CO) is a simple but important heteronuclear diatomic molecule. But only few studies on DEA to CO exist in literature using modern techniques like velocity map imaging or momentum imaging. DEA to CO have two possible channels for O^- production with energy threshold 9.62 and 10.88 eV and are shown by equation 1 and 2 respectively.

$$e^- + \mathrm{CO}(^1\Sigma^+) \rightarrow \mathrm{CO}^{-*} \rightarrow \mathrm{O}^-(^2P) + \mathrm{C}(^3P) \quad (1)$$

$$e^- + \mathrm{CO}(^1\Sigma^+) \rightarrow \mathrm{CO}^{-*} \rightarrow \mathrm{O}^-(^2P) + \mathrm{C}^*(^1D) \quad (2)$$

From the angular distribution measurements over a limited angular range Hall *et al.* [1] concluded a TNI state with Π symmetry is responsible for both the processes. In contrast, Tian *et al.* [2] reported that for 10 eV incident electron energy Σ and Π two states are involved but, for 10.6 eV three TNI states, Π, Δ and Φ, with coherent interference between them are present. To address these issues we have studied DEA to CO [3] using a recently developed velocity slice imaging (VSI) spectrometer similar to that was reported by Nandi *et al.* [4] with few modifications. VSI technique is a well-established method for simultaneous measurement of the kinetic energy and angular distribution of the fragment anions produced due to DEA over the entire 2π angle.

From the systematic kinetic energy distribution measurements we have identified both the processes. To determine the symmetry of the TNI states we have measured angular distributions of the O^- ions produced due to both the processes separately. We observed a Σ state is primarily responsible for both the processes. In addition a minor contrbution from a Π state also exists for process 1 and the contribution increases with increasing incident electron energy. In a recent R-matrix calculation Dora *et al.* [5] also observed only a Σ resonant state. From the angular distribution measurements we also have observed that there is no need to invoke the coherent interference between different states as suggested by Tian *et al.* In a recent experimental study Gope *et al.* [6] also observed similar results and strongly supported our conclusions.

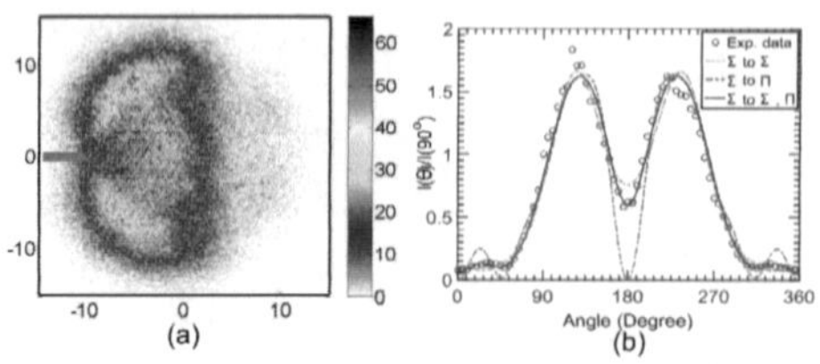

Figure 1. (a) The VSI of O^- ions for 11 eV beam energy and (b) angular distribution of the ions created due to process I for 11 eV beam energy along with fitted curves.

References

[1] R. I. Hall *et al.* 1977 *Phys. Rev. A* **15** 599

[2] S. X. Tian *et al.* 2013 *Phys. Rev. A* **88** 012708

[3] P. Nag and D. Nandi 2015 *Phys. Chem. Chem. Phys.* **17** 7130

[4] D. Nandi *et al.* 2005 *Rev. Sci. Instrum.* **76** 053107

[5] A. Dora *et al.* 2016 *Eur. Phys. J. D* **70** 197

[6] K. Gope *et al.* 2016 *Eur. Phys. J. D* **70** 134

[1]E-mail: pamir1118@iiserkol.ac.in
[2]E-mail: dhananjay@iiserkol.ac.in

IOP Conf. Series: Journal of Physics: Conf. Series **875** (2017) 062008 doi:10.1088/1742-6596/875/7/062008

Dipolar dissociation dynamics in electron collisions with carbon monoxide

Dipayan Chakraborty [1], **Pamir Nag** [2] **and Dhananjay Nandi** [3]

Indian Institute of Science Education and Research Kolkata, Mohanpur 741246, India

Synopsis Dipolar dissociation (DD) dynamics of carbon monoxide with electron collision has been studied using velocity slice imaging (VSI) technique. Threshold of the process has been obtained from the ion yield curve. Using the time sliced images kinetic energy and angular distribution have been obtained providing the detailed dynamics involved in the DD process. Two ion-pair states have been identified from the angular distribution measurements.

Low-energy (<100 eV) electron-molecule collision study is an important tool to unlock many unsolved problems in various field of science. Unlike dissociative electron attachment (DEA), dipolar dissociation (DD) process is a non-resonant process where the incident electron transfers (partial or complete) energy to the molecule and leaves it to a super excited state known as ion-pair state. Such ion-pair state then dissociates forming ion-pair as :

$$CO + e^- \rightarrow CO^* + e^- \rightarrow \begin{cases} C^+ + O^- + e^- \\ C^- + O^+ + e^- \end{cases} \quad (1)$$

Due to low C^- cross section, only O^- dissociation channel is focused here. The threshold value obtained from ion-yield curve is 19.8 eV, which is in good agreement with the previous reports [1]. The velocity sliced images have been taken at five different incident electron energies 25, 30, 35, 40 and 45 eV. Presence of lower and higher kinetic energy (KE) ions throughout the entire energy range of studied confirms two dissociation mechanism [2]. Low KE ions originate through indirect process where the molecule initially excited to a Rydberg state, that crosses the ion-pair state near ion-pair dissociation limit. With increasing electron energy same Rydberg state is accessed throughout and thus low KE energy ions. Higher ion KE peak for 25 eV incident electron energy centred at 1.2 eV whereas, an unchanged KE behaviour around 1.5 eV is observed with increasing incident energy. The initial increase in kinetic energy indicates the direct excitation to the different repulsive part of the ion-pair states in the Franck Condon (FC) region. But as the incident electron energy increases, the states are out of the FC transition window and by partial energy transfer an unchanged KE peak observed. Another possibility is due to involvement of another higher Rydberg state. The position of the ion-pair state can also be located using the KE information. According to conservation of momentum 1.5 eV O^- ions will be accompanied by 2 eV C^+ ions, thus the total KE release sholud be 3.5 eV. The truncated shape seen at the KE distribution for 25 eV incident energy clearly indicates that the ion-pair state enters the FC region around that energy. To know the symmetry of associated ion-pair states angular distribution of O^- ions for both KE regions are fitted using the formula given by Van Brunt [3]. We conclude that Σ and Π ion-pair states are involved for both dissociation processes. To quantify the result detailed theoretical potential energy curve calculations are needed.

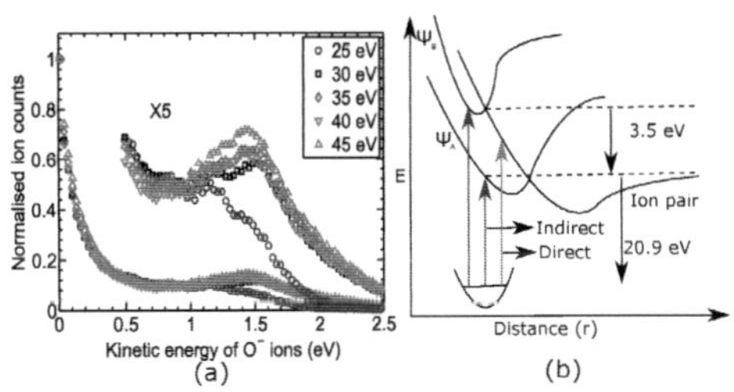

Figure 1. (a) KE distribution of O^- ions and (b) Schamatic diagram to understand the dissociation processes.

References

[1] W. W. Lozier 1934 *Phys. Rev.* **46** 268

[2] D. Chakraborty *et al.* 2016 *Phys. Chem. Chem. Phys.* **18** 32973

[3] R. J. Van Brunt 1974 *J. Chem. Phys.* **60** 3064

[1]E-mail: dc14rs010@iiserkol.ac.in
[2]E-mail: pamir1118@iiserkol.ac.in
[3]E-mail: dhananjay@iiserkol.ac.in

IOP Conf. Series: Journal of Physics: Conf. Series **875** (2017) 062009 doi:10.1088/1742-6596/875/7/062009

New insights in low-energy electron-fullerene interactions

Alfred Msezane† [1], **Zineb Felfli**† [2]
† Department of Physics and CTSPS, Clark Atlanta University, Atlanta, Georgia 30314, USA

Synopsis Regge pole calculated low-energy electron scattering total cross sections (TCSs) for ground and first excited states of fullerenes are generally robust; for C_{20} they resemble those of atomic Au. Here we explore the variation of R, the ratio of second to first Ramsauer-Townsend minima of excited states TCSs, from C_{20} through C_{70}. We find that for C_{20} R is close to that for Au (> 1.) while for C_{24} R is about unity. For C_{70} R is less than 0.5 demonstrating strong departure from atomic behavior due to significant polarization interaction in C_{70}; which also induces long-lived metastable anions in C_{70}.

The discovery of the now most studied C_{60} fullerene [1] inspired vigorous searches for new fullerenes, leading to the discovery of the smallest fullerene thought to exist, C_{20}. Its electron affinity (EA) was measured to be 2.25 eV [2]. Other small fullerenes of interest here are C_{24}, C_{26}, C_{28} and C_{44}, some of which have measured EAs available.

This investigation has been motivated by: 1) Amusia concluded that fullerenes can be viewed as "big atoms" [3]; 2) Rich resonances have been observed to characterize low-energy electron-fullerene interactions; 3) In [4] the Regge pole calculated low-energy electron elastic total cross sections (TCSs) for ground states C_{60} through C_{92} were found to be characterized by dramatically sharp resonances located at the second Ramsauer-Townsend (R-T) minima of their TCSs. The energy positions of these sharp resonances were generally in outstanding agreement with the measured EAs for the investigated fullerenes and 4) The calculated TCSs revealed polarization induced long-lived metastable fullerene anions [4].

The departure from atomic behavior is studied through the ratio R focusing on the green curve of TCSs for C_{20} through C_{70}. In the Fig. the first R-T minima for both C_{20} and Au are lower than the second, indicative of atomic behavior. For C_{26}, R is about unity and for C_{70} it is < 0.5, demonstrating complete departure and conversion of the 0.466 eV resonance to a long-lived metastable anion in C_{70}.

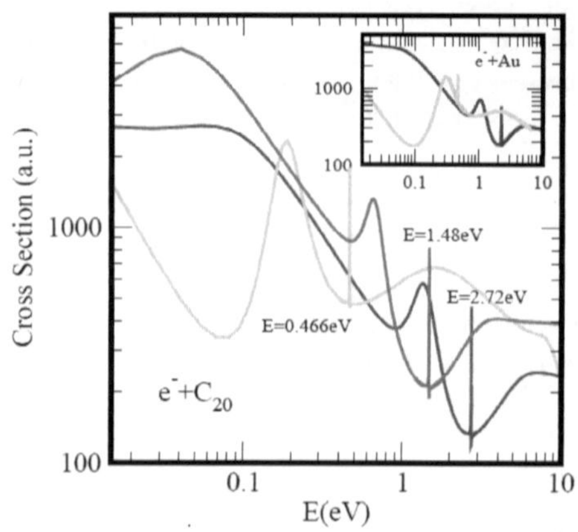

Figure 1: Total cross sections (a.u.) for electron elastic scattering versus E (eV) for C_{20}. The purple, pink and green curves represent results for the ground, induced metastable and first excited states, respectively; the insert represents TCSs for atomic Au. The dramatically sharp resonances correspond to fullerene anions formation during the collisions.

Research is supported by U.S. DOE, OBES, OS

References

[1] H. W. Kroto *et al.* 1985 *Nature* **318** 162
[2] H. Prinzbach *et al.* 2000 *Nature* **407** 60
[3] M. Ya Amusia 2013 *Chem Phys* **414** 168
[4] Z. Felfli and A. Z. Msezane 2017 *J. Phys. B* (Submitted).

[1] E-mail: amsezane@cau.edu
[2] E-mail: zfelfli@cau.edu

IOP Conf. Series: Journal of Physics: Conf. Series **875** (2017) 062010 doi:10.1088/1742-6596/875/7/062010

Binding energies of fullerene and complex atomic negative ions

Alfred Msezane[† 1], **Zineb Felfli** [† 2]

[†] Department of Physics and CTSPS, Clark Atlanta University, Atlanta, Georgia 30314, USA

Synopsis A robust potential which embeds fully the vital core polarization interaction has been used in the Regge-pole method to explore low-energy electron scattering from C_{60}, Eu and Nb through the total cross sections (TCSs) calculation. From the characteristic dramatically sharp resonances in the TCSs manifesting negative ion formation in these systems, we extracted the binding energies (BEs) for the C_{60}^-, Eu^- and Nb^- anions. Our BE for C_{60}^- ground state is in outstanding agreement with the measured electron affinity (EA) of C_{60} [1]. For anionic Eu^- and Nb^- our BEs are compared with the measured EAs [2-4].

Progress toward a fundamental theoretical understanding of the mechanism underlying low-energy electron scattering from heavy and complex atoms, including fullerenes, leading to stable negative ion formation has been very slow. The presence of many intricate and diverse electron configurations characteristic of low-energy electron interactions in these systems leads to computational complexity yielding uncertain electron affinities (EAs). The representation of the fullerene cage potential continues to plague theoretical studies as well.

Our Regge pole approach [5], completely new in the field of clusters/fullerenes, should be considered as an investigation toward a fundamental understanding of low-energy electron collisions with fullerenes and complex atoms. The Regge-pole method is appropriate for investigating fullerene negative ion formation as resonances since Regge poles, singularities of the S-matrix, rigorously define resonances. Fully embedded in the Regge pole method are the crucial electron-electron correlations and the vital core polarization interaction. These effects are mostly responsible for the existence and stability of typical negative ions.

Figure 1 exhibits the calculated e-C_{60} TCSs (the inset shows e-Nb TCSs for comparison). Indeed, the TCSs are characterized by Ramsauer-Townsend minima, shape resonances and dramatically sharp resonances corresponding to fullerene anions formation during the collisions. The BE at 2.66 eV (blue curve) agrees excellently with the EA [1] while for Nb the measured EA[4] agrees with our excited state BE of 0.902 eV and not with the ground state BE of 2.48 eV. This has created a quandary.

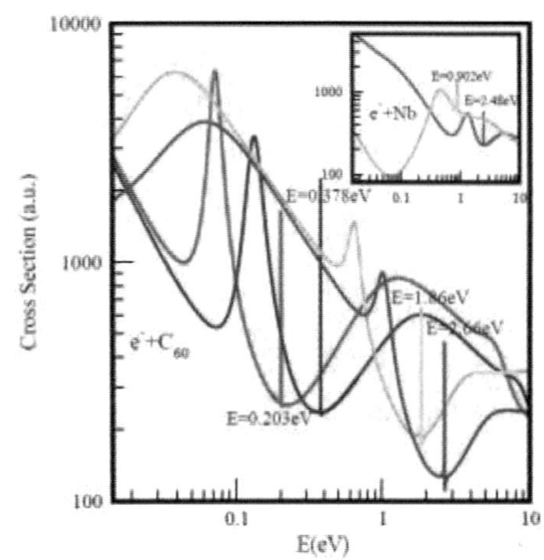

Figure 1: TCSs (a.u.) for e-C_{60} elastic scattering. The blue curve represents the ground state TCS; the other curves correspond to long-lived metastable and excited states. The inset represents TCSs for atomic Nb for comparison.

Research is supported by U.S. DOE, OBES, OS

References

[1] D. –L. Huang *et al* 2014 *J. Chem Phys.* **140** 224315
[2] S. -B. Cheng and A. W. Castleman 2015 *Sci. Rep.* 5 12414
[3] V. T. Davis and J. S. Thompson 2004 *J. Phys. B* 37 1961
[4] Z. Luo *et al.* 2016 *Phys. Rev. A* **93**, 020501(R) 93 020501(R)
[5] D. Sokolovski *et al.* 2007 *Phys. Rev. A* 76 012705

[1] E-mail: amsezane@cau.edu
[2] E-mail: zfelfli@cau.edu

IOP Conf. Series: Journal of Physics: Conf. Series **875** (2017) 062011 doi:10.1088/1742-6596/875/7/062011

Resonances in low-energy electron elastic scattering from Fullerenes

Zineb Felfli[†1] **and Alfred Msezane**[†2]

[†] Department of Physics and CTSPS, Clark Atlanta University, Atlanta, Georgia 30314, USA

Synopsis The benchmarked Regge-pole method on the calculated low-energy electron elastic scattering total cross section (TCS) for C_{60} through the measured electron affinity (EA) [1,2] is used to calculate TCSs for selected fullerenes, from C_{54} through C_{240} in the electron impact energy range $0.02 \leq E \leq 10.0$ eV. From the characteristic dramatically sharp resonances representing long-lived ground state fullerene negative ion formation we extracted the binding energies (BEs) and compared them with the measured EAs, obtaining outstanding agreement.

The EA provides a stringent test of theory when the calculated and measured EAs are compared. A strong motivation for the fundamental investigations of low-energy electron elastic scattering from the selected fullerenes is the availability of high quality measured EAs [1-7]. The Regge pole calculated TCSs are found to be characterized generally by Ramsauer–Townsend (R-T) minima, shape resonances and dramatically sharp resonances manifesting stable negative ion formation. The extracted BEs for the resultant anions agree excellently with the measured EAs of the considered fullerenes giving great credence to the Regge pole method.

Table 1 displays the outstanding capability of the Regge pole method to calculate BEs of fullerene anions over a wide range of fullerenes. The method can also be used for complex atoms Fig. 1 displays the typical TCSs for electron-C_{76} scattering, exhibiting rich resonances.

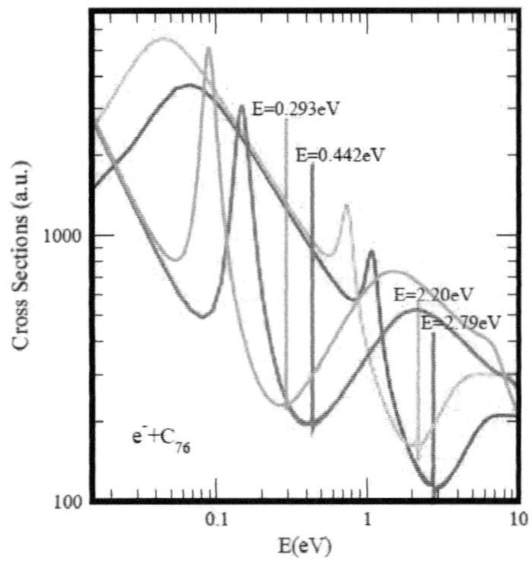

Fig. 1: TCSs for electron- C_{76} scattering.

Table 1: Comparison between the measured EAs (eV) and the present BEs (eV) for the investigated fullerenes, C_{54} through C_{240}.

System	EA-Expt. (eV)	BE-Present (eV)
C_{60}	2.666±0.001[2] 2.6835±0.0006[1]	2.66
C_{70}	2.72±0.05[4] 2.676±0.001 [2] 2.765(0.01)[3]	2.70
C_{74}	3.28±0.07[4] 3.28±0.07 [5]	3.21
C_{76}	2.88 ± 0.05[4] 2.89±0.05[5]	2.79
C_{78}	3.01±0.07[4] 3.10±0.06[5]; 3.10±0.01[6]	2.98
C_{54}	--	3.14
C_{80}	3.17 ± 0.06[4] 3.17±0.06 [5]	3.28
C_{82}	3.14±0.06[4] 3.14±0.06[5]	3.15
C_{84}	3.05±0.08[4] 3.14(6)[5]; 3.185±0.01[6] 3.16[7]	2.94
C_{86}	≥ 3.0 [4]	2.92
C_{90}	≥ 3.0 [4]	3.06
C_{92}	≥ 3.0 [4]	3.09
C_{180}		2.54
C_{240}		2.41

Research supported by U.S. DOE, OBES, OS

References

[1] D.-L. Huang *et al.* 2014 *J. Chem. Phys* **140** 224315
[2] C. Brink *et al.* 1995 *Chem. Phys. Lett.* **233** 52
[3] X. B. Wang *et al* 2006 *Phys Rev. Lett.* **96** 143002
[4] O. V. Boltalina *et al.* 1993 *J. Rapid Commun. Mass. Spectrom.* **7** 1009
[5] O. V. Boltalina *et al.* 1996 *Chem. Phys. Lett.* **256** 253
[6] X.–B. Wang *et al* 2007 *J. Phys. Chem. C* **111** 17684
[7] O. V. Boltalina *et al* 1997 *J. Phys. Chem. A* **101** 9561

[1] E-mail: zfelfli@cau.edu

[2] E-mail: amsezane@cau.edu

Integral cross sections of the dipole-allowed excitations of nitrogen studied by fast electron scattering and X-ray scattering

Ya-Wei Liu[1], **Long-Quan Xu, Dong-Dong Ni, Xin Xu, Xin-Chao Huang, Lin-Fan Zhu**[2]

Hefei National Laboratory for Physical Sciences at Microscale, Department of Modern Physics, University of Science and Technology of China, Hefei, Anhui 230026, People's Republic of China

Synopsis The integral cross sections (ICSs) of the electron scattering of nitrogen molecule are of great importance to the understanding of many different aspects of atmospheric physics. Based on the generalized oscillator strengths (GOSs) of these dipole-allowed transitions crosschecked by the high-energy electron scattering and high-resolution X-ray scattering, their ICSs are obtained systematically from the threshold to 5000 eV for the first time with the aid of the *BE*-scaling method, and the corresponding ICSs at the moderate and high energies are the only available data to the best of our knowledge.

As the primary atmosphere constituent of Earth and some satellites such as Titan and Triton, the inelastic cross sections of nitrogen by electron scattering, especially the ICSs, are of great importance to the understanding of many different aspects of atmospheric physics, such as the determination of the sample electron spectra for different electron environments at different altitudes of the atmosphere of Earth and the heating mechanisms of Titan's upper atmosphere. So the determination of the ICSs of nitrogen has been a subject of continuously increasing interests both experimentally and theoretically.

Recently, based on the scaled plane-wave Born models Kim has developed the *BE*-scaling method to calculate the ICSs of the dipole-allowed excitations of atoms and molecules [1,2]. Based on our present and previous GOSs of nitrogen measured by the high-energy electron scattering [3] and high-resolution X-ray scattering [4], the ICSs of the dipole-allowed excitations of nitrogen are obtained systematically from the threshold to 5000 eV for the first time [3], and the corresponding ICSs at the moderate and high energies are the only available data to the best of our knowledge.

It is clear from Fig.1 that the present ICSs for b(1-3) and b(4)+b'(0) are in good agreement with the recent ones of Malone *et al.* [5], while the ICSs of Malone *et al.* [6] at 60 eV and Ajello *et al.* [7] at 100 eV are lower than the present ones. As for the ICSs of Zipf and Gorman [8], they are larger than the other results below 60 eV and lower than our results at $E_0 >$ 60 eV although it is expected that the present *BE*-scaling results are more reliable at higher incident electron energies.

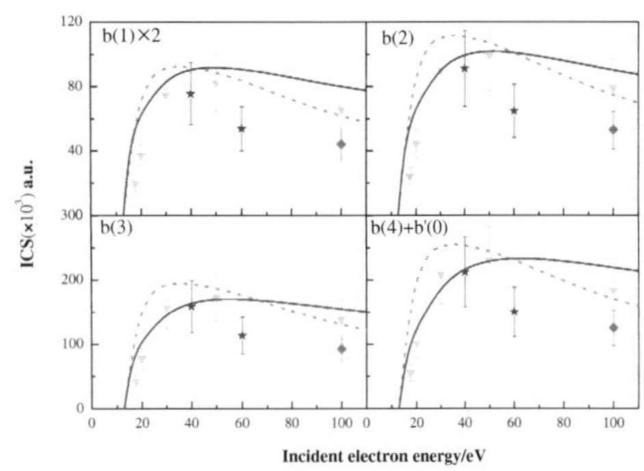

Fig.1 The ICSs for b(1-3) and b(4)+b'(0). The blue solid line is the present *BE*-scaling result [3]. The green triangle is the result of Malone *et al.* [5]; the violet star is the result of Trajmar *et al.*[6]; the red diamond is the result of Ajello *et al.* [7]; the pink dash line is the result of Zipf and Gorman [8] updated by Malone *et al.* [5].

In summary, the present ICSs of nitrogen is accurate, which can serve as the basics input parameters of the theoretical models to simulate the processes in planetary atmosphere and explain the astronomical observations.

References

[1] Y.K. Kim, 2001 *Phys. Rev. A.* 64, 032713.
[2] H. Tanaka, M.J. Brunger, *et al.*, 2016 *Rev. Mod. Phys.* 88, 025004.
[3] Y.W. Liu, L.Q. Xu *et al.*, 2017, *J. Geophys. Res.* (accepted).
[4] Y.G. Peng, X. Kang, K.Yang *et al.*, 2014 *Phys. Rev. A* **89**, 032512
[5] C.P. Malone, P.V. Johnson *et al.*, 2012 *Phys. Rev. A.* 85, 062704.
[6] S. Trajmar, D.F. Register and A. Chutjian, 1983 *Phys. Rep.* 97, 219.
[7] J.M. Ajello, M.H. Stevens *et al.*, 2007 *Geophys. Res. Lett.* 34, L24204.
[8] E.C. Zipf and M.R. Gorman, 1980 *J. Chem. Phys.* 73, 813.

[1]E-mail: liuyawei@ustc.edu.cn
[2]E-mail: lfzhu@ustc.edu.cn

IOP Conf. Series: Journal of Physics: Conf. Series **875** (2017) 062013 doi:10.1088/1742-6596/875/7/062013

Positive ion mass spectrometry for fragmentation of anthracene by low energy electron impact

M. Dunne*, M. L. Gradziel*, and P. J. M. van der Burgt* [1]

* Department of Experimental Physics, National University of Ireland Maynooth, Maynooth, Co. Kildare, Ireland

Synopsis. We have studied the fragmentation of anthracene molecules induced by low energy electron impact. A reflectron time-of-flight mass spectrometer has been used to mass resolve and detect positively charged fragments. Using computer controlled data acquisition, mass spectra have been measured as a function of electron impact from 5 to 100 eV. The mass spectra show clear indications of double and triple ionization. Ion yield curves and appearance energies for most of the positive ions have been extracted from the data.

Polycyclic aromatic hydrocarbons have environmental relevance, because they are formed as by-products in the combustion of organic materials. They are believed to be a major component of the interstellar medium, because they are considered to be responsible for the emission features that dominate the infrared spectra of many galactic and extragalactic sources [1].

In order to obtain information about electron-induced fragmentation patterns of anthracene ($C_{14}H_{10}$), we have measured mass spectra of positive ions of anthracene, with electron energies ranging from 5 to 100 eV in steps of 0.5 eV.

A resistively heated oven generates a beam of anthracene molecules. The oven is mounted in an expansion chamber and the forward section of the beam effusing from a capillary in the oven passes through a skimmer into the collision chamber, where the beam is crossed by a pulsed electron beam (0.3 μs, 8 kHz). Positively charged fragments are extracted into a reflectron time-of-flight mass spectrometer. LabVIEW based data acquisition techniques are used to measure mass spectra as a function of electron impact energy. More details about the experiment and the data analysis can be found in [2-4], which present results obtained for electron impact on some of the nucleobases.

Ion yield curves for most fragment ions of anthracene have been obtained by fitting groups of adjacent peaks in the mass spectra with sequences of normalized Gaussians. Appearance energies have been determined by fitting a threshold function of the form $f(E) = c(E - E_0)^p$ (for $E > E_0$) to the ion yield curves. This function is convoluted with a Gaussian to account for the electron beam energy resolution.

We observe clear indications of multiple ionization in the mass spectra. The triply-charged parent is weakly present at 59.3 u, just separated from the 60-65 u group of fragments, and the appearance energy is 45.5 ± 0.5 eV. In the 83-91 u group of fragments, half-integer peaks are clearly present. The threshold for double ionization is not entirely clear. The 89 u ion has an appearance energy of 20.2 ± 0.7 eV and a second onset at 23.6 ± 0.6 eV.

Mass spectra, ion yield curves and appearance energies will be presented at the conference and will be compared with other research on anthracene and other polycyclic aromatic hydrocarbons (eg. [5-7]).

References

[1] A. G. G. M. Tielens 2013 *Rev. Mod. Phys.* **53**, 1021
[2] P. J. M. van der Burgt 2014 *Eur. J. Phys. D* **68**, 135
[3] P. J. M. van der Burgt, F. Mahon, G. Barrett, and M. L. Gradziel 2014 *Eur. J. Phys. D* **68**, 151
[4] P. J. M. van der Burgt, S. Finnegan, and S. Eden 2015 *Eur. J. Phys. D* **69**, 173
[5] S. Denifl, B. Sonnweber, J. Mack, L.T. Scott, P. Scheier, K. Becker and T.D. Märk 2006 *Int. J. Mass Spectrom.* **249–250**, 353
[6] G. Reitsma, H. Zettergren, S. Martin, R. Brédy, L. Chen, J. Bernard, R. Hoekstra and T. Schlathölter 2012 *J. Phys. B: At. Mol. Opt. Phys.* **45**, 215201
[7] R. Brédy, C. Ortéga, M. Ji, J. Bernard, L. Chen, G. Montagne and S. Martin 2013 *Phys. Scr.* **T156**, 014042

[1] E-mail: peter.vanderburgt@nuim.ie

IOP Conf. Series: Journal of Physics: Conf. Series **875** (2017) 062014 doi:10.1088/1742-6596/875/7/062014

Positive ion mass spectrometry for fragmentation of 5-fluorouracil by low energy electron impact

M. A. Brown* and P. J. M. van der Burgt* [1]

* Department of Experimental Physics, National University of Ireland Maynooth, Maynooth, Co. Kildare, Ireland

Synopsis. Fragmentation of 5-fluorouracil in the gas phase has been studied using low energy electron impact. Positive ions have been mass resolved and detected using a reflectron time-of-flight mass spectrometer. Mass spectra have been measured as a function of electron impact from 5 to 100 eV, and ion yield curves and appearance energies for most of the positively charged fragments have been extracted.

The halo-uracils are important in radiation treatment therapy as radiosensitizers, because cancer tissues doped with these sensitizer molecules will be destroyed preferentially under exposure with ionizing radiation. Because secondary electrons formed along the tracks of the ionizing radiation play an important role in radiation damage processes on the molecular scale, low-energy electron impact studies of radiosensitizers are of relevance [1, 2].

We have measured mass spectra of positive ions for electron impact on 5-fluorouracil, with electron energies ranging from 5 to 100 eV in steps of 0.25 eV. A beam of 5-fluorouracil was generated by a resistively heated oven mounted in an expansion chamber, and the forward section of the beam effusing from a capillary in the oven passed through a skimmer into the collision chamber, where the beam was crossed by a pulsed electron beam (0.5 μs, 8 kHz). Ions have been mass resolved and detected using a reflectron time-of-flight mass spectrometer. LabVIEW based data acquisition techniques have been used to accumulate mass spectra as a function of electron impact energy. Details about the experiment and the data analysis can be found in [3-5], which present results obtained for electron impact on cytosine, thymine and adenine.

Ion yield curves for most fragment ions of 5-fluorouracil have been obtained by fitting groups of adjacent peaks in the mass spectra with sequences of normalized Gaussians. Appearance energies have been determined by fitting a threshold function of the form $f(E) = c(E - E_0)^p$ (for $E > E_0$) to the ion yield curves. This function is convoluted with a Gaussian to account for the electron beam energy resolution.

Mass spectra, ion yield curves and appearance energies will be presented at the conference and will be compared with other research on uracil and the 5-halo-uracils (see e.g. [6-8]).

References

[1] L. Sanche 2005 *Eur. J. Phys. D* **35**, 367
[2] E. Alizadeh, T. M. Orlando and L. Sanche 2015 *Annu. Rev. Phys. Chem.* **66**, 379
[3] P. J. M. van der Burgt 2014 *Eur. J. Phys. D* **68**, 135
[4] P. J. M. van der Burgt, F. Mahon, G. Barrett, and M. L. Gradziel 2014 *Eur. J. Phys. D* **68**, 151
[5] P. J. M. van der Burgt, S. Finnegan, and S. Eden 2015 *Eur. J. Phys. D* **69**, 173
[6] S. Denifl, S. Ptasińska, B. Gstir, P. Scheier and T.D. Märk 2004 *Int. J. Mass. Spectrom.* **232**, 99
[7] M. Imhoff, Z. Deng and M. A. Huels 2007 Int. J. Mass. Spectrom. **262**, 154
[8] E. Itälä, D. T. Ha, K. Kooser, E. Rachlew, M. A. Huels and E. Kukk 2010 *J. Chem. Phys.* **133**, 154316

[1] E-mail: peter.vanderburgt@nuim.ie

IOP Conf. Series: Journal of Physics: Conf. Series **875** (2017) 062015 doi:10.1088/1742-6596/875/7/062015

Differential Measurements of Ionization and Excitation of H_2 by 250 eV Positrons

O. G. de Lucio* [1]**and R. D. DuBois**† [2]

*Instituto de Física, Universidad Nacional Autónoma de México, Apdo. Postal 20-364, México DF, México
†Dept of Physics, Missouri University of Science and Technology, Rolla, MO 65409, USA

Synopsis Ejected electron-recoil ion and scattered projectile-recoil ion coincidences are measured for 250 eV positron impact on molecular hydrogen. The electron detector is also sensitive to photon emission. Various combinations of these coincidence spectra are used to extract differential information about ionization and excitation of H_2 .

For some time we have been studying differential ionization for positron impact with the main goals being to obtain kinematic complete information and to make direct comparisons between positron and electron impact data[1,2]. Here, we report a preliminary analysis of differential ionization and excitation of H_2 resulting from 250 eV positron impact. These data were obtained by measuring ejected electron-recoil ion and scattered projectile-recoil ion coincidences. Alone, or in combination, these coincidence data provide doubly and triply differential information about ionization of H_2. However, because the electron and recoil detectors are also sensitive to photons, differential information about excitation and ionization-excitation processes can also be obtained.

The position-sensitive electron channelplate, located perpendicular to the beam direction, was sensitive to ~17% of the full solid angle. The recoil detector was sensitive to ions extracted from the beam-target overlap region plus photon emission within ~1.5% of the full solid angle. Projectiles scattered between approximately ±5° (theta angle) and ±2.4° (phi angle) were energy analyzed and detected by a position-sensitive channelplate. Multi-hit electronics initiated coincidence events for both channelplates with a delayed signal from the recoil detector providing a common stop.

The figures show a preliminary interpretation of some of the results. The upper figure shows detection of either electrons or photons (the hit 1 curve) or electrons associated with photons (the hit 2 curve) by the electron detector in coincidence with detection of recoil target or background ions. The lower figure curves show the same coincidences with the additional restrictions that a scattered projectile was detected and the recoil ion was H_2^+.

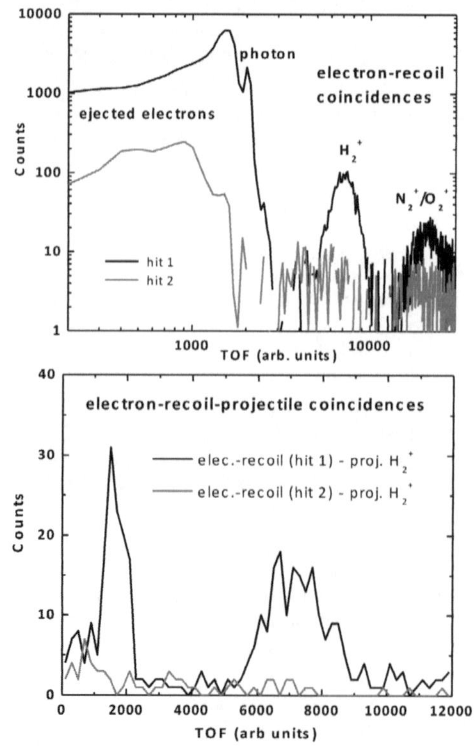

Figure 1. Coincidence spectra associated with ionization, excitation, and ionization-excitation of H_2 and background gases. Upper figure: all projectile scattering; lower figure: restricted to small scattering angles and H_2^+ production.

Acknowledgements: Dr. de Lucio is supported by UNAM-PAPIIT contracts TA100213 and IN116916.

References

[1] O. G. de Lucio et al. 2006 Phys. Rev. Lett. 97 243201.
[2] O. G. de Lucio et al. 2016 Phys. Rev. A 93 032710.

[1] E-mail: olucio@ciencias.unam.mx
[2] E-mail: dubois@mst.edu

IOP Conf. Series: Journal of Physics: Conf. Series **875** (2017) 062016 doi:10.1088/1742-6596/875/7/062016

Investigation of low energy electron interaction with silane molecules

Arvind Kumar Jain

Department of Physics, M. L. N. College, Yamuna Nagar, 135 001, Haryana, India

Synopsis Total and momentum transfer cross sections are calculated for the scattering of electrons from SiH_4 molecules in electron energy region of 0.1 – 20 eV. A parameter–free spherical optical potential (SOP) approach in the fixed nuclei approximation is applied. The present calculated results are compared with the available other calculations and the experimental measurements. The quantitative features of the scattering parameters (such as Ramsauer - Townsend minimum and shape resonance structure) as observed in experiments, are well reproduced in this study.

Knowledge of scattering cross sections for interaction of electrons with silane (SiH_4) molecules are very important in many fields of research including industries, space science, surface modification technique and even in fundamental chemistry. In the silicon hydrides, particularly silane (SiH_4) is the most important constituent used frequently in low temperature plasmas. It is also required in the fabrication of micro electronic devices and semi conducting component [1]. They are also used as surface coating agents for surface modification and as coupling agent for adhesion promotion between dissimilar materials [2]. In the glass industry [3], SiH_4 is used for the deposition of silicon - based layers, especially for car heat reflective windshields manufacturing that allow light but not heat to penetrate the glass.

Many groups have investigated the total cross sections for electron collisions with a large number of polyatomic molecules [4, 5]. In the present study, our aim here is to carry out *ab initio* calculations for the total (TCS) and the momentum transfer cross sections for electron scattering from SiH_4 molecules in the energy range 0.1 – 20 eV. A variable phase approach using parameter free spherical optical potential (SOP) model has been employed here. The projectile - target interaction is represented by a sum of three local and real terms, namely the static, the exchange and polarization potentials. All these potentials can be generated easily once the target charge density $\rho(\boldsymbol{r})$ is known. We determine $\rho(\boldsymbol{r})$ from a single-centre wave functions with enough terms in the expansion of each bound orbital. $\rho(\boldsymbol{r})$ is expanded in terms of symmetry – adapted functions belonging to the totally symmetry $^1A_1$ irreducible representation of the molecular point group. Present calculated TCS are compared with the available theoretical calculations and experimental measurements in the energy region 0.1 – 1.0 eV (see Figure 1). The quantitative feature such as Ramsauer – Townsend minimum is well reproduced in the present study. The details of the results will be presented in the conference.

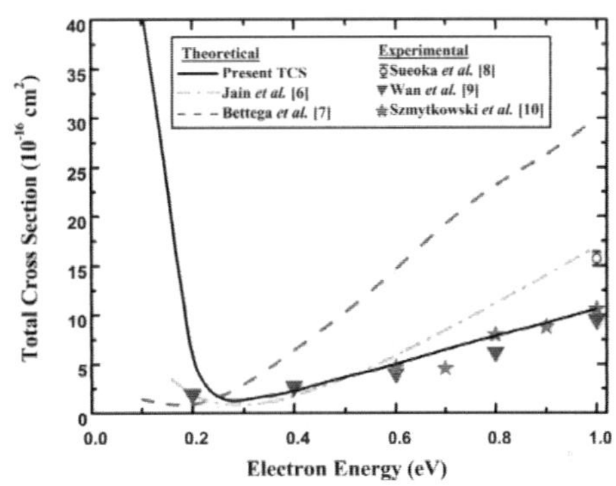

Figure 1. TCS for *e* – SiH_4 scattering in the energy range 0.1 – 1.0 eV.

References

[1] B. Bruneau *et al.* 2014 *J. Appl. Phys.* **115** 084901.
[2] H. Chen *et al.* 2015 *Adv. Powder Tech.* **26** 618.
[3] T. Yavuz and O. Eraslan 2016 *J Adv Prostho-dont.* **8** 75.
[4] M. A. Khakoo *et al.* 2016 *Phys. Rev. A* **93** 012710.
[5] G. Kaur *et al.* 2015 *Phys. Rev. A* **91** 022702.
[6] A. K. Jain *et al.* 1987 *J. Phys. B* **20** L389.
[7] M. H. F. Bettega *et al.* 2003 *Phys. Rev. A* **68** 012706.
[8] O. Sueoka *et al.* 1994 *J. Phys. B* **27** 1453.
[9] H.-X. Wan *et al.* 1989 *J. Chem. Phys.* **91** 7340.
[10]Cz. Szmytkowski *et al.* 1997 *J. Phys. B* **30** 4363.

E-mail: arvindjain61@gmail.com

IOP Conf. Series: Journal of Physics: Conf. Series **875** (2017) 062017 doi:10.1088/1742-6596/875/7/062017

Laser-assisted electron momentum spectroscopy of H_2^+

Andrew A. Bulychev[*1], **Konstantin A. Kouzakov**[†2]

[*] Laboratory of Theoretical Physics, Joint Institute for Nuclear Research, Moscow Region, 141980 Dubna, Russia
[†] Department of Nuclear Physics and Quantum Theory of Collisions, Faculty of Physics, Lomonosov Moscow State University, 119991 Moscow, Russia

Synopsis Electron impact ionization of the hydrogen molecular ion in the presence of laser radiation is studied theoretically. The kinematical regime of high impact energy and large momentum transfer is considered. The influence of a laser field on the target is taken into account using a quasistatic states model, whereas its effect on the incoming and outgoing electrons is described by the Volkov functions. For the field-dressed target, the molecular bond oscillation factor, extracted from the electron momentum distribution, demonstrates a significantly different behavior compared to the case of the unperturbed target.

Electron momentum spectroscopy (EMS) is a well-established technique that utilizes an $(e,2e)$ ionization to study electron momentum distributions in atoms and molecules [1]. Recent advances in the EMS have allowed to combine the $(e,2e)$ EMS spectrometer with a laser system, which has opened the door to several novel applications of this method. One of them is the laser-assisted EMS [2], where the EMS measurements are conducted on the atomic and molecular targets in the presence of laser radiation. This allows one to probe the laser effects on the electronic structure of the investigated target.

While the laser-assisted EMS has not been implemented experimentally yet, the theoretical analysis of its potential has already been carried out for the simplest atomic targets, such as the H and He atoms. In the present work, we extend this method to molecular targets, performing a theoretical analysis for the case of the laser-assisted EMS of the H_2^+ ion. The EMS kinematics involves high impact energy and large momentum transfer in such a way that the $(e,2e)$ ionizing collision is very close to the free electron-electron collision. This validates the use of the binary-encounter and first Born approximations for the S matrix of the process. In order to discard the possible photoionization and photodissociation effects and to neglect the laser influence on the nuclear motion, the case of a low-frequency laser field of relatively weak intensity is inspected. The laser field action on the target electron is taken into account within a quasistatic states approach, which goes beyond the usual perturbation theory.

In the case of a laser field with frequency $\hbar\omega = 1.55\,\mathrm{eV}$ and intensity $I = 1.55\times10^{11}\,\mathrm{W/cm^2}$, we found that the laser-induced modification of the target electron momentum distribution can be seen most clearly by comparing the bond oscillation profiles [3] corresponding to the field-dressed and unperturbed states of the target. Even for the moderate field strength, the bond oscillation interference factor exhibits notable sensitivity to the employed model of the dressed target state [4] (see Fig. 1).

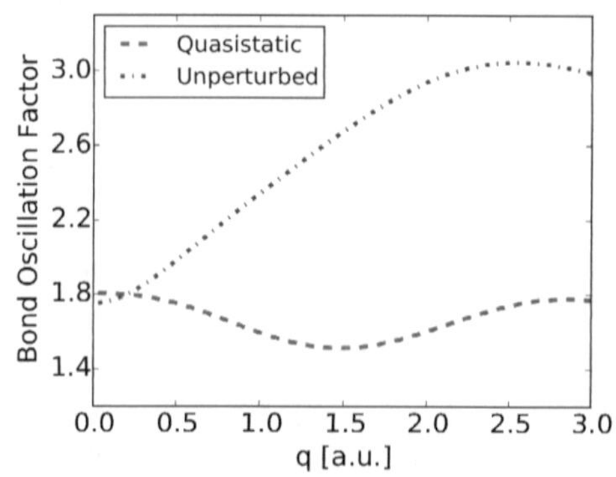

Figure 1. The bond oscillation factor for the cases of the quasistatic and unperturbed target states when the collision is accompanied by emission of one photon.

The work of A.A.B. was supported by the Russian Foundation for Basic Research (Grant No. 16-32-00428).

References

[1] E. Weigold, I. E. McCarthy 1999 *Electron Momentum Spectroscopy* (Kluwer Academic/Plenum Publishers, New York)

[2] K. A. Kouzakov, Yu. V. Popov, M. Takahashi 2010 *Phys. Rev. A* **82**, 023410

[3] M. Yamazaki, H. Satoh, N. Watanabe, D. B. Jones, M. Takahashi 2014 *Phys. Rev. A* **90**, 052711

[4] A. A. Bulychev, and K. A. Kouzakov 2017 *Eur. Phys. J. D* **71**, 23

[1]E-mail: bulychev@theor.jinr.ru
[2]E-mail: kouzakov@srd.sinp.msu.ru

N_2 dissociative ionization by electron impact

Y. Zhang*, X. Wang* [1], L. F. Zhu†, D. Lu*, R. Hutton*, Y. Zou*, and B. Wei* [2]

* Institute of Modern Physics, Department of Nuclear Science and Technology, Fudan University, Shanghai 200433, China

† Hefei National Laboratory for Physical Sciences at Microscale, Department of Modern Physics, University of Science and Technology of China, Hefei, Anhui 230026, China

Synopsis Dissociative single, double and triple ionization of N_2 induced by 200, 400 and 1000 eV electron impact are studied using the cold target recoil-ion momentum spectroscopy (COLTRIMS) technique. Kinetic energy releases (KERs) are determined for dissociation channels ($N_2^+ \rightarrow N^+ + N$, $N_2^{2+} \rightarrow N^{2+} + N$) and Coulomb explosion channels ($N_2^{2+} \rightarrow N^+ + N^+$, $N_2^{3+} \rightarrow N^{2+} + N^+$). It is observed that as the electron-impact energy increasing, different channels behave quite differently. Through analyzing the electronic states of the precursor molecular ion in each channel, the present KER measurements show that ionization from different valance orbitals plays a very important role in the dissociation of molecules.

Nitrogen, as a diatomic molecule with a triple bond, is in abundance in planetary atmospheres, *e.g.* Titan's upper atmosphere. The dissociative ionization dynamics or kinematics of N_2 following the interaction with electrons has been the research focus for many years. In collisions with electrons, N_2 molecules might dissociate subsequent to ionization, thereby resulting in the formation of fragments accompanied by the kinetic energy release (KER). In this work, taking advantage of the COLTRIMS technique [1], we report the dissociative single, double and triple ionization kinematics of N_2 molecules impact by 200, 400 and 1000 eV electrons, covering another electron-impact energy range.

The KER spectra of different ions or channels resulting from N_2 impact by 200, 400 and 1000 eV electrons respectively are shown in Figure 1. The N^+ yield from the dissociative channel $N^+ + N$ can be mainly separated into two parts – from the $F^2\Sigma_g^+$ state with an increasing relative cross section and the H band of N_2^+ with a decreasing relative cross section as the electron-impact energy increases, which can be explained by the ionization from the inner valance orbital of nitrogen. The charge symmetric decay (CSD, $N_2^{2+} \rightarrow N^+ + N^+$) channel is mainly originated through two electrons removal from the outer valance orbitals of N_2, while the charge asymmetric dissociation (CAD, $N_2^{2+} \rightarrow N^{2+} + N$) channel is formed through one hole in $2\sigma_g$ orbital and one hole in an outer valance orbital.

The present KER measurements of N_2 show that ionization from different valance orbitals plays a very important role in the dissociation of molecules. For nitrogen, ionization from the inner valence orbital $2\sigma_g$ gives rise to the abundance change of N^+ ions or neutral N atoms in the channel $N_2^+ \rightarrow N^+ + N$ and on the other hand determines the charge distribution of N_2^{2+}, i.e. the dissociation channel (CSD or CAD). Moreover, although the response of the KER distributions of different dissociation channels to the increase of the electron-impact energy is very different, the contributors or peak energies of the KER distributions at different impact energies are not changed for each channel. This phenomenon indicates that for a Coulomb explosion channel or a non-Coulomb-explosion channel, the KER distributions depend much more on the properties of the molecular ions, i.e. electronic states with different potential energy curves and branching ratios, rather than the electron-impact energy.

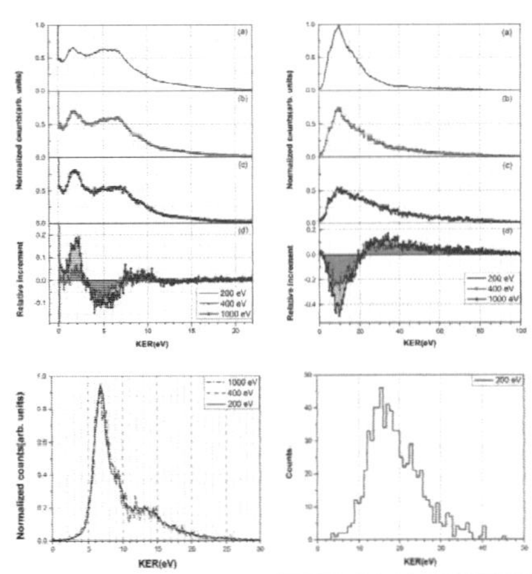

Figure 1. Upper panel: The normalized KER spectra of N^+ (left) and N^{2+} (right); Lower panel: The KER spectra of the channel $N_2^{2+} \rightarrow N^+ + N^+$ (left) and $N_2^{3+} \rightarrow N^{2+} + N^+$ (right).

References

[1] Y. Zhang *et al.* 2014 *Nucl. Instr. and Meth. B* **337**, 39.

[1] E-mail: xinchengwang@fudan.edu.cn
[2] E-mail: brwei@fudan.edu.cn

A space-charge-effect-compensated electron monochromator for electron-impact multi-coincidence measurements

T. Okumura, Y. Mori, H. Akasaka, K. Hosaka, M. Kitajima[1], and N. Kouchi

Department of Chemistry, Tokyo Institute of Technology, Meguro-ku, Tokyo 152-8551, Japan

Synopsis A new experimental apparatus was developed for electron-impact multi-coincidence measurements at high energy resolution. A new monochromator which was designed to take the space charge effect into account was developed. It is shown that the new monochromator is able to produce a monochromatized electron beam with high current and a high energy resolution which is sufficient to carry out the multi-coincidence measurements.

Dissociative ionization of molecules by low-energy electron impact is a fundamental process of great importance which produces reactive fragments driving pursuing reactions [1]. A detailed information on the electron-impact dissociative single ionizaiton of a molecule AB as

$$\mathrm{AB} + \mathrm{e_i^-}(E_\mathrm{i}) \longrightarrow \mathrm{AB}^+ + \mathrm{e_s^-}(E_\mathrm{s}) + \mathrm{e_e^-}(E_\mathrm{e}) \\ \hookrightarrow \mathrm{A} + \mathrm{B}^+(E_\mathrm{ion}), \quad (1)$$

where E_j $(j = \mathrm{i, s, e, or\ ion})$ is the kinetic energy of the particle, can be studied by coincidence detection of the three particles, $\mathrm{e_s^-}, \mathrm{e_e^-}$, and B^+. Only recently such experiments have been achieved [2] although insufficient energy resolution prevented complete determination of the dissociation dynamics. In the present study, we developed a new experimental apparatus for the coincidence measurements of $\mathrm{e_s^-}$ and $\mathrm{e_e^-}$ together with B^+ produced in the reaction (1) with a higher energy resolution.

The present apparatus utilizes an electron monochromator which suffices both high resolution and high current for the monochromatized electron beam. The new monochromator was designed taking into consideration of the space charge effect (SCE), the effect resulted from mutual repulsion of electrons in the beam.

Before constructing the new monochromator, we have studied the SCE in the electron monochromator using a typical spherical type apparatus. For the study of the SCE, we measured transmitted electron current of a hemispherical energy selector as a function of injecting electron current I_in, which is one of the key parameter determining the strength of the SCE. Trajectories of electrons in the selector under the SCE were also calculated and compared with the measurements. From the comparison, two main effects of the space charge were found: i) the SCE enlarges the focusing deflection angle of the selector; ii) the SCE enhances the energy dispersion of the selector. The results imply that the electron monochromator can produce the electron beam with a small energy width in the high current condition when the effect i) is compensated.

Following the above idea, we constructed the two-stage cylindrical-type electron monochromator with the deflection angle adjusted so that the SCE could enhance the energy dispersion without degrading the focusing property. In order to test the performance of the new monochromator, we measured the electron current I_out and the energy width ΔE_M of the beam transmitting the selector as a function of I_in. As shown in Fig. 1, both of the values of I_out and ΔE_M change drastically due to the strong SCE. It can also be found that an optimal condition exists where the maximum of I_out and the minimum of ΔE_M coincide due to the achievement of the focus of the electron beam in the strong SCE condition. According to Fig. 1, the incident electron beam with $I_\mathrm{out} \approx 30$ nA and $\Delta E_M \approx 20$ meV is available with the present monochromator, which is sufficient to carry out the multi-coincidence measurements.

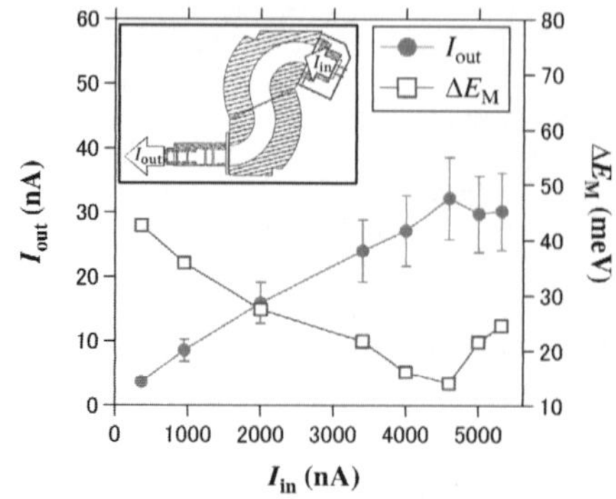

Figure 1. The injected electron current dependence of monochromatized electron beam current I_out and its energy width ΔE_M. The left vertical axis represents I_out, the right ΔE_M.

[1]E-mail: mkitajim@chem.titech.ac.jp

References

[1] For example, L. Campbell and M. J. Brunger 2016 *Int. Rev. Phys. Chem.* **35** 297

[2] X. Ren *et al.* 2012 *Phys. Rev. Lett.* **109** 123202

IOP Conf. Series: Journal of Physics: Conf. Series **875** (2017) 062022 doi:10.1088/1742-6596/875/7/062022

Molecular-frame EELS experiment on inner-valence ionization of N_2

Noboru Watanabe[1], **So Yamada, and Masahiko Takahashi**

IMRAM, Tohoku University, Katahira 2-1-1 Aoba-ku Sendai 980-8577, Japan

Synopsis We report molecular-frame electron scattering cross sections of N_2, obtained using an electron-ion coincidence technique. The vector correlation between the scattered electron and the fragment ion has been measured at incident electron energy of 1.4 keV for inner-valence ionization of N_2. The angular distributions of the electron scattering cross sections thus obtained exhibit clear momentum transfer dependence, indicating the significant change of the ionization dynamics due to contributions from electric multipole transitions.

In the past five decades, angle resolved electron energy loss spectroscopy (EELS) has evolved as a powerful means for investigating electron excitation processes of atoms and molecules. The momentum transfer dependence of the electron scattering cross section provides deep insights into the nature of electron excitation processes of interest [1]. EELS studies on molecules have, however, long been restricted to investigating targets with a random spatial orientation; the spherical averaging over the molecular orientation results in enormous loss of information on the electron-molecule collision dynamics. To remove ambiguities in the analysis of the spherically averaged EELS cross sections, we have recently developed a method to perform EELS experiments in a molecular frame, which involves a coincidence detection of the scattered electron and fragment ion [2]. In this work we have applied the method to the study of inner-valence ionization of N_2. In the inner-valence region, a number of two-hole one-particle ionic states appear. Since the major mechanisms responsible for transitions to such ionic states are configuration interactions in the target molecule, the inner-valence ionization offers an ideal opportunity for investigating ionization dynamics of molecules beyond the independent-particle model.

An electron-ion coincidence apparatus developed in our laboratory consists of an electron gun, a molecular beam source, a hemispherical electron analyzer, and an ion momentum imaging spectrometer equipped with a position sensitive detector. A pulsed electron beam is crossed with gaseous targets expanded from a nozzle and electrons scattered at a particular angle of θ are energy analyzed using the hemispherical electron spectrometer. Once a scattered electron is detected, a homogeneous ion-extraction field is applied to the scattering region to push towards fragment ions into the ion spectrometer. The recoil-momentum of the ion is determined from its time of flight and arrival position at the position sensitive detector.

By using the apparatus we have performed molecular-frame EELS experiments on N_2 at incident electron energy of 1.4 keV. Fig. 1 shows the EELS cross sections thus obtained for the ionization to the F $^2\Sigma_g^+$ two-hole one-particle state at energy loss of 38 - 40 eV. In the figure the EELS cross sections are plotted as a function of the angle ϕ_K of momentum transfer vector $\boldsymbol{K}$ from the molecular axis. It exhibits clear ϕ_K dependence and the angular distribution considerably varies with the increase in K. The result indicates that contributions from electric multipole transitions lead to the significant change of the ionization dynamics.

In the presentation we will show the results for transitions to other ionic states and discuss how information on ionization mechanisms is manifested in the angular distribution and its momentum transfer dependence.

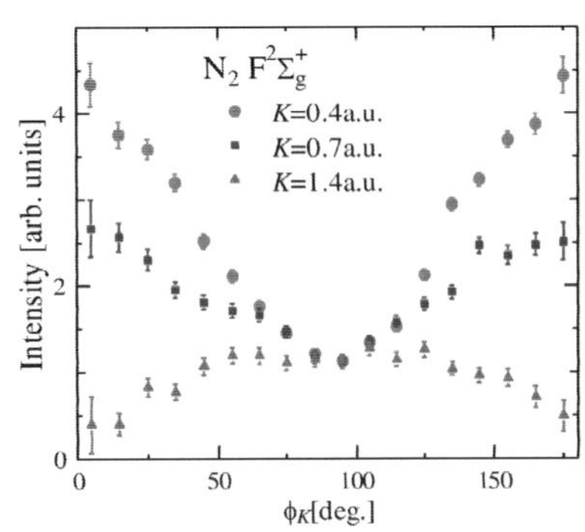

Figure 1. Molecular frame EELS cross section of N_2 for the transition to the F $^2\Sigma_g^+$ state.

References

[1] M. Inokuti 1971 *Mod. Phys.* **43** 297
[2] N. Watanabe *et al.* 2015 *J. Phys. Conf. Ser.* **635** 072030

[1] E-mail: noboru@tagen.tohoku.ac.jp

IOP Conf. Series: Journal of Physics: Conf. Series **875** (2017) 062023 doi:10.1088/1742-6596/875/7/062023

Absolute cross sections for silver clusters (Ag_n, n=1-4) by electron impact

Paresh Modak[1] and Bobby Antony[2]

[*] Department of Applied Physics, IIT (ISM), Dhanbad, India,

Synopsis The modified addidtivity formula [1] for homonuclear cluster is employed to calculate the elastic, total and ionization cross section of silver cluster form ionization threshold to 20 keV and is compared with available experimental data [2,3]. The absorption, polarization, exchange and correlation effects are introduced in an optical potential through SCOP formalism [4]. The elastic and total cross sections are reported for the first time.

The ionization of particles by electron impact is one of the basic processes in mass spectrometry. Besides, this interaction plays an important role in many practical applications, e.g. gas discharges, plasmas, radiation chemistry, planetary upper atmospheres etc. Hence, a large number of theoretical and experimental studies have been done for decades, to calculate/measure electron induced ionization cross section in gas phase for atoms and molecules. However, the cross section data for clusters, especially for metal clusters, is lacking.

The present study gives theoretical insight into evaluation of elastic, total and ionization cross section for silver clusters (Ag_n, n=1-4) by the impact of electron from ionization threshold to 10 keV. The spherical complex optical potential (SCOP) [4] method is used to calculate absolute elastic and total cross sections and then ionization contribution procedure [4] is employed to derive the contribution of ionization from the inelastic cross section. The energy of low ling excited state [5] is considered as threshold value for the present calculations. The modified additivity formula [1] employed presently for the clusters is,

$$\sigma_e(X_n) = n^a \sigma_e(X) \quad (1)$$

This modified rule is based on geometrical concept for homonuclear cluster. Where $\sigma_e(X)$ denotes the cross section for atoms X, n represents the cluster size and a is the hard sphere packing fraction.

Table 1: Ratio of ionization cross section of neutral silver atoms to its clusters for 46 eV.

Cluster	Present	Franzerb [2]	D.Koch [3]
Ag	1	1	1
Ag2	1.58	1.53±20%	1.3±10%
Ag3	3.74		3.8±10%
Ag4	2.52		2.2±10%

[1] E-mail: paresh@ap.ism.ac.in

[2] E-mail: bobby.iit.ism@gmail.com

In general, present calculations shows good agreement within the reported uncertainty of Franzreb *et al.* [2] throughout their energy range and that of D. Koch *et al.* [3] for a=2/3 till about 20 eV. The overestimation of the present calculation is expected, since the modified additivity formula is designed to calculate total cross section and experiments were done for partial cross section only.

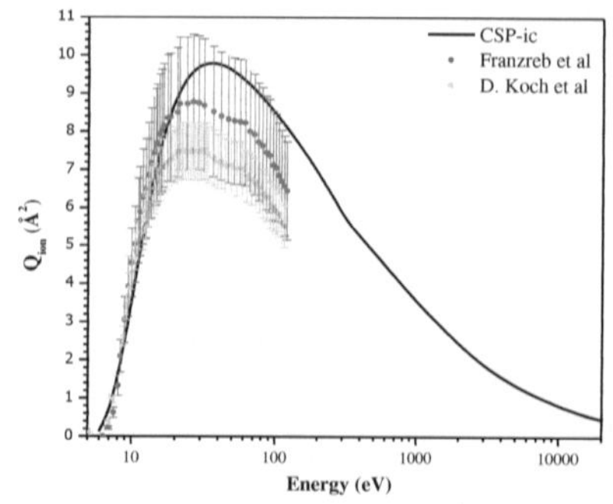

Figure 1. Ionization cross section (Q_{ion}) for silver dimer.

Fig.1. displays the Q_{ion} for silver dimer (n=2) up to 20 keV with reported experimental results. The present calculation shows higher value of cross section around the peak. The present calculation gives qualitatively good agreement within experimental error of previous data.

References

[1] M. Kappes 1988 *Chem. Rev.* 88
[2] K. Franzreb, A. Wucher and H. Oechesner 1991 *Z. Phys. D* 19
[3] D. Koch, M. Wahl and A. Wucher 1994 Z. Phys. D 32
[4] P. Modak, S. Shing, J. kaur and B. Antony 2016 *Int. J. Mass Spectrom.* 409
[5] R. S. Freund, R. C. Wetzel, R. J. Shul and T. R. Hayes 1990 *Phys. Rev. A* 41

IOP Conf. Series: Journal of Physics: Conf. Series **875** (2017) 062024 doi:10.1088/1742-6596/875/7/062024

Experimental scaling of plane-Born cross sections for the electron-impact excitation to the $b^1\Pi_u$ state of N_2 molecule

A. Tanaka*, H. Kato*, M. Hoshino*[1], and H. Tanaka*

*Department of Physics, Sophia University, Tokyo 102-8554, Japan.

Synopsis We report measurements on differential cross sections (DCSs) for electron-impact excitation to the optically-allowed $b^1\Pi_u$ state in molecular nitrogen in the impact energy range of 20 – 400 eV. Absolute excitation DCSs normalized by the relative flow technique were transformed to generalized oscillator strengths (GOSs) as a function of momentum transfer squared. Integral cross sections (ICSs) for the transitions to the $b^1\Pi_u$ state were compared with the BE*f*-scaled plane-Born ICS as well as previous experimental studies.

We have been measuring the electron impact electronic excitation cross sections of atomic and molecular targets [1]. Molecular nitrogen, N_2, is a major constituent of the Earth's atmosphere (~80%), so to obtain a quantitative understanding of the atmospheric behavior of our planet, the role of electron-driven processes is an important component [2]. In the present work, we focused on the N_2 molecule as the target, and measured the absolute DCS and ICS for the optically-allowed transition to the $b^1\Pi_u$ state. The experimental ICSs are compared with the BE*f*-scaled ICS based on the plane-Born approximation [3] to verify the validities of the BE*f*-scaling for optically-allowed excitation in N_2 molecule.

The present measurements were performed at Sophia University using a crossed-beam apparatus of a conventional electron spectrometer in the incident electron energies of 20 – 400 eV with total energy resolution of ~40 meV. The scattering angle range was $1° \leq \theta \leq 130°$. The true 0° scattering angle with respect to the incident electron beam was determined from the symmetry of the scattering intensities for the He 2^1P excitation. The absolute scale of the DCSs for the excitation to the $b^1\Pi_u$ state were obtained by the relative flow technique using the well-known elastic and 2^1P excitation DCSs of the He atom [4, 5]. Uncertainties in DCSs and ICSs were estimated ~25% and ~30%, respectively.

In this study, we measured angle-resolved electron energy loss (EEL) spectra in the EEL range of 12.0 – 15.0 eV, where it includes the excitation to the $b^1\Pi_u$ state. The EEL spectra of N_2 have shown rather complicated structures with many of the vibrational sublevels in an excited state overlapping with other vibrational sublevels in different excited states [2]. Thus, spectral deconvolution of the measured EEL spectra are necessary in order to obtain the absolute excitation DCSs, because the energy resolution of the electron spectrometer employed in the present measurements is ~40 meV that is not enough to resolve all of the vibrational sublevels in the final electronic states. The detailed of deconvolution procedure has been reported in ref. [2]. Excitation energies and Franck-Condon factors or relative intensity ratio between initial and final vibrational sublevels were referred in the literatures [6].

The GOSs as a function of the momentum transfer squared, K^2, were transformed from the measured absolute excitation DCSs and the scattering angles at each incident energy. Fitting the experimental GOSs at higher incident energies with the Vriens formula [7], where the Born approximation is satisfied, we obtained the optical oscillator strength (OOS) and apparent Born ICS for the excitation to the $b^1\Pi_u$ state. The present OOS is compared with the quasi-photon absorption measurement [8].

Finally, BE*f*-scaled Born ICS for $b^1\Pi_u$ excitation were estimated from the apparent Born ICS using the excitation and binding energies in the $b^1\Pi_u$ excited state of N_2 for the wide energy range. The comparison of the present BE*f*-scaled ICS with previous experimental data [9, 10] for this excitation, shows good agreement in the low energy region.

References

[1] H. Tanaka *et al.* 2016 *Rev. Mod. Phys.* **88** 025004
[2] H. Kato *et al.* 2010 *Phys. Rev. A* **81** 042717
[3] Y.-K. Kim 2001 *Phys. Rev. A* **64** 032713
[4] L. Boesten and H.Tanaka 1992 *At. Data Nucl. Data Tables.* **52** 25
[5] M. Hoshino *et al.* 2013 *J. Chem. Phys.* **139** 184301
[6] M. A. Khakoo *et al.* 2008 *Phys. Rev. A* **77** 012704
[7] L. Vriens. 1967 *Phys. Rev.* **160** 100
[8] W. F. Chan *et al.* 1993 *Chem. Phys.* **170** 81
[9] C. Malone *et al.* 2012 *Phys. Rev. A* **85** 062704
[10] S. Trajmar *et al.* 1983 *Phys. Rep.* **97** 219

[1]E-mail: masami-h@sophia.ac.jp

Theoretical investigation of water ionization by electron impact at low energies

István Tóth[*1], **Ladislau Nagy**[*2]

[*] Faculty of Physics, Babeş-Bolyai University, 400084 Cluj-Napoca, Romania

Synopsis Triple differential cross sections (TDCS) are calculated for the ionization of water by electron impact. The TDCSs are determined in the scattering plane and in the plane perpendicular to the incident beam direction. The obtained theoretical cross sections are compared with very recent experimental measurements and other theoretical calculations.

The ionization of water molecules has been studied intensely in the last years [1, 2, 3, 4, 5, 6, 7, 8]. This process is of interest for several research fields, like medicine or astrophysics. One possible application is related to radiative treatment methods, since organic tissue consists of a large percentage of water. It has been shown that low energy electrons have a particularly damaging effect on DNA strands.

Previously we have studied the ionization of water for different kinematical and geometrical conditions. We have calculated [7] TDCSs in the scattering plane for electron impact energies of 250 eV for the asymmetric ejection of the final state electrons. Further, the $3a_1$ orbital has been investigated [8] at low impact energies both in the scattering and the perpendicular planes and also for some geometrical arrangements between these two extreme configurations. In this case only symmetric ejection of electrons was considered.

In the current work we extend our investigation of water ionization for a different set of kinematical conditions and geometrical arrangements in order to further test the validity of our calculation method. The TDCSs are determined for a relatively low impact energy (81 eV) both in the scattering and perpendicular planes. The calculated cross sections represent the sum of the $1b_1$ and $3a_1$ TDCSs in order to compare with the experimental data [1]. In the experiment, these orbitals were not resolved, due to the limited binding-energy resolution of the experimental setup. Nevertheless, the cross sections are also calculated separately for the above mentioned orbitals. We investigate the asymmetric ejection of the final state electrons, the scattering angle of the projectile being set at -6° and -10°, respectively, relative to the incident beam direction. For both scattering angles, we consider ejected electrons at 5 eV and 10 eV, respectively. The ejected electron is the slower one. Beside the experimental data [1], the obtained results are compared with other theoretical calculations presented in the same paper.

In the calculations we use a distorted-wave approach for the description of the free states in the system, while the molecular orbitals are described by multi-center Gaussian-type orbitals. The distorted waves are calculated in the spherically averaged potential of the molecule or the molecular ion. Two schemes are considered for the final state particles. In the first one the ejected electron moves in the averaged field of the nuclei and residual electrons, while the faster electron experiences a potential which is totally screened by the slower one. In the second configuration we neglect this screening and both electrons move in the same potential of the molecular ion.

In the calculations we take into account the post collision interaction (PCI) effects between the final continuum electrons using the Coulomb distortion factor of [9]. Although the two electrons have different energies, these are relatively low values, therefore PCI effects may still be important for the correct description of the process.

Further details about the theoretical models and results will be presented at the conference.

References

[1] X. Ren *et al.* 2017 *Phys. Rev. A* **95** 022701

[2] K. L. Nixon *et al.* 2010 *J. Phys. B: At. Mol. Opt. Phys.* **43** 035201

[3] C. Y. Lin *et al.* 2014 *Phys. Rev. A* **89** 012703

[4] C. Champion *et al.* 2006 *Phys. Rev. A* **73** 012717

[5] D. S. Milne-Brownlie *et al.* 2004 *Phys. Rev. A* **69** 032701

[6] M. Sahlaoui *et al.* 2013 *J. Phys. B: At. Mol. Opt. Phys.* **46** 115206

[7] I. Tóth *et al.* 2012 *Eur. Phys. J. D* **66** 21

[8] I. Tóth *et al.* 2014 *Eur. Phys. J. D* **68** 369

[9] S. J. Ward *et al.* 1994 *Phys. Rev. A* **49** 1049

[1]E-mail: istvan.toth@phys.ubbcluj.ro
[2]E-mail: lnagy@phys.ubbcluj.ro

IOP Conf. Series: Journal of Physics: Conf. Series **875** (2017) 062028 doi:10.1088/1742-6596/875/7/062028

Absolute elastic differential cross sections for PF_3 molecule by electron impact: A comparative study with XF_3 (X = B, C, N and CH) molecules

N. Hishiyama*, M. Hoshino*[1], F. Blanco†, G. García‡ and H. Tanaka*

* Department of Physics, Sophia University, Tokyo 102-8554, Japan.

† Departamento de Fisica Atomica, Molecular y Nuclear, Facultad de Ciencias Fisicas, Universidad Complutense de Madrid, E-28040 Madrid, Spain

‡ Instituto de Fisica Fundamental, Consejo Superior de Investigaciones Científicas, 28006 Madrid, Spain

Synopsis We report absolute electron elastic scattering differential cross sections (DCSs) from PF_3 molecules. The crossed-beam method was used in conjunction with the relative flow technique (using helium as the reference gas) to obtain absolute values. Very good agreement between the experimental DCSs and those calculated with the IAM-SCAR method has been found for the higher impact energy (above 50 eV). We have also compared the measured DCSs for PF_3 with experimental results for other fluorine compounds, XF_3 (X = B, C, N and CH), as well as with those derived from the calculated atomic fluorine cross sections multiplied by a factor of 3.

We found for the first time an atomic-like behavior in the elastic scattering of electrons from the halomethanes, CH_3F, CH_3Cl, CH_3Br, and CH_3I [1], by comparing with that for the corresponding noble gases, Ne, Ar, Kr, and Xe, respectively. This comparison suggested that the halogen atoms dominated the angular distributions of the elastic scattering for high impact energies (above 50 eV). Similar behavior were observed for isoelectronic systems such as GeH_4 and Kr [2].

More recently, we have shown this atomic-like effect in the electron elastic scattering from fluorine compounds such as XF_3 (X = B, C, N, and CH) [3] and YF_4 (Y = C, Si, and Ge) [4] for energies above 50 eV. Furthermore, the angular distribution of elastically scattered electrons from XF_3 and YF_4 molecules for energies above 50 eV have been found to be well reproduced by the calculated DCS for atomic-fluorine multiplied by a factor of 3 and 4, respectively. These results suggested that the elastic scattering for high impact energies is virtually dominated by the outer fluorine atoms surrounding the central atom. This also evidenced that atomic-like effects persist in electron–molecule scattering systems.

In this study, we have measured the elastic DCSs for PF_3 molecules and results have been compared with the IAM-SCAR calculations [5] as well as with the previously reported for XF_3 molecules [3] to verify the atomic-like effect in the fluorine compound molecules.

The experiment was performed at Sophia University using an original crossed-beam apparatus [6]. The total energy resolution was about 50 meV (FWHM). The relative flow technique was used to obtain absolute DCS values. Estimated uncertainties of the DCSs are about 15%.

[1]E-mail: masami-h@sophia.ac.jp

Figure 1 shows the measured angular distribution of the elastic DCS for PF_3 at 100 eV impact energy together with the present IAM-SCAR calculated values and those corresponding to XF_3 molecules. The present experimental results also confirm the atomic-like effect of the elastic scattering DCSs for the PF_3 molecule.

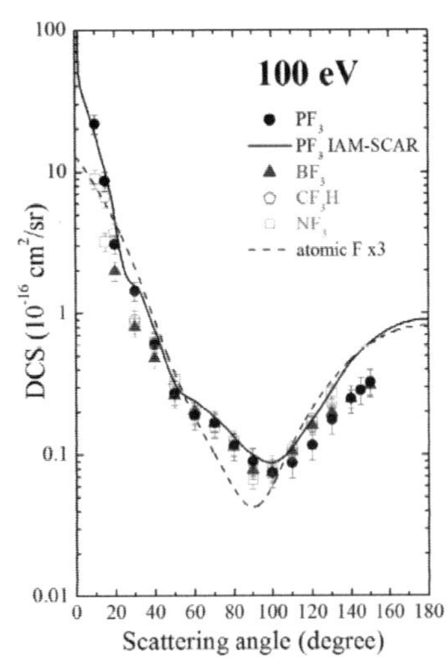

Figure 1. Angular distributions of DCSs for XF_3 (X = B, CH, N, and P) molecules at 100 eV impact energy, together with the IAM-SCAR calculation for PF3 and DCS of fluorine atom multiplied by a factor of 3.

References

[1] H. Kato *et al.* 2010 *J. Chem. Phys.* **132** 074309
[2] M. A. Dillon *et al.* 1993 *J. Phys. B* **26** 3147
[3] M. Hoshino *et al.* 2015 *J. Chem. Phys.* **143** 024313
[4] H. Kato *et al.* 2012 *J. Chem. Phys.* **136** 134313
[5] F. Blanco and G. García 2007 *Phys. Lett. A* **360** 707
[6] H. Tanaka *et al.* 1998 *Phys. Rev. A* **57** 1798

IOP Conf. Series: Journal of Physics: Conf. Series **875** (2017) 062029 doi:10.1088/1742-6596/875/7/062029

Ionization of NH_3 by Electron and Photon Impact

Carlos Mario Granados–Castro[*1], **Alessandro Genoni**[†], **Lorenzo Ugo Ancarani**[†2]

[*] Institute for Physics, Martin-Luther-Universität Halle-Wittenberg, 06120 Halle, Germany
[†] Equipe TMS, UMR CNRS 7565, Université de Lorraine, 57078 Metz, France

Synopsis We study single ionization by electron and photon impact from the outer valence orbital $3a_1$ of NH_3. The wave function describing the ejected electron is expanded in an appropriate set of Generalized Sturmian Functions. The cross sections obtained with two different molecular target descriptions are compared with each other, and with experimental data.

Photoionization and ionization by electron impact are two fundamental processes which test our ability to describe the target electronic structure, the role of electron-electron interactions and the collision mechanisms. Thanks to diverse experimental and theoretical advances, in the last years the focus on molecular targets has grown. Most of the theoretical studies, however, have been limited to small molecules such as H_2 or N_2. The extension to molecules of traditional atomic approaches is limited by the complexity of the problem (multicenter formulation and generally highly noncentral). Recent $(e,2e)$ measurements on polyatomic molecules motivated the development of theoretical methods to describe ionization cross sections in their most differential form.

We propose to use the Sturmian approach, using generalized Sturmian functions (GSFs) [1] to study single ionization of molecules. The method was applied successfully to study single photoionization of atoms and small molecules [2, 3], as well as double ionization of atoms induced by photon [4], electron [5] or proton [6] impact. In the present investigation, in a one-active electron approach, the scattering wave function is expanded in a set of GSFs that have an appropriate asymptotic outgoing-type behavior; this property makes the method rather efficient. Moreover, the scattering amplitude can be extracted from the asymptotic behavior of the scattering solution (essentially the expansion coefficients), without the need of calculating a transition matrix element.

We study single ionization by electron and photon impact from the outer valence orbital $3a_1$ of NH_3. To describe the initial state wave function, we use either the one-center wave function calculated by Moccia [7], or the one obtained by a DFT calculation. The molecular potential is either the one obtained through a static exchange approximation [8] or the one of the DFT results. The comparison between the two allows us to separate out the importance of the molecular structure quality from the numerical description of the molecular single continuum.

[1]E-mail: carlos.granados@physik.uni-halle.de
[2]E-mail: ugo.ancarani@univ-lorraine.fr

We compare our results with experimental data. For photoionization we compare with the cross section reported in Ref. [9], while for electron impact we focus on triple differential cross sections (TDCS) and compare with the measurements presented in Ref. [10]; an example is shown in Fig. 1.

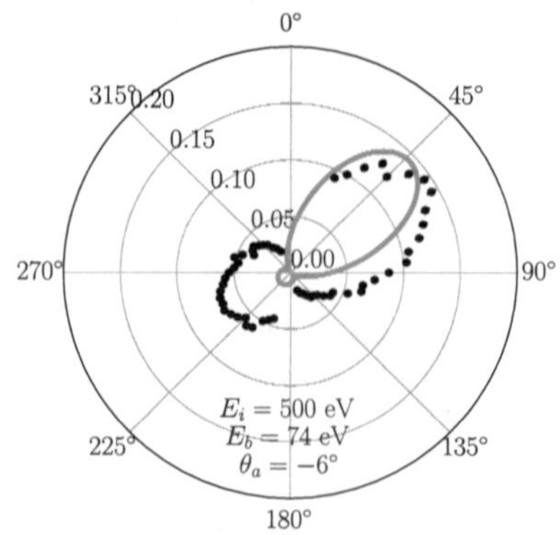

Figure 1. TDCS for $NH_3\ (3a_1^{-1})$. The kinematic conditions are: incident electron energy $E_i = 500$ eV, ejected electron $E_b = 74$ eV. The scattered electron is detected at $\theta_a = -6°$. Experimental data from [10].

References

[1] D. M. Mitnik *et al.* 2011 *Comp. Phys. Commun.* **182** 1145; G. Gasaneo *et al.* 2013 *Adv. Quantum Chem.* **67** 153
[2] C. M. Granados–Castro *et al.* 2016 *Adv. Quantum Chem.* **73** 3
[3] C. M. Granados–Castro 2016 Ph.D. thesis, Université de Lorraine, Metz
[4] J. M. Randazzo *et al.* 2015 *Eur. Phys. J. D* **69** 189
[5] M. J. Ambrosio *et al.* 2015 *J. Phys. B: At. Mol. Opt. Phys.* **48** 055204
[6] M. J. Ambrosio *et al.* 2015 *Phys. Rev. A* **92** 042704
[7] R. Moccia 1964 *J. Chem. Phys.* **40** 2176
[8] L. Fernández–Menchero and S. Otranto 2010 *J. Chem. Phys.* **82** 22712
[9] M. S. Banna *et al.* 1987 *Chem. Phys.* **114** 157
[10] R. El Mir *et al.* 2015 *J. Phys. B: At. Mol. Opt. Phys.* **48** 175202

IOP Conf. Series: Journal of Physics: Conf. Series **875** (2017) 062030 doi:10.1088/1742-6596/875/7/062030

Dissociation dynamics of transient anion formed via electron attachment to sulphur dioxide (SO_2)

Krishnendu Gope * [1], Vaibhav S. Prabhudesai * [2], Nigel J Mason† [3] and E. Krishnakumar* [4]

* Dept. Nuclear and Atomic Physics, Tata Institute of Fundamental Research, Colaba, Mumbai 400005, India
† Dept. of Physical Sciences, the Open University, Walton Hall, Milton Keynes, MK7 6AA, UK

Synopsis: We have investigated the dynamics of dissociative electron attachment to sulfur dioxide (SO_2) using velocity slice imaging (VSI) technique. The symmetric dissociation of the transient parent anion results in S^- formation whereas its asymmetric dissociation results in the O^- and SO^- channels. We clearly see the competition between bending and stretch mode of vibrations of the parent anions in S^- formation whereas the stretch modes dominate in the other two ions production.

Very little is known about the excited anion states of sulphur dioxide (SO_2) in spite of their importance in atmospheric chemistry and industrial pollution. Dynamics of dissociative electron attachment (DEA) may provide some information about the excited states of the anions formed by electron attachment. We have studied the DEA dynamics for SO_2 in terms of angular distribution and kinetic energy distribution of the fragment anions using velocity slice imaging [1].

DEA to SO_2 is known to produce O^-, S^- and SO^-. Among all of the observed ions, the most dominant are O^- which peaks at 4.6 eV and 7.2 eV, S^- peaking at 4.2 eV, 7.4 eV and 9 eV and the SO^- peaking at 4.8 eV and 7.3 eV. We have recorded the momentum images for these ions across all these peaks.

Since the three-body fragmentation thresholds to form S^- or O^- are above 9 eV, the observed formation of these ions is due to two-body dissociation. The momentum images obtained for S^- channel show very little kinetic energy release. This implies that the excess energy appears as the internal excitation of the O_2 fragment. This excitation is found to increase with electron energy indicating the interplay between the bending and stretch mode of vibrations in the parent anion. For electron energies beyond 4.8 eV, O_2 can also be formed with electronic excitation. However, even for that channel the excess energy appears as the vibrational excitation of O_2.

The KER spectra obtained for the O^- channel clearly indicates internal excitation of SO. This internal excitation may appear in its rotational and vibrational motion. The rotational excitation of SO is a consequence of the asymmetric two body fragmentation of the resonant state. It is found that the internal excitation of the SO increases with the electron energy. Our analysis of the results indicates the dominance of both symmetric and asymmetric stretch modes of vibration over the bending mode in the parent anion. We draw similar conclusion for the SO^- channel as well. Though both O^- and SO^- channels have very similar dynamics, we note a clear distinction between the two. In the O^- channel the SO fragment can store considerable amount of energy as internal excitation, while in the SO^- channel this is limited by the electron affinity of SO which is 1.1 eV only. This may explain the lower value of the ratio of absolute cross sections of SO^- and O^- at the 7.3 eV peak as compared to that at 4.8 eV.

References

[1] D. Nandi *et al.* 2005 *Rev. Sci. Instrum.* **76** 053107

[1] E-mail: gope.krishnendu@gmail.com
*[2] E-mail: vaibhav@tifr.res.in
[3] E-mail: n.j.mason@open.ac.uk
[4] E-mail: ekkumar@tifr.res.in

IOP Conf. Series: Journal of Physics: Conf. Series **875** (2017) 062032 doi:10.1088/1742-6596/875/7/062032

Electron Impact Ionization of CH_4 for Different Momentum Transfers

Carlos Mario Granados–Castro[*1], **Lorenzo Ugo Ancarani**[†2]

[*] Institute for Physics, Martin-Luther-Universität Halle-Wittenberg, 06120 Halle, Germany
[†] Equipe TMS, UMR CNRS 7565, Université de Lorraine, 57078 Metz, France

Synopsis The electron impact single ionization of the outer valence orbital $1t_2$ of methane is investigated theoretically using an expansion on Generalized Sturmian Functions. Triple differential cross sections, calculated for several coplanar asymmetric geometries, are compared with other theoretical models and with two sets of relative experimental data (incident energy of 500 eV and 250 eV). An analysis with respect to the momentum transfer is presented. A double peak structure in the cross section binary region, a clear signature of the p-nature of the molecular orbital, is predicted for given kinematical conditions.

We study the electron impact ionization of CH_4 from the outer valence orbital $1t_2$. We calculate triple differential cross sections (TDCSs) for coplanar asymmetric geometries, and compare them with those obtained with other theoretical results (in an absolute scale), and with the relative scale measurements reported in Refs. [1] and [2], The binary to recoil ratio is analyzed as a function of the momentum transfer. Some of our results were already presented in Ref. [3].

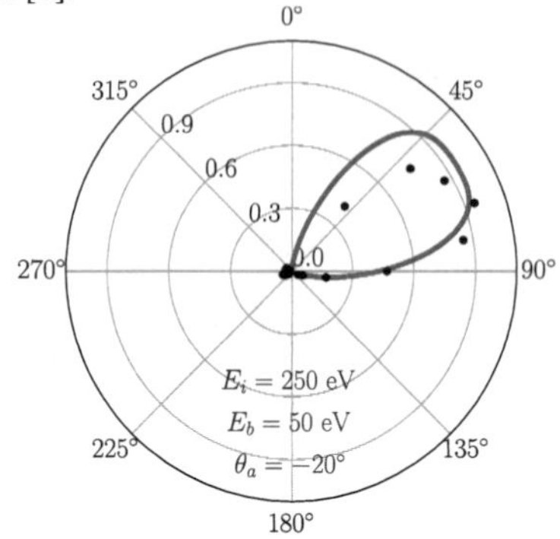

Figure 1. TDCS for $CH_4\left(1t_2^{-1}\right)$. Kinematic conditions: incident electron energy $E_i = 250$ eV, ejected electron energy $E_b = 50$ eV. Scattered electron detected at $\theta_a = -20°$. Experimental data from Ref. [2].

The TDCSs are obtained using a Sturmian approach with generalized Sturmian functions (GSFs) [4]. Such approach has been used to study, among others, single photoionization of atoms and small molecules [5]. In order to simplify the scattering problem, we use as molecular potential the one described through a static exchange approximation [7], make the one-center expansion, and take the initial state wave function from Ref. [6]. In a first Born approximation, the scattering wave function describing the ejected electron is expanded in a set of GSFs with appropriate Coulomb asymptotic conditions; this allows us to extract the scattering amplitude directly from the expansion coefficients, without the need of calculating a transition matrix element.

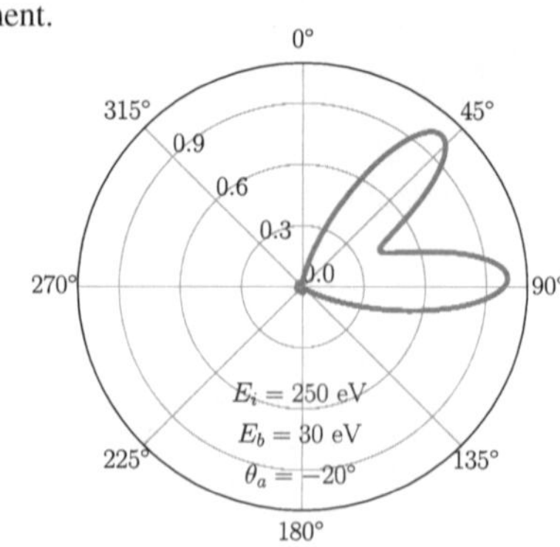

Figure 2. Same as Fig. 1, bur for ejected electron with $E_b = 30$ eV.

In Fig. 1 we present an example of one of the kinematical conditions reported in Ref. [2]. The binary peak is clearly dominant. As an extension (and a complement) to such experimental results, we consider other parameters; in the case presented in Fig. 2, for example, we predict a double peak structure in the cross section binary region, which is a clear signature of the p-nature of the molecular orbital.

References

[1] A. Lahmam–Bennani *et al.* 2009 *J. Phys. B: At. Mol. Opt. Phys.* **42** 165201
[2] Işik *et al.* 2016 *J. Phys. B: At. Mol. Opt. Phys.* **49** 065203
[3] C. M. Granados–Castro and L. U. Ancarani 2017, *Eur. J. Phys. D* in press
[4] D. M. Mitnik *et al.* 2011 *Comp. Phys. Commun.* **182** 1145; G. Gasaneo *et al.* 2013 *Adv. Quantum Chem.* **67** 153
[5] C. M. Granados–Castro *et al.* 2016 *Adv. Quantum Chem.* **73** 3
[6] R. Moccia 1964 *J. Chem. Phys.* **40** 2164
[7] L. Fernández–Menchero and S. Otranto 2010 *J. Chem. Phys.* **82** 22712

[1]E-mail: carlos.granados@physik.uni-halle.de
[2]E-mail: ugo.ancarani@univ-lorraine.fr

IOP Conf. Series: Journal of Physics: Conf. Series **875** (2017) 062033 doi:10.1088/1742-6596/875/7/062033

Direct evidence of Interatomic Coulombic Decay in electron impact ionization of Ne dimer

S. Yan* [1]**, P. Zhang*, X. Ma*** [2]**, S. Xu*, B. Li*, X. L. Zhu*, W. T. Feng*, D. M. Zhao* and L. L. Shen***†

* Institute of Modern Physics, Chinese Academy of Sciences, Lanzhou 730000, China
† University of Chinese Academy of Sciences, Beijing 100049, China,

Synopsis The unambiguous evidence of the existence of Interatomic Coulombic Decay (ICD) of Ne dimer in (e, 2e) experiment with impact energy of 380 eV is observed. Our result suggests that the ICD electron accounted for more than 80% of low energy electron yields.

The collision between the radiation particle and biological tissue will produce lots of secondary electrons, which will interact with neighboring molecule further, and lead to secondary charge and energy transfer between molecules and their environment. The (e, 2e) study of small clusters can be used to simulate these secondary processes.

One of energy transfer mechanisms is termed Interatomic Coulombic Decay (ICD) [1, 2], which is identified in numerous X ray and heavy ion impact experiments. However, the ICD investigation by electron impact is few. And the corresponding evidence is only obtained based on electron energy spectrum subtraction [3]. In order to get the ICD evidence directly, we performed the fragmentation experiment of Ne dimers induced by 380 eV electron impact using the Reaction Microscope in the Institute of Modern Physics, CAS [4].

By detecting the Ne^+/Ne^+ ions and the emitted electron in coincidence, the momenta and energy of all charged products, as well as the Kinetic Energy Release (KER), are obtained. As shown in Figure 1, the relationship of the electron energy and the KER is presented in a two dimensional map, in which a diagonal island at electron energy from 0 to 2 eV and KER from 4 to 6 eV is observed.

According to the energy conservation law, if a 2s electron of one Ne atom is ionized, and the transition energy from 2p to 2s is transferred to ionize the electron of the other Ne atom, the sum energy of the ICD electron and KER will be a constant (5.5 eV). In figure 1, this constant is presented as a dashed line. Obviously, most events in the diagonals island locate around this line. This means that the detected ICD arises from initial states Ne($2s^{-1}$)-Ne($2p^6$). The present result provides the most direct evidence of ICD in (e, 2e) experiment so far.

In addition, by comparing the electron energy spectrum of Ne dimer with that of Ne atom, a significant enchantment of electron yield in low energy range is observed in the ICD of neon dimer. It is concluded that, for the low energy area from 0 eV to 2 eV, the ICD electrons account for more than 80% of electron yields.

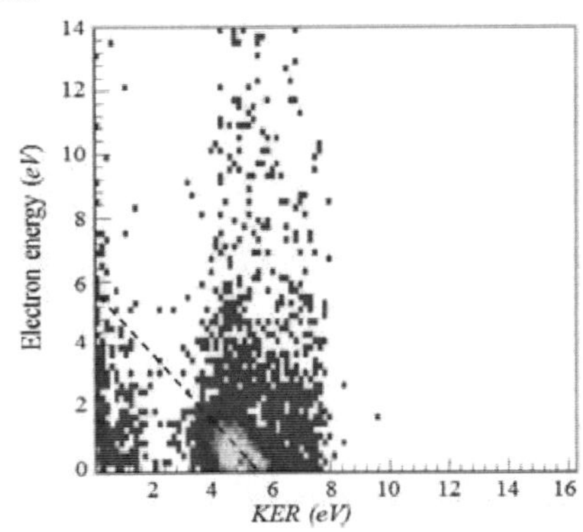

Figure 1. Correlation map between electron energy and KER. Black dashed line: position of ICD initial state Ne($2s^{-1}$) -Ne($2p^6$).

This work is supported by NSFC under Grants Nos. U1532129, 11304325, and U1332128.

References

[1] L. S. Cederbaum *et al.* 1997 *Phys. Rev. Lett.* **79** 4778
[2] T. Jahnke *et al.* 2004 *Phys. Rev. Lett.* **93** 163401
[3] X. Ren *et al.* 2016 *Nat. Commun.* **6** 139
[4] X. Ma *et al.* 2011 *Phys. Rev. A.* **83** 052707

[1] E-mail: yanshuncheng@impcas.ac.cn
E-mail: x.ma@impcas.ac.cn

IOP Conf. Series: Journal of Physics: Conf. Series **875** (2017) 062034 doi:10.1088/1742-6596/875/7/062034

Electron attachment studies with 2,3-dimethoxy-5-methyl-1,4-benzoquinone

F. Ferreira da Silva[* 1], J. Ameixa[* 2], J. Khreis[† 3], and S. Denifl[† 4]

[*] Institut für Ionenphysik und Angewandte Physik, Leopold Franzens Universität Innsbruck, Technikerstrasse 25, 6020 Innsbruck, Austria

[†] Laboratório de Colisões Atómicas e Moleculares, CEFITEC, Departamento de Física, Faculdade de Ciências e Tecnologia, Universidade NOVA de Lisboa, Campus da Caparica, 2829-516 Caparica, Portugal

Synopsis We demonstrate that low-energy electron (< 3 eV) interaction with 2,3-dimethoxy-5-methyl-1,4-benzoquinone (DMB) in the gas phase leads to a wide assortment of anionic fragments. DMB is the functional head-group of ubiquinone electron carrier involved in the electron transport chain in mithochondrial membrane. This study is a key contribution to a more comprehensive understanding of the electron transport in the mithochondrial membrane by ubiquinone.

The electron transport chain is the final stage of aerobic respiration and plays the final role in reducing ADP to ATP by oxidative phosphorylation in mithocondria membrane. In this process, ubiquinone acts as an electron carrier [1]. Remarkably, no other molecule can perform this function. Briefly, ubiquinone is composed by a quinone derivative head-group which works as an electron carrier as well as an hydrophobic-tail formed by isopropenoid subunits [2]. In order to understand the fundamental processes of electron transport by ubiquinone, we have performed studies on electron attachment to 2,3-dimethoxy-5-methyl-1,4-benzoquinone (DMB -$C_9H_{10}O_4$) in the gas phase. This compound represents the functional unit of ubiquinone. The measurements were carried out utilizing a high resolution double focusing mass spectrometer in reversed Nier-Johnson geometry. The incident electron current was set to 10 μA and the energy resolution at 0 eV is ~ 1 eV and the mass resolution is about $m/\Delta m$= 1000. At electron energies < 3 eV we observe the formation of the parent anion, $C_9H_{10}O_4^-$ as well as the following fragments: $C_9H_9O_4^-$, $C_8H_6O_4^-$, $C_7H_4O_4^-$. The highest anionic yield corresponds to the fragment $C_8H_6O_4^-$ formed by the loss of a hydrogen atom and a methyl group. The most relevant recorded ion yields are presented in Figure 1. Additionally, we report more six anionic fragments formed by DEA at higher energies. The corresponding peak resonance energies are listed in Table 1. Therefore, gas-phase DMB exhibits a highly reactive behavior upon interaction with electrons close to 0 eV.

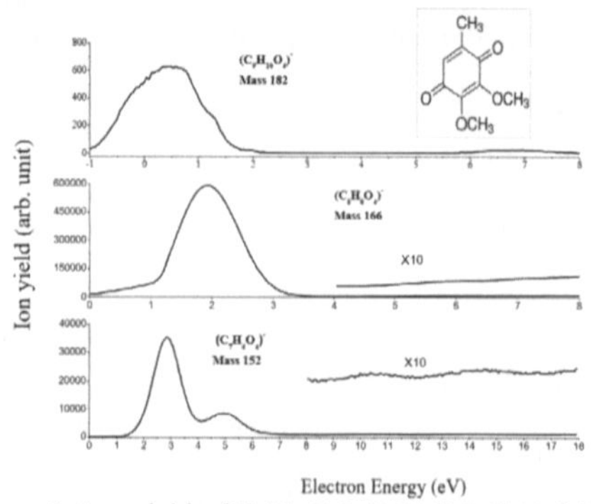

Figure 1. Ion yield of $C_9H_{10}O_4^-$, $C_8H_6O_4^-$, $C_7H_4O_4^-$ for electron attachment to DMB as a function of electron energy. Also shown is molecular structure of DMB (mass 182 u).

Table 1. Resonance energies of anions formed upon electron attachment to DMB.

Mass (Da)	Anion	Resonance Energies (eV) 1st	2nd	3rd	4th
182	$(DMB)^-$	0	-	-	-
166	$(DMB\text{-}H\text{-}CH_3)^-$	1.9	-	-	-
152	$(DMB-2CH_3)^-$	2.8	4.9	10.7	14.5
137	$(C_6H_7O_4)^-$	3.2	7.3	-	-
125	$(C_5HO_4)^-$	2.9	5.9	-	-
124	$(C_5O_4)^-$	3.1	7.1	8.8	-
123	$(C_6H_3O_3)^-/(C_7H_7O_2)^-$	7.0	9.1	8.8	-
110	$(C_6H_6O_2)^-$	7.0	9.9	-	-
97	$(C_4HO_3)^-$	6.4	-	-	-

References

[1] G. Karp, 2010 *Cell and Molecular Biology: Concepts and Experiments* pp. 185 ed Wiley (New York)

[2] G. Lenaz *et al.* 2009 *Biochim. Biophys. Acta - Bioenerg.* **1787** 563–73

[1] E-mail: f.ferreiradasilva@fct.unl.pt
[2] E-mail: jusuf.khreis@uibk.ac.at
[3] E-mail: j.ameixa@campus.fct.unl.pt
[4] E-mail: Stephan.denifl@uibk.ac.at

IOP Conf. Series: Journal of Physics: Conf. Series **875** (2017) 062035 doi:10.1088/1742-6596/875/7/062035

$Pt(CO)_2Cl_2$ fragmentation upon low energy electron interactions

F. Ferreira da Silva* [1], **R. Thorman**†, **H. Lu**‡, **L. McElwee-White**‡, **and O. Ingólfsson**†

* Laboratório de Colisões Atómicas e Moleculares, CEFITEC, Departamento de Física, Faculdade de Ciências e Tecnologia, Universidade NOVA de Lisboa, 2829-516 Caparica, Portugal
† Department of Chemistry and Science Institute, University of Iceland, Dunhagi 3, 107 Reykjavik, Iceland
‡ Department of Chemistry, University of Florida, Gainesville, FL 32611-7200, US

Synopsis: In the present communication we present low energy electron interactions with $Pt(CO)_2Cl_2$, a FEBID precursor. Electron impact ionization as well as dissociative electron attachment studies are discussed.

Focused electron beam induced deposition (FEBID) [1-3] of metallic materials is a promising technique for controlled metal deposition on surfaces. FEBID allows precise size and thickness control of the deposit and the fabrication of three-dimensional structures on planar as well as non-planar substrates. Hence, it is a technique with promising complementary capabilities as compared to current industrial nanofabrication methods. In FEBID, precursor molecules, often organometallic compounds containing the metal to be deposited, are continuously dosed at or close to a substrate surface where they are decomposed by exposure to a high-energy electron beam. An enormous amount of low energy electrons are formed inducing further decomposition that may occur through dissociative electron attachment (DEA), dissociative ionization (DI), neutral dissociation (ND) or dipolar dissociation (DD) [4]. A novel precursor, *cis*-platinum dicarbonyl dichloride (*cis*-$Pt(CO)_2Cl_2$) has been designed specifically for use in FEBID of Pt nanostructures. This precursor was synthesized in the laboratory of Lisa McElwee-White at the University of Florida and has been tested in electron induced decomposition under ultra-high vacuum (UHV) in the laboratory of Howard Fairbrother at Johns Hopkins University (JHU) [5].

Electron impact ionisation studies have been carried out showing a rich dissociation leading to bare platinum cation. Appearance energies (AE) for all detected cations are presented in Table 1.

Table 1. Appearance energies

Cation	AE	Error (%)
$Pt(CO)_2Cl_2$	11.07	9
$PtCOCl_2$	11.25	11
$Pt(CO)_2Cl$	12.11	3
$PtCl_2$	14.22	2
PtCl	17.81	2
PtCOCl	15.48	5
PtCO	17.18	2
PtC	26.66	2
Pt	19.91	3

Dissociative electron attachment leads predominantly to the loss of a carbonyl (CO) group at ~0 eV electron impact energy (Figure 1). Loss of two CO ligands is due as well to the excess of energy via a resonant process at 3.5 eV.

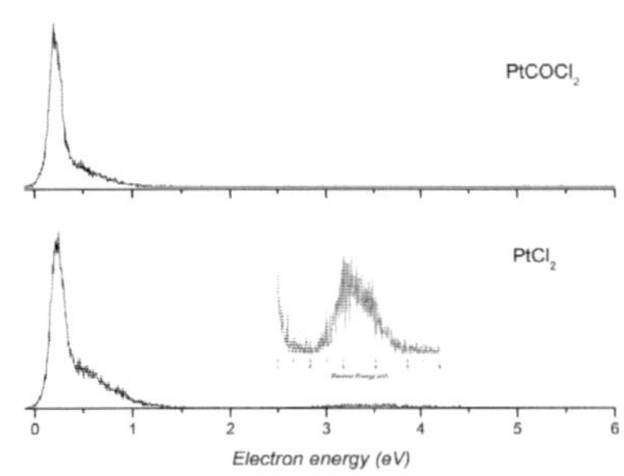

Figure 1. Ion yield for $PtCOCl_2$ and $PtCl_2$ for electron attachment of $Pt(CO)_2Cl_2$ as a function of electron energy.

References

[1] S. Chattopadhyay *et al.* 2006 *Critical Reviews in Solid State and Materials Sciences.* **31** 15-53.
[2] I. Utke *et al.* 2008 *Journal of Vacuum Science & Technology B: Microelectronics and Nanometer Structures.* **26**, 1197.
[3] W. F. van Dorp *et al.* 2008 *Journal of Applied Physics.* **104**, 081301.
[4] R. M. Thorman *et al.* 2015 *Beilstein J. Natotechnol.* **6**, 1904-1926.
[5] J. A. Spencer *et al.* 2016 *J. Am. Chem. Soc.* **138**, 9172-82.

[1] E-mail: f.ferreiradasilva@fct.unl.pt

IOP Conf. Series: Journal of Physics: Conf. Series **875** (2017) 062036 doi:10.1088/1742-6596/875/7/062036

Elastic Differential Cross Sections for Electron Scattering with Dichloromethane

E. Lange*, **K. Krupa*‡**, **J. Ameixa***, **A. S. Barbosa†**, **D. F. Pastega†**, **P. Limão-Vieira***, **M. H. F. Bettega†**, **F. Blanco§**, **G. García‡**, **F. Ferreira da Silva*1**

* Laboratório de Colisões Atómicas e Moleculares, CEFITEC, Departamento de Física, Faculdade de Ciências e Tecnologia, Universidade NOVA de Lisboa, 2829-516 Caparica, Portugal
† Departamento de Física, Universidade Federal do Paraná, 81531-990, Curitiba, Paraná
§ Departamento de Física Atómica, Molecular y Nuclear, Universidad Complutense de Madrid, Avenida Complutense, 28040 Madrid, Spain
‡ Instituto de Física Fundamental, Consejo Superior de Investigaciones Científicas, Serrano 113-bis, 28006 Madrid, Spain

Synopsis In the present study joint experimental and theoretical elastic differential cross sections for electron scattering from dichloromethane in the incident electron energy region 7 to 50eV are discussed.

Dichlorotomethane (CH_2Cl_2) is a relevant atmospheric and environmental molecule, where its high volatility results from the constant use in chemical industries [1], in biomass production [2] as well as from oceanic emissions [3]. Once in the stratosphere, the main sink mechanism has been attribute to photolysis leading to chlorine radical formation. Such radical at tropospheric altitudes further acts as a catalyser to ozone fragmentation into ClO.
Photolysis [4], photoabsorption [5] and DEA [6] studies can be found in the literature, but only a few related with electron scattering processes and even those are in the high energy regime [7] or strictly theoretical approaches [8, 9].

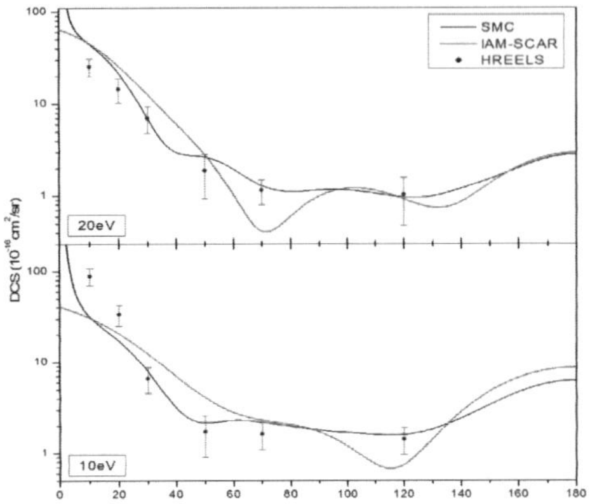

Figure 1. Elastic DCS for electron scattering from CH_2Cl_2 at 10 and 20 eV electron impact energy.

Here we present a comprehensive joint experimental and theoretical study on differential cross sections (DCSs) for elastic electron scattering from CH_2Cl_2 molecule for incident electron energies 7, 10, 20, 30 and 50 eV. The experimental DCSs were obtained in a High Resolution Electron Energy Loss Spectrometer (HREELS) [10] with an energy resolution of 120 meV (FWHM). The theoretical calculations were performed with two different methodologies: the Schwinger Multichannel Method (SMC) implemented with pseudopotentials [11] and the Independent Atom Method with Screening Corrected Additivity Rule (IAM-SCAR)[12]. The SMC method presents a better description of the elastic scattering at lower electron impact energies (up to 20eV), where the Born Closure correction for long range potentials gives a good description of the dipolar cross section dependence for smaller scattering angles. The IAM-SCAR method shows a better description for electron impact energies above 20eV.

The excellent agreement between experimental DCSs and both theoretical approaches leads to a good description of the shapes and angular distribution of the cross sections.

[1]E-mail: f.ferreiradasilva@fct.unl.pt

References

[1] K. F. Alcantara *et al.* 2011 *J. Phys. B* **44** 165205

[2] J. M. Lombert *et al.* 1999 *J. Geophys. Res.* **104** 8373

[3] M. A. K. Khalil in *Reactive halogen compounds in the atmosphere* Springer; 1999. **4E** p. 45-79

[4] M. Yen *et al.* 1993 *J. Chem. Phys.* **99** 123

[5] A. Mandal *et al.* 2014 *J. Quant. Spec. Rad.* **149** 291

[6] G. A. Gallup *et al.* 2011 *J. Chem. Phys.* **135** 134316

[7] C. G. P. Karwasz *et al.* 1999 *Phys. Rev. A* **59** 1341

[8] A. P. P. Natalense *et al.* 1999 *Phys. Rev. A* **59** 879

[9] R. Naghma *et al.* 2014 *J. Elec. Spec. Rel. Phen.* **193**48

[10] F. MotteTollet *et al.* 1992 *J. Chem. Phys.* **97** 7314

[11] R. F. da Costa *et al.* 2015 *Eur. Phys. J. D* **69** 159

[12] F. Blanco *et al.* 2013 *Eur. Phys. J. D* **67** 199

A combined gas phase and surface study on electron induced decomposition of the heteronuclear FEBID precursor; $CpFe(CO)_2Mn(CO)_5$

Oddur Ingólfsson* [1], **Rachel M. Thorman***, **Ilyas Unlu**†, **Julie Spencer**†, **Kelsea R. Johnson**‡, **Lisa McElwee-White**‡, **Howard Fairbrother** † [4]

* Science Institute and Department of Chemistry, University of Iceland, Dunhaga 3, 107 Reykjavík, Iceland
† Department of Chemistry, Johns Hopkins University, Baltimore, Maryland 21218, United States
‡ Department of Chemistry, University of Florida, Gainesville, Florida 32611-7200, United States

Synopsis Here we percent a combined gas phase and surface study on electron induced decomposition of the heteronuclear FEBID precursor; $CpFe(CO)_2Mn(CO)_5$. Dissociative electron attachment and dissociative ionization of this compound in the gas phase is explored and discussed in context to controlled deposition experiments where the same compound is exposed to high-energy electrons, when adsorbed on a surface.

In recent years, a number of gas phase studies on the interaction of low energy electrons with organometallics have been conducted in conjunction with surface studies on deposit formation from thin layers of the same compounds adsorbed on surfaces and exposed to high-energy electrons [1]. The aim of these studies has primarily been to evaluate the potential of these compounds as precursor molecules in focused electron beam induced deposition (FEBID) [2] and to unravel the mechanisms behind deposit formation [1]. In this context, the combination of gas phase and surface experiments is advantageous as the former can deliver information on the extent and energy dependency of individual processes, while the latter elucidates the surface interactions and exposure to secondary electrons with an energy distribution similar to what is expected in FEBID instruments. The use of heteronuclear precursors is an attractive alternative to mixed gas or multiple gas inlet systems for the fabrication of alloy nanostructures, and recent experiments with $HFeCo_3(CO)_{12}$ [3] achieved a typical metal purity of 80% with 1:3 FeCo ratio, clearly demonstrating the viability of this approach.

To follow this path and explore further heteronuclear precursor candidates, we have conducted gas phase studies on dissociative ionization and dissociative electron attachment to the heteronuclear organometallic $CpFe(CO)_2Mn(CO)_5$ [5], a compound specifically synthesized for this purpose. Furthermore, the combination of the multiply coordinated cyclopentadienyl group, which is known to be a poor leaving group (see ref. 4 and refs therein), and the more labile CO ligands offers an excellent opportunity to compare these ligands in a heteronuclear architecture.

The two most intense fragments produced in DEA to by $CpFe(CO)_2Mn(CO)_5$ are both formed via a low-lying resonance leading to a maximum in the ion yield near 0 eV. These channels are the loss of one CO, and the cleavage of the iron-manganese bond to form $Mn(CO)_5^-$. Hence, both these channels are associated with a single bond rupture. Further CO loss is also observed in DEA, covering the formation of $[M - nCO]^-$ over the range from n = 2 to n = 6. These channels are, however, orders of magnitude less efficient than the primary single bond rupture channels. No cyclopentadienyl loss is observed in DEA to this compound. In dissociative ionization, the fragmentation pattern observed are very different to these observed in DEA. At about 70 eV, the DI spectra is dominated by fragments formed through the loss of 5 and 6 CO ligands, the formation of the iron cyclopentadienyl fragments FeCp and $FeCp(CO)_2^+$ and the bare metal Fe^+ and Mn^+. Hence, the fragmentation through DI is much more extensive than through DEA.

Recently, surface science experiments were performed in the Fairbrother lab, wherein $CpFe(CO)_2Mn(CO)_5$ was physisorbed onto gold and highly ordered pyrolytic carbon (HOPG) substrates under ultra-high vacuum (UHV) and irradiated with varying electron dose, while the surface composition was monitored using X-ray photoelectron spectroscopy (XPS). These studies show that the primary deposition pathway is through carbonyl loss, while no cyclopentadienyl loss is observed.

In the current contribution we discuss the efficiency of the DI vs. the DEA channels in conjunction with the decomposition pathways and deposit formation from thin layers of $CpFe(CO)_2Mn(CO)_5$ at surfaces.

Acknowledgments: This work was supported by Icelandic Centre for Research (RANNIS) and the University of Iceland Research Fund and conducted within the framework of the COST Action CM1301 (CELINA)

References

[1] R. M. Thorman, *et al.* 2015 *Beilstein J. Natotechnol.* **6**, 1904-1926
[2] I. Utke *et al.* 2008 *J. Vac. Sci. Technol. B:* **26**, 1197.
[3] M. Porrati, et al. 2015 Nanotechnol. **26**, 475701
[4] R. M. Thorman *et al.* 2017 *Submitted.*

[1] E-mail: odduring@hi.is

IOP Conf. Series: Journal of Physics: Conf. Series **875** (2017) 062040 doi:10.1088/1742-6596/875/7/062040

A combined experimental and theoretical study on dissociative ionization and dissociative electron attachment to the heteronuclear FEBID precursor; $HFeCo_3(CO)_{12}$

Oddur Ingólfsson* [1], **Ragesh Kumar T P *** [2], **Ragnar Bjornsson *** [3], **Sven Christian Barth** [†4]

[*] Science Institute and Department of Chemistry, University of Iceland, Dunhaga 3, 107 Reykjavík, Iceland
[†] TU Wien, Institute of Materials Chemistry, Getreidemarkt 9/BC/02, A-1060 Wien, Austria,

Synopsis Here we percent a combined experimental and theoretical study on dissociative ionization (DI) and dissociative electron attachment (DEA) to the heteronuclear FEBID precursor; $HFeCo_3(CO)_{12}$ with the emphasis on the DEA process, the nature of the attachment and the dissociation dynamics.

Low energy electron interaction with organometallics has received appreciable attention in recent years [1], specifically in relation to the use of these compounds as precursors in focused electron beam induced deposition (FEBID) [2,3]. For the fabrication of alloy nanostructures through FEBID, the use of heteronuclear precursors is an attractive alternative to mixed gas or multiple gas inlet systems. To date, however, the use of heteronuclear organometallics in FEBID and related studies on such potential precursor molecules are limited. In fact, the only FEBID work with heteronuclear precursors we are aware of are that with $HFeCo_3(CO)_{12}$ [4] and unpublished work with $H_2FeRu_3(CO)_{13}$ [5]. In the first case, i.e., $HFeCo_3(CO)_{12}$ a typical metal purity of 80% with 1:3 FeCo ratio was reached, clearly attesting the feasibility of using bimetallic precursors for pure alloy deposition.

Motivated by the successful deposition of alloy structures using $HFeCo_3(CO)_{12}$ and the role of low energy secondary electrons in FEBID (see e.g. ref [1] and refs. therein), we have studied the gas phase decomposition of this compound through dissociative electron attachment and dissociative ionization.

The two main channels observed in DEA to $HFeCo_3(CO)_{12}$ are the loss of the apical Fe-group, mainly as $Fe(CO)_4^-$, and sequential CO loss [5]. The first of these two channels agrees well with quantum chemical calculations conducted at the BP86/def2-TZVP level of theory [6]. The calculations offer a consistent picture of our experimental observations with weakening of the bonding of the apex $Fe(CO)_3$ group to the Co_3 plane and appreciable spin density at the apical iron, reflecting the observed efficiency of the $Fe(CO)_4^-$ formation. The sequential CO loss, the second dominating channel, comprises the loss of 1 and 2 CO close to 0 eV up to the loss of all 12 CO ligands, which appears at around 20 eV incident electron energy - about 12 eV above its ionisation energy. To elucidate this remarkable attachment of electrons about 12 eV above the ionisation limit, we conducted calculations of the state densities of this compound in comparison to model compounds both with and without the metal core [7]. The calculations reveal a high density of metal-CO antibonding orbitals about 3 eV above the HOMO, while the density of antibonding metal-metal orbitals is low. The accommodation of the large amount of excess energy associated with the attachment process is thus ascribed to a quasi-continuum of resonant states associated with multiple electronic excitations within the metal-CO anti bonding "band".

In dissociative ionization, the fragmentation pattern observed is similar to DEA in the sense that the loss of the iron-centred apex and the sequential CO loss are the dominating channels. However, in DI the intensity ratios within the respective channels are very different from that in DEA. Here, the apex loss appears through formation of $Fe(CO)_n^+$ with n = 1-5, all with comparable intensity, and the main CO loss channels are associated with the loss of 5 to 12 CO; the bulk of the intensity in DEA is on the loss of 2 CO ligands.

In the current contribution, we discuss the mechanisms behind the observed DEA processes in conjunction with our calculations, as well as the efficiency of the DI vs. the DEA channels in conjunction with the decomposition pathways and deposit formation from thin layers of $HFeCo_3(CO)_{12}$ at surfaces. These will be presented in a separate poster.

Acknowledgments: This work was supported by Icelandic Centre for Research (RANNIS) and the University of Iceland Research Fund and conducted within the framework of the COST Action CM1301 (CELINA).

References

[1] R. M. Thorman, *et al.* 2015 *Beilstein J. Natotechnol.* **6**, 1904-1926
[2] I. Utke *et al.* 2008 *J. Vac. Sci. Technol. B:* **26**, 1197.
[3] W. F. van Dorp *et al.* 2008 *J. Appl. Phys.* **104**, 081301.
[4] M. Porrati, et al. 2015 Nanotechnol. **26**, 475701
[5] R. K. TP *et al.* 2017 *Submitted.*
[6] R. K. TP *et al.* 2016 *Eur. Phys. J. D* ***70***, 163
[7] R. K. TP *et al.* 2017 *Submitted.*

[1] E-mail: odduring@hi.is

Electron induced fragmentation and deposit formation from nano-meter thin surface layers of $HFeCo_3(CO)_{12}$ adsorbed on gold surfaces.

Oddur Ingólfsson* [1], **Ragesh Kumar T P***, **Ilyas Unlu**[†], **D. Howard Fairbrother**[†] **and Sven Barth**[‡¶]

[*] Science Institute and Department of Chemistry, University of Iceland, Dunhagi 3, 107 Reykjavík, Iceland
[†] Department of Chemistry, Johns Hopkins University, Baltimore, Maryland 21218, United States
[‡] TU Wien, Inst. of Materials Chemistry, Getreidemarkt 9/BC/02, A-1060 Wien, Austria.

Synopsis In the current contribution, we present electron induced fragmentation and deposit formation from nano-meter thin surface layers of $HFeCo_3(CO)_{12}$ adsorbed on gold surfaces and exposed to high-energy (400 eV) electrons from a flood gun.

There is an increase in demand for developing and improving the efficiency of techniques that are capable of fabricating sub-10 nm structures in a controlled manner. Focused electron beam induced deposition (FEBID) [1,2] is a technique with this potential and the capability of fabricating direct 3D-nano-structures on top of almost any substrate irrespective to its geometry. In FEBID, a high-energy electron beam is focused onto a substrate on which precursor molecules are adsorbed at constant partial pressure. The precursor molecules dissociate under exposure to the electron beam, apparently through interaction with low energy secondary electrons produced through the interaction of the primary beam with the surface [3]. Ideally the precursor decomposes completely and volatile ligands are pumped away while the non-volatile fragments (metal) deposits to grew the desired nanostructures. However, FEBID is still facing challenges such as lateral broadening of structures and impurities in the deposit from incomplete decomposition of the precursor molecules.

$HFeCo_3(CO)_{12}$ is the first single-source heteronuclear precursor molecule used in FEBID to fabricate bimetallic nanostructures and it was shown that 80% pure FeCo deposits could typically be attained [4], a purity much higher than normally attainable. In an attempt to better understand the mechanisms behind the decomposition of this precursor we recently conducted a gas phase study on the energy dependency of dissociative electron attachment (DEA) and dissociative ionization (DI) of this molecule [5, 6]. It shows rich fragmentation through both DEA and DI and These experiments give valuable information on the potential fragmentation mechanisms involved in its deposition process, however, no information on the role of the surface interaction can be obtained in these experiments.

In the current contribution, we present electron induced fragmentation and deposit formation from nano-meter thin surface layers of $HFeCo_3(CO)_{12}$ adsorbed on gold surfaces and exposed to high-energy (400 eV) electrons from a flood gun. Here, desorbing fragments are monitored by mass spectrometry, while the electron dose dependence of the deposit composition, is monitored by means of X-ray photo electron spectroscopy (XPS). As expected the predominant desorption, monitored with mass spectrometry, is that of CO. Correspondingly a rapid reduction of the CO content is observed at the surface. This is followed by decomposition of the remaining CO and an apparent iron oxide formation. Hence, based on our preliminary analysis of the development of the electron dose dependency of the XPS peak areas, and peak positions, electron induced decomposition of $HFeCo_3(CO)_{12}$ proceeds in two steps. In the first step, with electron dose of $<5x10^{16}e^-/cm^2$, the peak area analysis shows an average 9 CO lost per molecule. The second step proceeds considerably slower and constitutes decomposition of the remaining CO and the apparent conversion iron into iron oxide. Furthermore, by increase the substrate temperature from the initial -60 ^0C to RT (25 ^0C), all remaining CO is removed from the surface. This observation is consisting with the high purity of FeCo deposit in FEBID from $HFeCo_3(CO)_{12}$

In the current contribution, we compare the electron induced reaction of $HFeCo_3(CO)_{12}$ adsorbed on gold surfaces with previous gas phase studies and FEBID using $HFeCo_3(CO)_{12}$ as precursor

Acknowledgments: This work was supported by Icelandic Centre for Research (RANNIS) and the University of Iceland Research Fund. RKTP acknowledges financial support through a PhD grant from the University of Iceland Research Fund and a from CELINA for a shot term scientific mission.

References

[1] I. Utke *et al.* 2008 *J. Vac. Sci. Technol. B:* **26**, 1197.
[3] M. Thorman *et al.* 2015 *Beilstein J. Nanotechnol*, **6**, 1904-1926, and references therein.
[4] F. Porrati et al. 2015 *Nanotechnol.* **26**, 475701.
[5] T P. R *et a*l. 2016 *EPJ D*, **70**, 163.
[6] T P. R *et al*, 2017 *Submitted*

E-mail: odduring@hi.is

IOP Conf. Series: Journal of Physics: Conf. Series **875** (2017) 062042 doi:10.1088/1742-6596/875/7/062042

Dissociative electron attachment to 3-bromopyruvic acid

F. Ferreira da Silva* [1], **S. Denifl**[†], **I. Bald**[‡], **and J. Kopyra**[•]

[*] Laboratório de Colisões Atómicas e Moleculares, CEFITEC, Departamento de Física, Faculdade de Ciências e Tecnologia, Universidade NOVA de Lisboa, 2829-516 Caparica, Portugal
[†] Institut für Ionenphysik und Angewandte Physik, Leopold Franzens Universität Innsbruck, Technikerstrasse 25, 6020 Innsbruck, Austria
[‡] Institute of Chemistry—Physical Chemistry, University of Potsdam, Potsdam, Germany
[•] Department of Chemistry, Siedlce University, 3 Maja 54, 08-110 Siedlce, Poland.

Synopsis: Fragmentation of gas phase 3-bromopyruvic acid ($BrCH_2COCO_2H$) caused by capture of low energy electrons (0-12 eV) was studied. Ion yields of negative ions formed via dissociative electron attachment were recorded.

Many cancer cells have aberrant energy metabolism therefore they consume more glucose than normal cells and rapidly convert it to lactate. Hexokinase type II (HK II) is the key enzyme for maintaining increased glycolysis in cancer cells where it is overexpressed. 3-bromopyruvate (3-BrPA), has an ability to inhibit HK II, hence it can cause cell death in cancer tissues.

Mitochondria are important in the control of both cell survival and cell death. Mitochondrial dysregulations, such as loss of trans-membrane potential, accumulation of reactive oxygen species (ROS), membrane permeability transition and release of pro-apoptotic factors lead to apoptosis [1]. 3-BrPA is a strong alkylating agent. It inhibits both HK II and mitochondrial oxidative phosphorylation, leading to a decrease of ATP formation which will lead to cell death [2]. Although 3-BrPA has been described as a good anti-cancer agent [3], the underlying molecular mechanism of cell death induced is still not clear.

In the present study we describe dissociative electron attachment (DEA) mechanisms leading anionic fragmentation of 3-BrPA. The peak resonances listed in the Table 1 were extracted from the recorded ion yields.

Table 1. DEA resonances

Anion	Peak resonance
3-BrPA-H	0 eV; 1 eV
3-BrPA-Br	0 eV; 2 eV
$BrCH(OH)_2$	0 eV
Br	0 eV; 2 eV; 5 eV
COOH	2.2 eV; 4.5 eV
COO	2.2 eV; 4.5 eV
OH	5 eV
O	5.5 eV

Dissociative electron attachment to 3-BrPA induces rich anionic fragmentation, prevailing 0 eV channels. The measurements have been carried out at the Siedlce University by means of crossed molecular beam apparatus coupled with trochoidal monochromator together with quadrupole mass spectrometer.

References

[1] Y. Higuchi Y 2004 *J Cell Mol Med* **8**, 455–464
[2] Y. H. Ko *et al.* 2004 *Biochem Biophys Res Commun* **324**, 269–275
[3] Y. H. Ko *et al.* 2001 *Cancer Lett* **173**, 83–91

[1] E-mail: f.ferreiradasilva@fct.unl.pt

IOP Conf. Series: Journal of Physics: Conf. Series **875** (2017) 062043 doi:10.1088/1742-6596/875/7/062043

Low energy ELEctron driven chemistry for the advantage of emerging NAno-fabrication methods (ELENA); a Marie Skłodowska-Curie Innovative Training Network

Oddur Ingólfsson* [1]

* Science Institute and Department of Chemistry, University of Iceland, Dunhagi 3, 107 Reykjavík, Iceland

Synopsis The current contribution relates to a recently endorsed Marie Skłodowska-Curie Innovative Training Network on Low energy ELEctron driven chemistry for the advantage of emerging NAno-fabrication methods; ELENA. Here we discuss the objectives of ELENA relating to the understanding the low energy electron induced processes underpinning two emerging nanotechnology methods; focused electron beam induced deposition (FEBID) and extreme ultra violet lithography (EUVL).

ELENA; Low energy ELEctron driven chemistry for the advantage of emerging NAno-fabrication methods, is a recently endorsed Marie Skłodowska-Curie, Innovative Training Network. The network focuses on the, fundamental, electron driven processes underpinning two emerging nano-fabrication methods; Focused Electron Beam Induced Processing FEBIP [1,2,3] (with an emphasis on deposition FEBID) and Extreme Ultraviolet Lithography (EUVL) [4,5]. Exploitation of these methods has the potential to improve the manufacturing process of existing devices, improve the performance of specific nano-devices through better material control and higher spatial resolution in their fabrication, and enable the fabrication of new devices. Improved EUVL can provide cost efficient high-volume microchip production for the fabrication of more-advanced integrated circuits housing higher component densities than is currently achievable [6]. FEBID on the other hand is ideally suited for direct, one step production of high aspect ratio components and may be integrated in the production of functional devices where the 3D shape of components is critical [1].

For FEBID the objectives of ELENA embrace the development of new precursor molecules for ultimate composition control of the deposits and for the development of in situ purification protocols. Hence, a core objective of ELENA is the design and development of protocols for precursor molecules that lead to purer deposits and higher resolution. With regards to FEBID, ELENA's objective is to achieve the desired control of the deposit composition by understanding the low energy electron induced chemistry involved and to acknowledge this chemistry in the design of new precursor molecules such that pure deposits with well controlled composition may be generated on a routine basis. Similarly for EUV resists the objective is to gain control over the low energy induced chemistry within these resists and to use that chemistry to control adverse and limiting factors such as the mean free path of low energy electrons and acid diffusion within the resist. Hence, to control the low energy electron induced chemistry within the resists to improve the performance of EUVL with regards to resolution, sensitivity and line width roughness (RLS). This is only achievable by understanding the low energy electron induced chemistry involved and to acknowledge this chemistry in the design of new FEBID precursors and EUVL resist material.

In the current contribution we discuss the objectives of ELENA and the means to achieve these.

References

[1] I. Utke, *et al.* 2008 *J. Vac. Sci. Technol., B* **26**, (4), 1197.
[2] W. F. van Dorp *et al.* 2008 *J. Appl. Phys.* ***104***, 081301.
[3] S. Randolph *et al.* 2006 *Crit. Rev. Solid State Mater. Sci.* **31**, (3), 55
[4] Saša Bajt *et al.* 2008 *Surf. Sci. Rep.* (2) 73-99
[5] B. Wu *et al.* 2007 *J. Vac. Sci. Technol. B* **25** (6), 1743-1761
[6] C. Wagner *et al.* 2010. *Ind. Persp. Nature Photonics*, **4**, 24

[1] E-mail: odduring@hi.is

Interference effect in e-emission spectrum from a molecular (N_2) double slit in collisions with fast electrons

Madhusree Roy Chowdhury*, C. R. Stia†, O. A. Fojon†, R. D. Rivarola† and Lokesh C. Tribedi*[1]

* Tata Institute of Fundamental Research, Colaba, Mumbai 400005, India
† Institute de Fisica Rosario (CONICET-UNR), Universidad Nacional de Rosario, 2000 Rosario, Argentina

Synopsis Double differential cross sections (DDCS) of electrons ejected in ionization of N_2 by fast electrons have been measured. Signature of interference effect was revealed in the DDCS ratio (N_2/2N). Interference oscillation was also confirmed from the asymmetry parameter, determined from the DDCS of two complementary angles. Traces of second order effect was also observed from the asymmetry parameter.

Ionization of diatomic molecules by electrons, heavy ions and photon to investigate the Young type interference effect has been a subject of study for more than a decade, although initially proposed by Cohen and Fano in 1966 [1]. Its application towards fundamental quantum mechanical issues as well as structural and collisional aspects involving di- and multi-atomic molecules are well known. Several experimental measurements have been performed for the simplest diatomic molecule H_2 using different probes and in all cases the observation of interference oscillations were consensual [2-4]. This work was extended further for multi electronic targets like N_2 and O_2. Though oscillations were observed in individual orbital for both the targets upon photoionization [5], but, for heavy ions, there exists an ambiguity in the evidence of interference oscillations [6,7]. According to a theoretical prescription, interference effect may not be readily observed in case of heavy ion collision with N_2 because of the cancellation of phase mismatched oscillations contributing from each individual orbitals [8]. For fast electrons as projectile, the Coulomb perturbation strength is much small, resulting in negligible multiple ionization. Hence it is expected that the interference effect may be observed much clearly in case of electrons than heavy ions.

In this work, we report the measurement of DDCS of secondary electrons emitted from ionization of N_2 in collision with 7 keV electrons. The DDCS have been measured for electron energies between 1 and 550 eV and emission angles from 15^0-145^0 using a hemispherical electrostatic analyzer. The atomic DDCS were calculated using B1 model. The experimental-to-theoretical DDCS ratios (N_2/2N) revealed clear signature of interference oscillations [9,10] which was fitted by the Cohen-Fano model. The difference in the frequency of oscillation between forward and backward angles causes an oscillation in the forward-backward asymmetry parameter (α(k), Fig.1).

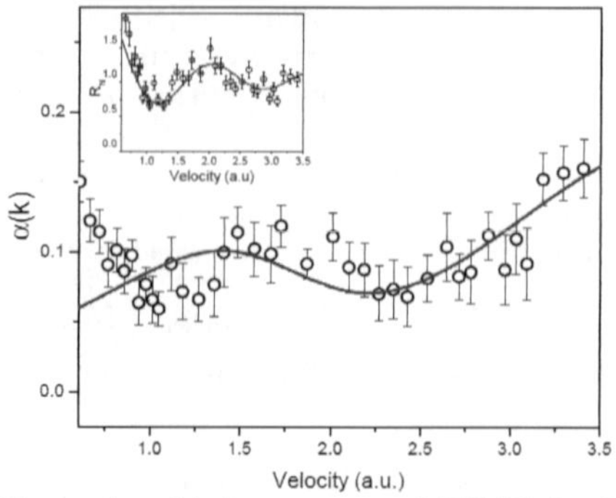

Figure 1: α(k) from measured DDCS for 35^0 and 145^0. Inset : α(k) divided by first order fitting reveals 2nd order effect.

References

[1] H. D. Cohen and U. Fano 1966 *Phys. Rev.* **150** 30
[2] N. Stolterfoht *et al.* 2001 *PRL* **87** 023201
[3] D. Misra *et al.* 2004 *Phys. Rev. Lett.* **92** 153201
[4] S. Chatterjee *et al* 2009 *J Phys B* **42** 065201
[5] M. Ilchen *et al.* 2014 *Phys. Rev. Lett.* **112** 023001
[6] J. Baran *et al.* 2008 *Phys. Rev. A* **78** 012710
[7] S. Nandi *et al.* 2015 *Eur Phys J D* **69** 192
[8] C. A. Tachino *et al.* 2012 *J. Phys. B* **45** 025201
[9] M. Roy Chowdhury *et al.* 2016 *PRA* **94** 052703.
[10] M. Roy Chowdhury *et al. EPJD* (submitted)

[1]E-mail: lokesh@tifr.res.in

IOP Conf. Series: Journal of Physics: Conf. Series **875** (2017) 062045 doi:10.1088/1742-6596/875/7/062045

Angle-resolved $e^- + C_{60}$ elastic scattering cross section versus Ramsauer minima in partial elastic scattering cross sections

V. K. Dolmatov[*1], **M. Ya. Amusia**[†‡2], **L. V. Chernysheva**[‡3]

[*] Department of Physics and Earth Science, University of North Alabama, Florence, Alabama 35632, U.S.A.
[†] Racah Institute of Physics, Hebrew University, 91904 Jerusalem, Israel
[‡] A. F. Ioffe Physical-Technical Institute, 194021 St. Petersburg, Russia

Synopsis Ramsauer-type minima in $e^- + C_{60}$ partial electron elastic scattering cross sections, brought about by C_{60} polarization, and their exposition to light in the angle-differential scattering cross section are revealed and detailed.

Work [1], devoted to the study of electron elastic scattering off endofullerenes $A@C_{60}$, predicted the emergence of Ramsauer-type minima in partial scattering cross sections σ_ℓ with $\ell = s$, p, d and f, due to polarization of $A@C_{60}$ by scattering electrons. The energy positions of the minima were somewhat different from one another. For this reason, they were unresolved in the total scattering cross section $\sigma_{\rm total} = \sum \sigma_\ell$. However, there are good reasons to expect that the minima should be resolved in the angle-differential scattering cross section $\frac{d\sigma}{d\Omega}$, since the position of a minimum depends on the scattering angle. This constitutes the primary aim of the present work, namely, to investigate the extent to which Ramsauer minima in σ_ℓs might be exposed to light in $\frac{d\sigma}{d\Omega}$.

In the present paper, for the sake of simplicity, we focus on electron elastic scattering off an empty C_{60}, $e^- + C_{60}$. As in [1], we model C_{60} by an attractive spherical potential $U_{\rm c}(r)$ of certain depth U_0, inner radius r_0 and thickness Δ. The impact of the polarization potential $V_{\rm pol}(r)$ of C_{60} on scattering electrons is taken into account in the framework of a static dipole-polarization approximation: $V_{\rm pol}(r) \approx -\alpha/[2(r^2+b^2)^2]$. Here, α is the static polarizability of C_{60} and b is a parameter of the order of r_0. Our model, thus, accounts for the motion of scattering electrons in the field of an effective potential $U_{\rm eff}(r) = U_{\rm c}(r) + V_{\rm pol}(r)$.

Calculated partial σ_ℓs and $\frac{d\sigma}{d\Omega}$ at 80° (the calculation accounted for fourteen partial waves, up to $\ell_{\max} = 13$), as a case study, are depicted in Fig. 1 along with corresponding experimental data [2] borrowed from [3].

First, Fig. 1 demonstrates a reasonable qualitative and semi-quantitative agreement between theory and experiment. The utilized in the present work simple approximation is, thus, proven to be surprisingly usable.

Second, one can see from Fig. 1 that the positions of the Ramsauer minima in σ_s ($\varepsilon \approx 0.25$ eV), σ_d ($\varepsilon \approx 0.99$ eV) and σ_f ($\varepsilon \approx 1.48$ eV) approximately (for reasons) match the positions of the first, second and third lower energy minima in $\frac{d\sigma}{d\Omega}$, respectively.

It is, thus, shown in the present work that (a) the simple model utilized in the present paper is usable and (b) the angle-differential cross section $\frac{d\sigma}{d\Omega}$ allows one to expose to light the existence of Ramsauer minima in electron scattering of partial electronic waves.

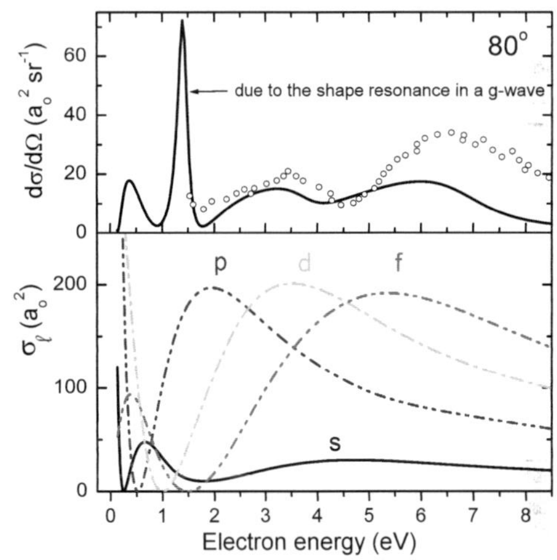

Figure 1. Top panel, calculated $\frac{d\sigma}{d\Omega}$ ($\theta = 80^{\rm o}$, open circles - experiment [2]) and, bottom panel, partial σ_ℓs for $e^- + C_{60}$ elastic scattering. Parameters of the utilized model (in atomic units): $r_0 = 5.26$, $U_0 = -0.26$, $\Delta = 2.91$, and $\alpha = 850$.

V.K.D. acknowledges the financial support in the form of a travel grant from COAS, UNA.

References

[1] V. K. Dolmatov *et al.* 2017 *Phys. Rev. A* **95** 012709
[2] H. Tanaka *et al.* 1994 *J. Phys. Soc. Japan* **63** 485
[3] F. A. Gianturco *et al.* 1999 *J. Phys. B* **32** 2181

[1]E-mail: vkdolmatov@una.edu
[2]E-mail: amusia@012.net.il
[3]E-mail: larissa.chernysheva@mail.ioffe.ru

Total cross section for low-energy electron scattering from formic acid, (HCOOH), molecules

Paweł Możejko[*1], **Alicja Domaracka**[†2], **Mateusz Zawadzki**[*], **Elżbieta Ptasińska-Denga**[*] **and Czesław Szmytkowski**[*]

[*] Department of Atomic, Molecular and Optical Physics, Faculty of Applied Physics and Mathematics, Gdańsk University of Technology, ul. G. Narutowicza 11/12, 80-233 Gdańsk, Poland

[†] Centre de Recherche sur les Ions, les Matériaux et la Photonique, Normandie University, ENSICAEN, UNICAEN, CEA, CNRS, CIMAP, F-14000 Caen, France

Synopsis Total cross section (TCS) for low-energy electron scattering from formic acid molecules has been measured using electrostatic electron spectrometer working in linear transmission mode. Two local maxima centered around 1.7 eV and 7.8 eV have been observed and associated with resonant scattering processes.

Reliable, accurate experimental data concerning electron collisions with atoms, molecules or their clusters are of great importance in many fields of pure and applied science, ranging from space sciences, astrobiology, radiation physics and chemistry to gaseous electronics, energy conversion and electron impact-induced reactions on surfaces. In this work, we present our experimental studies on electron collisions with formic acid, (HCOOH), molecules.

The absolute total cross section (TCS) for electron scattering from HCOOH molecules has been measured using the electrostatic 127° electron spectrometer working within the linear transmission configuration [1] for impact energies ranging from 0.5 eV up to 300 eV. In this experimental setup the target sample in the scattering cell is irradiated with a monoenergetic electron beam (fwhm $\simeq$ 50 meV). Those electrons which emerge from the reaction region through a cell exit orifice are energy discriminated with a planar retarding field filter and, finally, electrons are collected with a Faraday cup detector. The total cross section $\sigma(E)$ at given electron impact energy E has been obtained according to the Bougue–de Beer–Lambert attenuation formula:

$$I_n(E) = I_0(E)\exp\left[-nL\sigma(E)\right]. \qquad (1)$$

Here, $I_n(E)$ and $I_0(E)$ are the measured intensities of the electron beam passing the distance L through the reaction volume in the presence or absence of the target vapor, respectively; n is the number density of the target molecules in the scattering cell.

The TCS obtained in the present experiments, for impinging electron energies ranging from 1 to 30 eV, is shown in figure 2 together with experimental total [2] and elastic [3] cross sections. In the presented energy range, TCS function is characterised by two resonant like features located near 1.7 eV and 7.8 eV, respectively. Both of them are connected with dissociative electron attachment processes which lead to production of three anionic fragments: $HCOO^-$ at 1.3 eV, OH^- at 7.8 eV and O^- at 7.5 eV and 9 eV [4].

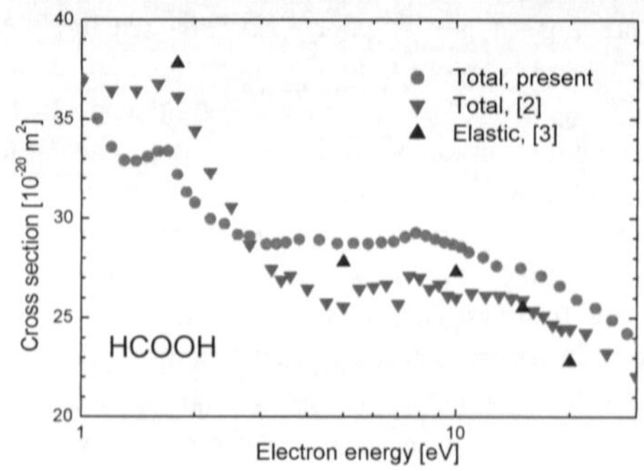

Figure 1. Comparison of the experimental cross sections for electron collisons with HCOOH molecule

Below 1.5 eV TCS function is increasing with energy decreasing, which is typical behaviour for electron collisions with polar molecules (dipole moment of formic acid μ_D = 1.41D). While the data of Kimura *et al* [2] shows the same energetic dependence, with both resonant features clearly visible, as in the present results, some discrepancies in magniudes of both data sets can be easily observed.

This work has been partially supported by the Polish Ministry of Science and Higher Education.

References

[1] Cz. Szmytkowski and P. Możejko 2001 *Vacuum* **63** 549 **635** 001001

[2] Kimura, M., O. Sueoka, A. Hamada, and Y. Itikawa 2000 *Adv. Chem. Phys.* **111** 537

[3] V. Vizcaino, M. Jelisavcic, J. P. Sullivan and S. J. Buckman 2006 *New J. Phys.***8** 85

[4] A. Pelc, W. Sailer, P. Scheier, N.J. Mason and T.D. Märk 2002 *Eur. Phys. J.* D **20** 441

[1]E-mail: paw@mif.pg.gda.pl
[2]E-mail: Alicja.Domaracka@ganil.fr

Cross sections calculations for electron scattering from dimethylamine, $NH(CH_3)_2$, molecule

Paweł Możejko*[1], Bożena Żywicka*[2] and Alicja Domaracka†[3]

* Department of Atomic, Molecular and Optical Physics, Faculty of Applied Physics and Mathematics, Gdańsk University of Technology, ul. G. Narutowicza 11/12, 80-233 Gdańsk, Poland

† Centre de Recherche sur les Ions, les Matériaux et la Photonique, Normandie University, ENSICAEN, UNICAEN, CEA, CNRS, CIMAP, F-14000 Caen, France

Synopsis The total cross section for single electron-impact ionization and the integral elastic cross section for electron scattering from dimethylamine have been calculated using the binary-encounter-Bethe model and the independent atom method, respectively.

In the present work we have theoretically studied elastic electron scattering from dimethylamine, ($NH(CH_3)_2$), and positive ionization of this molecule by electron impact.

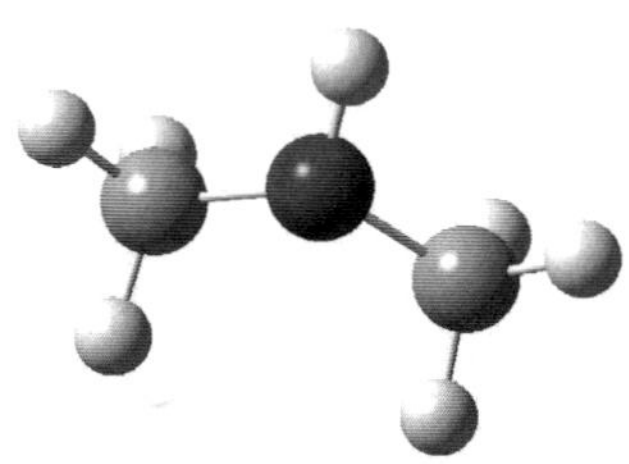

Figure 1. Geometry of the dimethylamine molecule.

The integral elastic cross section has been calculated using the independent atom method [1] for electron energies ranging from 30 eV to 3 keV. The interaction between scattered electron and dimethylamine molecule has been described by a static-polarization model potential. The total cross section for the single electron-impact ionization has been derived using the binary-encounter-Bethe (BEB) model [2] for energies between the ionization threshold up to 3 keV. All physical quantities necessary in the BEB calculations have been obtained for the ground state of the geometrically optimized molecule with the Hartree-Fock method using Gaussian code [3] and Gaussian 6-311G+(d,p) basis set. The valence orbital energies have been calculated with OVGF method implemented in Gaussian.

In figure 2 the calculated elastic and ionization cross sections are shown. Predicted total cross section, estimated as the sum of elastic and ionization cross sections is also shown. Although this approach seems to be rather crude, we expect that the total cross section estimated this way will be quite satisfactorily to within $\pm10\%$. This expectation is justified by the results of our previous calculations which reproduced the experimental intermediate-energy data successfully for a variety of complex molecular targets [4].

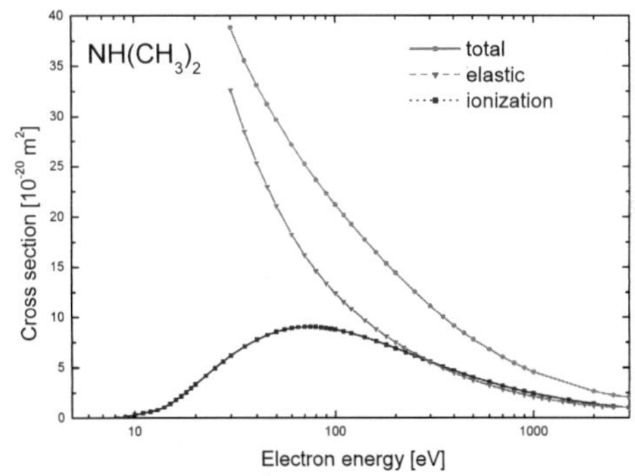

Figure 2. Calculated cross sections for electron impact ionization and elastic scattering for dimethylamine molecule. Total cross section is obtained as the sum of these partial cross sections.

This work has been partially supported by the Polish Ministry of Science and Higher Education. The authors acknowledge the Academic Computer Centre in Gdansk (TASK) for the Computational Grant.

References

[1] N. F. Mott and H. S. W. Massey 1965 *The Theory of Atomic Collisions* Oxford University Press

[2] W. Hwang, Y. K. Kim and M. E. Rudd 1996 *J. Chem. Phys.* **104** 2956

[3] M. J. Frisch *et al* 2003 *GAUSSIAN 03, Revision B.05* Pittsburgh: Gaussian

[4] Cz. Szmytkowski, A. Domaracka, P. Możejko and E. Ptasińska-Denga 2007 *Phys. Rev.* A **75** 052721

[1]E-mail: paw@mif.pg.gda.pl
[2]E-mail: boshena@mif.pg.gda.pl
[3]E-mail: Alicja.Domaracka@ganil.fr

IOP Conf. Series: Journal of Physics: Conf. Series **875** (2017) 062049 doi:10.1088/1742-6596/875/7/062049

Dissociation of 2-oxopropanoic acid by low energy electrons

Mateusz Zawadzki[*†1], **Jaroslav Kočišek**[*], **Juraj Fedor**[*]

[*] J. Heyrovský Institute of Physical Chemistry vvi, Academy of Sciences of the Czech Republic, Dolejškova 3, 182 23 Praha, Czech Republic

[†] Atomic Physics Division, Department of Atomic, Molecular and Optical Physics, Faculty of Applied Physics and Mathematics, Gdańsk University of Technology, ul. Gabriela Narutowicza 11/12, 80-233 Gdańsk, Poland

Synopsis Electron attachment to molecular target can lead to a variety of negative fragment ions. We experimentally probe the main dissociative electron attachment (DEA) channels in 2-oxopropanoic acid ($CH_3 - CO - COOH$) - an important trace component in the Earth's atmosphere. In our studies we utilise nearly mono-energetic electrons at electron energies from 0 eV up to 15 eV.

Atmospheric gases directly influence the global radiative balance by interaction and scattering of incoming radiation. 2-oxopropanoic acid[2] is a substantial atmospheric organic compound that originates from oxidation of isoprene emitted from trees. Apart from a tree-covered environment, pyruvic acid also comes from biomass burning, metabolism of animals and plants. However, in urban environment the primary origin of pyruvic acid comes from vehicular emissions [1, 2].

Dissociation of pyruvic acid by photon impact has been already studied in detail [3]. However, there is a lack of experimental studies dealing with interactions of electrons with pyruvic acid molecule, in particular concerning dissociative electron attachment. Free electrons are abundant species in the upper atmospheres and therefore interactions of electrons with pyruvic acid may play an important role in the atmospheric chemistry. Particularly, the dissociative attachment process, which is effective in bond breaking at subexcitation energies.

In our studies we used the mass spectrometer with trochoidal electron monochromator (MS-TEM). The MS-TEM consist of an electron beam of well defined incident energy, a target chamber where electrons collide with the target gas molecules, and quadrupole filter for analysing ions created in the dissociative attachment process [4].

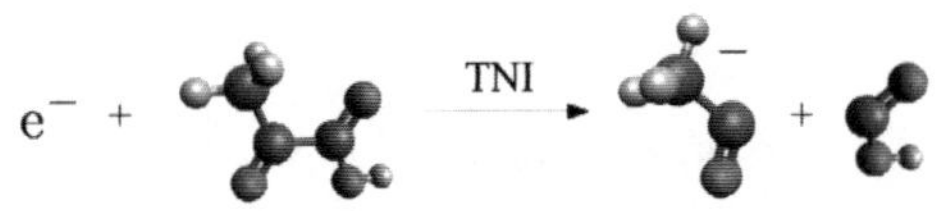

Figure 1. An example of fragmentation via possible DEA channel; with TNI being the transient anion formed upon resonant free electron attachment.

We experimentally probe the main dissociative electron attachment (DEA) channels in pyruvic acid. Electron attachment to pyruvic acid leads to a variety of negative fragment ions. Detailed energetics of formation of the DEA channels will be discussed. The example of such molecular fragmentation is shown in Fig. 1.

The most dominant main DEA channels for pyruvic acid (Eqs. 1-3):

$$e^- + CH_3COCOOH \xrightarrow{TNI} CH_3COCO + OH^- \quad (1)$$
$$CH_3^- + COCOOH \quad (2)$$
$$CH_3CO^- + COOH \quad (3)$$

These fragments OH^-, CH_3^-, CH_3CO^- arise from the cleavage of a single bond and corresponding resonances are observe within 3-8 eV. Other fragments can be described using more complex fragmentation reactions. Additionally, the evidence of dimers formation is also reported. Particularly, presence of dimers in the sample lead to a completely different behaviour of the dehydrogenated $(M\text{-}H)^-$ anion. The fragmentation pattern and the energy range completely change upon the formation of hydrogen bonding. Dramatic energy shift towards lower energies (near 0 eV) is assigned to spontaneous proton transfer along the hydrogen bond during the electron attachment to the dimer.

Acknowledgments: The work was supported by the Czech Science Foundation Grant no. 16-10995Y.

References

[1] H. Khwaja 1995 *Atmos. Environ.* **29** 127.

[2] K. F. Ho, S. C. Lee, J. J. Cao, K. Kawamura, T. Watanabe, Y. Cheng and J. C. Chow 2006 *Atmos. Environ.* **40** 3030.

[3] R. N. Rosenfeld and B. Weiner 1983 *J. Am. Chem. Soc.* **105** (11) 3485.

[4] M. Stepanovic, Y. Pariat and M. Allan 1999 *J. Chem. Phys.* **110** 1137.

[1]E-mail: m.zawadzki@mif.pg.gda.pl

IOP Conf. Series: Journal of Physics: Conf. Series **875** (2017) 062050 doi:10.1088/1742-6596/875/7/062050

Electron impact study of H_2 and D_2 continuum radiation

Michal Ďurian, Marián Danko, Juraj Országh and Štefan Matejčík[1]

Faculty of Mathematics, Physics and Informatics, Comenius University in Bratislava, Slovakia

Synopsis We report our recent experimental study regarding H_2 and D_2 continuum radiation ($a^3\Sigma_g^+ \rightarrow b^3\Sigma_u^+$) in the spectral range of 200 – 650 nm, excited by electron impact at near-threshold impact energies. In contrast to earlier studies, present results indicate that H_2 and D_2 continuum radiation is detectable in the spectral range above 500 nm. To supplement this, we present excitation-emission photon efficiency curves (PEC) of D_2 measured at 650 nm, with precautions taken to rule out other possible signal sources.

Hydrogen and deuterium range among the most widely studied molecules in modern science [1, 2]. High-quality kinetic data of processes involving H_2 and D_2 are crucial for modeling of fusion and other plasmas. We are using the electron induced fluorescence (EIF) technique, where precise control of electron kinetic energy and high photon detection sensitivity allows obtaining high-quality spectral and kinetic data. The experiment is capable of measuring emission spectra initiated by electron impact and corresponding photon efficiency curves. These can be subsequently calibrated to obtain excitation-emission cross-sections.

In this abstract we present results of our recent experiments involving H_2 and D_2 continuum radiation spectra originating from their $a^3\Sigma_g^+ \rightarrow b^3\Sigma_u^+$ transition. The hydrogen continuum spectrum is a superposition of narrower continua corresponding to excitation of different vibrational states in the upper electronic state and their radiative decay to the lower unbound state [1, 3].

The EIF apparatus is described in detail in [4]. In present experiment we have used electron beam with electron energy resolution of 0.5 eV FWHM at few hundreds nA of electron current. Photon emission was measured using optical system comprising of Cornerstone 260 grating monochromator, focusing optics and Hamamatsu H4220P PMT module, with spectral range from 200 to 670 nm. Spectral sensitivity calibration was performed using tungsten filament emission, taking gray body radiation model above 500 nm and using known H_2 continuum radiation shape for wavelengths above 200 nm [5].

Fluorescence spectra of H_2 and D_2 were recorded in 200 – 650 nm spectral range. Distinct spectral features, such as the continuum maximum and Fulcher bands were visible. Uniqueness of these spectra lies in observation of the continuum radiation up to 650 nm, which is much higher than commonly reported 500 nm limit. Spectral sensitivity of the PMT module did not allow to extend the measurements to longer wavelengths.

To prove the origin of the 650 nm radiation is in the $a^3\Sigma_g^+ \rightarrow b^3\Sigma_u^+$ transition, we have measured excitation-emission PEC of D_2 continuum at this wavelength and compared its shape to PEC measured at 230 nm (Figure 2). 420 nm cut-off optical filter was used to cancel higher order diffraction from the monochoromator grating.

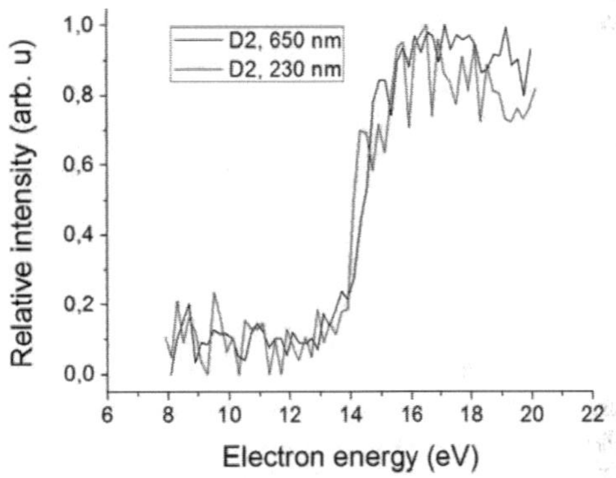

Figure 1. Photon efficiency curves of D_2 at 230 nm and 650nm (measured with optical filter).

Acknowledgment: This work was supported by the Slovak Research and Development Agency, project Nr. APVV-0733-11 and the grant agency VEGA, project Nr. 1/0417/15. This project has received funding from the European Union's Horizon 2020 research and innovation programme under grant agreement number 633053.

References

[1] B. P. Lavrov et al. 1999 *Phys. Rev. E* **59** 3526
[2] M. Ajello, D. E. Shemansky 1993 *ApJ* **407** 820
[3] H. M. James et al. 1939 *Phys. Rev.* **55** 184
[4] M. Danko et al. 2013 *J. Phys. B* **46** 045203
[5] G. K. James et al. 1998 *J. Geophys. Res.* **103** 20113

[1]E-mail: stefan.matejcik@fmph.uniba.sk

IOP Conf. Series: Journal of Physics: Conf. Series **875** (2017) 062051 doi:10.1088/1742-6596/875/7/062051

Experimental and theoretical progress in time-resolved (e, 2e) electron momentum spectroscopy of photodissociation dynamics of acetone at 195 nm

Shotaro Nakayama [1], Masakazu Yamazaki, and Masahiko Takahashi [2]

Institute of Multidisciplinary Research for Advanced Materials, Tohoku University, Sendai 980-8577, Japan

Synopsis We report on experimental and theoretical progress in time-resolved (e, 2e) electron momentum spectroscopy of photodissociation dynamics of the deuterated acetone molecule at 195 nm. Experimentally, we are conducting a series of measurements to have improved experimental results, compared to the earlier ones, by using a pump laser with higher intensity. Theoretically, we have examined the predicted minimum energy reaction path to investigate whether associated (e, 2e) calculations meet the experimental results. A noticeable difference between the experiment and the calculations has been found at around binding energy of 10 eV, suggesting that the observed difference may originate, at least partly, in ever-unconsidered non-minimum energy paths.

Through many years of intensive studies, electron momentum spectroscopy (EMS) has been established as a powerful means to measure spectral electron momentum density or to look at individual molecular orbitals in momentum space [1, 2]. Extension of the applicability of EMS to transient species is one of the challenges to be tackled, as the change of the electron motion is the driving force behind any chemical reactions. For this purpose, we have recently developed time-resolved EMS (TR-EMS) by combining EMS with a pump-probe technique using femtosecond laser (120 fs) and picosecond electron (1 ps) pulses [3]. The potential ability of TR-EMS has already been demonstrated through orbital imaging of the highest occupied molecular orbital of deuterated acetone (acetone-*d*6) in its short-lived (13.5 ps), second singlet excited (S_2) state [4], an EMS study on the electronic structure of the relatively long-lived (86 ns), first singlet excited (S_1) state of toluene [5], and an attempt at a product vibrational analysis of a photo-induced chemical reaction that occurs as a deactivation process of acetone-*d*6 in the S_2 state [6].

The purpose of this study is to obtain information about dynamics of the photo-induced chemical reaction of S_2 acetone-*d*6 as a result of a joint experimental and theoretical investigation. Experimentally, we are conducting TR-EMS measurements with improved statistics, compared to the earlier ones [4, 6], by using a pump laser with higher intensity. Our idea here is that although the current time-resolution is limited to ±35 ps owing to velocity mismatch effects, information about electron dynamics during the chemical reaction may be left as a trace in the measured TR-EMS spectra. Indeed, we have calculated associated TR-EMS spectra and compared those with the experiments.

The S_2 acetone-*d*6 is known to undergo a three-body dissociation that eventually gives CO and two CD_3 within ~20 ps, firstly via rapid dissociation of acetone-*d*6 to acetyl (CD_3CO) and CD_3 after electronic relaxation to its S_1 state and secondly via subsequent dissociation of CD_3CO to CO and CD_3 [7]. We have found, for instance, that the experimental TR-EMS binding energy spectrum measured at a delay time of 0 ps exhibits substantially larger intensities at around 10 eV, compared to the associated theoretical spectrum that was generated by summing up EMS spectra of the initial, intermediate, and final reaction products calculated at their equilibrium geometries with appropriate weight factors determined by their lifetimes. Furthermore, the observed difference in intensity between the experimental and calculated spectra can be reduced by considering contributions of the dissociating acetyl radical in a bent molecular geometry that is far different from the equilibrium geometries along the minimum energy reaction path. We are planning to perform molecular dynamics trajectory simulations, under professional guidance, to resolve the difference.

References

[1] E. Weigold and I. E. McCarthy 1999 *Electron Momentum Spectroscopy* (New York: Kluwer)
[2] M. Takahashi 2009 *Bull. Chem. Soc. Jpn.* **82** 751
[3] M. Yamazaki, Y. Kasai, K. Oishi, H. Nakazawa, M. Takahashi 2013 *Rev. Sci. Instrum.* **84** 063105
[4] M. Yamazaki, K. Oishi, H. Nakazawa, C.Y. Zhu, M. Takahashi 2015 *Phys. Rev. Lett.* **114** 103005
[5] M. Yamazaki, Y. Tang, M. Takahashi 2016 *Phys. Rev. A* **94** 052509
[6] M. Yamazaki, H. Nakazawa, C.Y. Zhu, M. Takahashi 2015 *J. Phys. Conf. Ser.* **635** 012010
[7] W.K. Chen, Jr-W. Ho, P.Y. Cheng 2003 *Chem. Phys. Lett.* **380** 411

[1] E-mail: seagals@mail.tagen.tohoku.ac.jp
[1] E-mail: masahiko@tagen.tohoku.ac.jp

IOP Conf. Series: Journal of Physics: Conf. Series **875** (2017) 062052 doi:10.1088/1742-6596/875/7/062052

Introducing a phase factor for the two-electron continuum representation

A. S. Zaytsev*, **L. U. Ancarani**[†,1] **and S. A. Zaytsev**[*,2]

[*] Pacific National University, Khabarovsk, 680035, Russia
[†] Équipe TMS, UMR CNRS 7565, Université de Lorraine, 57078 Metz, France

Synopsis We propose an approach to describe three-body Coulomb continuum wave functions in the entire space. It makes use of expansions on Convoluted Quasi Sturmian (CQS) basis, modified by an adequate phase factor as to possess an asymptotic form very close to the formal one. To show numerically that they constitute a suitable set of basis functions we consider the two-electron continuum case (e.g., electron impact double ionization of helium): a considerable improvement of the partial transition amplitudes' convergence rate is indeed observed.

The continuum spectrum of three charged particles in Coulomb interaction is notoriously difficult to describe. When solving atomic or molecular ionization problems, imposing the cumbersome boundary conditions to the wave function constitutes one of the primary mathematical and numerical difficulty. Besides, the long range nature of the Coulomb interaction implies solving Schrödinger's equation on relatively large spatial domains, which eventually translates into using large basis sets and a high computational cost. Ideally, such a domain should be extended up to the boundary of the asymptotic region (Ω_0) where all interparticle distances are large.

Only few *ab initio* methods (see, e.g., the review [2] and Introduction of [3]) have been and are being developed for constructing three-body continuum wave functions. In this contribution, we would like put forward a numerical approach in which the solution is obtained in the entire space so that the ionization amplitude is extracted from the expansion coefficients. The key idea is to use an expansion on a basis set of functions whose asymptotic behavior is as close as possible to the formal one in the Ω_0 region [1].

The proposed basis set contains two ingredients. First, it uses two-particle functions, named Convoluted Quasi Sturmian (CQS) in Ref. [3]; these behave asymptotically as a six-dimensional outgoing (incoming) spherical wave, meaning that they already possess intrinsically some three-body features. While promising, truncated expansions on CQS functions failed to converge satisfactorily with increasing basis size. The reason behind it is that they miss out an important term in the large hyperradius ($\rho = \sqrt{r_1^2 + r_2^2}$) domain: a Coulomb logarithmic phase which corresponds to the interelectronic potential. This brings us to the second ingredient, which consists in introducing – from the outset – an appropriate phase factor into the basis set. The modified CQS functions possess then an asymptotic behavior closer to the formal one, and lead to a considerable convergence improvement in numerical results.

For the sake of simplicity, we consider the two-electron continuum that occurs, for example, in the electron-impact double ionization of helium. To first order, the process can be described by a three-body Schrödinger driven equation. We seek a solution in the form of a truncated partial wave expansion on modified CQS. The desired phase is obtained as to cancel out at best the Coulomb potential $1/r_{12}$.

As we focus on the region Ω_0 we consider, as a numerical test case, typical experimental kinematical conditions with two electrons escaping with 10 eV each. We first investigate the compatibility of introducing a phase factor when using truncated expansions with a two-channel case with total angular momentum $L = 0$. We construct a suitable formula for the phase, and show that it indeed accelerates significantly the convergence of the calculated amplitudes as the number of basis functions is increased. We then consider total angular momentum $L = 1$, which provides the dominant contribution to the cross sections [4]. It is found that the introduction of the phase factor (with just two parameters) is again effective in obtaining satisfactory convergence of the partial transition amplitudes on a basis set of reasonable size. Thus, we expect the CQS basis combined with the proposed phase method to provide an effective tool for the study of full $(e, 3e)$ processes as well.

References

[1] S. P. Merkuriev and L. D. Faddeev, *Quantum Scattering Theory for Several Particle Systems* (Kluwer Academic, Dordrecht, 1993).
[2] I. Bray *et al.* 2012 *Phys. Rep.* **520** 135
[3] A. S. Zaytsev, L. U. Ancarani, S. A. Zaytsev 2016 *Eur. Phys. J. Plus* **131** 48
[4] M.J. Ambrosio, F.D. Colavecchia, G. Gasaneo, D.M. Mitnik, D. M. and L.U. Ancarani 2015 *J. Phys. B* **48** 055204

[1]E-mail: ugo.ancarani@univ-lorraine.fr
[2]E-mail: zay1962@mail.ru

IOP Conf. Series: Journal of Physics: Conf. Series **875** (2017) 072001 doi:10.1088/1742-6596/875/8/072001

Electron impact secondary electron emissions from atomic and molecular solid targets

A. K. F. Haque*[1], M. A. Uddin[*], A. K. Basak[*], B. C. Saha[†], M. Maaza[§], M. A. R. Patoary[*], M. M. Haque[*], and M. Ismail Hossain[*]

[*]Department of Physics, University of Rajshahi, Rajshahi-6205, Bangladesh.
[†]Department of Physics, Florida A&M University, Tallahassee, FL-32307, USA,
[§]University of South Africa (UNISA), Muckleneuk Ridge, PO Box392, Pretoria, South Africa

Synopsis The Sternglass theory [Sternglass, Phys. Rev. 108, (1957) 1] for fast-ion-induced secondary electron emission, which is proportional to the stopping powers, has been modified to calculate the electron impact secondary electron yield from both elemental and compound targets with atomic number Z = 4 - 92 for incident energy range $5 \le E_i \le 10^5$ eV. This modification includes the use of a realistic stopping power expression that involves calculations of the effective atomic electron number, effective mean excitation energies and realistic electron density distribution of the target atoms along with the effective charge of incident electron.

Secondary electrons (SEs) are those emitted when an elemental or compound solid targets irradiated by projectiles - electrons, ions and photons. Knowledge of secondary electron yield (SEY) lies at the heart of many diversified fields. In recent years its renewing interest grows enormously, in both research and applications including the studies of radiation effects in materials, plasma-surface interaction scanning electron microscopy (SEM) microscope images, atomic structure of a given target and also many other applications.

The theory for the SE emission (SEE) has been a subject of continuous interest and variation since its discovery, resulting in the publications of a large number of theoretical works on this topic. The objective of the present study is to propound formalism capable of furnishing reasonably accurate electron impact SEY. The SEY is proportional to the stopping power of the target material. Encouraged by our recent calculation [1] involving the electron impact stopping powers (ESPs) of material media, we use the same ESPs formula to propose a new semi analytical model for SEY in the framework of Sternglass theory [2], embodying a modified factor essential for the best fit of the data. The present work investigates the performance of our proposed *easy-to-use* model for the calculation of the electron impact SEY for both elemental and compound solids. To the best of our knowledge, still there is not a single model available in the literature capable of reproducing the experimental SEY data for large number of species over such a wider incident energy domain. In Fig. 1 we compare our results for elemental and compound targets and found good agreement with experiments.

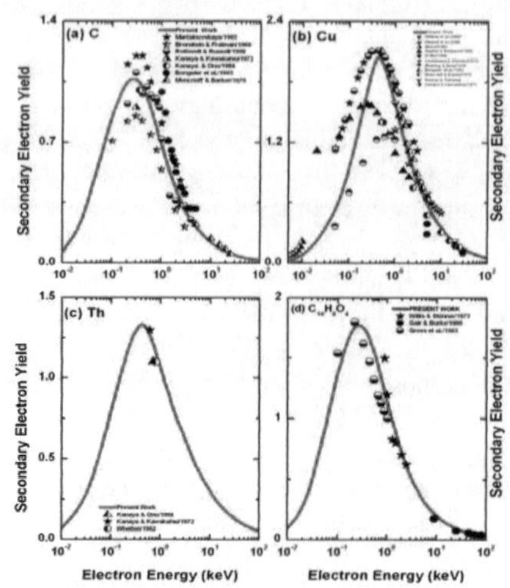

Figure 1. Electron impact SEY of (a) C, (b) Cu, (c) Th and (d) $C_{10}H_8O_4$.

References

[1] A.K.F. Haque *et al.* 2016 Vacuum, 132, 123
[2] E. Sternglass, 1957 Phys. Rev. 108(1), 1

[1]E-mail: fhaque2001@yahoo.com

IOP Conf. Series: Journal of Physics: Conf. Series **875** (2017) 072002 doi:10.1088/1742-6596/875/8/072002

Diffraction in matter-antimatter binding: Positronium formation from C_{60}

Himadri Chakraborty* [1], Anzumaan Chakraborty*, and Paul-Antoine Hervieux† [2]

* Department of Natural Sciences, D.L. Hubbard Center for Innovation and Entrepreneurship, Northwest Missouri State University, Maryville, Missouri 64468, USA

† Université de Strasbourg, CNRS, Institut de Physique et Chimie des Matériaux de Strasbourg, 67000 Strasbourg, France

Synopsis Due to the dominant electron capture from its hull, C_{60} acts like a spherical diffractor inducing resonances in the positronium (Ps) formation as a function of the positron impact energy. Results open new avenue of Ps spectroscopy with nanomaterials and push the barrier to motivate level-differential measurements.

Following the impact of positrons with matter, an exotic electron-positron binary, positronium (Ps), transiently forms until their annihilation – a process of great interests in astrophysics, materials, QED, and antimatter gravity.

Ps formation studies accessed materials that include a breathtaking landscape ranging from atoms, molecules, polymers to solids, liquids, surfaces/films, metal-organic-frameworks and embedded nanostructures. However, straddling the line between atoms and condensed matters are nanoparticles that not only have hybrid properties of the two extremes, but also exhibit outstanding spectroscopic behaviors. Here we take the pilot stride toward probing the Ps formation from gas phase fullerene molecules [1].

The structure of a C_{60} molecule is described by the local-density functional approximation augmented by the LB94 exchange-correlation [2]. The positron impact on C_{60} leading to the Ps formation is treated by the continuum distorted-wave-final-state approximation [3].

The cross section of the Ps formation in its $1s$ and $2s$ states by selective captures from various molecular levels of C_{60} are calculated. The shapes of the positron scattering potential by the residual C_{60} ion and the wavefunctions of C_{60} and Ps levels confine the capture process within the C_{60} molecular width. Strong constructive interferences take place as a function of the recoil momentum (Q). When the odd integer multiple of the half-wavelength of *effective* continuum wave in Q fits the C_{60} cavity, systems of peaks (bright-spots) in the energy domain are formed which subsequently results into a single centroid fringe pattern via a dephasing mechanism across the width [1]. The calculated resonances in the ratio of the HOMO-to-HOMO-1 capture are shown in Figure 1 as a function of Q, along with a schematic representation of the underlying diffraction process. As a proof of the prediction, the separation between two neighboring resonances is found to proportionally reduce going from C_{60} to the larger molecule C_{240} in our extended calculations.

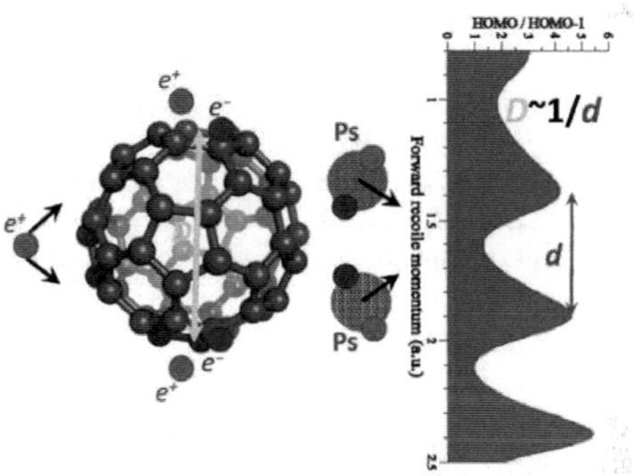

Figure 1. Schematic of a C_{60} diffractor in Ps formation. Resonances in the ratio of cross sections of HOMO and HOMO-1 captures are presented.

The Ps formation off fullerenes thus unearths unique physics and spotlights on a novel direction of Ps spectroscopy with nanosystems that promises to be a great topical field. Of particular interest is the impetus that the current results generate to break new grounds towards target-state differential measurements. The work is supported by the U.S. National Science Foundation Grant PHY1413799.

References

[1] P.-A Hervieux *et al.* 2017 *Phys. Rev A* 95 020701(R)
[2] J. Choi *et al.* 2017 *Phys. Rev. A* 95 023404
[3] O. A. Fojón *et al.* 2001 *J. Phys. B* 34 4279

[1] E-mail: himadri@nwmissouri.edu
[2] E-mail: hervieux@unistra.fr

IOP Conf. Series: Journal of Physics: Conf. Series **875** (2017) 072003 doi:10.1088/1742-6596/875/8/072003

Electron transmission through macroscopic metallic capillaries

D. Borka*, K. Tőkesi†, and C. Lemell‡1

* Atomic Physics Laboratory, Vinča Institute of Nuclear Sciences, University of Belgrade, Belgrade, Serbia
† Institute for Nuclear Research, Hungarian Academy of Sciences (ATOMKI), Debrecen, Hungary
and ELI-ALPS, ELI-HU Non-profit Kft., Szeged, Hungary
‡ Institute for Theoretical Physics, Vienna University of Technology, Vienna, Austria

Synopsis We perform classical transport simulations to model the transmission of electrons through metallic capillaries. Excellent agreement between measured and simulated energy spectra of transmitted electrons is found.

We present trajectory simulations of the transmission of electrons with primary kinetic energies below 1 keV through macroscopic metallic capillaries. In contrast to insulating materials no local charge patches may build up to deflect subsequent projectiles. Electrons will therefore come in close contact with the capillary wall. A classical transport model with stochastic scattering events simulates such impact events.

The initial conditions for the CTMC simulation are chosen to match typical experimental conditions. Within the capillary electrons follow straight-line trajectories, interaction with its own image charge is only indirectly accounted for (deflection at vacuum-metal boundary). Within the material, projectiles are deflected due to elastic or inelastic scattering events and may, possibly after additional impacts on the capillary wall, eventually exit the capillary. If secondary electrons are generated along the trajectory of the primary we follow their trajectories as well with the position of the inelastic scattering event as their starting place. The initial momentum is given by the energy loss of the primary electron, the initial direction of motion is chosen randomly. Finally, the energy of all electrons (primaries and secondaries) reaching the exit opening of the capillary determines the total energy spectrum of our simulation which can be compared directly with experimental data (e.g. [1]).

Near the elastic peak the spectrum is dominated by trajectories with a small number of impacts on the inner surface of the capillary (see figure for 200 eV electron transmission through a platinum capillary [2]). Despite the small number of trajectories undergoing more than 2 impact events on the inner wall of the capillary (see figure) their contribution to the total spectrum has to be accurately accounted for in order to reach good agreement between experimental and simulated spectra (red solid line as opposed to black dashed line representing the energy spectrum for a single impact event). We also find that the electron-energy spectra are largely insensitive to the divergence of the incoming electron beam due to the broad exit angle distribution after impact on the surface.

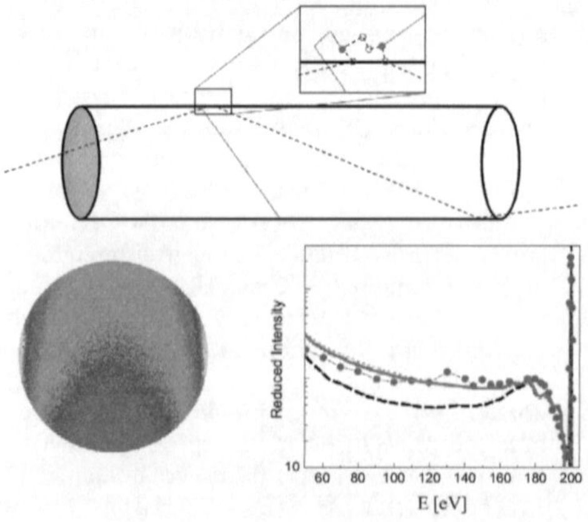

Figure 1. Schematic diagram of simulated system: an electron enters the capillary through the shaded entrance plane and will hit the capillary wall. There, elastic and inelastic scattering events (inset) may cause secondary electron emission and an eventual re-escape of the primary particle. If an electron exits the capillary its energy is recorded (bottom right, data from [1]). Depending on its entry point it will have suffered one (red area), two (green area), or more scattering events (other colors) along its trajectory (bottom left).

References

[1] A. R. Milosavljević *et al.* 2015 *Nucl. Instr. Meth. Phys. Res. B* **354** 86

[2] D. Borka *et al.* 2017 *Nucl. Instr. Meth. Phys. Res. B*, in print

1E-mail: lemell@concord.itp.tuwien.ac.at

IOP Conf. Series: Journal of Physics: Conf. Series **875** (2017) 072004 doi:10.1088/1742-6596/875/8/072004

The Monte Carlo simulation of electron transmission through Al_2O_3 nanocapillary

A. X. Yang, B. H. Zhu, X. M. Chen, J. X. Shao*

School of Nuclear science and Technology, Lanzhou University, Lanzhou, China

Synopsis The detailed simulation was conducted to study the 900 eV electron transmission through insulating nanocapillaries and to identify the transmission probability and guiding effect. We get, the uneven negative and positive charge distributed on the inner wall of the capillary made the transmission probability fall off rapidly and slightly effected the guiding effect.

The study of electron through nanocapillary and tapered capillary have attracted many attentions. Electron do not change their state in capillary and second electron production depend on the energy, angle of primaries, material of capillary. In Ref. [1], simulation results get charge up of surface plays only a minor role in guiding process in Mylar. And for Al_2O_3 [2], the transmission of electron decreases as a function of time, in which charging-up attract the trajectories.

In this paper, the detailed simulation was conducted to study the 900 eV electron transmission through insulating nanocapillaries and to identify the transmission probability and guiding effect. The diameter of Al_2O_3 nanocapillary was 60 nm and the current of incident electron was 4.0 pA/mm. In simulation, the secondary electron production and trajectories were considered.

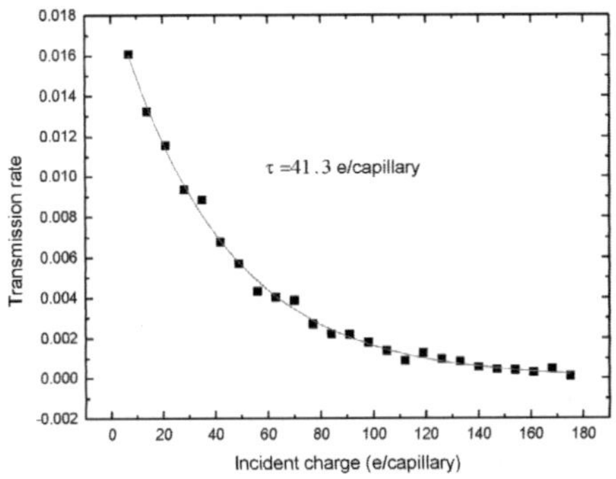

Figure 1. The transmission rate change with incident charge for 0.2 deg.

The transmission of electron decreases as exponential decay. In capillary, the negative and positive charge are both exist. The negative charges are low energy secondary electron from opposite wall and positive charges produce from secondary production.

For incident angle of 0.2 deg., the positive charge distribution is uneven, which makes the transmission fall off rapidly. For different incident angle, the transmission angles out of the capillary were slightly attracted to the incident side by the stronger positive charge patch. Our results were also consistent with the experiments of our laboratory (not present).

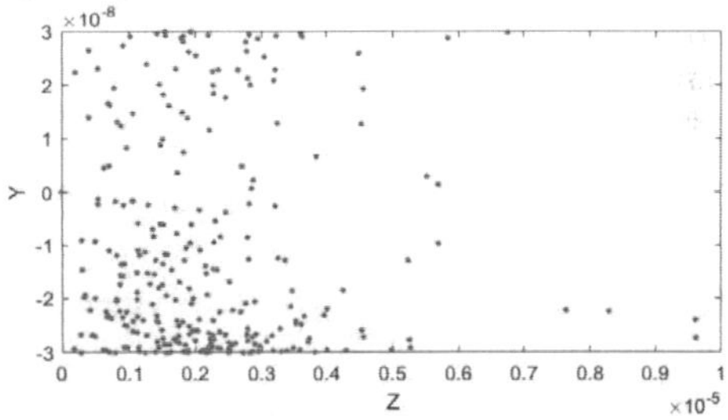

Figure 2. Positive charge distribution in the capillary for incident 0.2 deg.

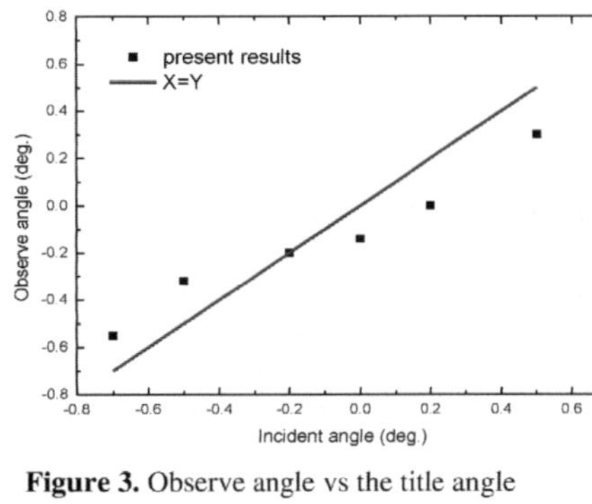

Figure 3. Observe angle vs the title angle

References

[1] K. Schiessl *et al.* 2009 *Phy. Rev. Lett.* **102** 163201

[2] A. R. Milosavljevic *et al.* 2011 *Nucl. Instr. Meth, Phys. B* **279** 190

E-mail: yangax08@lzu.edu.cn

E-mail: shaojx@lzu.edu.cn

IOP Conf. Series: Journal of Physics: Conf. Series **875** (2017) 072005 doi:10.1088/1742-6596/875/8/072005

The investigation of quasi-characteristic radiation of electrons channeled along the charged axes in the crystals of zinc blende

M.V. Maksyuta* [1], **V.I. Vysotskii*** [2], **S.V. Efimenko*** [3], **V.V. Syshchenko**† [4], **A.I. Tarnovsky**† [5], **A.Yu. Isupov**‡ [6]

* Department of Radio Physics, Electronics and Computer Systems, Taras Shevchenko National University of Kyiv, 64/13, Volodymyrska Street, Kyiv 01601, Ukraine
† Belgorod State University, 85, Pobedy Street, Belgorod 308015, Russian Federation,
‡ Laboratory of High Energy Physics, Joint Institute for Nuclear Resarch, Dubna 14980, Russian Federation

Synopsis For slight relativistic electrons moving along the main charged [100] and [110] axes in the crystals of zinc blende there was made a numerical calculation of the transverse energy levels and corresponding wave functions. It was done the calculation and the comparison of quasi-characteristic radiation spectra for these axes.

In [1] it was shown that at electron channeling along the main charged [110] axes in the crystals of zinc blende type the potential wells become very non-unimodal due to a close position of oppositely charged axes. Note that potential wells in [100] axes in these crystals are unimodal. One may suppose that such structure differences in the interaction potentials must cause the essential difference of quasi-characteristic radiation (QCR) spectra.

In [2] there were investigated the peculiarities of slight relativistic electrons axial channeling at Lorentz –factors $\gamma = 2$, 6 and 10 in ZnS, ZnSe and ZnTe crystals which are characterized by a various correlation between nuclei charges of positively and negatively charged ions. These investigations showed that the channeling along the charged [110] axes leads to great spectral intensities in broad spectral ranges and, in the long run, to great total intensities of QCR.

The given paper deals with the consideration of the channeling peculiarities of slight relativistic electrons in main charged [110] axes. The principle emphasis is made on a comparative analysis and a comparison of these results with the data of analogous theoretical and experimental investigations got at the study of the orientation motion of electrons in other crystallographic directions. For example, in Figure 1 QCR spectra originating at the channeling of electrons with Lorentz-factor $\gamma = 10$ along [100] axes (the curve 1, $G_{\max} \approx 1\ \text{cm}^{-1}$) and along [110] axes (the curve 2, $G_{\max} \approx 40\ \text{cm}^{-1}$) are compared.

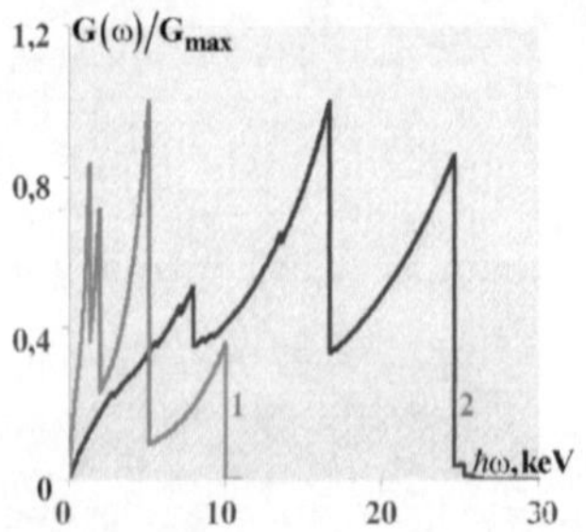

Figure 1. The spectral distribution of QCR energy by an electron per unit path at its motion at zero angle to the axes [100] – the curve 1 and [110] – the curve 2 of ZnS crystal (calculated in relative units at $\gamma = 10$ and zero angular dispersion).

It is shown for the first time that the considered channeling regime allows to forecast the creation of much more effective sources of radiation. Besides, the paper studies the channeling of positively charged particles in non-unimodal directions.

References

[1] N. Maksyuta *et al* 2015 *Abstract of the XXIX International Conference on Photonic, Electronic and Atomic Collisions*, ICPEAC – Abstracts Available Online, TU – 152.
[2] N.V. Maksyuta *et al* 2016 *Theses of the Reports of VII International Conference Charged & Neutral Particles Channeling Phenomena,* Sirmione-Desenzano del Garda, Italy 68

[1] E-mail: maksyuta@univ.kiev.ua
[2] E-mail: vivysotskii@gmail.com
[3] E-mail: svye@ukr.net
[4] E-mail: syshch@yanex.ru
[5] E-mail: artur_tarnovsky@mail.ru
[6] E-mail: isupov@moonche.jinr.ru

IOP Conf. Series: Journal of Physics: Conf. Series **875** (2017) 082001 doi:10.1088/1742-6596/875/9/082001

Lifetimes of ultralong-range Strontium Rydberg molecules in a dense BEC

F. Camargo* , J. D. Whalen* , R. Ding*, T. C. Killian*, F. B. Dunning*[1], J. Pérez-Ríos†, S. Yoshida‡, and J. Burgdörfer‡

* Department of Physics & Astronomy, Rice University, Houston, Texas, USA
† Department of Physics & Astronomy, Purdue University, West Lafayette, Indiana, USA
‡ Institute for Theoretical Physics, Vienna University of Technology, Vienna, Austria, EU

Synopsis The lifetimes and decay channels of ultralong-range strontium Rydberg molecules that contain tens to hundreds of ground-state atoms within the electron orbit are examined by monitoring the time evolution of the Rydberg population using field ionization.

Ultralong-range Rydberg molecules excited in a BEC can be used to explore collective phenomena in quantum degenerate gases such as the creation of polarons[1]. Their use as a probe, however, requires that their lifetimes be sufficient to allow interactions to produce measurable effects. Here we examine the lifetimes of strontium Rydberg molecules in a BEC and the reactions responsible for their destruction.

^{84}Sr atoms cooled in a magneto-optical trap are loaded into an optical dipole trap where they are subject to evaporative cooling to create a BEC with a peak density of ~4x10^{14} cm^{-3}. Two-photon excitation is used to create Rydberg molecules with values of principal quantum number n in the range $49 \leq n \leq 72$. A typical excitation spectrum is included in Fig. 1, expressed as a function of detuning from the atomic line. The sharp peak at zero detuning results from excitation of thermal non-condensed atoms present in the trap, the broad feature extending to the red results from molecular excitation, the larger the detuning the larger the number of atoms present in the molecule. The evolution of the Rydberg population is monitored by field ionization. Analysis of the data indicates that the destruction of Rydberg molecules results from the same processes as identified in earlier studies with rubidium[2], namely associative ionization, leading to formation of Sr_2^+ molecules, and L-changing reactions.

Figure 1 shows the time evolution of the total Rydberg population. The initial rapid decrease in the Rydberg population results from associative ionization. The slower rate of loss seen at later times results from the decay of L-changed Rydberg atoms. The measured molecular lifetimes, ~3-10 μs, limit the timescales over which studies involving Rydberg molecules in cold, dense gases can be undertaken, and reduce the coherence time in such measurements.

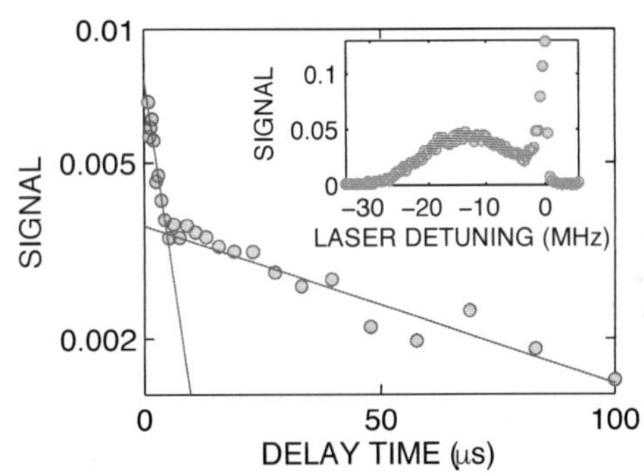

Figure 1. Time evolution of the Rydberg population following excitation of n=60 Rydberg molecules at a detuning of -21.6 MHz in a BEC. The inset shows the n=60 excitation spectrum.

Research supported by the AFOSR, the NSF, the Robert A. Welch Foundation, and the FWF (Austria)

References

[1] F. Camargo et al. to be published
[2] M. Schlagmüller et al. 2016 *Phys. Rev. X* **6** 031020

[1] E-mail: fbd@rice.edu

IOP Conf. Series: Journal of Physics: Conf. Series **875** (2017) 082002 doi:10.1088/1742-6596/875/9/082002

Rydberg atom scattering in K(12p)-CH_3NO_2 collisions: role of transient ion pair formation

M. Kelley, S. Buathong, and F. B. Dunning[1]

Department of Physics & Astronomy, Rice University, Houston, Texas, USA

Synopsis Studies of K(12p)-CH_3NO_2 collisions reveal unusually strong Rydberg atom scattering which is attributed to the formation of transient $K^+..CH_3NO_2^-$ ion-pair states.

Collisions between low-*n* Rydberg atoms and attaching targets can lead to formation of weakly-bound ion-pair states through electron transfer reactions of the type

$$X^{**} + AB \rightarrow X + AB^{-*} \rightarrow X^+..AB^- \quad (1)$$

where AB^{-*} is a short-lived excited state populated by initial electron capture, AB^- is a long-lived metastable anion formed by intramolecular vibrational relaxation, and $X^+..AB^-$ is an ion-pair state. Ion-pair states represent a novel class of long-range molecules which, as shown by recent measurements, posses unusual physical and chemical properties [1].

Earlier studies of reaction (1) have focused on targets that form valence-bound anions. Low-*n* Rydberg collisions with polar targets can also lead to formation of dipole-bound negative ions. We explore the role of dipole binding in forming ion-pair states using CH_3NO_2 whose dipole moment is sufficient to support a dipole-bound state but that can also form long-lived valence-bound anions. As shown in Fig. 1a, pulses of K(12p) atoms are photo-excited near the center of a gas cell where they interact with the target gas. A fraction of the neutral reaction products exit through a slit and enter an analysis region where they are subject to a transverse, pulsed electric field. The resulting negative particles are detected by a position sensitive detector that records their arrival times and positions. The particles are observed to be electrons produced by field ionization of scattered Rydberg atoms. No long-lived ion-pair states are seen. The electron arrival position distributions point to strong Rydberg atom scattering with a collision cross section comparable to the geometric size of the Rydberg atom itself.

Rydberg scattering is attributed to ion-ion scattering through formation of transient ion-pair states with lifetimes ≥10 ps via transitions between the covalent K(12p)+CH_3NO_2 and ionic $K^+..CH_3NO_2^-$ terms in the quasi-molecule

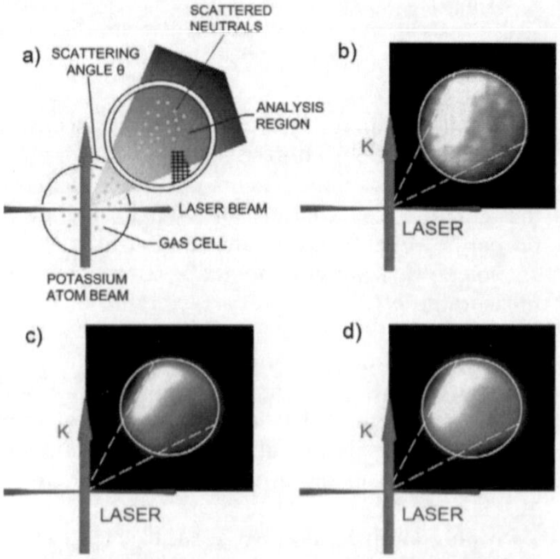

Figure 1. a) Schematic of the experimental approach. b) Measured spatial distribution of scattered Rydberg atoms following a 50 μs flight time. c), d) Results of model calculations that assume ion-pair lifetimes of 10 and 100 ps, respectively.

formed during collision. The $CH_3NO_2^-$ lifetime is limited by electron detachment induced by the field of the K^+ ion, the detached electron remaining bound to the K^+ ion in a Rydberg state. The large reaction rate observed is consistent with theoretical calculations of resonant quenching, i.e., *n*, ℓ changing, in Rydberg collisions with CH_3NO_2. [2]

Research supported by the Robert A. Welch Foundation under Grant No. C-0734.

References

[1] C.H. Wang, M. Kelley, S. Buathong, and F.B. Dunning 2014 *J. Chem. Phys.* **140** 234306
[2] A.A. Narits, E.S. Mironchuk, and V.S. Lebedev 2016 *J. Phys. B : At. Mol. Opt. Phys.* **49** 124001

[1]E-mail: fbd@rice.edu

IOP Conf. Series: Journal of Physics: Conf. Series **875** (2017) 082003 doi:10.1088/1742-6596/875/9/082003

Optical fields to control ultracold atomic/molecular collisions

A. Orbán*, O. Dulieu*, and N. Bouloufa-Maafa*[1]

* Laboratoire Aimé Cotton, CNRS, Univ. Paris-Sud, ENS Paris-Saclay, Univ. Paris-Saclay, 91400 Orsay, France,

Synopsis Ways to suppress inelastic or reactive processes between ultracold particles using electromagnetic fields are investigated.

Research focusing on the formation of ultracold atomic and molecular quantum gases is a continuously expanding field due to its envisioned applications such as quantum-controlled chemistry or quantum simulation.

The aim of our theoretical work is to find ways to suppress inelastic or reactive processes between colliding particles in ultracold quantum gases. Using a laser field detuned to the blue of a relevant transition, we propose to couple the initial colliding particle state to a repulsive excited one, thus preventing the particles to come close to each other [1].

We intend to apply this "blueshielding" technique for ultracold quantum gases, close to quantum degeneracy. In this case, due to the extremely small collisional energies (10-100 nK) it is possible to address a single repulsive channel with the laser field, thus ensuring a full control of the suppression efficiency.

The selection of a well-defined molecular state requires the knowledge of the molecular structure down to the hyperfine scale in particular for the electronically excited molecular states. We have developed an asymptotic model [2] relying on an asymptotic description of the molecular hyperfine structure, extended to the short range.

We will present our results in case of the $^{39}K^{133}Cs$ molecule for the first and second excited dissociation limits. We have also performed a full treatment of molecular rotation for these asymptotes. We will compare the possibilities of blue-shielding of ^{39}K and ^{133}Cs collision (heteronuclear molecular case) to the case of Na + Na collision (homonuclear molecular case) [3]. As a first step we consider rotational molecular states for shielding.

Prospects for the generalization to the shielding of molecule-molecule collisions will also be addressed.

[1]E-mail: Nadia.Bouloufa@u-psud.fr

References

[1] S. Bali, D. Hoffmann, T. Walker 1994 *Europhys. Lett.* **27** 273

[2] A. Orbán. R. Vexiau, O. Kriegelsteiner, H.-C. Nägerl, O. Dulieu, A. Crubellier, N. Bouloufa-Maafa 2015 *Phys. Rev. A* **92** 265

[3] K.-A. Suominen 1996 *J. Phys. B: At. Mol. Opt. Phys.* **29** 5981

IOP Conf. Series: Journal of Physics: Conf. Series **875** (2017) 082004 doi:10.1088/1742-6596/875/9/082004

Quantum-State Resolved Study of the Ultracold K+KRb Reaction

J. F. E. Croft[*1], **C. Makrides**[†2], **M. Li**[†3], **A. Petrov**[†4], **S. Kotochigova**[†5], **B. K. Kendrick**[‡6], **N. Balakrishnan**[*7]

[*] Department of Chemistry, University of Nevada, Las Vegas, NV 89154, USA
[†] Department of Physics, Temple University, Philadelphia, Pennsylvania 19122, USA
[‡] Theoretical Division (T-1, MS B221), Los Alamos National Laboratory, Los Alamos, NM 87545, USA

Synopsis We report a numerically exact quantum mechanical treatment of the K+KRb reaction and discuss both universal and statistical aspects of its reaction dynamics. It is shown that the total rate is universal in character - validating the use of universal models in other similar reactions. However, the rotational distribution is shown to be statistical in character - which we argue is due to the chaotic nature of the reaction complex.

Chemistry's ultimate goal is the absolute control of quantum states of both reactants and products. This is only achievable in the ultracold regime where precise control over initial states including, nuclear spin orientation is possible. In a landmark experiment, research groups at JILA were able to produce an ultracold gas of $^{40}K^{87}Rb$ molecules at nano Kelvin temperatures where even the nuclear spins were oriented [1]. Rate coefficients for exothermic reactions between KRb+KRb and KRb+K were also reported. It was shown that by merely flipping a nuclear spin the reaction could be turned on and off.

An explicit quantum calculation of reaction dynamics for such systems poses a formidable computational challenge, as the computational costs scale with the cube of the number of channels. For these reasons, while the pioneering experiments [1] on KRb reactions were performed over six years ago, no accompanying scattering calculations have been performed at the state-resolved level, until now [2].

Here, we report explicit quantum calculations of the K+KRb→K_2+Rb reaction with resolution of rotational and vibrational level of the product K_2 molecule. Figure 1 provides a comparison of rate coefficients from the present study along with the experimental result of Ospelkaus et al. [1] and predictions of a universal model. It is seen that the computed results are within a factor of two of the experiment and in excellent agreement with the prediction of the universal model suggesting its validity. The experimental result corresponds to a translational temperature of about 250 nK and only s-wave contributes at this temperature. The figure also shows that the total rate is insensitive to three-body forces in the interaction potential. The vibrationally resolved rates indicate that the reaction populates vibrational levels $v' = 0-2$ at comparable rates. The high density of states allows a statistical description of the reaction dynamics. An analysis of the rotational population reveals a Poisson distribution.

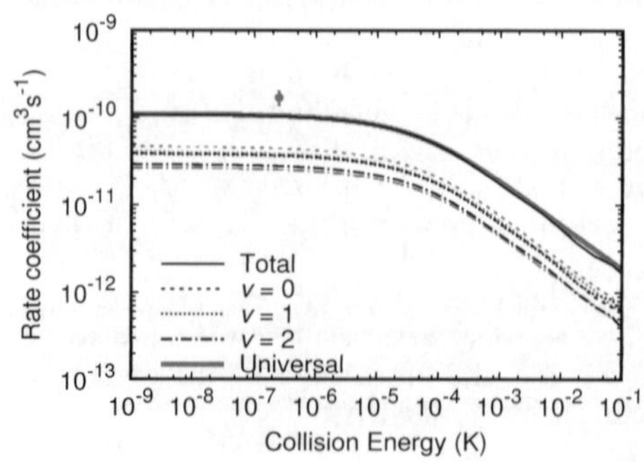

Figure 1. Rate coefficient for K+KRb($v = 0, j = 0$) →$K_2(v')$+Rb reaction as a function of the translational temperature. The red curves are based on a pairwise potential that does not include three-body forces and the black curves are from an interaction potential that includes three-body forces.

References

[1] S. Ospelkaus *et al.* 2010 *Science* **327** 853.

[2] J. F. E. Croft *et al.* 2017 *Nature Comm.* (submitted).

[1]E-mail: j.f.e.croft@gmail.com
[2]E-mail: makrides@umd.edu
[3]E-mail: tug38145@temple.edu
[4]E-mail: alexsandernp@gmail.com
[5]E-mail: skotoch@temple.edu
[6]E-mail: bkendric@lanl.gov
[7]E-mail: naduvala@unlv.nevada.edu

IOP Conf. Series: Journal of Physics: Conf. Series **875** (2017) 082005 doi:10.1088/1742-6596/875/9/082005

Anisotropic blockade using pendular long-range Rydberg molecules

M. T. Eiles[*1] **H. Lee**[*2] **J. Pérez-Ríos**[*3] **and C. H. Greene**[*4]

[*]Department of Physics and Astronomy, Purdue University, West Lafayette, Indiana 47907, USA

Synopsis Ultralong-range "butterfly" Rydberg molecules possess large dipole moments which lead to long-range anisotropic interactions between molecules. Their properties are studied using a Hamiltonian that fully includes the electronic and nuclear spin degrees of freedom. Using this more accurate and complete theoretical description, we investigate the intermolecular interaction when the molecules are prepared in pendular states. The anisotropy of this interaction manifests itself strongly in the density of created molecules.

The recent observation of "butterfly" Rydberg molecules [1] has confirmed theoretical predictions made well over a decade ago [2]. The high experimental resolution of the photoassociation spectrum, along with the additional measurements of the bond lengths and dipole moments of these molecules, necessitates a more sophisticated formulation of the theory including nuclear and electronic spin degrees of freedom to quantitatively understand this system.

We have constructed a Hamiltonian which incorporates these effects, including the dependence of the electron-atom P-wave interaction on the different ^3P_J scattering phase shifts. This has been neglected in nearly all previous treatments, and our approach clarifies its derivation and demonstrates its importance [3]. This full theory has been used to calculate accurate potential curves and electronic multipole moments for both Rb and Cs, and has revealed an improved picture of recent experimental observations. We report on how this ^3P_J dependence of the Hamiltonian mixes states of different molecular symmetries, reducing their dipole moments in good agreement with experiment [1].

With the improved theoretical calculations of these molecular properties in hand, we have studied the long-range intermolecular interactions. Since even weak external fields tightly align these molecules, putting them deep within the so-called "pendular" regime, butterfly molecules are ideal candidates for exploring many-body physics in dipolar gases. We have calculated the potential energy surface, dominated at long range by the anisotropic dipole-dipole interaction but also including higher-order multipole and van der Waals terms, between aligned molecules in a quasi-one-dimensional trap at an angle θ relative to the applied field. By varying θ the dipole-dipole contribution can be tuned to zero at the magic angle $\theta_{M,n}$, which provides a clear experimental signature of this anisotropy in the density of molecules since the strong interactions "blockade" nearby excitations. This straightforward experimental scheme outlines a path towards the study of polaron/impurity physics in a mixed system, angulon/pendulon interactions, or the formation of crystal-type phases with Rydberg molecules.

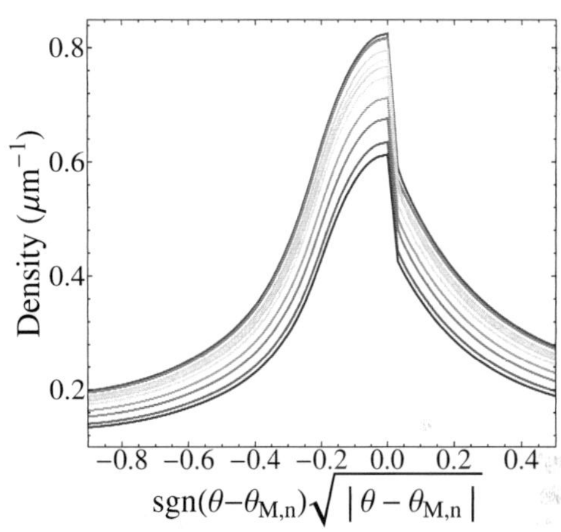

Figure 1. Simulated density as a function of the angle between the field and trap axes. As n increases from 20-40 the color changes from blue to red.

[1]E-mail: meiles@purdue.edu
[2]E-mail: lee2232@purdue.edu
[3]E-mail: jperezri@purdue.edu
[4]E-mail: chgreene@purdue.edu

References

[1] Niederprüm T *et al* 2016 *Nat. Comm.* **10** 1038

[2] Hamilton E L *et al* 2002 *J. Phys. B: At. Mol. Opt*, **35** 199; Chibisov M I *et al* 2002 *J. Phys. B: At. Mol. Opt*, **35** 193.

[3] Eiles M T and Greene C H 2016, arXiv:1611.04508.

[4] Eiles *et al* 2017, arXiv:1702.00341.

Ultracold molecules strongly coupled to a nanophotonic crystal: an universal platform for ultracold chemistry experiments

J. Pérez-Ríos[*1], **M. Kim**[*2] **and C.-L. Hung**[*3].

[*] Department of Physics and Astronomy, Purdue University, West Lafayette IN 47907, USA

Synopsis We develop a novel platform for the study of ultracold chemical reactions and quantum information protocols based on the strong coupling between ultracold molecules and a nanophotonic device.

The possibility of controlling and manipulating atoms as well as internal degrees of freedom of molecules nowadays is the main driving force in the field of atomic, molecular and optical (AMO) physics as well as in chemical physics. In particular, such degree of control is the first step towards quantum memories, quantum logic spectroscopy, and state selective chemistry, among other applications. However, despite the promising applications of photonics devices, the study of ultracold molecules in the presence of photonics devices has remained largely unexplored. Indeed, the presence of internal degrees of freedom in ultracold quantum emitters coupled to photonic modes may lead to unprecedented coherent control of chemical reactions, robust non-destructive measurements techniques, and state-selective strong coupling mechanisms

We propose a novel method for deterministic production and manipulation of ultracold molecules (temperature < 1mK) in their rovibrational ground state. In particular, we seek to enhance PA conversion efficiency to ground state molecules by making use of the paradigmatic waveguide and cavity quantum electrodynamics (QED) [1]. In the proposed experiment, a nanophotonic structure will be devised to trap magneto-optically pre-cooled atoms, efficiently guide the photoassociation light to excite trapped atoms into electronically excited bound states, and selectively enhance the subsequent radiative decay channel into a single molecular quantum state leading to > 99% conversion efficiency and >1 MHz production rate of ultracold molecules in a given rovibrational state. Most importantly, exactly the same photon following spontaneous emission will be collected by the nanophotonic structure, and can be detected and analyzed to inquire the information about the final molecular state. Taking the advantage of strong light-molecule coupling, the nanostructure can also be used to guide light and perform subsequent quantum control, such as coherently driving transition between rovibrational states or inducing controlled ultracold chemical processes.

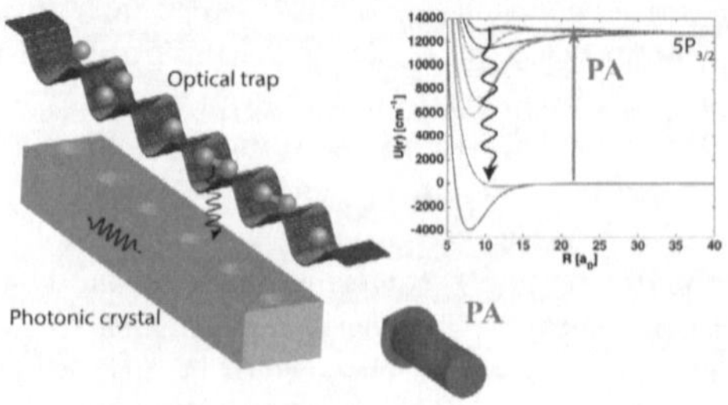

Figure 1. Photoassocition of Rb_2 in the presence of a photonic crystal. The photoassociation (PA) laser is blue detuned from the $5P_{3/2}$ atomic line of Rb, leading to the formation of molecules in an excited electronic state, and decaying in a time $\sim$ 10 ns to a broad distribution of vibrational states associated with the $a^3\Sigma$ state. However, the presence of the photonic crystal highly suppress all these decay modes except one, corresponding to the decay to the ground vibrational state, which is on resonance with the guiding mode of the photonic crystal.

References

[1] A. Goban *et al.* 2015 *Phys. Rev. Lett.* **115** 063601

[1]E-mail: jperezri@purdue.edu
[2]E-mail: kim2257@purdue.edu
[3]E-mail: clhung@purdue.edu

IOP Conf. Series: Journal of Physics: Conf. Series **875** (2017) 082007 doi:10.1088/1742-6596/875/9/082007

Heteroisotopic Feshbach resonances in collisions of cold Ca and Ca^+

Marko Gacesa[*1] **and Robin Côté**[†2]

[*] NASA Ames Research Center, Moffett Field, CA 94035, USA
[†] Department of Physics, University of Connecticut, CT 06268, USA

Synopsis We investigate ultracold collisions of heteroisotopic pairs of Ca with Ca^+ mediated by an external magnetic field. Based on the nuclear spin of Ca, namely $I = 7/2$ for ^{43}Ca and $I = 0$ for other isotopes, we identify different scattering dynamics and characterize magnetic Feshbach resonances in ^{43}Ca+^{43}Ca$^+$ and ^{43}Ca+$^{(40,42)}$Ca$^+$ collisions. Our results suggest that interisotopic Feshbach resonances could be used to control charge-exchange in ultracold atom-ion mixtures of Ca and its ion.

Recent progress in experiments involving small ensembles of trapped ultracold atoms and ions cooled below the critical mass ratio is for the first time allowing direct studies of atom-ion collisional dynamics and controlled chemistry at temperatures in the milli- and microkelvin regime[1]. In addition, such hybrid systems are suitable for studying atom-ion interactions, many-body physics of cold neutral samples doped with charges, quantum simulation, and quantum information processing. Such studies rely on detailed descriptions of atom-ion interactions and the means to control them via external electromagnetic fields. As in case of neutral atoms[2], Feshbach resonances allow the pairwise atom-ion interactions to be tuned over a wide range and offer means to control charge exchange rate in collisions.

In this study, we theoretically investigate the properties of magnetic Feshbach resonances and their impact on charge-exchange rates in collisions of Ca with Ca^+. Specifically, we consider cold collisions between ^{43}Ca (nuclear spin $I = 7/2$) and $^{(40,42)}$Ca$^+$ ($I = 0$) in a magnetic field. A key component of our model are non-Born-Oppenheimer couplings between the potential energy curves of different symmetry due to the mass polarization term present for the collisions between different isotopes [3]. These couplings give rise to magnetic Feshbach resonanances between different isotopes that could be used to resonanly enhance charge-exchange rate in a collision between the particles in selected hyperfine states.

We find and characterize Feshbach resonances by numerically solving a system of time-independent coupled-channel equations for the collision energy $E = 10\ \mu$K, where the atom-ion interactions are described by *ab-initio* electronic interaction potentials smoothly extended to the long-range region [4]. The electronic potentials for ^{43}Ca+^{42}Ca$^+$ with all corrections included are given in Fig. 1. The resulting scattering length for the incoming hyperfine channel $(\alpha\beta)$=(23) as a function of the magnetic field B is shown in Fig. 1 (inset) (see Refs. [1, 5] for details related to the notation). A Feshbach resonance occurs at about 70 G. The charge-exchange rate at ultracold temperature is related to the scattering length as $K_{\rm cx}(B) = 4\pi {\rm Im}(a)/\mu$, implying that a large variation of the rate is possible. Similar results are found in other channels. While the exact positions of the resonances are impossible to predict without an accurate spectroscopic study, the Feshbach resonances appear to be much broader than for Be+Be$^+$[5], suggesting Ca+Ca$^+$ as a more suitable candidate system for further investigation.

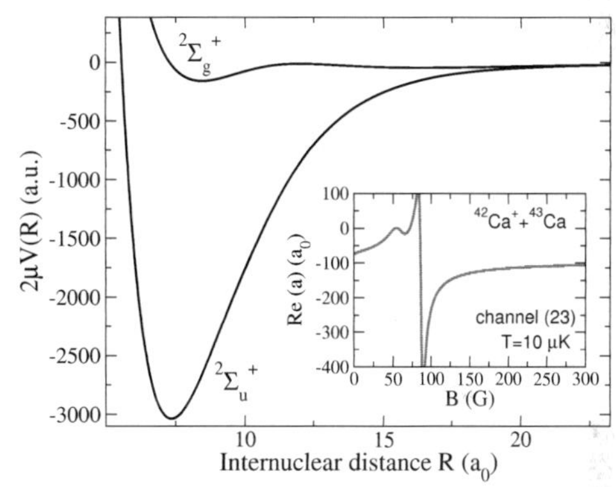

Figure 1. Electronic potential energy curves for $(^{43}Ca^{42}Ca)^+$ with non-Born-Oppenheimer couplings included. Inset: Scattering length as a function of the B field for the scattering channel $(\alpha\beta) = (23)$. Feshbach resonance occurs at about 70 G.

References

[1] Côté, R. 2016 *Adv. At. Mol. Opt. Phy.* **65** 67.

[2] Chin, C. et al. 2010 *Rev. Mod. Phys.* **82** 1225

[3] Zhang, P., Bodo, E., Dalgarno, A. 2009 *J. Phys. Chem. A* **113** 15085

[4] Banerjee, S. et al. 2012 *Chem. Phys. Lett.* **542** 138

[5] Gacesa, M., Côté, R. 2016 arXiv:1612.02555

[1]E-mail: marko.gacesa@nasa.gov
[2]E-mail: robin.cote@uconn.edu

IOP Conf. Series: Journal of Physics: Conf. Series **875** (2017) 082008 doi:10.1088/1742-6596/875/9/082008

Long-range dispersion interactions between excited states of K and rare-gas atoms

Deng-Hong Zhang, Ya-Bin Xu, Jun Jiang[1], Li Jiang, Lu-You Xie and Chen-Zhong Dong[2]

Key Laboratory of Atomic and Molecular Physics and Functional Materials of Gansu Province, College of Physics and Electronic Engineering, Northwest Normal University, Lanzhou 730070, P. R. China.

Synopsis The long range dispersion coefficients between the ground state of rare gas atoms and low-lying states of K are calculated by using relativistic method, in which the spin-orbital interactions are included.

The long-range interaction potential plays an important role in the current research in the field of cold atoms[1-3]. such as Bose-Einstein condensation, ultra-cold collisions, and ultra-cold photo-association spectra. Precisely definition of the interaction potential between alkali metal and rare gas atoms has attracted much interest due to its signification in cold atom physics in recent years. For example, the stability and structure of Bose-Einstein condensation (BEC)[1] depend on the sign (and magnitude) of the scattering length, and the scattering length depends on the precise values of the dispersion constants. The long-range van der Waals dispersion coefficients between two atoms, is one of the more important parts of long-range potential.

In this work, the long-range interaction between low-lying states of K and the ground state of rare gas atoms are calculated in JJ coupled states, in which the spin-orbital interactions are included. The energy levels and transition arrays that contribute to the dispersion parameters are computed by using the relativistic semiempirical-core-potential method (RCICP)[4]. The rare gas oscillator strengths distributions are derived by using calculations of rare-gas polarizabilities and dispersion coefficients to tune Hartree-Fock single-particle energies and expectation values [5].

The present dispersion coefficients between the ground state of Ar and low-lying states of K are listed in Table 1 and compared with others available values [5]. The dispersion coefficients between the ground state of K and the ground state of rare gas atoms have been calculated in our previous paper [6]. The present results are in good agreement with the configuration interaction plus core polarization (CICP) results [5].

This work was supported by National Natural Science Foundation of China (Grant No. 11564036, U1332206, 11464042, U1330117, U1331122, U153014), the Young Teachers Scientific Research Ability Promotion Plan of Northwest Normal University (NWNU-LKQN-15-3).

References

[1] J. Weiner, *et al.* 1999 *Rev. Mod. Phys.* **71** 1
[2] K. M. Jones *et al.* 2006 *Rev. Mod. Phys.* **78** 483
[3] T. Kohler *et al.* 2006 *Rev. Mod. Phys.***78** 1311
[4] J. Jiang *et al.*2016 *Phys. Rev.A* **94** 062514
[5] J. Mitroy *et al.* 2007 *Phys. Rev. A* **76** 032706
[6] J. Jiang, *et al.* 2015 *Atomic Data and Nuclear Data Tables* **101** 158

Table 1 The dispersion coefficient (in a.u.) between the ground state of Ar and low-lying states of K. The numbers in the round brackets denote powers of ten.

	MT	C_6		C_8		C_{10}	
		Present	Other[7]	Present	Other[7]	Present	Other[7]
$4s_{1/2}$	1/2	299.37	299.3	22265	22280	2.1300(6)	2.132(6)
$4p_{1/2}$	1/2	578.59		76428		1.3286(7)	
$4p_{3/2}$	3/2	470.44	471.7	11232	11220	8.0594(5)	8.022(5)
	1/2	693.81		1.4340(5)		2.6173(7)	
$5s_{1/2}$	1/2	1784.6	1786	6.1844(5)	6.192(5)	2.5250(8)	2.529(8)
$3d_{5/2}$	5/2	707.73	712.0	-31943	-31900	-6.7078(5)	-6.713(5)
	3/2	1035.8		2.0671(5)		1.8168(7)	
	1/2	1199.9		5.4485(5)		2.1838(8)	

[1] E-mail: phyjiang@yeah.net

[2] E-mail: dongcz@nwnu.edu.cn

IOP Conf. Series: Journal of Physics: Conf. Series **875** (2017) 082009 doi:10.1088/1742-6596/875/9/082009

Emerging novel phases of Bose-Einstein Condensate for various topology

Priyam Das *, [1], Ayan Khan †, [2] and Prasanta K. Panigrahi ‡, [3]

*Department of Physics, Indian Institute of Technology Delhi, New Delhi – 110016, India
†Department of Physics, Bennett University, Greater Noida - 201310, Uttar Pradesh, India
‡Department of Physical Sciences, Indian Institute of Science Education and Research – Kolkata, Mohanpur – 741246, India,

Synopsis: In this work, we realize the soliton's negative mass regime, when a Bose-Einstein condensate is subjected to a harmonic confinement. We also discuss that this system is favorable for the formation of a bound state when an additional species is considered. When a time modulated optical lattice potential is introduced, the sinusoidal lattice modes lead to the generation of nonlinear resonances. A classical dynamical phase transition is also identified in this case through a superfluid to the insulating phase transition.

We study Bose-Einstein condensate (BEC) for different geometrical confinement through mean-field description. Interestingly we observe several novel consequences arising from the interplay of confinement and interaction. When BEC is subjected to the harmonic confinement, we observe that the solitonic excitations in the system initiates generation of negative mass regime, as is shown in Fig.1. We also realize that this geometry is favorable for bound state formation when we introduce another species of BEC. Introduction of a time modulated optical lattice potential in the system, drastically changes the dynamics of the system. In this case, we observe the emergence of nonlinear resonances and classical dynamical phase transition, through the density distribution and energy spectrum. When the harmonic trap is switched off, the system undergoes a nonlinear compression for the static OL potential. In a harmonic trap, the center of mass undergoes periodic oscillation (Fig. 2a), as a consequence of Kohn mode. When the harmonic trap is made time dependent, it breaks the periodicity and the center of mass no longer obeys the Kohn theorem, as can be seen from Fig.2b. Possible application of such geometry to information storage and retrieval is also discussed.

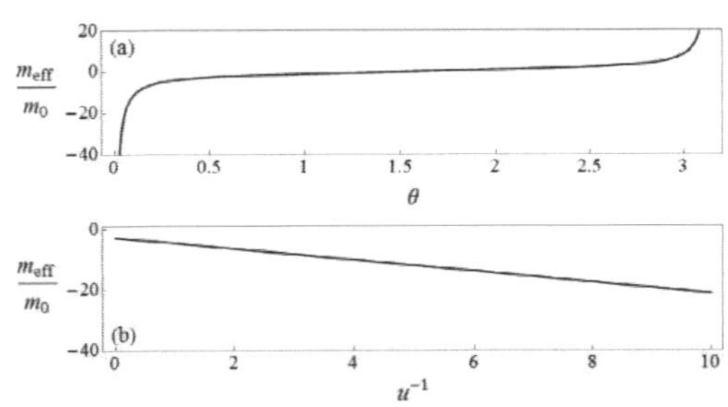

Fig.1: The effective mass profile (m_{eff}) is shown as a function of (a) Mach angle (θ), showing negative mass regime at the low momenta regime. (b) Shows the negative mass regime as a function of inverse of the soiton's velocity (u^{-1}).

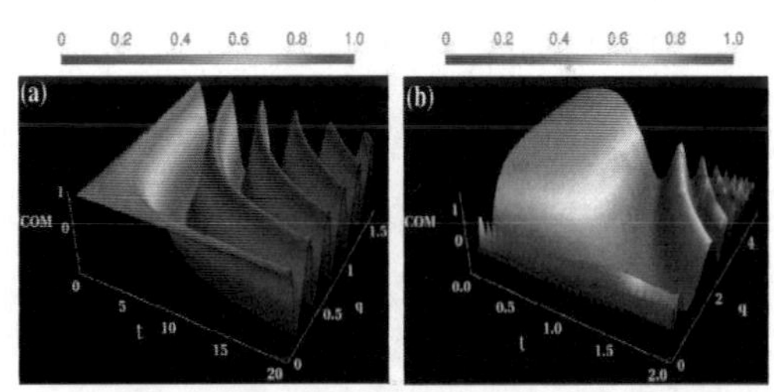

Fig.2: (a) The center of mass profile is depicted in presence of a regular harmonic confinement, showing that the COM oscillates with trap frequency. (b) The COM profile described by the Bessel function, where the harmonic trap varies exponentially with time.

[1] E-mail : daspriyam3@gmail.com
[2] E-mail : ayan.khan@bennett.edu.in
[3] E-mail : panigrahi.iiser@gmail.com

References

[1] Priyam Das *et. al.*, 2016 *Eur. Phys. J. D* 70 113.

[2] Priyam Das *et. al.*, 2015 *Laser Phys.* 25 125501.

IOP Conf. Series: Journal of Physics: Conf. Series **875** (2017) 082010 doi:10.1088/1742-6596/875/9/082010

Elementary excitation of spin-orbit coupled Bose-Einstein condensates in optical lattice

Xu-Dan Chai and Ju-Kui Xue*

Key Laboratory of Atomic & Molecular Physics and Functional Materials of Gansu Province, College of Physics and Electronics Engineering, Northwest Normal University, Lanzhou 730070, China

Synopsis The phonon mode excitations of spin-orbit (SO) coupled Bose-Einstein condensates (BECs) trapped in an one-dimensional optical lattice are studied and the sound speed of the system is obtained analytically.

The elementary excitation and collective mode, which are the low-energy excitation of the system and can be precisely measured, play important roles for simulating more complicated and correlated condensed matter systems with Spin-orbit (SO) coupling [1]. Motivated by the rapid experimental and theoretical progress in these areas, great deal of works have been done to explore the properties of the SO coupled BECs. Especially, the SO coupled BECs in optical lattice has become a subject of immense interest theoretically and experimentally [2]. However, the analytical description of the low-energy excitation of the SO coupled BECs in optical lattice is still missing.

Under the mean field theory, by using the perturbation method [3], the sound speed of SO coupled BECs trapped in an one-dimensional optical lattice for zero momentum and plane wave phases regimes are derived analytically, respectively. The most obvious phenomenon is the softening of the phonon,i.e., the vanishing of sound speed in optical lattice. Without lattice potential, the softening of phonon mode occurs only at the phase transition point exactly, which is not affected by the atomic interaction strength and Raman coupling when the SO coupling is strong. However, considering the optical lattice, the softening of phonon modes takes place in a regime near the phase transition point. Interestingly, the regime is widened as lattice strength and SO coupling increase or atomic interaction decreases. The suppression of the sound speed by lattice strength relies on atomic interaction, Raman coupling, and SO coupling. Furthermore, we find that the sound speed in plane wave phase regime and zero momentum phase regime behaves very different characters for Raman coupling and SO coupling changing. In zero momentum phase regime, sound speed monotonically increases/decreases with Raman coupling/SO coupling. However, in plane wave regime, sound speed can either increase or decrease with Raman coupling and SO coupling, which strongly depends on the atomic interaction. Physically, phonon mode softening indicates the the appearance of the dynamical instability and the disappearance of the superfluidity. Specially, the presence of the optical lattice enhances this instability.

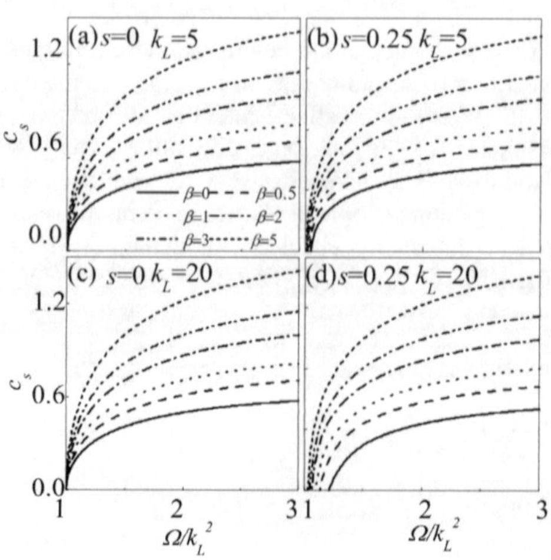

Figure 1. Sound speed against Raman coupling for different atomic interaction in zero momentum phase regime

This work is supported by the National Natural Science Foundation of China under Grants No. 11274255, 11305132, and 11475027.

References

[1] Y. Li *et al* 2012 *Phys. Rev. Lett.* **108** 225301

[2] G. I. Martone *et al* 2012 *Phys. Rev. A* **86** 063621

[3] A. X. Zhang *et al* 2009 *Phys. Rev. A* **80** 043617

*E-mail: xuejk@nwnu.edu.cn

Phase transition of soft-core bosons in disordered optical lattice

Ji-Ming Gao and Ju-Kui Xue*

Key Laboratory of Atomic & Molecular Physics and Functional Materials of Gansu Province, College of Physics and Electronics Engineering, Northwest Normal University, Lanzhou 730070, China

Synopsis The mean-field phase diagram of two-dimensional soft-core extended Bose-Hubbard model in the presence of spatial disorder is determined.

The interplay between disorder and interactions in ultracold atoms has been a topic of intense theoretical investigations in the past years. In the presence of disorder, ultracold atoms in deep optical lattice are described by the Bose-Hubbard(BH) model and its extensions with disorder. It was found that disorder in BH model generates a novel Bose-glass(BG) phase, which always intervenes between the Mott-insulator(MI) and the Superfluid(SF) states. For the extended BH model, there may be possible a supersolid state, which also intervenes between the incompressible solid and SF states in the parameter space. The introduction of disorder into the extended BH model leads to the competition between the BG and SS phases, which may influence the existence of the SS phase. For hard-core bosons in square lattice, external spatial disorder always destroys the long rang density wave order and makes it no chance to become a SS state[1]. QMC and disordered mean field theory studies demonstrate that the SS are stable against weak spatial disorder on the simple cubic lattice.[2]. To our knowledge, the phase digram of soft-core bosons with nearest-neighbor interactions and disorder in two-dimensional(2D) optical lattice has not been studied, and the fate of the SS state in such system is still unknown.

We use inhomogeneous mean-field theory[3] to study the effects of spatial disorder in the soft-core Bose-Hubbard model on a 2D square lattice. The mian results are summarized in Figure 1, rendering a rich phase diagram. We find five distinct phases: MI, Checkerboard(CB) solid, Disorder Solid(DS), SS and SF. In contrast to hard-core bosons in square lattice, we find that the solid and SS phases are stable against the spatial disorder. And the spatial disorder yields a sequence of DS phases which intervene between the incompressible lobes. Especially, it is remarkable that the regions of DS and SS phases can be tuned by varying the disorder strength. For weak nearest-neighbor interactions, the disorder induces the solid and SS phase shrinking to smaller hopping regions. However, for strong nearest-neighbor repulsions, the competition between nearest-neighbor interactions and disorder stabilizes the SS phase and makes it extending to large hopping region. Therefore, our results suggest that the prospect of observing the SS phase is much higher when loading dipolar Bose gases into disordered optical lattices.

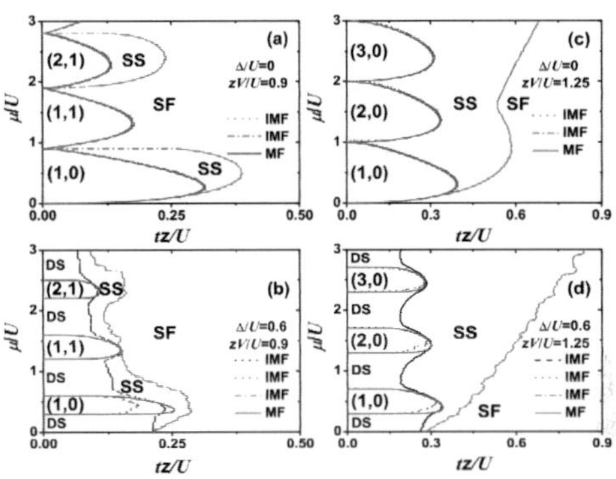

Figure 1. (Color online) The ground-state phase diagrams as a function of the chemical potential μ/U and hopping tz/U.

This work is supported by the National Natural Science Foundation of China under Grants No. 11274255, 11305132, and 11475027.

References

[1] K. Bernardet*et al* 2002 *Phys. Rev. B.* **66** 054520

[2] B. M. Kemburi*et al* 2012 *Phys. Rev. B.* **85** 020501

[3] A. E. Niederle*et al* 2013 *New. J. Phys.* **15** 075029

*E-mail: xuejk@nwnu.edu.cn

IOP Conf. Series: Journal of Physics: Conf. Series **875** (2017) 082012 doi:10.1088/1742-6596/875/9/082012

Internal cooling of cold Rb_2^+ ions with cold Rb atoms

Maurice Raoult[*1], **Humberto da Silva Jr**[*], **and Olivier Dulieu**[*]

[*] Laboratoire Aimé Cotton, CNRS, Univ. Paris-Sud, ENS Paris-Saclay, Univ. Paris-Saclay, 91400 Orsay, France

Synopsis The cooling of the internal degrees of freedom of cold Rb_2^+ ions induced by collisions with cold Rb atoms is modeled using a close-coupling approach. Due to the strong anisotropy of the potential energy surface, the process is predicted to be very efficient. Possible experimental implementation will be discussed.

The simultaneous trapping of cold atoms and cold ions in hybrid traps has opened new avenues for studying interactions in cold, dilute gas ensembles. In particular, molecular ion formation processes [1] have been investigated as a premise to ultracold chemistry. The molecular ions are generally obtained in a distribution of rotationally and vibrationally levels in their ground electronic state [2] which still remain to be cooled down to the lowest quantum level as required in many applications of single quantum state preparation.

Here we theoretically study the internal cooling of Rb_2^+ ions by ultracold Rb atoms, in the context of the recent experiment of Ref. [3].

A potential energy surface (PES) in standard atom-diatom Jacobi coordinates $V(R,r,\theta)$ has been calculated using a full configuration interaction approach, where the valence electrons of the individual atoms are represented by a relativistic effective core potential (ECP), corrected with core polarization potential (CPP) depending on the electronic angular momentum [4]. Thus, the Rb_3^+ system reduces to a two-electron problem, where the Hartree-Fock and the excitation determinants are computed in an atom-centered Gaussian basis set.

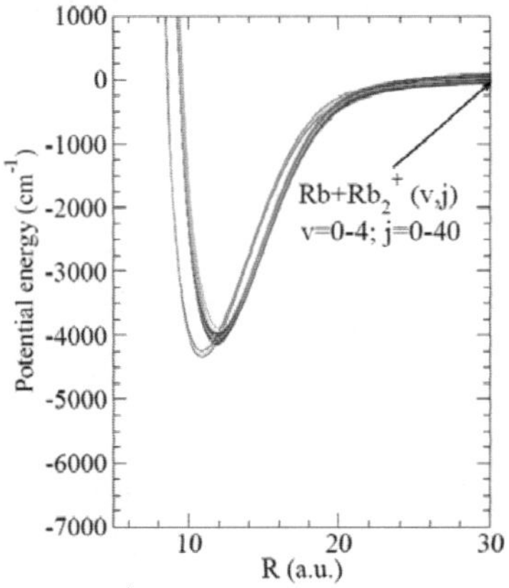

Figure 1. Rb-Rb_2^+ diabatic potential curves.

This PES is used to determine the interaction matrix in the asymptotic free rotor basis set $d^j_{m,0}(\theta)\chi_v(r)$ (rotational Wigner function $d^j_{m,0}(\theta)$ and vibrational wavefunction $\chi_v(r)$ of the diatom). The diagonal elements of the interaction matrix $V_{vj,v'j'}(R)$ (diabatic potentials) are presented in Fig.1. In Fig. 2 are shown the corresponding adiabatic potentials resulting from the diagonalization of $V_{vj,v'j'}(R)$. The huge difference between the diabatic and adiabatic potentials reflects the strong short-range interaction between the collisional channels, suggesting an efficient rovibrational de-excitation. The results of our preliminary close-coupling calculations confirm this expectation, in agreement with Ref. [5].

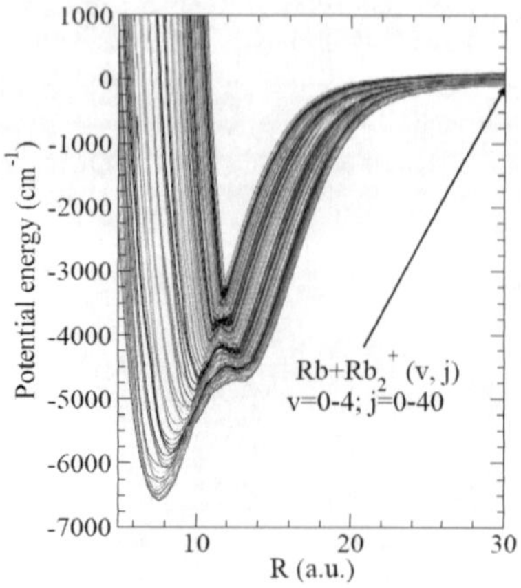

Figure 2. Rb-Rb_2^+ adiabatic potential curves.

References

[1] F. H. J. Hall *et al*, 2015 *Phys. Rev. Lett.* **107** 243202
[2] H. da Silva Jr *et al.* 2015 *New J. Phys.* **17** 045015
[3] S. Jyothi *et al.* 2016 *Phys. Rev. Lett.* **117** 213002
[4] M. Aymar and O. Dulieu 2005 J. Chem. Phys **122** 204302
[5] T. Stoecklin et al. 2016 *Nature Comm.* **7** 11234

[1] E-mail: maurice.raoult@u-psud.fr

IOP Conf. Series: Journal of Physics: Conf. Series **875** (2017) 082013 doi:10.1088/1742-6596/875/9/082013

Photodissociation of cold $RbBa^+$ ions produced in a hybrid cold atom-ion trap

Humberto da Silva Jr [*], Maurice Raoult[*], Amir Mohammadi[†], Johannes Hecker Denchlag[†], and Olivier Dulieu[*1]

[*] Laboratoire Aimé Cotton, CNRS, Univ. Paris-Sud, ENS Paris-Saclay, Univ. Paris-Saclay, 91400 Orsay, France
[†] Institut für Quantenmaterie and Center for Integrated Quantum Science and Technology, Universität Ulm, Germany

Synopsis The formation and destruction processes of cold $RbBa^+$ molecular ions are modelled in the context of their experimental observation in hybrid setups merging cold atom and cold ion traps.

Investigating atom-ion interactions in hybrid setups merging a cold atom trap and a cold ion trap has revealed to be one of the novel developments of research on ultracold matter. Beside elastic or inelastic binary collisions, charge exchange, or three-body collisions, the formation of cold molecular ions has been directly observed by one group [1], while they are predicted to be created in most experiments of this kind [2].

Here we present a theoretical investigation about the possible destruction of $RbBa^+$ molecular ions by a photodissociation mechanism, due to the many lasers involved in the experimental setup of Ref. [3]. The potential energy curves (PEC) and all the relevant permanent and transition electric dipole moments (PEDM and TEDM, respectively) of $RbBa^+$ are calculated in a semiempirical methodology which has been already used in our group to describe a number of diatomic systems [4]. Both atoms are represented as a one-electron system, where the valence electron is moving in the field of a relativistic effective core potential (ECP) with a core polarization potential (CPP) correction. The $RbBa^+$ electronic structure is thus reduced to an effective two-electron problem. The re-maining two valence electrons are considered to calculate the Hartree-Fock and the excitation determinants, in an atom-centered Gaussian ba-sis set, through the usual self-consistent field methodology. Finally, a full configuration inter-action (FCI) is achieved.

With this data we computed the state-to-state absorption cross section, $\sigma_{\Lambda vj}(\nu)$, of a $RbBa^+$ rovibrational level (v,j) in the Λ electronic state induced by a laser with frequency ν. Following Ref. [3] we focused on photodissociation by a 1064 nm laser towards accessible electronic states Λ', reaching continuum states with angular momentum j' and energy $\varepsilon=\hbar k$. The corresponding process is illustrated in Figure 1.

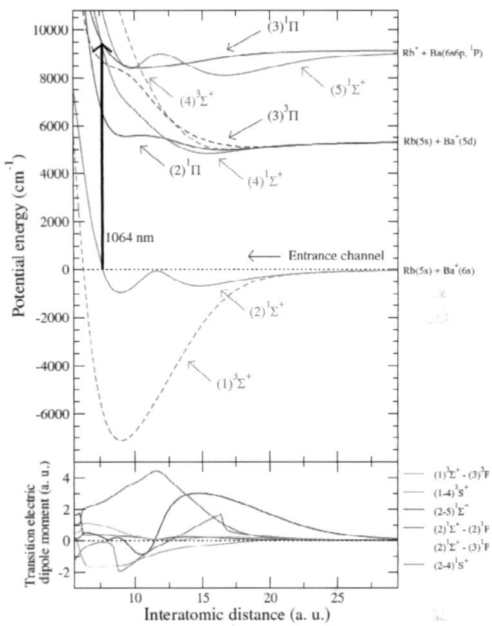

Figure 1. Computed $RbBa^+$ potential energy curves and transition dipole moments. The suggested transition (black arrow) corresponds to an initial weakly-bound molecular ion assumed to be created by three-body collision (see Ref. [3])

Preliminary results indicate that the lifetime of the molecular ions is strongly limited by the presence of laser light in the experiment, in agreement with the findings of Ref. [5].

References

[1] F. H. J. Hall *et al*, 2015 *Phys. Rev. Lett.* **107** 243202
[2] H. da Silva Jr *et al.* 2015 *New J. Phys.* **17** 045015
[3] A. Krükow *et al.* 2016 Phys. Rev. A **94** 030701 (R)
[4] M. Aymar *et al* 2011 *J. Chem. Phys.* **135** 064305
[5] S. Jyothi *et al.* 2016 *Phys. Rev. Lett.* **117** 213002

[*] E-mail: olivier.dulieu@u-psud.fr

IOP Conf. Series: Journal of Physics: Conf. Series **875** (2017) 092001 doi:10.1088/1742-6596/875/10/092001

APPA R&D — BMBF collaborative research at FAIR

S. Schippers*[1], Th. Stöhlker†,‡ for the APPA collaborations

* I. Physikalisches Institut, Justus-Liebig-Universität Gießen, 35392 Giessen, Germany
† Helmholtz-Institut Jena, 07743 Jena, Germany
‡ GSI Helmholtzzentrum für Schwerionenforschung, 64291 Darmstadt, Germany

Synopsis The APPA R&D research collaboration at the international Facility for Antiproton and Ion Research (FAIR) in Darmstadt, Germany, is introduced and its activities are briefly summarized pointing out, in particular, the experimental program at the low-energy storage ring CRYRING which will become available for users already in 2018.

The research collaboration APPA R&D [1] comprises the German university groups who have set out to perform scientific research at the future international accelerator complex FAIR (Facility for Antiproton and Ion Research) under the umbrella of APPA (Atomic, Plasma Physics and Applications). The FAIR-installations (Fig. 1) are currently under construction at the site of the GSI Helmholtz Center for Heavy Ion Research in Darmstadt, Germany. APPA is one of the four research pillars of FAIR hosting the international research collaborations BIOMAT (biophysics and material research), FLAIR (physics with low-energy antiprotons), HEDgeHOB/WDM (plasma physics), and SPARC (atomic physics) who focus on investigations of matter under extreme conditions such as strong fields, high densities, high pressures, and high temperatures [2].

The APPA R&D research collaboration pursues coordinated research projects in the area of accelerator based experiments with heavy ions at the future FAIR-installation. Central issues are i) further development of the the experimental infrastructure, in particular, research and development for enhancing the scientific capabilities of the existing installations and of the future accelerator and detector systems including the respective base technologies, and ii) set-up of the APPA experiments of the modules 0-3 of the modularized start version of FAIR.

FAIR is expected to operate fully in 2024/2025. However, the existing GSI accelerators including the existing Experimental Storage Ring (ESR) are available already now as well as the low-energy storage ring CRYRING, a Swedish in-kind contribution to FAIR, which is currently being commissioned. CRYING will receive cooled and decelerated highly charged ions from ESR, thus, permitting, e.g., precision experiments that will challenge state-of-the-art calculations within the frame-work of quantum electrodynamics (QED). In addition to ESR, a separate injector can also provide ions to CRYRING, such that it can be operated and used for experiments even when ESR is serving other purposes. First user experiments will start in 2018. From then on, a rich and diverse experimental program covering topics from atomic physics, nuclear physics, material sciences, and accelerator physics will be pursued [3].

APPA R&D is funded by the German Federal Ministry for Education and Research (BMBF) within the collaborative-research frame work ('Verbundforschung').

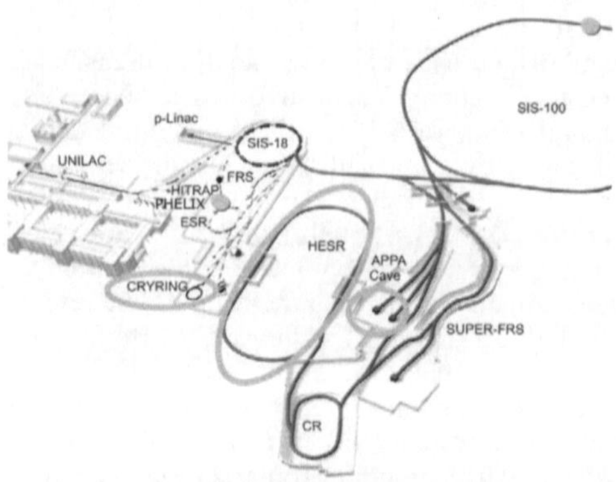

Figure 1. Experimental facilities at the Modularized Start Version (MSV) of FAIR [2]. The existing ion-beam lines and the new FAIR facilities are marked in blue and red, respectively. Sites where major APPA activities will take place are marked in yellow. Experiments in the low-energy storage ring CRYRING [3] will start in 2018. According to the current planning, FAIR will fully operate in 2024/2025.

References

[1] http://appa-rd.fair-center.eu

[2] Th. Stöhlker *et al.* 2015 *Nucl. Instrum. Meth. B* **385** 680

[3] M. Lestinsky *et al.* 2016 *Eur. Phys. J ST* **225** 797

[1]E-mail: stefan.schippers@physik.uni-giessen.de

IOP Conf. Series: Journal of Physics: Conf. Series **875** (2017) 092002 doi:10.1088/1742-6596/875/10/092002

On the beam emittance/brilliance of highly charged ions extracted from the Frankfurt 14GHz Electron-Cyclotron-Resonance-Ion-Source in standard and enhanced (MD) mode

J. C. Müller*[1], R. Dörner*, F. King*, L. Schaechter+, K. E. Stiebing*[2]

* Institute for Nuclear Physics, Goethe Universitaet, Max von Laue Str. 1, D-60438 Frankfurt am Main, Germany
+ National Institute for Physics and Nuclear Engineering, 77425 Bucharest, Romania

Synopsis Electron Cyclotron Resonance Ion Sources (ECRIS) provide only moderate emittances as compared to other types of ion sources. Besides considerable effort to improve the total output of ECRIS´s by dedicated modifications (gas mixing, wall coating etc.), it is important to also determine the influence of such improvements on the beam quality. In this contribution dedicated experiments on the influence of the MD method [1] on the emittance/brilliance of extracted ion beams from the Frankfurt 14GHz ECRIS are presented.

Electron-Cyclotron-Resonance-Ion-Sources (ECRIS) are widely used as injectors of highly charged ions into linear accelerator structures. Substantial experimental and theoretical effort has been made to improve this type of ion source with regard to their high charge state performance. One of these methods is the use of a special layer with a well-defined metal-dielectric transition at the inner wall of the plasma chamber (MD-method) [1], which has been tested in detail in a series of experiments at the 14 GHz ECRIS of the IKF in Frankfurt (see Figure 1).

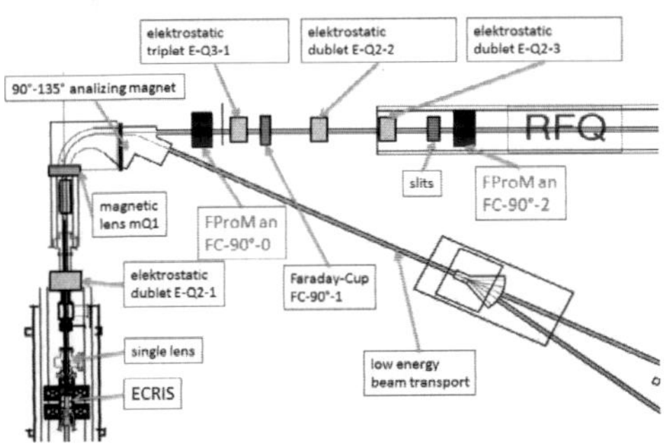

Figure 1. Infrastructure of the Frankfurt 14GHz ECRIS and RFQ installation, FProM: Frankfurt profile and emittance monitor.

This method strongly increases the output of highly charged ions and allows reducing the x-ray load for a given intensity [2]. Dedicated measurements have shown that the effect is due to an increase of density and temperature of the plasma electrons as compared to the standard stainless steel walls [3].

To effectively inject the ion beam into subsequent accelerator structures and/or into a transport channel to the experiments a sufficiently small emittance is needed. In order to provide information on the influence of the MD-method on these beam parameters dedicated experiments have been performed in the frame of a master thesis [4].

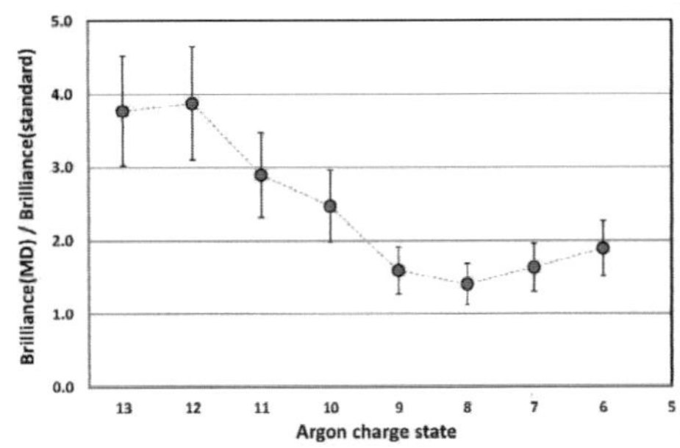

Figure 2. relative brilliance of argon ions extracted from the ECRIS in MD configuration as a function of the charge state, normalized to the brillaince for the standard source.

In Fig. 2 the relative brilliance of extracted argon ions as a function of the charge state from the ECRIS in MD configuration is presented and demonstrates the strong improvement of the brilliance particularly for high charge states. These results will be discussed in detail.

References

[1] L. Schaechter *et al* 1999 *Rev.Sci.Instr.* **70** 1367
[2] L. Schaechter *et al* 2000 *Rev.Sci.Instr.* **71** 918
[3] L. Schaechter *et al* 2002 *Rev.Sci.Instr.* **73** 4172
[4] J. C. Müller 2015 "*Emittanzmessungen an der Frankfurter Elektron Zyklotron Resonanz Ionenquelle*" Master Thesis IKF

[1] E-mail: janmueller@atom.uni-frankfurt.de

[2] E-mail: stiebing@em.uni-frankfurt.de

IOP Conf. Series: Journal of Physics: Conf. Series **875** (2017) 092003 doi:10.1088/1742-6596/875/10/092003

Measurements of the kinetic-energy-release (KER) of vibrational cooled HeH^+ and H_3^+ at the electrostatic Frankfurt low energy storage ring (FLSR)

J. C. Müller[*,1], R. Dörner[*], F. King[*], L. Ph. H. Schmidt[*], M. S. Schöffler[*], K. E. Stiebing[*,2]

[*] Institute for Nuclear Physics, Goethe Universitaet, Max von Laue Str. 1, D-60438 Frankfurt am Main, Germany

Synopsis The KER for electron capture of vibrational cooled HeH^+ and H_3^+ ions at 20 keV from residual gas atoms has been measured in the Frankfurt Low Energy Storage Ring (FLSR). At a vacuum in the order of few 10^{-11} mbar, this residual gas consists to 99% of H_2 molecules. For the identification of the recoil products of this reaction, a recoil spectrometer (with an MCP-detector with position and time sensitive read out) was installed at one of the focus points (IP) in the FLSR. The planned extension of this set up by a gas target to a full COLTRIMS reaction microscope will be discussed.

The Frankfurt Low Energy Storage Ring (FLSR) [1] is a fully electrostatic device designed for the storage of ions with a total energy of up to 50 keV. As no magnetic fields are involved, the ion mass of the stored ions is not limited by the ion optics of the ring. FLSR has racetrack geometry with a circumference of 14.7 m resulting in a revolution time of e.g. 4.5 μs for 50 keV protons. The four-fold super symmetry of the ring provides four straight sections with regions of enhanced ion density (interaction points (IP)) for carrying out experiments and/or beam diagnostics. For the detection of neutralized atoms/molecules two position and time sensitive MCP-detector systems are installed at the 0°- and 180°- port of the ring (see Figure 1).

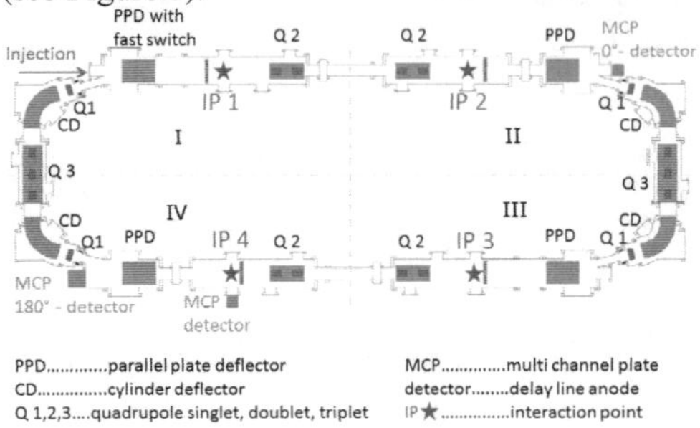

Figure 1: Schematic view of FLSR.

In Figure 2 results are presented from an exper-iment on the dissociative recombination of vi-brational cooled HeH^+ ions on an electron target [2]. For this experiment an electron gun was installed at IP 4. As this reaction channel does not offer the possibility to identify the reaction by measuring a recoil ion we have substituted the electron gun by a recoil ion spectrometer in order to identify the competing capture channel of the reaction. The results of this new experiment will be reported.

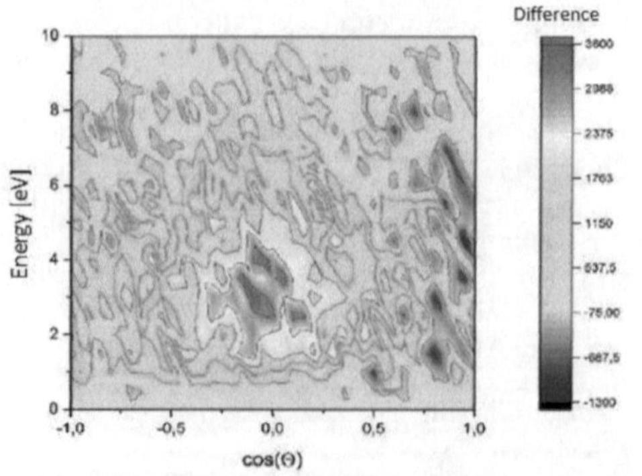

Figure 2: 3D plot of the KER to the momentum in electron beam direction ($\cos(\Theta) = p_x / p_{ges}$) [2]

In order to complete this set up also for user defined target gas species, it is planned to install a gas target at IP-4 of FLSR, providing a full COLTRIMS reaction microscope for future experiments at the ring. The design of this set up will be presented.

References

[1] K. E. Stiebing *et al* 2010 *Nucl. Instr. and Meth.* **A614,**10
[2] F.King 2015 *"Experimente zum dissoziativen Elektroneneinfang in raumfeste Molekülionen im Frankfurter Speicherring"* PHD-thesis IKF

[1] E-mail: janmueller@atom.uni-frankfurt.de
[2] E-mail: stiebing@em.uni-frankfurt.de

IOP Conf. Series: Journal of Physics: Conf. Series **875** (2017) 092004 doi:10.1088/1742-6596/875/10/092004

Multiple ionization effects on L X-ray emission in Bi by proton impact

Sunita Sharma

Department of Chemistry, M. L. N. College, Yamuna Nagar, 135 001, Haryana, India

Synopsis L X-ray intensity ratios for Bismuth (Bi) as a function of proton energy has been calculated in the energy range 260 – 400 keV. An important phenomenon i.e. Multiple Ionization (MI) effect, which influenced the X-ray production process included in the calculations. These are based on the ECPSSR model using the data basis of different workers. Theoretically calculated values are compared with the experimental measurements and show fairly a well agreement over the entire energy region except at lower energy edge. The present results clearly illustrate the significance of multiple ionization effect and offer relatively good opportunity to bring the theoretical values closer to the experimental measurements.

In atomic and molecular physics, interaction of charged particles with matter has been the subject of extensive study both experimentally and theoretically [1, 2]. In ion-atom collisions various processes can occur like creation of vacancy, filling up of vacancy and lasts with X-ray emission. Analytical technique like PIXE relies on X-ray production cross sections. An accurate knowledge of cross sections has been an increasing requirement, as it plays the great role for solution of many analytical problems [3]. It provides elemental information qualitatively as well as quantitatively over a broad range of elements. The accuracy of cross sections depends upon ionization mechanism along with other parameters viz. Fluorescence yield and Coster-Kronig (CK) probabilities. These parameters are highly affected by the multiple ionization (MI). In view of this, theoretical model with different effects viz. ECPSSR + UA, ECPSSR + MI are employed for calculation of ionization, X-ray production cross sections and their line ratios. Much of this work has been with the heavy ions [4, 5]. But in case of proton, this effect is still largely unexplored. So an effort is made to understand this effect for Bi (Bismuth) due to proton impact within the energy range 260 – 400 keV. Bi is chosen as it is used in the manufacturing of plastics, the synthesis of methanol and as a catalyst. It has wide application in production of low melting alloys used in protective coatings.

We have presented intensity ratios which are calculated with multiple ionization (MI) and without multiple ionization (WOMI) using data sets of different workers [6-9] along with measured values [10] (see figure 1). It clearly shows that how the inclusion of MI effect substantially refines the agreement of calculated and measured values. The detailed results will be presented and discussed at the conference.

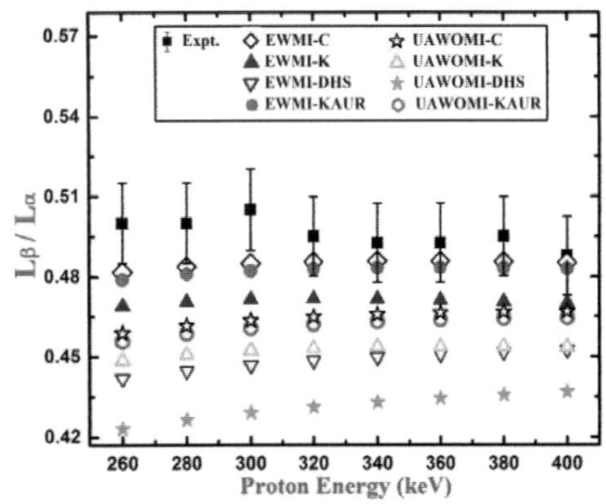

Figure 1. L_β/L_α for Bismuth as a function of proton energy.

E-mail: sunitamlnc@gmail.com

References

[1] Y. P. Singh *et al.* 2000 *Phys. Rev.* A **63** 012713
[2] J. Miranda *et al.* 2000 *At. Data Nucl. Data Tables* **100** 651
[3] C. Fourdrin *et al.* 2016 *J. Microchem.* **126** 446.
[4] C. Ciortea *et al.* 2003 *Nucl. Instr. Meth.* B **205** 789
[5] R. Mehta *et al.* 1995 *J. Phys.* B: *At. Mol. Opt. Phys.* **28** 1187
[6] G. Kaur *et al.* 2016 *J. Quant. Spectrosc. Radiat. Transfer* **87** 3368
[7] J. L. Campbell 2009 *At. Data Nucl. Data Tables* **95** 115
[8] M. O. Krause 1979 *J. Phys. Chem. Ref. Data* **8** 307
[9] S. Puri *et al.* 1993 *X-Ray Spectrom.* **22** 358
[10] H. Mohan *et al.* 2014 *Nucl. Instr. Meth.* B **332** 103

IOP Conf. Series: Journal of Physics: Conf. Series **875** (2017) 092005 doi:10.1088/1742-6596/875/10/092005

Competition between radiative and Auger decay processes of doubly excited Li-like C, N and O ions

N. Numadate[1], Y. Uchikura, K. Shimada, T. Akutsu and H. Tanuma

Department of Physics, Tokyo Metropolitan University, 1-1 Minami-Ohsawa, Hachioji, Tokyo 192-0397, Japan

Synopsis We performed charge exchange collision experiments of metastable He-like C, N and O ions (1s2s $^3S_1$) with neutral target gases, and observed soft X-rays from doubly excited Li-like C, N and O ions produced by single electron capture. These emissions corresponded to $1s^22s\ ^1S_0$ - 1s2s ($^3S_1$) *np* transitions according to calculations for transition energies by the Cowan code.

Radiative and Auger transitions are dominant and competitive decay processes of doubly excited and inner-shell excited states of atoms and molecules above their ionization potentials. These two processes have attracted much attention, because X-rays and Auger electrons from the superexcited states are of great importance for diagnostic of fusion and astrophysical plasmas [1], investigation of decay dynamics of molecules [2] and radiation induced damage on DNA [3]. Now there are very few papers about the radiative decay of superexcited atoms of light elements, because the Auger decay is much faster than the radiative decay in the light atoms.

The doubly excited states can be produced by double electron capture in atomic collisions. We produced doubly excited Li-like C, N and O ions [1s2s ($^3S_1$) *nl*] by single electron capture between metastable He-like ions (1s2s $^3S_1$) and He or O_2 gas. Resonance and inter-combination transitions ($1s^22s$ - 1s2s*np*) were observed by a grazing incidence spectrometer and a silicon drift detector (SDD). In order to identify spectral peaks, we calculated transition energies and rates of the Li-like ions by the Cowan code.

Table 1 shows calculation results of the Li-like oxygen ion by the Cowan code. In our experimental condition, the resonance transitions (^2S - ^2P) were mainly observed due to their high decay rates. Figure 1 shows soft X-ray spectra of the Li-like oxygen ions observed by the SDD. Two peaks at approximately 560 and 640 eV corresponds to radiations from the oxygen ions. A broad tail in low energy region is background signal derived from the SDD. Chen *et al.* calculated the Auger and radiative decay rates of Li-like C, N and O ions, and it was found that the Auger decay rate was only one order of magnitude higher than the radiative decay rate in the Li-like ions [4].

We discuss the competition between the radiative and Auger decay processes in the doubly excited Li-like C, N and O ions by comparing observed spectra with calculation results of each radiative and Auger decay rate.

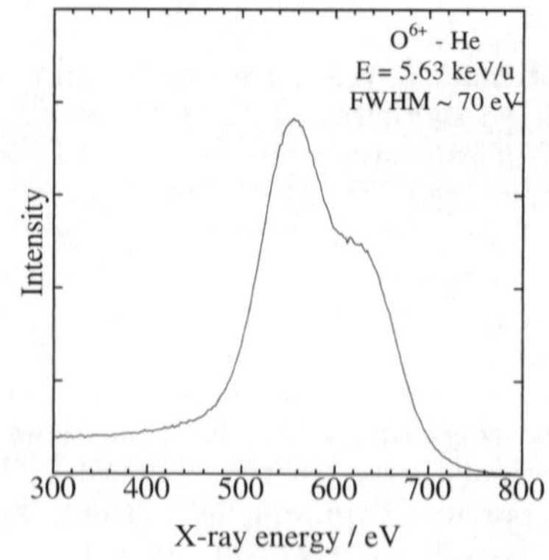

Figure 1. Soft X-ray spectrum of Li-like oxygen ions observed by the SDD. Energy resolution is approximately 70 eV.

Table 1. Calculated transition energies and radiative decay rates for states of 1s2s (^3S) *np* configurations of Li-like oxygen ion.

Lower state	Upper state		Energy (eV)	Rate (s^{-1})
$1s^22s\ ^2S$	$1s2s(^3S)2p$	$^4P_{1/2}$	554.0	5.17E+07
		$^4P_{3/2}$	554.0	2.61E+08
		$^2P_{1/2}$	567.8	6.24E+11
		$^2P_{3/2}$	567.8	1.15E+12
	$1s2s(^3S)3p$	$^4P_{1/2}$	640.1	4.11E+07
		$^4P_{3/2}$	640.1	2.11E+08
		$^2P_{3/2}$	641.6	2.43E+12
		$^2P_{1/2}$	641.6	1.22E+12

References

[1] A. Burgess 1964 *Astrophys. J.* **139** 776.
[2] M. Ukai 1996 *J. Electron Spectrosc. Rel. Phenom.* **79** 423.
[3] B. Boudaïffa *et al.* 2000 *Science* **287** 1658.
[4] M. H. Chen *et al.* 1983 *Phys. Rev.* A **27** 544.

[1] E-mail: numadate-naoki@ed.tmu.ac.jp

Single ionization of Helium at 0.5 - 2 MeV proton impact: On the quest for projectile coherence effects

J. Gatzke[†], F. Navarrete[¤], M. Ciappina[‡,§], H. Gatzke[†], O. Chuluunbaatar[ϒ,¶], S. A. Zaytsev[♣], A. A. Bulychev, K. A. Kouzakov[♦], A. Galstyan, M. Waitz[†], H.-K. Kim[†], T. Bauer[†], A. Laucke[†], S. Eckart[†], G. Kastirke[†], J. Müller[†], M. Ritzer[†,■], E. Bloch[†*], M. Richter, K. Fehre[†], M. Kunitski[†], Ch. Müller[†], J. Voigtsberger[†], J. Rist[†], K. Pahl[†], M. Honig[†], M. Pitzer[†], M. Weller[†], I. Vela Pérez[†], J. Hoehl[†], G. Nalin[†], S. Grundmann[†], H. Maschkiwitz[†], C. Janke[†], S. Zeller[†], C. Goihl[†], Y. Herrman[†], D. Trabert[†], T. Jahnke[†], L. Ph. H. Schmidt[†], Yu. V. Popov[¶,♦], R. Dörner[†] R. O. Barrachina[¤], and M. S. Schöffler[†1]

[†] Institut für Kernphysik, Goethe University, Max-von-Laue-Straße1, 60438 Frankfurt, Germany
[*] Univ. Bordeaux, CNRS CEA, CELIA, UMR5107, F-F33405 Talence, France
[■] Experimental Physics IV, University of Kassel, Heinrich-Plett-Straße 40, D-34132 Kassel, Germany
[¶] Joint Institute for Nuclear Research, Dubna, Russia
[ϒ] National University of Mongolia, UlaanBaatar, Mongolia
[♣] Pacific State University, Khabarovsk, Russia
[♦] Lomonosov Moscow State University, Russia
[¤] Centro Atómico Bariloche e Instituto Balseiro (Comisión Nacional de Energía Atómica and Univ. Nac. de Cuyo), Bariloche, Río Negro, Argentina
[‡] Max-Planck-Institut für Quantenoptik, Hans-Kopfermann-Straße 1, D85748 Garching, Germany
[§] Institute of Physics of the ASCR, ELI-Beamlines project, Na Slovance 2, 182 21 Prague, Czech Republic

Synopsis: We investigated single ionization in proton-helium-collisions at 0.5, 1 and 2 MeV for various beam properties (divergence/coherence length).

Single ionization, induced by a fast charged particle has been investigated since a long time. While electron impact experiments – at least on a helium target – agreed well with theory in lowest order perturbation theory, impacting heavy particles, such as protons or highly charged ions never agreed well with theory. The typical electron angular distribution shows two peaks, the pronounced binary and the recoil peak, and a nodal structure in-between them. In a collision experiment with C^{6+} projectiles at 100 MeV/u [1], the node was mostly filled, and the disagreement with state-of-the-art theories, started an avalanche of discussions and further experiments. 15 years later, the issue is still not solved and two explanations are heavily debated: insufficient momentum resolution and the influence of transversal beam coherence. According to [2] the macroscopic geometrical preparation of the beam, e. g. its divergence, should influence the microscopic scattering behavior and therefore for example the electron's angular distribution.

In a series of single ionization experiments, we investigated p+He collisions at impact energies from 0.5 to 2 MeV, produced by a Van-de-Graaff accelerator. We applied the COLTRIMS reaction microscope technique [3], optimized for highest momentum resolution, to determine the momentum vectors of all emitted particles in coincidence. For the highest projectile energy (2 MeV), we manipulated the beam divergence by utilizing an electro static quadrupole lense. The electron angular distribution in the scattering plane is shown in Figure 1, both with lense being turned on and off. Within the experimental error bars we don't observe any difference.

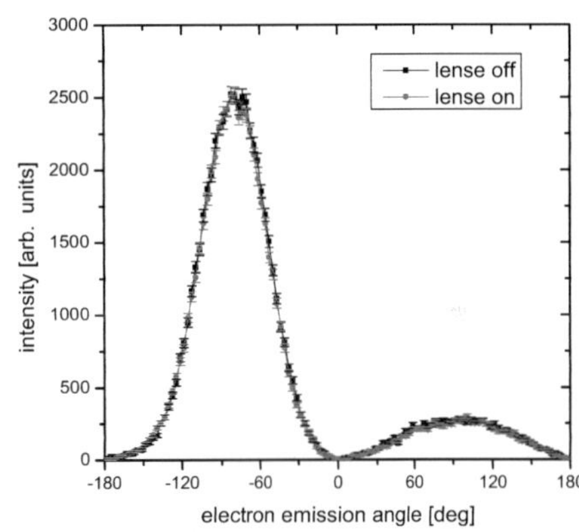

Figure 1. Electron emission angle in the scattering plane for 2 MeV p+He, E_e=6.5±3.5 eV, q=0.75±0.25 au, with projectile lense turned on (red circles) or turned off (black squares).

References

[1] Schulz et al., Nature, **422**, 48, (2003)
[2] Schulz et al., Phys. Rev. A, **76**, 032712, (2007)
[3] Gassert et al., Phys. Rev. Lett **116**, 073201, (2016)
[4] Dörner et al., Phys. Rep. 330, 95 (2000)

[1] E-mail: schoeffler@atom.uni-frankfurt.de

IOP Conf. Series: Journal of Physics: Conf. Series **875** (2017) 092007 doi:10.1088/1742-6596/875/10/092007

Effective solid angle correction factors for long-lived Auger states populated in low-Z ion collisions with gas targets

E. P. Benis[‡], S. Nanos[‡], I. Madesis[*†], A. Laoutaris[*†], T. W. Gorczyca[§], T. J. M. Zouros[1 *†]

[‡]Department of Physics, University of Ioannina, GR 45110 Ioannina, Greece
[*]Department of Physics, University of Crete, P.O. Box 2208, GR 71003 Heraklion, Greece
[†]Tandem Accelerator Laboratory, INPP, NCSR Demokritos, GR 15310 Ag. Paraskevi, Greece
[§]Department of Physics, Western Michigan University, Kalamazoo, Michigan 49008, USA

Synopsis: Effective solid angle correction factors for two high resolution electrostatic spectrometers - a hemispherical and a tandem parallel plate – are determined using SIMION Monte Carlo simulations. Results are compared to experimental correction factors directly determined from Auger spectra produced by 1s ionization in collisions of 6.6 MeV C^{2+} ($1s^22s^2\ {}^1S$, $1s^22s2p\ {}^3P$) mixed state ion beams with H_2 targets. Excellent agreement between simulation and measurement is found.

A persistent problem in high resolution Auger projectile spectroscopy is the determination of the contribution from metastable states due to their inherent long lifetime. As an example we refer to the accurate evaluation of the $1s2s2p\ {}^4P/{}^2P$ ratio of K-Auger cross sections, whose observed non-statistical production by electron capture into He-like ions, is a field of differing interpretations, awaiting further investigation [1-3].

Here, we present SIMION Monte Carlo type simulations of the response function and detection solid angle of two analyzers: A single stage hemispherical spectrograph with injection lens and position sensitive detector [4], and a two-stage parallel plate analyzer [5], both used extensively in zero-degree Auger projectile electron spectroscopy. At 0^0 observation the spectrometers lie in the direct path of the ion with the excited metastable projectile states (lifetimes ~1-10^3 ns) Auger decaying all along its path towards the spectrometer (and even inside the spectrometer). Thus, the overall detection solid angle of the electron emission varies due to the moving source position, resulting in a considerable correction to the measured electron yield [4]. These effects, particular for Auger emission from fast moving projectile ions, are also included in our simulations.

Our SIMION model calculations are also checked experimentally utilizing collisions of Be-like carbon beams produced in mixed ($1s^22s^2\ {}^1S$, $1s^22s2p\ {}^3P$) states which also give rise to $1s2s2p\ {}^4P$ states, but free of cascade repopulation effects since they are formed by needle ionization of the $1s$ electron of the $1s^22s2p\ {}^3P$ state [5]. A typical spectrum obtained recently with our hemispherical spectrograph is shown in Fig. 1. The necessary determination of the metastable beam fraction is obtained from the spectrum of Fig. 1 utilizing the prompt $1s2s^2\ {}^2S$ and $1s(2s2p^3P)\ {}^2P$. lines whose ratio is *known* [5]. The effective solid angle correction factor for the hemispherical spectrograph is then obtained via the determination of the also *known* 1s2s2p ${}^4P/{}^2P$- ratio of K-Auger electron yields [5]. Our simulations are in excellent agreement with experimental results, thus providing an empirical consistency check for future determinations of these correction factors in other spectrographs and geometries.

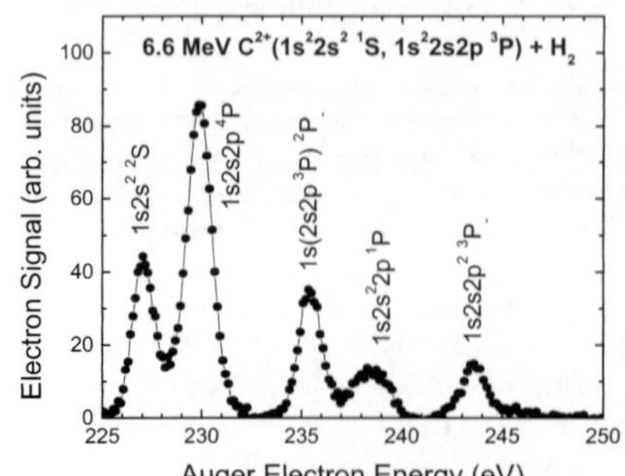

Figure 1. Auger spectra from 6.6 MeV C^{2+} collisions with H_2 targets. C^{2+} beam is delivered in the mixed $1s^22s^2\ {}^1S$ ground and $1s^22s2p\ {}^3P$ metastable states. The dominant Auger lines are depicted in the figure.

References

[1] T. J. M. Zouros, B. Sulik, L. Gulyás, and K. Tőkési 2008 *Phys. Rev. A* 77 050701
[2] D. Röhrbein, T. Kirchner, and S. Fritzsche *2010 Phys. Rev. A* **81** 042701
[3] E. P. Benis and T. J. M. Zouros 2016 *J. Phys. B* **49** 235202
[4] S. Doukas *et al 2015 Rev. Sci. Instrum.* **86** 043111
[5] D. H. Lee *et al.* 1992 *Phys. Rev. A* **46** 1374

[1]E-mail: tzouros@physics.uoc.gr

IOP Conf. Series: Journal of Physics: Conf. Series **875** (2017) 092008 doi:10.1088/1742-6596/875/10/092008

Atomic structure calculations of He and Li-like Ti ions for interpreting astrophysical spectra

Gajendra Singh[*1], **A K Singh**[*2],

[*]Department of Applied Sciences, MSIT, affiliated to USICT, GGSIPU, New Delhi, India

Synopsis We are working on atomic structure calculations with main focus on the ions relevant to Astrophysics and upcoming fusion reactors like ITER. Main work comprises of first generating atomic data of ions by theoretical calculations by Grasp2k code and then use it for plasma diagnostic studies by identifying the relevant transitions.

Study of physical processes in astrophysical and fusion plasmas consists of detail analysis of high resolution atomic spectra obtained from such plasmas. The X-ray spectra from Ti, Cu, Ni and Fe K-shell and L-shell ions are particularly important for astrophysics as they are in wavelength range covered by telescopes on board space observatories like Chandra and XMM-Newton. To conduct the astrophysical plasma diagnostic studies of Active galactic nuclei (AGN), solar corona and other similar astronomical entities, a large number of accurate transition data both from theory and experiment are indispensable. Here we have calculated similar atomic data in use for these studies by multi-configuration Dirac-Fock (MCDF) formalism used in Grasp2K code. Calculations consists of ground and few low level excited states in He-like and Li-like Ti ions and some important transitions connecting these levels.

Table 1. Calculated Energy levels for He and Li-like Ti ions.

config.	Term	This work (cm^{-1})	NIST (cm^{-1})
Ti^{20+}			
$1s^2$	$^1S_0$	0	0
1s2s	$^3S_0$	37923300	37923880
1s2p	$^3P_{0,-1,-2}$	38114400	38114760
		38124911	38125260
		38180261	38180620
1s2p	$^1P_{-1}$	38308448	38308340
Ti^{19+}			
$1s^22s$	$^2S_{1/2+}$	0	0
$1s^22p$	$^2P_{1/2-,3/2-}$	324401	323521
		386569	385666
1s2s2p	$^4P_{1/2-,3/2-,5/2-}$	37758694	37759000
		37774736	37774000
		37817431	NA

Our results are listed in Table 1, where we have presented our calculated level energies of He-like and Li-like Ti. Here $1s^2$ state is used as the ground state configuration and 1s2s, 1s2p as excited state configurations are used for He-like Ti ions, whereas for Li-like Ti ions $1s^22s$ is used as the ground and $1s^22p$, 1s2s2p as excited states configurations. We compared our results of the level energies with those of NIST database [1]. It may here be pointed out that the NIST energy levels were extrapolated using experiments and COWAN code. As we can see, nearly all the computed values of ours are in good agreement with the NIST compilation; except at one point in Li-like system. In fact this level i.e. $^4P_{5/2-}$ belonging to 1s2s2p configuration is not even listed in NIST database.

Remembering that Li-like ions of all the cosmically abundant elements like Ti, Fe, Ni and Cu etc. are actually used in astrophysical plasma diagnostic studies of active galactic nuclei or Solar corona, the presence of above noted level in our estimate turns out to be significant addition to the existing knowledge. It may here be pointed out that exclusion of an important level such as $^4P_{5/2}$ will adversely affect the plasma diagnostic studies while calculating the electron temperature $G(T_e)$ and electron number density $R(n_e)$. Where $G(T_e)$ is calculated as the ratio of forbidden plus inter-combination to the resonance and $R(n_e)$ as the ratio of forbidden to the inter-combination transitions.

Considering the importance of including one additional energy level $^4P_{5/2}$ in the existing NIST database, which makes it complete, we had also computed all important radiative transition rates like E1, M1 and M2 for both the ions under study and these radiative rates soon will be communicated in a reputed journal for publication.

References

[1] T. Shirai *et al.* 2000 *J. Phys. Chem. Ref. Data* Monograph No. 8, 632 pages, Vol 29(2000)

[1]E-mail: gpskmc@gmail.com
[2]E-mail: drajayphd@gmail.com

Theoretical study on projectile coherence effects in fully differential ionization cross sections

Ladislau Nagy*, **Sándor Borbély***, **Ferenc Járai-Szabó***[1]

* Faculty of Physics, Babeş Bolyai University, 400084 - Cluj-Napoca, Romania

Synopsis In order to clarify the origin of the projectile coherence effects we present a theoretical, ionization amplitude based study of ionization of helium by 100 MeV/u C^{6+}, 1 Mev proton, and 75 keV proton projectiles.

After performing fully differential cross section (FDCS) measurements for the ionization of helium [1] significant difference between experimental data and theoretical continuum distorted wave (CDW) results have been observed. These, and other similar theoretical calculations were not able to reproduce the experimentally observed structures in the perpendicular plane. However, our semiclassical calculations [2] and the convolution of the first Born approximation with elastic scattering [3] described fairly well these structures. The discrepancy between these calculations and the relationship with the experimental data has been explained, by the importance of the projectile coherence effects [4]. Significant differences have been observed in the FDCSs for projectiles causing the same perturbation, but having different transversal coherence lengths. If the coherence length was larger than the atomic dimensions, the results were closer to the CDW calculations, which assumes a plane wave for the projectile, while for smaller coherence widths the results were closer to the convoluted [3] and our semiclassical results.

Recently, we have shown in two extreme cases the FDCS dependence on the projectile coherence [6] by assuming classical particle and plane wave description of the projectile. Later, a more realistic, finite wave packet description has been proposed [7]. In the meantime, another fully differential study yielded good agreement between experiment and theory but no conclusive evidence regarding the projectile coherence effect was reported [5]. Accordingly, the topic is not closed yet, and surely it is intensively studied by the scientific community.

In order to clarify the effects which lead to the FDCS differences for coherent and incoherent projectiles, we present a theoretical, ionization amplitude based study for ionization of helium by 100 MeV/u C^{6+}, 1 Mev proton, and 75 keV proton projectiles. These studies are focusing on the averaged ionization amplitudes over the impact parameter's azimuthal angle, and the shape of the projectile wavepacket. These quantities are shown in Fig. 1 for the ionization of helium by 75 keV proton projectile, in a region where significant differences in FDCS have been detected for coherent and incoherent projectile beams [7]. In case of coherent projectile beam the whole impact parameter regime contributes to the FDCS, while in case of projectiles having shorter coherence length the FDCS is calculated by integrating ionization amplitudes over a shorter impact parameter region where only a single change of the signs of real and imaginary parts of the ionization amplitude is present.

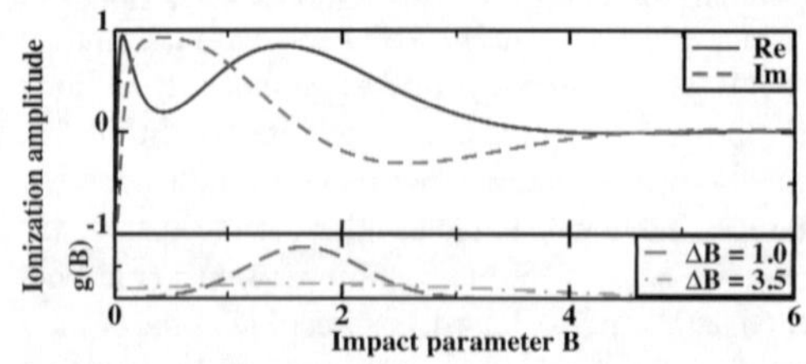

Figure 1. Top panel shows real and imaginary parts of the averaged ionization amplitude for ionization of helium by 75 keV proton projectile. The fixed x-component of the recoil-ion momentum $p_{rec} = 0.7$ au., the electron ejection angles are $\theta_{el} = 30°$, and $\phi_{el} = 90°$. The bottom panel shows the Gaussian functions which describes the shape of the coherent ($\Delta B = 1.0$) and incoherent ($\Delta B = 3.5$) projectile wavepackets.

The above described amplitude analysis is performed for many kinematical parameters and for different projectiles, too.

References

[1] M. Schultz *et al.* 2003 *Nature* **422** 48

[2] F. Járai-Szabó, L.Nagy 2009 *Nucl. Instrum. Methods B* **267** 292

[3] M. Schulz *et al.* 2007 *Phys. Rev. A* **76** 032712

[4] Wang X *et al.* 2012 *J. Phys. B: At. Mol. Opt. Phys.* **45** 211001

[5] H. Gassert *et al.* 2016 *Phys. Rev. Lett.* **116** 073201

[6] F. Járai-Szabó and L. Nagy 2015 *Eur. Phys. J. D* **69** 4

[7] T. Arthanayaka *et al.* 2016 *J. Phys. B: At. Mol. Opt. Phys.* **49** 13LT02

[1]E-mail: jferenc@phys.ubbcluj.ro

IOP Conf. Series: Journal of Physics: Conf. Series **875** (2017) 092010 doi:10.1088/1742-6596/875/10/092010

Radiative Double Electron Capture (RDEC) in F^{9+} + He, Ne Collisions

D. S. La Mantia[*1], **P. N. S. Kumara**[*], **A. Kayani**[*], **A. Simon**[†], **J. A. Tanis**[*]

[*] Department of Physics, Western Michigan University, Kalamazoo, MI 49008-5252 USA
[†] Department of Physics, University of Notre Dame, Notre Dame, IN 46556-5670 USA

Synopsis Preliminary cross sections for RDEC, as well as total single and double electron capture, were measured for 40 MeV F^{9+} + He and Ne targets. Emitted x rays and charge-changed particles were measured using coincidence techniques.

Radiative electron capture (REC) is a one-step process by which a target electron is captured to a projectile with the emission of an x ray and can be considered the time-inverse of photoionization. Double photoionization and the effect of electron correlation can be investigated through the process of radiative double electron capture (RDEC), in which two electrons are captured with the simultaneous emission of a single photon. Preliminary RDEC cross sections, as well as those for total single and double capture, for F^{9+} + He and Ne were measured.

This work was performed using the tandem Van de Graaff accelerator at Western Michigan University. A 40 MeV beam of fully-stripped fluorine ions was collided with helium (40μ) and neon (25μ) targets inside a differentially pumped cell. A Si(Li) x-ray detector was positioned at 90° to the beamline on the gas cell. The ion beam was charge state analyzed using a dipole magnet and detected using silicon surface-barrier particle detectors. The x-ray and particle detector data were collected using an event-mode data acquisition system to assign the measured x rays to the corresponding charge-changed particles.

Previous RDEC experiments using high Z, high energy projectiles on gas targets [1][2] were performed and did not find evidence for the process. Theoretical predictions [3] suggest mid-Z projectiles in low energy collisions will give larger RDEC cross sections. Successful observations of RDEC were performed at WMU using bare oxygen [4] and fluorine [5] projectiles incident on thin carbon foil targets.

The events that can result in RDEC are the capture of two electrons to the projectile K shell ($1s^2$) or the capture of one electron to the K and L shells ($1s^1 2s^1$). X rays coincident with single (Q-1) and double (Q-2) electron capture for F^{9+} + Ne (25μ) can be seen in Fig. 1. In the Q-1 spectrum, beyond the characteristic x rays for F, REC dominates. In the Q-2 spectrum the counts at the high energy end of the REC region can be due to RDEC from target KK $\rightarrow$ projectile KL transitions, with the higher energy counts being due to target KK $\rightarrow$ projectile KK transitions.

Preliminary analysis gives $\sigma_{\mathrm{RDEC}}^{\mathrm{Ne}} \sim 110$ mb (10^{-27} cm^2), which is in reasonable agreement with the theory of Mistonova and Andreev [6]. For He a single event was seen in the RDEC region, corresponding to $\sigma_{\mathrm{RDEC}}^{\mathrm{He}} \sim 3$ mb. At least an additional two months of continuous beam time is needed to obtain reliable statistics for each target.

*Supported in part by NSF Grant No. PHY-1401429.

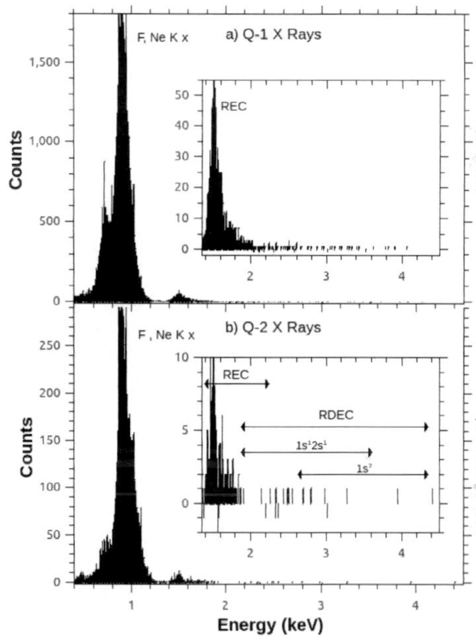

Figure 1. Spectra for x rays coincident with (a) single and (b) double capture for 40 MeV F^{9+} + Ne collisions.

References

[1] Bednarz *et al.* 2003 Nucl. Instrum. Meth. B **205** 573
[2] Winters *et al.* 2013 Phys. Scr. **T156** 014048
[3] Nefiodov *et al.* 2005 PLA **346** 158
[4] Simon *et al.* 2010 PRL **104** 123001
[5] Elkafrawy *et al.* 2016 PRA **94** 042705
[6] Mistonova and Andreev. 2013 PRA **87** 034702

[1] E-mail: david.s.lamantia@wmich.edu

IOP Conf. Series: Journal of Physics: Conf. Series **875** (2017) 092011 doi:10.1088/1742-6596/875/10/092011

Time evolution of the population distribution in charge exchange collisions

Kento Shimada*1, Naoki Numadate*, Yoshiyuki Uchikura*, Takuto Akutsu*, Ling Liu†, Jianguo Wang†, and Hajime Tanuma*2

* Department of Physics, Tokyo Metropolitan University, Hachioji, Tokyo 192-0397, Japan,
† Institute of Applied Physics and Computational Mathematics, Beijing 100088, China

Synopsis: We have calculated the population distributions of many electronic excite states as a function of time in collisions of bare, hydrogen-like and helium-like C, N and O ions with neutral targets, H, He, and H_2. Using the population distribution, we have calculated theoretical emission spectra in the collision system. Comparing the theoretical spectra with the observation, we have estimated the detection efficiency of the spectrometer and discuss the singlet-triplet ratio in charge exchange collisions of hydrogen-like ions.

Both the soft X-ray emission from comets and the X-ray background signals observed with X-ray satellite observatories have been understood as the transitions of highly charged ions in the solar wind that underwent charge-exchange collisions with neutral gases around the comets in the heliosphere [1, 2, 3]. For the quantitative analyses of future observed X-ray spectra with high resolution micro-calorimeters, the accurate charge exchange cross sections and emission cross sections are required by astrophysicists.

We have measured electron capture cross sections and emission cross sections in collisions of highly charged ions produced with a 14.25 GHz electron cyclotron resonance ion source with neutral gas targets. In the emission cross section measurements, we have used a collision cell 50 mm long and only the emission from the center of the cell has been observed with a windowless silicon drift detector for soft X-ray and a grazing incident spectrometer for the ex-treme ultra-violet (EUV) region. In the experiments, we kept the target gas pressure low enough to avoid double collisions in the cell. However, the contribution of the intermediate lifetime states produced upstream of the observation center must be investigated. Furthermore, the detection efficiency of the EUV spectrometer, which strongly depends on the wavelength, must be estimated to obtain emission cross sections.

In this study, we have calculated the time dependence of the population for many electronic excited states of incident ions in a collision cell filled by a target gas with very low density to keep single collision conditions. For this calculation, we have used single electron capture cross sections found by the AOCC (Atomic Orbital Close Coupling) method [4] and Einstein's A-coefficients stored in the CHIANTI atomic database [5]. Using cross sections and transition rates, we assumed the following simultaneous rate equations:

$$\frac{\mathrm{d}[\mathrm{X}^{(q-1)+}(\mathscr{S})]}{\mathrm{d}t} = \sigma_{\mathscr{S}}\, v[\mathrm{X}^{q+}][\mathrm{Y}] + \sum_{\mathscr{S}'}\left\{A_{\mathscr{S}',\mathscr{S}}[\mathrm{X}^{(q-1)+}(\mathscr{S}')]\right\} - \left\{\sum_{\mathscr{S}''}A_{\mathscr{S},\mathscr{S}''}\right\}[\mathrm{X}^{(q-1)+}(\mathscr{S})],$$

where $[\mathrm{X}^{(q-1)+}(\mathscr{S})]$ means the population of the $\mathscr{S}$ state of the incident ion after the electron capture, [Y] is the density of the target gas, v is the collision velocity, $A_{i,j}$ is the optically transition rate from i to j, and $\sigma_{\mathscr{S}}$ means state-selective capture cross section to the $\mathscr{S}$ state. In the AOCC calculation, it is difficult to distinguish the fine-structure with different total angular moment J. Therefore, we assumed the statistical weight ratios of $2J+1$ for each capture cross section. After the calculation of the population evolution of the incident highly charged ions, we have also calculated the emission spectra.

We have calculated the population evolution and the emission spectra in collisions of bare, hydrogen-like, and helium-like C, N, and O ions with H, He, and H_2 target gasses. And, by comparing of the observed emission spectrum with the theoretical spectrum, we could estimate the detection efficiency of the EUV spectrometer if the theoretical capture cross sections are appropriate. However, singlet-triplet ratio in collisions of hydrogen-like ions are not well known at present. In this study, we will discuss the singlet-triplet ratio in the capture cross sections.

References

[1] S. L. Snowden *et al.* 1994 *Astrophys. J.* **424** 714-728
[2] C. M. Lisse *et al.* 1996 *Science* **274** 205-209
[3] T. E. Cravens, 2002, *Science,* **296(5570)** 1042-1045
[4] L. Liu, *private communication*
[5] P. R. Young *et al.* 2016 *J. Phys. B* **49(7)** 074009

1E-mail: shimada-kento1@ed.tmu.ac.jp
2E-mail: tanuma-hajime@tmu.ac.jp

IOP Conf. Series: Journal of Physics: Conf. Series **875** (2017) 092012 doi:10.1088/1742-6596/875/10/092012

MeV ion beam extraction into air with a glass capillary filled with He

M. Asamura*[1], K. Ishii†[2], and H. Ogawa†

* Graduate School of Humanities and Sciences, Nara Women's University, Nara, 630-8506, Japan
† Department of Physics, Nara Women's University, Nara, 630-8506, Japan

Synopsis In order to obtain an intense ion beam in air, we have developed the He capillary, which is a normal glass capillary filled with low pressure He gas. The energy spectra of an ion beam transmitted via the He-capillary with and without He have been measured. An enhancement of the peak intensity and suppression of low energy components have been observed for the He- capillary.

Developments of in-air-analysis methods with MeV ion beam would have applications in various fields such as biology, nano-technology and archaeology. In particular, a technique to extract an ion beam into air through a tapered glass capillary has been developed by Nebiki *et al.* They have used this method to perform in-air-PIXE (particle induced X-ray emission) for sea sludge [1]. This technique was so simple and effective to obtain an in-air micro beam that it is now used in several laboratories. We have also developed several in-air-analysis techniques using glass and metal capillaries [2] and reported transmission properties of ion beams through several capillaries [3]. It is concluded that an ion beam diameter in air predouced via a capillary is determined by the inner diameter of the capillary and the beam energies and intensities are spread by atmospheric gas inside and outside the capillary. In this work, we have newly developed a capillary termed a He-capillary, which is a normal glass capillary filled with low pressure He gas. It is expected that an ion beam spread in the capillary will be suppressed, since the energy loss of an ion beam in a He atmosphere is smaller than that in air.

We have measured the energy spectra of an ion beam transmitted via a He capillary as a function of the He pressure in the capillary by using a SSD (silicone semiconductor detector). Figure 1 shows the energy spectra of a 3 MeV proton beam transmitted via the He-capillary filled with and without He gas. As is shown in Figure, the peak intensity is more enhanced in the capillary filled with He compared to no He, although the peak energy is shifted lower. Furthermore, low energy components of the energy spectra with He are suppressed copared to no He.

In the presentation, we will introduce the He- capillary and discuss transmission properties of ion beams via the capillary in detail.

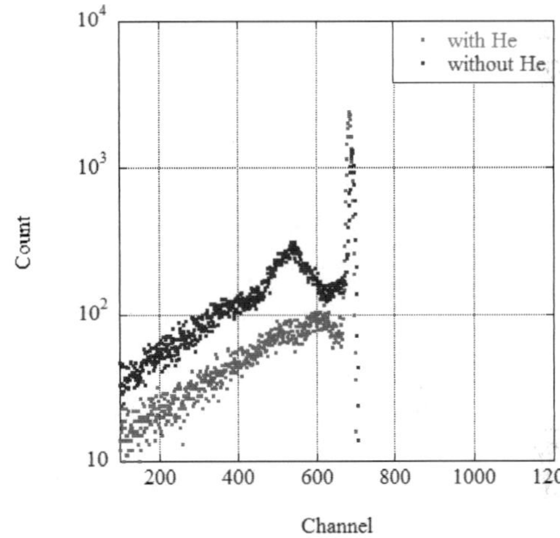

Figure 1. Energy spectra of 3.0 MeV proton beam via the He-capillary with and without He.

References

[1] T. Nebiki, M. H. Kabir and T. Narusawa 2006 *Nucl. Instr. Meth. B* **249** 226
[2] N. Fujita, K. Ishii and H. Ogawa 2011 *Nucl. Instr. Meth. B* **269** 1023. ; K. Ishii, N. Fujita and H. Ogawa 2011 *Nucl. Instr. Meth. B* **269** 1026
[3] N. Fujita, A. Yamaki, K. Ishii and H. Ogawa 2013 *Nucl. Instr. Meth. B* **315** 332

[1]E-mail: nam_amanura@cc.nra-wu.ac.jp
[2]E-mail: ishii@cc.nara-wu.ac.jp

IOP Conf. Series: Journal of Physics: Conf. Series **875** (2017) 092013 doi:10.1088/1742-6596/875/10/092013

SPARC experiments with highly charged ions at the HESR of FAIR

O. Kovalenko*[1], Yu. A. Litvinov*, Th. Stöhlker†

*GSI Helmholtzzentrum für Schwerionenforschung, Planckstrasse 1, 64291 Darmstadt, Germany
†Helmholtz-Institut Jena, Fröbelstieg 3, 307743 Jena, Germany

Synopsis One of the aims of the SPARC collaboration [1] at FAIR is to perform precision atomic physics experiments with highly charged heavy ions at the High Energy Storage Ring (HESR). An internal target is indispensably an integral part for many such experiments. Ions with different charge states, which are obtained as a result of interaction of an ion beam with the target, need to be effectively separated and detected. In this work we present ion optical studies unambiguously showing the feasibility of SPARC experiments at the HESR.

The SPARC collaboration at FAIR [1] aims at atomic physics research with highly charged ions (HCI) in energy domains previously not accessible for precision experiments [2,3]. Here, the High Energy Storage Ring (HESR) will allow for storing stochastically and/or electron cooled HCIs up to energies of ~5 A GeV.

The missing dipole concept in the HESR enables an installation of internal target stations [4] in two arcs (ROI 1 and 2) [5] (see Figure 1).

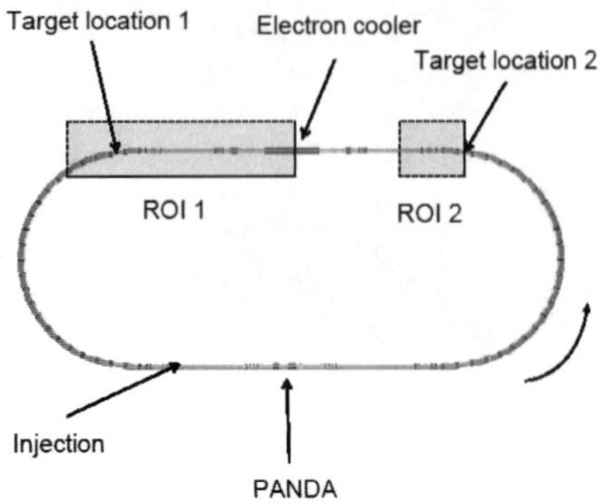

Figure 1. Two target locations in the HESR and the corresponding regions of interest (ROI) can be seen.

The experimental conditions in the ROI 1 region, which was initially proposed for the SPARC target location [6], have a complication: ions, recombined in the electron cooler and the rest gas along the long straight section, will produce an unwanted background. This situation is avoided in the ROI 2 case resulting in much cleaner experimental conditions. However, only one dipole magnet is then available for charge state separation [5]. Figure 2 shows a separation of the primary fully ionized $^{238}U^{92+}$ uranium beam (2-sigma emittance) and two charge states obtained after the reaction in the target (ROI 2). The ion optical computations prove that the resolution of the charge states at the particle detector, which is placed 3 m after the last defocusing quadrupole, is more than sufficient. For experiments with lighter beams, the separation improves further since ΔQ/Q increases [5]. As a conclusion, the experiments with internal target, placed in the missing dipole gap in ROI 2, are feasible [7]. Examples of the growing number of the proposed precision experiments in the HESR, which thus be enabled can be found in [8-14].

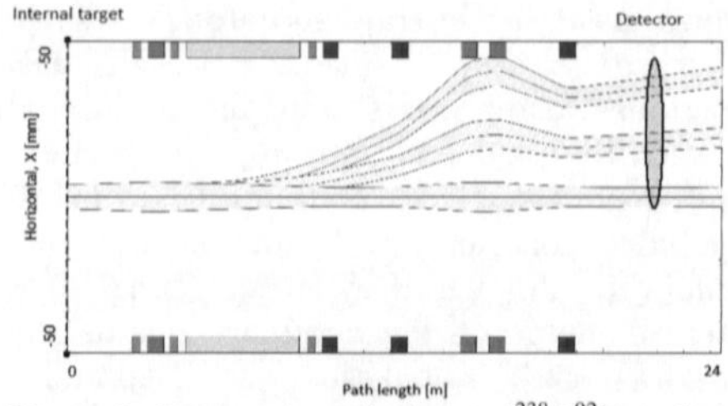

Figure 2. Propagation of primary ($^{238}U^{92+}$) and secondary charge exchange reaction products ($^{238}U^{91+}$, $^{238}U^{90+}$) after interaction with the target. The dipole magnets, focusing and defocusing quadrupoles are shown by cyan, red and blue colors, respectively [5].

References

[1] T. Stöhlker *et al.* 2011 *AIP Conf. Proc.* **1336** 132
[2] T. Stöhlker *et al.* 2015 *Nucl. Instr. Meth.* **B365** 680
[3] T. Stöhlker *et al.* 2013 *Phys. Scr.* **T156** 014085
[4] N. Petridis *et al.* 2015 *Phys. Scr.* **T166** 014051
[5] O. Kovalenko 2015 PHD Thesis, Heidelberg
[6] T. Stöhlker *et al.* 2014 *Hyperf. Interact.* **227** 45
[7] O. Kovalenko *et al.* 2015 *Phys. Scr.* **T166** 014042
[8] S. Hagmann *et al.* 2013 *Phys. Scr.* **T156** 014086
[9] F. Bosch 2013 *Prog. Part. Nucl. Phys.* **73** 84
[10] Y. Litvinov *et al.* 2013 *Nucl. Instr. Meth. B* **317** 603
[11] T. Stöhlker *et al.* 2015 *Phys. Scr.* **T166** 014025
[12] A. Gumberidze *et al.* 2015 *Phys. Scr.* **T166** 014076
[13] P. Hillenbrand *et al.* 2015 *Phys. Scr.* **T166** 014026
[14] W. Nörtershäuser 2015 *Phys. Scr.* **T166** 014020

*[1] E-mail: O.Kovalenko@gsi.de

IOP Conf. Series: Journal of Physics: Conf. Series **875** (2017) 092014 doi:10.1088/1742-6596/875/10/092014

Electromagnetic non-destructive detectors for storage rings

Shahab Sanjari[*1], **Yuri A. Litvinov**[*], **Thomas Stöhlker**[*†]

[*] GSI Helmholtzzentrum Darmstadt, [†] Helmholtz-Institut Jena

Synopsis Electromagnetic detectors are essential tools for the accelerator physicist. They can as well be designed to meet specific needs for performing precision atomic and nuclear physics experiments with stored highly-charged ions in storage rings. In this work we report on our ongoing detector developments at GSI.

The passage of particles through non-destructive electromagnetic detectors in storage rings induces a periodic signal that can be related to the particle's mass through its revolution frequency. Additionally the recorded signal can deliver information on the lifetime of the unstable nuclei if the data is processed through short time Fourier analysis. Furthermore, these so called Schottky signals deliver information on general properties of the particle beam such as momentum spread [1] and can even be used to observe stability of other elements in the storage ring such as the electron cooler [2] and power supplies.

Electromagnetic detectors designed for the detection of Schottky signals are based on a variety of geometries with their respective advantages and disadvantages for each application [3]. Of particular interest for precise determination of particle frequencies and intensities in storage rings are resonant cavities. Due to their high Q value, they can achieve high sensitivities, albeit at the cost of lower operation bandwidth. The usually large eigenmode frequency also allows for higher frequency and hence higher mass resolution power. An example of such a detector has been developed for the ESR storage ring at GSI facility [4] and its single ion sensitivity has been demonstrated [5].

Currently a R&D prototype of a similar longitudinal detector is under construction and is planned for test in the ESR as well as the future HESR and CR facilities of FAIR. The new detector will allow for variation of sensitivity during the operation. This feature would extend the scope of application of resonant cavity detectors in future storage ring experiments.

Designs based on resonant cavity structure are also suitable for detection of trasnversal displacement of the ion beams (please refer to [6] for a review). Here the electric dipole field is used to discern the beam position relative to the electric axis. Although this approach is successful in most cases, the small amplitude of the electric dipole field at the vicinity of the center of very large beam pipes (such as those found in the CR storage ring at FAIR) causes a low coupling amplitude and therefore low sensitivity. In order to overcome this issue, a design based on the monopole mode of a cavity with an elliptical cross section was proposed [7] and further investigated in [8]. Combined with an isochronous ion optical setting inside the storage ring, it is possible then to use the position information as an additional correction for Schottky mass measurements [9]. Currently R&D prototypes are planned to be tested in collaboration with S-DALINAC facility in Darmstadt before final designs are approved for the future CR at FAIR.

One of the inherent shortcomings of cavity based electromagnetic detectors is their rapid loss of sensitivity for low energy beams (i.e. very slow particles) due to the effect of transit time factor. However, there is a growing interest for experiments at lower energies motivated by precision studies in atomic, nuclear structures and astrophysics. A common practice is to use the so called slow or traveling wave detectors that allow for lower phase velocities [3]. A detector based on this approach is currently under developement for the newly installed CRYRING [10] facility at GSI.

This work has received funding from the European Research Council under the EU Horizon 2020 research and innovation program (N° 682841 ASTRUm).

References

[1] F. Caspers 2009 CERN-2009-005.407

[2] S. Sanjari *et al. GSI Sci. Rep. 2015* **p** 335

[3] G. R. Lambertson 1989 *ISBN* 978-3662137116

[4] F. Nolden *et al.* 2011 *NIM A* **v** 659 pp 69-77

[5] F. Bosch *et al.* 2013 *Phys. Lett. B.* **v** 1726, pp 638-645

[6] X. Chen 2014 *GSI Report* GSI-2014-00313

[7] S. Sanjari *et al.* 2015 *Phys. Scr.* 014060

[8] X. Chen *et al.* 2016 *NIM A* **v** 826 pp 39-47

[9] X. Chen *et al. Hyperfine Interact.* 05-2015

[10] M. Lestinsky *et al.* 2016 *EPJ ST* 255 5

[1]E-mail: s.sanjari@gsi.de

IOP Conf. Series: Journal of Physics: Conf. Series **875** (2017) 092015 doi:10.1088/1742-6596/875/10/092015

Proton and α capture studies for nuclear astrophysics at GSI storage rings

Jan Glorius*[1], C.Langer†, Z.Slavkovská†, Y.Litvinov*, R.Reifarth†, C.Brandau*,°, B.Brückner†, X.Chen‡, T.Davinson§, P.Erbacher†, S.Fiebiger†, T.Gassner*, A.Gumberidze*, G.Gyürky+, K.Göbel†,*, M.Heil*, R.Hess*, P.M.Hillenbrand*, O.Hinrichs†, B.Jurado#, C.Kozhuharov*, D.Kurtulgil†, G.Lane$, C.Lederer-Woods§, M.Lestinsky*, S.Litvinov*, B.Löher*, F.Nolden*, N.Petridis*, U.Popp*, M.Reed$, S.Sanjari*, H.Simon*, U.Spillmann*, M.Steck*, T.Stöhlker*,¶, T.Szücs+, B.Thomas†, M.Träger*, H.Törnqvist*, S.Torilov&, C.Trageser*, S.Trotsenko*, M.Volknandt†, M.Weigand†, C.Wolf† and P.J.Woods§

*GSI Darmstadt, Germany | †Goethe University Frankfurt, Germany | ‡Institute of Modern Physics, Lanzhou, China
§University of Edinburgh, UK | +ATOMKI Debrecen, Hungary | #CENBG, France | °JLU Gießen, Germany
$Australian National University, Australia | ¶HI Jena, Germany | &Saint Petersburg State University, Russia

Synopsis The $^{124}Xe(p,\gamma)$ reaction has been measured for the first time at energies around the Gamow window by using stored ions at the ESR facility. The desired beam energies below 10 MeV/u introduce new experimental challenges like windowless ions detection under UHV conditions, extremely short beam lifetimes and efficient beam deceleration and cooling, all of which have been successfully met.

In the nucleosynthesis of the so-called *p* nuclei radiative capture reactions like (p,γ) or (α,γ) play an important role to model the reaction network and to explain the stellar production yields in different explosive scenarios [1]. Most of the key reactions involve radioactive nuclei [2] and can be studied solely in inverse kinematics. The GSI facility offers the unique possibility to produce such exotic ions and to store them in the experimental storage ring, ESR, and eventually in the CRYRING. This setting allows one to use the limited intensities available for radioactive ions with maximum efficiency.

In a first step, fully-stripped ions are stored at beam energies below 10 MeV/u. Subsequently, nuclear reactions are introduced by colliding the stored ions with the internal jet target that consists of either hydrogen or helium gas for (p,γ) or (α,γ) reactions, respectively. Ions, which capture a proton or an α particle at the target, are separated from the stored beam in the next dipole magnet and are detected by UHV compatible double-sided silicon-strip detectors (DSSSD) with a 100% efficiency. Due to atomic interactions the lifetime of the highly charged beam is on the order of seconds. However, the dominant mechanism reflected by this is the well-known radiative electron capture (REC), which can be used for normalization by employing high resolution x-ray spectroscopy around the target.

The very first measurement in the ESR was performed with a beam of stable $^{124}Xe^{54+}$ ions decelerated to and stored at energies between 5.5 MeV/u and 8 MeV/u to study the reaction $^{124}Xe(p,\gamma)^{125}Cs$. The spacial resolution of the employed DSSSD allowed a clear identification of the (p,γ) signal sitting on a background of elastically scattered ions, as shown in Fig. 1. Similar signals could be identified for five different beam energies in the aforementioned energy range. The analysis of the data set is ongoing and preliminary results will be presented.

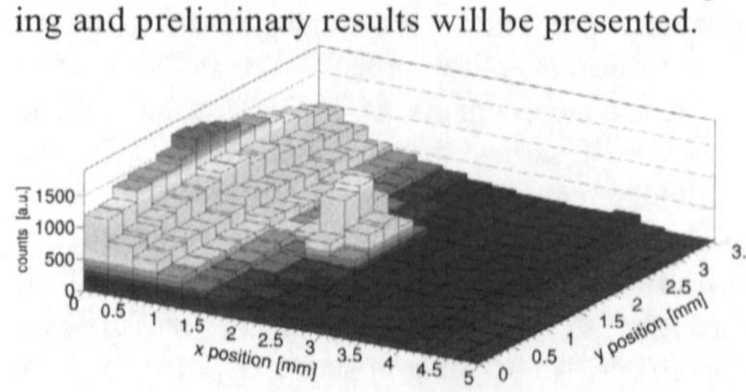

Figure 1. 2D map of the DSSSD data taken at 7 MeV/u. The peak contains the (p,γ) products sitting on a distribution of elastically scattered ions.

In the future, first reaction studies on radioactive nuclei are planned and will be carried out using the ESR setup described above. For energies below 4 MeV/u the newly installed CRYRING facility [3] is ideally suited to serve as a low-energy extension of the ESR. Corresponding experimental equipment is already being designed and will be ready for first experiments in 2018.

This project has received funding from the ERC (grant agreement No 682841 "ASTRUm").

References

[1] M. Pignatari *et al.* 2016 *Int. J. Mod. Phys. E* **25** 1630003
[2] T. Rauscher *et al.* 2016 *Mon. Not. R. Astron. Soc.* **463** 4153
[3] M. Lestinsky *et al.* 2016 *EPJ ST* **255** 5

[1] E-mail: j.glorius@gsi.de

Stopping and charge state revolution of low-energy ion beam in plasma

Y.Zhao*, †, 1, R.Cheng†, X.Zhou†, Y.Wang†, Y.Lei†, G. Xu*, J. Ren*, D.Hoffmann‡, S.Savin§, R.Gavrilin§, A.Golubev§, Z. Hu¶, F. Gao¶, Y.N. Wang¶ and G.Xiao*

*School of Science, Xi'an Jiaotong University, Xi'an 710049, China
† Institute of Modern Physics, Chinese Academy of Sciences, Lanzhou 730000, China
‡ Technique University Darmstadt, Darmstadt 64289, Germany
§Institute of Theoretical and Experimental Physics, Moscow, Russia
¶Dalian University of Technology, Dalian, China

Synopsis We have studied the energy loss and charge state revolution of ions in plasma. The beam energy ranged from 50-200keV/u. In future, the charge exchanging process of an ion in a plasma will be investigated.

Due to the strong non-linear effects and the special importance in ICF research, ion-beam and plasma interaction has attracted a lot of attention, especially for ion-beams in the low energy range and/or for plasma with high intensity. Herein we address the recent progress on the energy loss measurement in the case of hundreds of keV proton and helium beams passing a gas-discharging plasma target.

The experiments were carried out at the HV-ECR (High Voltage Electronic Cyclotron Resonance ion source) platform at IMP, where both proton and heavy ion beams with energy up to 320*q keV (q is the charge state of the ions) can be provided. After passing through the plasma target, the ion beam is analyzed by a bending magnet and reaches a fast-gated position-sensitive detector with minimum gate of 10ns (in experiments the gate was set as 100-1000ns to ensure sufficient statistics). The energy resolution of the setup is around 1%, depending on the beam size and the beam divergence (see details in references [1-3]).

Figure 1 shows the experimental result of the energy loss for a 400keV helium beam penetrating the hydrogen gas-discharge plasma target (with initial gas pressure of 3mbar, and discharging high-Voltage of 3kV) in terms of time after discharging, the energy losses calculated with different models are shown as well. As can be seen in the figure, the measurements of energy loss in plasma were much lower that the theoretical predictions.

It was also found that there were quite a few He^{1+} ions after the He^{2+} ion beam passed though the plasma, so that the effective charge state for the calculations should not simply take the nuclear charge 2. By fitting the data with an effective charge state of 1.7, the Bethe model fits well with the experimental data.

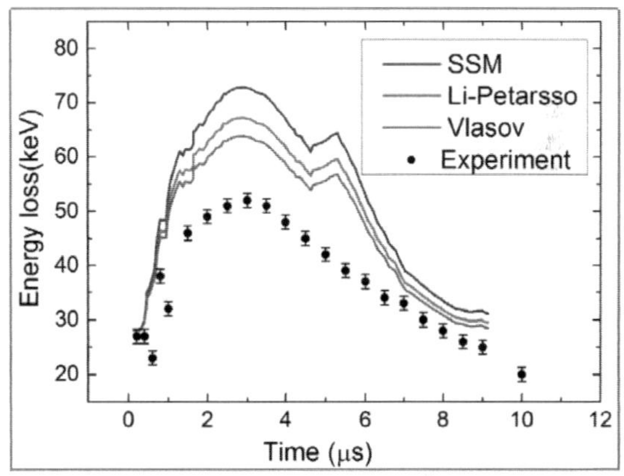

Figure 1. Energy loss for 400keV helium ion beam penetrating the hydrogen gas-discharge plasma target

In the future, we plan to measure the charge state distribution for the ion beam after pass though the plasma, and study the equilibrium time and the equilibrium charge state by varying the plasma density in a lower density regime.

This work has been supported by NSFC China, (No. U1532263, 11275241 and 11375034) and the Science Challenge Program on HED.

References

[1] Zhao Y. *et al* 2012 *Laser and Particle Beams* 30 679–706
[2] Cheng R. *et al* 2013 *Phys. Scr.* T156 014074
[3] Y. Zhao *et al* 2014 *High Power Laser Science and Engineering* 2 e39

[1]E-mail: zhaoyt@impcas.ac.cn

Storage test of a table-top electrostatic ion storage ring (μE-ring)

Jun Matsumoto[1], **Ryujiro Saiba, and Haruo Shiromaru**

Department of Chemistry, Tokyo Metropolitan University, Hachioji, Tokyo 192-0397, Japan

Synopsis A table-top electrostatic ion storage ring (μE-ring) with the circumference of about 0.8 m was developed. The first storage of ions, about 2 ms storage under vacuum of about 1 × 10^{-5} Pa, was achieved for 4 keV Ar^+ ions.

An electrostatic ion storage ring (E-ring) is a powerful device for investigation of slow reactions, such as metastable decay and radiative cooling of molecular ions. The first table-top E-ring was constructed by Lyon group [1]. We developed a second table-top E-ring (μE-ring), of which the size is reduced ten times from that of the existing ring at Tokyo Metropolitan University (TMU E-ring) [2], keeping the basic idea for beam manipulation unchanged.

The μE-ring consists of two 160 degree deflectors (160Def-1~2), four 10 degree deflectors (10Def-1~4) and four focusing and defocusing electrostatic quadrupole doublets (Q-1~4), as shown in figure 1. The circumference of an ion beam trajectory is about 0.8 m. All the electrodes are made of aluminum, mounted on a single rectangle plate of 480 mm × 200 mm, for precise alignment of the electrodes and efficient bake-out of the ring. Voltages of all the electrodes are provided from high-voltage modules controlled by a personal computer. Presently, the vacuum pressure is about 5 × 10^{-5} Pa with a 400 L/s turbo molecular pump.

Ion storage tests were performed by using Ar^+ ions at an energy of 4 keV extracted from an electron impact ion source. Typical current of the incident ion beam was about 100 nA. A pulsed voltage on 10Def-1 changes from injection to storage mode. Rotation period of the ions in the ring is about 6.3 μs. During ion storage, neutral argon atoms were produced due to collisions of Ar^+ ions with residual gas. When the neutralization occurs in the straight section (SS-2), the neutrals were detected by the secondary electron multiplier (SEM) in the forward direction of the SS-2.

Figure 2 shows neutral argon yield as a function of storage time. Background coming from the SEM was subtracted. The ions were stored for at least a few milliseconds (the ions turned round a few hundred times). In the early stage of storage, a fast decay appeared due to loss of ions with unstable trajectories. Then, a slow exponential decay with a lifetime of about 2 ms is observed, corresponding to collision with the residual gas.

The storage test has so far been conducted rather low vacuum. In near future, operation under better vacuum will be tested to store ions for much longer storage.

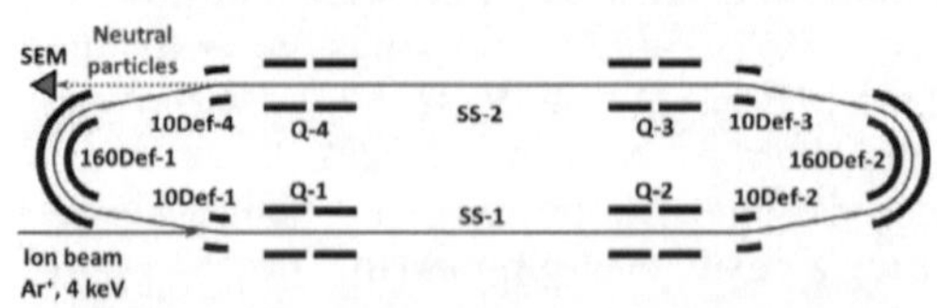

Figure 1. Schematic drawing of μE-ring. All electrodes are mounted on a single rectangle plate of 480 mm × 200 mm. 4 keV Ar^+ ion beam is injected from the lower left and makes a counterclockwise turn.

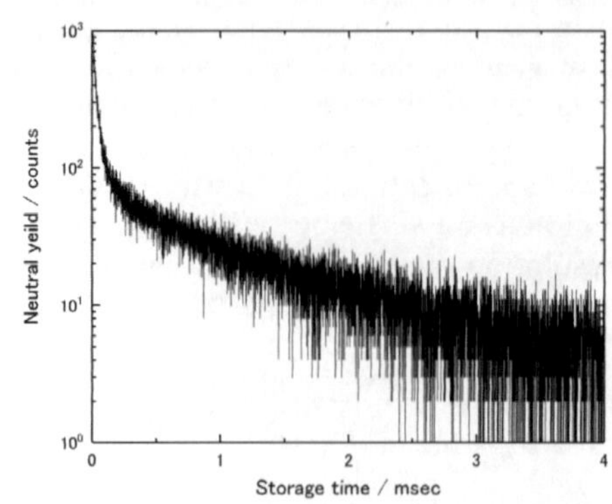

Figure 2. Neutral argon yield as a function of storage time. Background caused by the SEM was subtracted. Vacuum in the ring was about 5 × 10^{-5} Pa.

References

[1] J. Bernard *et al.* 2008 *Rev. Sci. Instrum.* **79** 075109.
[2] S. Jinno *et al.* 2007 *Nuecl. Instrum. Meth. A* **572** 568.

[1] E-mail: junmatsu@tmu.ac.jp

State-selective electron capture in He^+ + He collisions at intermediate energies

D. L. Guo*, X. Ma*[1], R. T. Zhang*, S. F. Zhang*, X. L. Zhu*, W. T. Feng*, Y. Gao*, B. Hai*†, M. Zhang*†, H. B. Wang*†, and Z. K. Huang*†

* Institute of Modern Physics, Chinese Academy of Sciences, Lanzhou 730000, China
† University of Chinese Academy of Sciences, Beijing 100049, China

Synopsis A combined experimental and theoretical study on single capture in 30- and 100-keV He^+ on He collisions was performed. We report on experimental state-selective cross sections and projectile angular-differential cross sections and compared with the theoretical results by using dynamic screening classical trajectory Monte Carlo method (dCTMC), and the significance of different electron-electron correlations are discussed. The structure observed in the projectile angular-differential cross sections is explained in terms of Fraunhofer-type diffraction of the He^+ projectile de Broglie wave.

The studies of charge exchange processes using fully stripped ions as projectiles have been extensively made in the past. Processes involving dressed projectiles therefore deserve special attention. In this work, we performed a combined experimental and theoretical investigation on the single capture at 30 and 100 keV impact energies for He^+ + He collisions.

In the present experiment, the recoil ion He^+ was recorded in coincidence with the scattered projectile He by using a reaction microscope. The state-selective cross sections and the projectile angular-differential cross sections have been extracted from the measurements [1]. We show the state selective cross sections as a function of the impact energy in Figure 1. It was found that the ground-state transfer process is dominant over excited-state transfer and transfer excitation processes. A comparison of the experimental results with our dCTMC calculations shows a reasonable agreement.

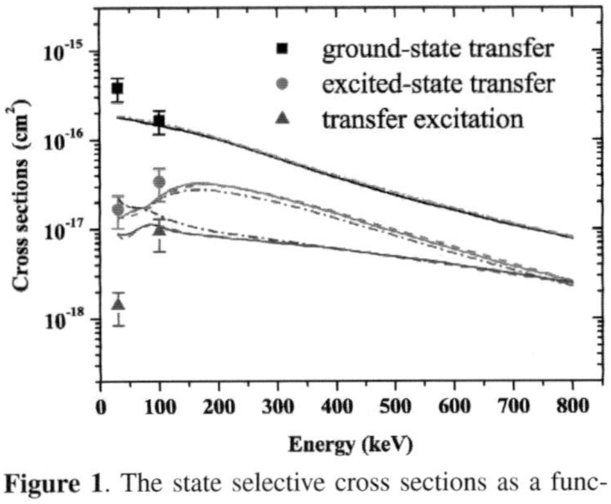

Figure 1. The state selective cross sections as a function of impact energy. Solid symbols: experiment; solid lines: dCTMC results; dash lines: dCTMC-c results; dash dot lines: CTMC-s results (see details in Ref. [1])

In Figure 2 we show the projectile angular distributions for ground-state transfer. With the impact parameter range obtained from dCTMC calculations, the positions of the first dark and first bright fringes expected from the Fraunhofer-type diffraction theory are indicated by arrows. The strikingly good agreement of the positions of arrows and the experimental results suggests that the oscillation structure appearing in the angular-differential cross sections originates from atomic-size Fraunhofer-type diffraction.

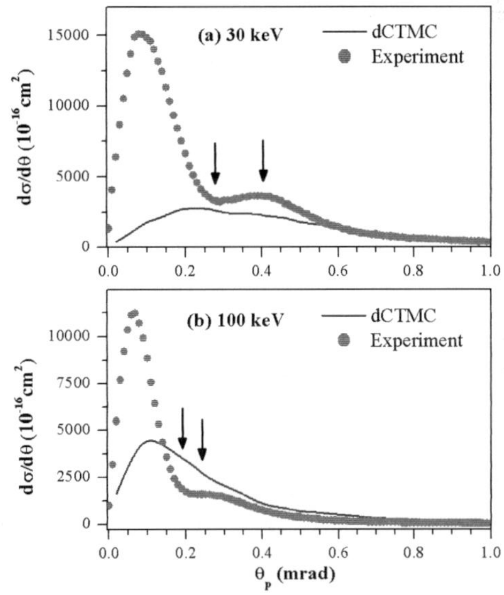

Figure 2. The projectile angular-differential cross sections for ground-state transfer.

This work was supported by the National Natural Science Foundation of China under Grants No. 11504387 and No.11574327.

References

[1] D. L. Guo *et al.* 2017 *Phys. Rev. A* **95** 012707

[1]E-mail: x.ma@impcas.ac.cn

IOP Conf. Series: Journal of Physics: Conf. Series **875** (2017) 092020 doi:10.1088/1742-6596/875/10/092020

Relativistic coupled-channel calculations of differential ionization cross sections

A. I. Bondarev[*,†,‡1], **Y. S. Kozhedub**[†], **I. I. Tupitsyn**[†], **V. M. Shabaev**[†], **and G. Plunien**[§]

[*] Center for Advanced Studies, Peter the Great St. Petersburg Polytechnic University, 195251 St. Petersburg, Russia
[†] Department of Physics, St. Petersburg State University, 199034 St. Petersburg, Russia
[‡] NRC "Kurchatov Institute" - ITEP, 117218 Moskow, Russia
[§] Institut für Theoretische Physik, Technische Universität Dresden, D-01062 Dresden, Germany

Synopsis A relativistic method for calculation of differential cross sections for ionization in ion-atom collisions is developed. It is applied to ionization of atomic hydrogen by antiproton impact, where several discrepancies in available predictions are resolved. The relativistic effects are studied for ionization of hydrogenlike xenon ion under the impact of carbon nuclei.

Recent progress in experimental technique of the recoil-ion and electron momentum spectroscopy [1] stimulates theoretical studies of the fully differential cross sections (FDCS) for ionization in ion-atom collisions.

In this contribution, we present a semiclassical relativistic method for calculation of the FDCS for ionization in ion-atom collisions. The method can be applied to collisions involving heavy targets, since the Dirac equation is used to describe electron dynamics. The active-electron approximation is implied for the target description, and the projectile is assumed to be a bare ion. The method is based on the finite-basis-set expansion of the wave function of the active electron. The basis functions are constructed within the dual-kinetically-balanced B-spline approach. The Fourier transform is employed to extract the FDCS for a given projectile momentum transfer.

As the first application, the method is utilized to calculate FDCS for antiproton-impact ionization of atomic hydrogen. This collisional system is very convenient for theoretical study, since it is a pure three particle system without charge-exchange and many-electron processes. An experimental investigation of the FDCS in this collision is not possible at the moment. Nevertheless, the ionization cross sections have already been extensively studied theoretically by various non-perturbative methods [2, 3, 4, 5]. Their results are in overall agreement. However, several ionization cross sections, predicted by these methods, considerably differ. Within our independent calculation [6], we can give preference to the results of certain approaches.

Furthermore, the impact-parameter dependencies of the total ionization probabilities from the K- and L-shells have been calculated for the 100 MeV/u C^{6+}-Xe^{53+} collision. In order to explore the relativistic effects induced by a large target charge, we as well carried out the calculation in the non-relativistic limit, where the standard value of the speed of light c was multiplied by a factor of 1000. The comparison of the results of both calculations is shown in the figure. Here the total ionization probabilities averaged over the magnetic quantum number of the initial state are plotted.

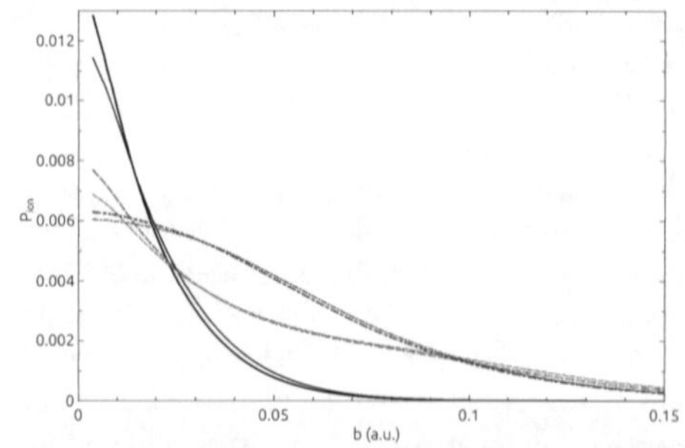

Figure 1. Impact-parameter dependence of the total ionization probability in the 100 MeV/u C^{6+}-Xe^{53+} collision for various initial states. Solid line, 1s; dashed line, 2s; dash-dotted line, 2p. Thick (thin) lines correspond to the results of the relativistic (non-relativistic) calculation.

This work was supported by SPSU-DFG (Grant No. 11.65.41.2017), DFG (Grant No. STO 346/5-1), RFBR-NSFC (Grant No. 17-52-53136), and by the Ministry of Education and Science of the Russian Federation (Grant No. 3.1463.2017). A.I.B. acknowledges support from the FAIR-Russia Research Center.

References

[1] J. Ullrich *et al.* 2003 *Rep. Prog. Phys.* **66** 2003

[2] M. McGovern *et al.* 2009 *Phys. Rev. A* **79** 042707

[3] I. B. Abdurakhmanov *et al.* 2011 *J. Phys. B* **44** 165203

[4] M. F. Ciappina *et al.* 2013 *Phys. Rev. A* **88** 042714

[5] I. B. Abdurakhmanov *et al.* 2016 *Phys. Rev. A* **94** 022703

[6] A. I. Bondarev *et al.* 2017 *Phys. Rev. A* **95** 052709

[1]E-mail: bondarev@spbstu.ru

IOP Conf. Series: Journal of Physics: Conf. Series **875** (2017) 092021 doi:10.1088/1742-6596/875/10/092021

Frontiers of atomic physics with highly charged heavy ions at HIAF

X. Ma*,[1], W. Q. Wen*, S. F Zhang*, Z. K. Huang*,†, H. B. Wang*,†, J. Yang*, X. L. Zhu*, D. Y. Yu*, R. Cheng*, X. Cai*, L. J. Mao*, J. C. Yang*, L. N. Sheng*, B. Wu*,†, X. H. Zhou*, H. S. Xu*, Y. J. Yuan*, J. W. Xia*, H. W. Zhao*, G. Q Xiao*, and W. L Zhan*

* Institute of Modern Physics, Chinese Academy of Sciences, Lanzhou, 730000, China
† University of Chinese Academy of Sciences, Beijing, 100049, China

Synopsis High Intensity heavy ion Accelerator Facility (HIAF) is a new project currently being under design and construction in China. HIAF will provide beams of stable and unstable heavy ions with high energies, high intensities and high quality. An overview of atomic phys-ics program using highly charged ions and radioactive heavy ions at HIAF is presented.

The High Intensity heavy ion Accelerator Facility (HIAF) is a new project which was approved by the Chinese government on 31 December 2015 [1]. The construction of HIAF is expected to be completed in 2024 and its commissioning is scheduled on 2025. The HIAF locates in the coast area in Huizhou city of Guangdong Province of South China.

As shown in Figure. 1, the HIAF project consists of ion-Linac (iLinac) with a length of 100m, a Booster Ring (BRing), a radioactive beam line of fragmentation type, a spectrometer ring (SRing), and several experimental terminals at low- and high-energy ends as well as setups for in-ring experiments. Some key parameters of each accelerator are summarized in Table 1. The rings will be equipped with electron cooler and stochastic cooling devices. The marked experimental terminals in figure 1 include, ① Nuclear structure spectrometer, ② Low energy irradiation, ③ RIBs beam line, ④ High precision spectrometer ring, ⑤ External target station ⑥ electron-ion recombination spectroscopy, ⑦ Ion-Ion Merging, ⑧ Warm Density Matter Physics, ⑨ High energy irradiation target. At low energy branch together with the nuclear structure terminals, a collinear laser spectrometer will also be built.

New possibilities for atomic physics experiments are foreseen at HIAF [2]. The in-ring experiments include a dedicated electron target for dielectronic recombination studies, an internal target for spectroscopy and relativistic collision dynamics experiment, a setup for laser cooling and laser spectroscopy studies, and a double-ToF detector system for precision mass measurements. Unstable ions far from stability will be used for precision spectroscopy at HIAF. In the meantime, the international FAIR project, which is being built in Germany, will offer unique experimental opportunities for atomic physics [3].

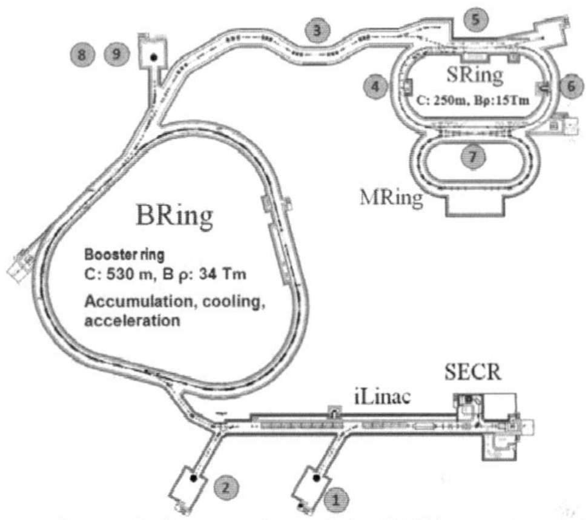

Figure 1: An overview of the HIAF complex.

Table 1 Key parameters of HIAF

	Ions	Energy	Intensity
SECR	U^{34+}	14 keV/u	0.05 pmA
iLinac	U^{34+}	17 MeV/u	0.028 pmA
BRing	U^{34+}	0.8 GeV/u	1.0×10^{11}ppp
SRing	U^{92+}	0.8 GeV/u	1.0×10^{11}ppp

References

[1] J. C. Yang *et al.* 2013 *Nucl. Instrum. Meth. B* **317** 263
[2] X. Ma *et al.* 2017 *Nucl. Instrum. Meth. B* submitted
[3] T. Stöhlker and Y.A. Litvinov 2015 *Phys. Scr.* **T166** 014025

[1] E-mail: x.ma@impcas.ac.cn

IOP Conf. Series: Journal of Physics: Conf. Series **875** (2017) 092022 doi:10.1088/1742-6596/875/10/092022

Molecular effects in M-shell ionization by slow light ions

M. Pajek*[1], D. Banaś*, Ł. Jabłoński* and T. Mukoyama†

* Institute of Physics, Jan Kochanowski University, 25-406 Kielce, Poland
† Institute for Nuclear Research of the Hungarian Academy of Sciences (ATOMKI), H-4026 Debrecen, Hungary

Synopsis The studies of M-shell ionization, or M-X-ray emission, by slow light ions indicate rather strong, up to a factor of 3-5, underestimation of the measured cross sections by the available theoretical calculations describing direct Coulomb ionization by charged particle impact. Guided by our earlier systematic measurements of M-shell ionization cross sections by ^{1,2}H and ^{3,4}He ions we demonstrate here that for the low energy ions, i.e. in the adiabatic regime, the electron binding energy does not saturates at the united atoms (UA) limit, commonly used in the calculations, but instead it has a smaller value corresponding to a transient molecule (MO) formed in collision. This effect, when described in the one-electron molecule approximation, can explain the observed discrepancies between the measured and calculated M-shell ionization cross sections.

Inner-shell ionization by charged particles is a process of both fundamental interest and in applications, namely, for the particle induced X-ray emission (PIXE) technique which needs accurate knowledge of the M-X-ray production cross sections by light ions. Based on our earlier systematic measurements of the X-ray production cross sections for dominating M-X-ray lines and the M-shell ionization cross sections for heavy atoms bombarded by hydrogen ^{1,2}H and helium ^{3,4}He ions in the energy range 0.1-1.0 MeV/amu [1,2], as well as the recent results reported by other authors [3-5], here we discuss the observed underestimation of the measured M-shell ionization cross sections by available calculations for the lowest energies. The existing theories treating the direct Coulomb ionization process, expected to dominate for asymmetric collisions, are based on the plane wave Born approximation (PWBA), semiclassical approximation (SCA) and the binary encounter approximation (BEA). The PWBA approach was further developed to include the higher-order effects and is known as the ECPSSR [6] or ECUSAR [7] theories. In the ECPSSR theory the electron binding effects is treated within the perturbed stationary state approximation (PSS), while in the ECUSAR the saturation of the binding effect at the united atom limit (UA) was introduced for the lowest ion energies.

Generally, the measured M-subshell X-ray production cross sections were found [1-5] to be systematically up to a factor of 3-5 higher than the predictions of the ECPSSR and ECUSAR theories at low energies. For this reason the present data on the M-subshell ionization by light ions were compared with the predictions of the SCA [8] and BEA [9] calculations in order to investigate a role of the electronic wave function and the electron binding effects. In particular, to calculate the M-subshell ionization cross sections different electronic wave functions, from nonrelativistic hydrogenic to relativistic Dirac-Fock types, were used within the SCA while in the BEA both the nonrelativistic and relativistic hydrogenic and the Hartree-Fock-Roothan wave functions (see Ref. [9]) were used.

The electron binding effect was taken into account within the extreme separated-atom (SA) and united-atom (UA) limits as well as the proposed molecular (MO) model describing the variation of the electron binding energy with respect of separation between the projectile and target atom in the adiabatic regime. We found that within the molecular picture the united atom limit cannot be reached, but instead slightly smaller molecular electron binding energy has to be used to calculate the electron binding effect. Consequently, the experimental results for the M-shell ionization by light ions can be interpreted in terms MO effects using the SCA calculations [8] in order to explain the observed discrepancies, in particular, for the lowest impact energies.

References

[1] M. Pajek *et al.*2006 *Phys. Rev. A* **73** 012709
[2] M. Pajek et al. 1999 *Nucl. Instr. Meth. B* **150** 33
[3] L.C. Phinney *et al.*, 2009 *J. Phys. B* **42** 085202
[4] D.D. Cohen *et al.* 2014 *Nucl. Instr. Meth. B* **318** 11
[5] S.J. Cipolla 2014 *Nucl. Instr. Meth. B* **330** 66
[6] W. Brandt and G. Lapicki 1981 *Phys. Rev. A* **23** 1717
[7] G. Lapicki 2002 *Nucl. Instr. Meth. B* **189** 8
[8] D. Trautmann and T. Kauer 1989 *Nucl. Instr. Meth. B* **12** 426
[9] T. Mukoyama 2015 *Nucl. Instr. Meth. B* **354** 155

[1]E-mail: pajek@ujk.edu.pl

IOP Conf. Series: Journal of Physics: Conf. Series **875** (2017) 092023 doi:10.1088/1742-6596/875/10/092023

Electron-positron pair creation in collisions of heavy bare nuclei: One-center approach

R. V. Popov[*,**] **A. I. Bondarev**[♯,*,**], **I. V. Ivanova**[*], **Y. S. Kozhedub**[*], **I. A. Maltsev**[*,**] [1], **V. M. Shabaev**[*] **I. I. Tupitsyn**[*], **X. Ma**[†], **G. Plunien**[††], **and Th. Stöhlker**[‡,§,¶]

[*] Department of Physics, St. Petersburg State University, 199034 St. Petersburg, Russia
[**] NRC "Kurchatov Institute" - ITEP, 117218 Moscow, Russia
[♯] Center for Advanced Studies, Peter the Great St. Petersburg Polytechnic University, 195251 St. Petersburg, Russia
[†] Institute of Modern Physics, Chinese Academy of Sciences, 730000 Lanzhou, China
[††] Institut für Theoretische Physik, Technische Universität Dresden, D-01062 Dresden, Germany
[‡] GSI Helmholtzzentrum für Schwerionenforschung GmbH, D-64291 Darmstadt, Germany
[§] Helmholtz-Institute Jena, D-07743 Jena, Germany
[¶] Institut für Optik und Quantenelektronik, Friedrich-Schiller-Universität, D-07743 Jena, Germany

Synopsis The probabilities and cross sections of electron-positron pair creation in low-energy collisions of heavy bare nuclei are calculated. The calculations are performed in the framework of the one-center approach.

Low-energy heavy-ion collisions provide a unique tool for investigation of quantum electrodynamics in presence of the supercritical electromagnetic field [1]. To date nonperturbative calculations of the electron-positron pair creation are mainly confined within the monopole approximation [2, 3], in which the spherically symmetric part of the two-center potential is considered in the center-of-mass frame. In the present work we perform nonperturbative calculations of the electron-positron pair creation in the frame centered at one of the nuclei. The calculation method is based on numerical solution of the time-dependent Dirac equation in the basis of eigenfunctions of the static one-center Hamiltonian. These eigenfunctions are obtained by using the dual kinetic balance method [4] with B-splines as the basis functions.

As the first step, we calculated the probabilities of the electron-positron pair creation in the monopole approximation, with an electron being captured into the $1s$ state of the target nucleus. In order to take into account electron capture by the projectile nucleus, the obtained values were multiplied by two. In Table 1, the obtained results for the U^{92+}–U^{92+} collision at 6.2 MeV/u collision energy are compared with the data of Ref. [3]. Despite the fact that Ref. [3] presents the probabilities of the pair creation with an electron captured into all bound states, this comparison is well justified because the ground state gives the dominant contribution to the pair creation probability. As one can see from Table 1, our results are in a reasonable agreement with the data in Ref. [3].

Within the approximation under consideration, we have also calculated the cross section of the electron-positron pair creation in the U^{92+}–U^{92+} collision for the relative ion velocity $v = 0.1c$, where c is the speed of light. The obtained value is several times smaller than the one given by a rough estimate in Ref. [5]. Now we are working on the calculations of the pair-creation probabilities and the cross sections beyond the monopole approximation. The results of these calculations will be presented at the conference.

[1]E-mail: i.maltsev@spbu.ru

Table 1. Probability of the electron-positron pair creation in the collision of bare uranium nuclei as a function of the impact parameter b. The projectile energy is 6.2 MeV/u and the nuclear trajectory is of the Rutherford type.

b (fm)	This work	Maltsev *et al.* [3]
0	1.14×10^{-2}	1.25×10^{-2}
5	9.60×10^{-3}	1.05×10^{-2}
10	6.48×10^{-3}	7.03×10^{-3}
15	4.08×10^{-3}	4.39×10^{-3}
20	2.54×10^{-3}	2.70×10^{-3}
25	1.57×10^{-3}	1.66×10^{-3}
30	9.86×10^{-4}	1.03×10^{-3}
40	3.94×10^{-4}	4.09×10^{-4}

This work was supported by RFBR-NSFC (Grant No. 17-52-53136), SPSU-DFG (Grant No. 11.65.41.2017), and DFG (Grant No. STO 346/5-1). R.V.P., A.I.B., and I.A.M. also acknowledge the support of the FAIR-Russia Research Center.

References

[1] W. Greiner, B. Müller, and J. Rafelski 1985 *Quantum Electrodynamics of Strong Fields*, Springer-Verlag, Berlin

[2] U. Müller *et al.* 1988 *Phys. Rev.* A **37** 1449

[3] I. A. Maltsev *et al.* 2015 *Phys. Rev.* A **91** 032708

[4] V. M. Shabaev *et al.* 2004 *Phys. Rev. Lett.* **93** 130405

[5] I. B. Khriplovich 2016, *Int. J. Mod. Phys.* A **31** 28n29 1645035

IOP Conf. Series: Journal of Physics: Conf. Series **875** (2017) 092024 doi:10.1088/1742-6596/875/10/092024

Pair creation in low-energy collisions of heavy nuclei beyond the monopole approximation

I. A. Maltsev[*,**,1], **I. I. Tupitsyn**[*], **Y. S. Kozhedub**[*], **R. V. Popov**[*,**], **G. Plunien**[†], **V. M. Shabaev**[*], **and Th. Stöhlker**[‡,§,¶]

[*] Department of Physics, St. Petersburg State University, Universitetskaya naberezhnaya 7/9, 199034 St. Petersburg, Russia
[**] SSC RF ITEP of NRC "Kurchatov Institute", Bolshaya Cheremushkinskaya 25, 117218 Moscow, Russia
[†] Institut für Theoretische Physik, Technische Universität Dresden, Mommsenstraße 13, D-01062 Dresden, Germany
[‡] GSI Helmholtzzentrum für Schwerionenforschung GmbH, Planckstrasse 1, D-64291 Darmstadt, Germany
[§] Helmholtz-Institute Jena, D-07743 Jena, Germany
[¶] Theoretisch-Physikalisches Institut, Friedrich-Schiller-Universität Jena, D-07743 Jena, Germany

Synopsis The creation of electron-positron pairs in slow collisions of heavy nuclei is considered beyond the monopole approximation. The nonperturbative calculations of pair-creation probabilities are performed with taking into account the full two-center potential of colliding nuclei.

In low-energy collisions of heavy ions, the electromagnetic fields of extremely high strength can be achieved. It gives an unique opportunity for investigation of quantum electrodynamics in supercritical regime, where the vacuum becomes unstable [1]. To date the nonperturbative calculations of pair creation in such collisions were mainly restricted to the monopole approximation, according to which only the spherically-symmetric part of the two-center potential is taken into account [2, 3, 4]. In the present work, we calculate the pair-creation probabilities beyond this approximation.

Our approach is based on the numerical solving of the two-center time-dependent Dirac equation in a finite basis set. The basis functions are obtained from B-splines according to the dual-kinetic-balance technique for the systems with axial symmetry [5]. The employed method allows us to take into account the full two-center potential. However, the rotation of the internuclear axis is neglected that enables to consider the Dirac equation only in two dimensions.

In Table 1, we present the obtained results for pair-creation probabilities with an electron captured in a bound state (bound-free pair creation). The calculations have been performed for the collision of bare thorium nuclei ($Z = 90$) at energy $E = 6$ MeV/u. The obtained values are compared with the corresponding ones calculated in monopole approximation using the method described in Ref. [4]. As one can see from the table, the results are close to each other. However, the two-center values are systematically larger than the monopole ones. One can also notice that the difference grows with the value of the impact parameter. It can be explained by the fact that the monopole approximation works better for smaller internuclear distances.

The pair creation with the electron in the continuum requires more cumbersome calculations. These calculations are currently underway and the results will be presented at the conference.

This work was supported by SPSU-DFG (Grant No. 11.65.41.2017), DFG (Grant No. STO 346/5-1) and FAIR-Russia Research Center. I.A.M. also acknowledges the support of the Dynasty foundation and the German-Russian Interdisciplinary Science Center (G-RISC).

Table 1. Probability of the bound-free pair creation in the collision of bare thorium nuclei as a function of the impact parameter b. The calculations have been performed for collision energy $E = 6$ MeV/u.

b (fm)	Two-center potential	Monopole potential
0	6.43×10^{-3}	6.18×10^{-3}
5	5.46×10^{-3}	5.21×10^{-3}
10	3.76×10^{-3}	3.52×10^{-3}
15	2.41×10^{-3}	2.20×10^{-3}
20	1.52×10^{-3}	1.35×10^{-3}
25	9.69×10^{-4}	8.29×10^{-4}

References

[1] W. Greiner, B. Müller, and J. Rafelski, *Quantum Electrodynamics of Strong Fields*, (Springer-Verlag, Berlin, 1985).

[2] U. Müller *et al.* 1988 *Phys. Rev. A* **37** 1449

[3] E. Ackad and M. Horbatsch 2008 *Phys. Rev. A* **78** 062711

[4] I. A. Maltsev *et al.* 2015 *Phys. Rev. A* **91** 03270

[5] E. B. Rozenbaum *et al.* 2014 *Phys. Rev. A* **89** 012514

[1]E-mail: i.maltsev@spbu.ru

IOP Conf. Series: Journal of Physics: Conf. Series **875** (2017) 092026 doi:10.1088/1742-6596/875/10/092026

Energy straggling cross section for antiproton-atom collisions

X.-M. Tong[*], **S. Borbély**[§], **I. Březinová**[†], **F. Lackner**[†], **S. Nagele**[†], **J. Feist**[¶], **L. Nagy**[§], **K. Tőkési**[‡], **N. Toshima**[*], **J. Burgdörfer**[†]

[*] Center for Computational Sciences, University of Tsukuba, Tsukuba, Ibaraki 305-8757, Japan
[§] Faculty of Physics, Babes-Bolyai University, 400084 Cluj-Napoca, Romania, EU
[†] Institute for Theoretical Physics, Vienna University of Technology, 1040 Vienna, Austria, EU
[¶] Departamento de Física Teórica de la Materia Condensada and Condensed Matter Physics Center (IFIMAC), Universidad Autónoma de Madrid, E-28049 Madrid, Spain, EU
[‡] Institute for Nuclear Research, Hungarian Academy of Sciences, Debrecen, Hungary, EU and ELI-ALPS, ELI-HU Non-profit Ltd., Dugonics tér 13, H-6720 Szeged, Hungary, EU

Synopsis We present benchmark simulations of the energy straggling cross section for antiprotons colliding with hydrogen and helium. We solve the one- and two-particle Schrödinger equation numerically within the framework of the straight-line impact-parameter approximation. Our results provide first evidence for the convergence of the *ab-initio* simulation toward the classical Bohr straggling cross section proposed a century ago.

One frequently employed key observable characterizing the collisions of charged particles with matter, both in the gas phase and condensed phase, is the stopping power or stopping cross section introduced more than a century ago by N. Bohr and intensely studied ever since. It characterizes the first moment of the energy loss distribution or, in case of atomic collisions, the collisional energy transfer to the target nucleus and the electronic excitation and ionization spectrum. Surprisingly, unlike the stopping cross section, the higher moments of the spectral distribution have not been systematically studied. In particular, *ab-initio* simulations of the straggling cross section characterizing the second moment of the energy loss distribution appear to be still missing.

We present first numerical results for straggling for gas phase collisions of antiproton with hydrogen and helium. Antiproton projectiles are of fundamental interest as they allow to probe the pure Coulomb excitation and ionization spectrum without admixture of the charge transfer channel. The main computational challenge is the accurate coverage of the large spectral width of the ionization spectrum at high impact energies. The ionization cross section peaks near threshold ($\varepsilon \approx 0$) while the binary encounter peak extends to $\varepsilon \approx 20$keV electron energy (e.g., for 10MeV impact energy). The higher moments sensitively depend on the tails of this wide distribution. We employ a straight-line impact parameter approximation for the $\bar{p}$. The time evolution of the electronic wave function is propagated by the split-operator method with a generalized pseudospectral method in the energy representation as described in Ref. [1]. In the present simulation we improve the method by projecting the ionized electron onto the Coulomb continuum wave function at given time intervals Δt which are determined by the fastest components of the electronic wave packet. We have tested the convergence against time step, space grid structure as well as the number of partial waves included. The latter is of particular importance for higher moments of the ionization spectrum. For example, at 5MeV impact energy, the total inelastic cross section converges, within 1% error, with inclusion of 30 partial waves. However, to reach the convergence for energy loss already 90 partial waves and for straggling even 150 partial waves are required. For energy loss cross section (Fig.1), our results are in reasonable agreement with the recent results by Lühr and Saenz [2]. The straggling cross section monotonically increases with impact energy and converges to the classical Bohr straggling cross section σ_{bs} at high energies. Extension to antiproton-helium collisions and the influence of correlations effects will be presented at the conference.

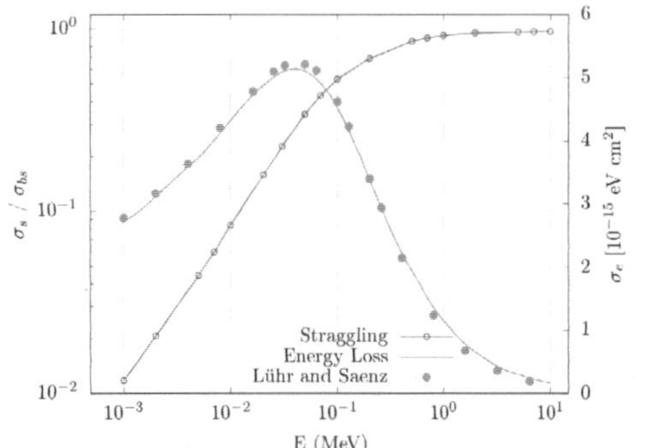

Figure 1. Energy loss σ_e (green) and straggling σ_s (purple) cross sections in units of the Bohr straggling cross section σ_{bs} of antiproton colliding with hydrogen.

References

[1] X. M. Tong *et al.* 2001 *Phys. Rev. A* **64** 022711
[2] A. Lühr and A Saenz, 2009 *Phys. Rev. A* **79** 042901

IOP Conf. Series: Journal of Physics: Conf. Series **875** (2017) 092027 doi:10.1088/1742-6596/875/10/092027

Upgrade of the Main Magnetic Focus Ion Trap in Giessen

Marc Keil*, Stefan Schippers*, Alfred Müller† and Alexander Borovik, Jr.*[1]

* I. Physikalisches Institut, Justus-Liebig-Universität Gießen, 35392 Giessen, Germany
† Institut für Atom- und Molekülphysik, Justus-Liebig-Universität Gießen, 35392 Giessen, Germany

Synopsis The Main Magnetic Focus Ion Trap installed in Giessen is being upgraded aiming at the extension of the range of producible ion species. The presently reported effort includes the installation of a system for extra fine-dosed introduction of gaseous parent materials into the trap region and the arrangement for the possibility of periodic dumping of the ion trap.

The Main Magnetic Focus Ion Trap (MaMFIT), a compact tool for spectroscopy of highly-charged ions [1], has been recently installed in Giessen and has already been employed for investigation of dielectronic recombination in highly-charged iridium ions [2]. The present construction, however, enables the experimental access almost exclusively to the ions of the sputtered cathode material, which, presently, is iridium. The main reason for this restriction is the process of evaporative cooling, in which heavier ions are cooled in Coulomb collisions with lighter ions resulting in the former being trapped for much longer time (and, thus, in higher charge states), while the latter mostly leave the trap after gaining additional kinetic energy before their high ionization stages can be achieved.

We report on an effort aimed at the enhancement of the versatility of the MaMFIT in Giessen thus enabling experimental access to ions of a wider spectrum of elements. It includes the arrangement for periodic dumping of the local ion trap by changing the potential on the drift tube thus preventing the trap from being filled with ions of the cathode material. For Electron Beam Ion Traps (EBIT), periodic time spans between the trap dumps are typically in the range from 0.1 to few seconds. In the case of MaMFIT, the electron current density in the small local trap is predicted to be higher than in typical EBITs. The corresponding required dumping time periods are, therefore, expected to be shorter. Another aspect of the presently reported upgrade is the system for supplying gaseous parent elements. When aiming at production of ions of injected elements, a sufficient amount of the parent gas should be introduced into the trap region, however, with smallest possible influence on the overall pressure in the setup. Employing the experience with the available ballistic gas inlet system installed on the EBIT at NIST [3], an analogous system adjusted for the present MaMFIT setup has been designed, built and installed. Fig. 1 shows the CAD model of the drift tube region of the present setup. With the described upgrade, the Giessen MaMFIT setup is currently being prepared for test measurements of spectra of highly-charged noble-gas ions.

Figure 1. The fragment of the CAD model of the MaMFIT setup in Giessen showing the drift tube arrangement together with the gas inlet system.

References

[1] V. P. Ovsyannikov 2014 *arXiv* 1403.2168.

[2] A. Borovik, Jr. et al., to be published.

[3] K. Fahy 2007 *Phys. Rev. A* **75** 032520.

[1]E-mail: Alexander.Borovik@physik,uni-giessen.de

IOP Conf. Series: Journal of Physics: Conf. Series **875** (2017) 092028 doi:10.1088/1742-6596/875/10/092028

COLTRIMS studies on singly ionizing ion-atom collisions: The roles of the projectile and target coherence

Konstantin A. Kouzakov[*1]

[*] Department of Nuclear Physics and Quantum Theory of Collisions, Faculty of Physics, Lomonosov Moscow State University, Moscow 119991, Russia

Synopsis A theoretical analysis of the so-called projectile and target coherence in COLTRIMS experiments on single ionization of atoms by fast ions is presented. Using time-dependent quantum scattering theory, it is shown that the projectile coherence has no serious impact on the measured fully differential cross sections. In contrast, the effects of the target coherence might be of greater importance in the discussed experiments.

Cold target recoil ion momentum spectroscopy (COLTRIMS) is a reaction microscope that allows high-precision measurement of fully differential cross sections (FDCS) of various atomic fragmentation processes. Of special interest are the COLTRIMS experiments on single ionization of helium by fast ionic projectiles [1, 2, 3], particularly due to the claim that for explaining their results one should account for the projectile-coherence effects.

As formulated by Wang et al. [2], in analogy to classical optics and in accordance with Huygens' principle, the projectile transverse coherence length is given by $\Delta x \approx \lambda L/2a$, where a and L are the width of the collimating slit and its distance to the target, respectively, and λ is the de Broglie wavelength of the projectile. If the projectile coherence length is larger (smaller) than the size of the target atom, the projectile is coherent (incoherent). For example, the transverse coherence length of the projectile beam in the 100-MeV/u C^{6+}+He experiment [1] was estimated as $\Delta x \approx 10^{-3}$ a.u. [2], thus suggesting that the C^{6+} projectiles were strongly incoherent in that experiment. From this fact it follows that the usual time-independent quantum scattering theory is not directly applicable, because this theory is based on the assumption that the projectile wave packet is much broader than the spatial extent of the target.

The proper theoretical framework for inspecting the effects of the projectile coherence is quantum scattering theory of wave packets. It treats a time-dependent collision of the projectile and target wave packets, as opposed to the time-independent formulation of scattering theory that deals with stationary plane waves. The time-independent formulation follows from the time-dependent one in the limit of the perfect localization of wave packets in momentum space. Suppose that the initial state of the ionic projectile in momentum space, as it is prepared in a COLTRIMS experiment, is given by the wave packet $\Phi_p(\mathbf{q}_p)$. The projectile coherence is directly related to the properties of this wave packet, namely the projectile becomes fully coherent when $|\Phi_p(\mathbf{q}_p)|^2 \to (2\pi)^3\delta(\mathbf{q}_p - \mathbf{k}_p)$, where $\mathbf{k}_p$ is the centroid of the wave packet. The same consideration applies to the target coherence, which is determined by the initial wave packet $\Phi_t(\mathbf{q}_t)$ that describes the state of the center of mass of the He atomic target in momentum space.

The performed theoretical analysis [4] involves general arguments based on quantum scattering theory of wave packets. The general formula for FDCS has been derived, where the projectile wave packet appears to be canceled. In other words, it does not support the claim that the projectile-coherence effects play an important role in atomic ionization by fast ions studied with COLTRIMS. In contrast, the incoherent effects can arise only due to the target center-of-mass wave packet. The detailed properties of this wave packet depend on experimental procedures employed for creation of a cooled atomic beam. This reflects the fact that it constitutes a part of experimental uncertainties pertinent to the COLTRIMS technique.

For accurate quantitative comparisons with experiment, one must further convolute the FDCS with a distribution of the target wave packets in a supersonic jet over the mean velocities, $\mathbf{v}_t = \mathbf{k}_t/M_t$, where M_t is the atomic mass, and with finite energy and angular resolutions of the detectors. In the case of weak collimation of the projectile beam, it is necessary to carry out an additional averaging due to the spread of the mean velocities $\mathbf{v}_p$ of the projectile wave packets. Thus, singling out the effects of the target center-of-mass wave packet in the measured FDCS is not a straightforward task in general.

References

[1] M. Schulz *et al.* 2003 *Nature* **422** 48

[2] X. Wang *et al.* 2012 *J. Phys. B* **45** 211001

[3] H. Gassert *et al.* 2016 *Phys. Rev. Lett.* **116** 073201

[4] K. A. Kouzakov 2017 *Eur. Phys. J. D* (in press)

[1]E-mail: kouzakov@srd.sinp.msu.ru

IOP Conf. Series: Journal of Physics: Conf. Series **875** (2017) 092029 doi:10.1088/1742-6596/875/10/092029

Molecular Orbital interpretation to the couplings in collisions of 2.5 and 3 MeV $Xe^{10+,12+}$-Au and Zr systems

Punita Verma*[1], Kajol Chakraborty*†, Ruchika Gupta*†, Sarvesh Kumar*, Gaurav Sharma‡, Deepak Swami‡, Samit K. Mandal†, C.P. Safvan‡.

* **Department of Physics, Kalindi College, University of Delhi, New Delhi, India**
† **Department of Physics and Astrophysics, University of Delhi, Delhi, India**
‡ **Inter University Accelerator Center, Aruna Asaf Ali Marg, New Delhi, India**

Synopsis Super heavy quasimolecules formed during slow, heavy-ion, heavy-atom collisions provide a tool for investigating the inner shell couplings. Correlation diagrams have been drawn for the Xe-Au, Zr collision systems which allow a satisfactory qualitative explanation for the experimentally observed enhanced cross sections.

Super-heavy quasi-atoms or quasi-molecules can be approached in relatively slow, heavy ion-heavy atom collisions which are slow compared to the orbital velocity of the innermost electrons K or L. A renewed interest has been generated recently in these super-heavy atomic systems with united atomic numbers far beyond existing matter (i.e.$Z_{UA}=Z_1+Z_2>100$, Z_1 and Z_2 being the atomic numbers of the projectile and target respectively). The normal Dirac equation for a point charge cannot be solved beyond $Z_{UA}=137$. By providing vacancies in the inner shells of the projectile inner shell couplings can be investigated by an observation of x-ray emission in the separated partners or in the quasi-molecule itself.

The Low Energy Ion Beam Facility (LEIBF) at Inter University Accelerator Centre (IUAC), New Delhi with its 10GHz ECR ion source was used to bombard 2.5 & 3 MeV $Xe^{10+,12+}$ ions on 640μg/cm^2 Au targets and 250μg/cm^2 Zr target on a carbon backing of 40μg/cm^2. X-Rays emitted were detected by a KETEK SDD and a CANBERRA LeGe detector.

The intensity ratios and cross sections have been measured for Xe and Zr L X-rays and Au M X-rays. Au M_α and Zr L_α have been observed with higher intensities as compared to Xe L_α X-rays (Figure 1). As the charge state of projectile was increased, enhanced intensities of target X-rays has been observed, an effect observed similarly by Ying *et al.*[1]. The results agree with previous known values[2]. These observations indicate an additional phenomena apart from well known electron capture or coulomb ionisation. Molecular Orbital approach is applicable to collision systems in the quasi-molecular regime <1 (adiabaticity parameter) which is the case for these collision systems.

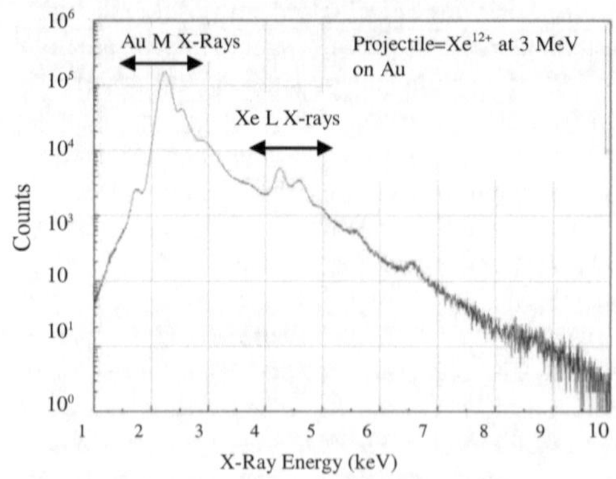

Figure 1. 3 MeV Xe^{12+} on Au target X-ray spectra

Diabatic correlation diagrams for Xe on Au and Xe on Zr collision systems indicate clearly the finite probability of direct ionization of 4f and 3d shells of the united atom. For Xe on Au collison system, the incoming vacancies in the outer shells of Xe may experience an easy transfer to Au M through the correlated Xe 3p levels and united atom (Z=133) 4d levels. The latter are further correlated to the Au 3p levels. High intensity ratios of Au M X-Rays observed thus seems to be obvious. Zr L X-rays exhibit similar behaviour. Either symmetric or antisymmetric systems in the low energy region will be investigated further for clarity of vacancy transfer mechanism at these low energies [3].

References

[1] Ying *et al.* 2008 *Sci. China Ser G-Phys Mech. Astron*, **51**, 1240
[2] Fink *et al.* 1973 *Proceedings of international conference on Inner shell ionization phenomena and future applications (Tennessee)* 396
[3] XU Jin-Zhang *et al.* 2007 *China Phys. Letters*, **25**, 1649

[1] E-mail: drpunitaverma.nature@gmail.com

Double differential cross sections for ion impact ionization of ammonia(NH_3) and methane (CH_4) molecules

M. Purkait

Department of Physics, Ramakrishna Mission Residential College, Narendrapur, Kolkata-700103, India

Keywords : Ion-molecule collision, ionization, cross sections

Synopsis We present a theoretical approach to calculate double-differential cross sections (DDCS) for single ionization of ammonia (NH_3) and methane (CH_4) molecules by the impact of proton at energies 0.25 MeV and 2 MeV respectively. The calculation are performed by the three-body formalism of three-Coulomb wave model (3C-3B). In the present formalism, the interaction of the projectile ion with the target is described by the Coulomb continuum wave functions in the initial channel. In the final channel, we have expressed the total wavefunctions as the product of pair-wise Coulomb wavefunctions among the ejected electron, projectile ion and the residual target ion, respectively.

Single ionization of atoms and molecules by bare ion impact is one of the more important challenges facing several fields of physics such as plasma physics, diagnostics for fusion in tokamaks, astro-physics, and irradiation of living matter etc. This process is also relevant to the study of the penetration of charged particles through matter. It is obvious that the interaction of heavy ions with multi-electronic molecular targets is very complex. For this reason, studies on the above mentioned processes should be such that they allow us to extract precise data without two many complications. Here, our attention is focussed on the calculation of double differential cross sections (DDCS) for the single ionization of CH_4 and NH_3 molecules by the impact proton in the framework of three body formalism of three Coulomb wave (3C-3B) model [1]. The transition matrix for single ionization may written as

$$T_{if} = \langle \psi_f^- | (V_{TP} + V_{Pe}) | \psi_i^+ \rangle$$

where ψ_i^+, the initial channel wavefunction is given by.

$$\psi_i^+ = \phi_i(\vec{r}_{Te}) \chi_i^+.$$

Here $\phi_i(\vec{r}_{Te})$ is the one-center molecular wavefunction [2] and χ_i^+ represent the Coulomb distortion. Here the initial bound state is distorted by the incoming projectile. The final state is characterised by the product of three coulomb wave function (3CW), called the BBK model [3]. We have displayed our computed results by angular distribution of DDCS for single electron ionization of NH_3 by a 250 keV proton in figure 1 at selective electron emission energies of 20, 50, 200 and 500 eV respectively.

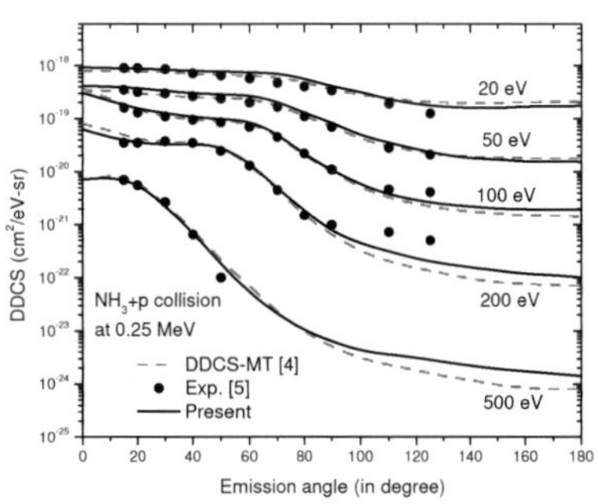

Figure 1. p-NH_3 collision at 0.25 MeV

We acknowledge the support by the Department of Science & Technology (DST), SERB, New Delhi, India under project No. SB/S9/LOP-0009/2013.

References

[1] A Mondal, T K Ghosh, C R Mandal and M Purkait 2016 *J. Phys.B: At.Mol. Opt. Phys.* ***49***, 245203

[2] C Champion, J Hassen and P A Hervieux 2004, *J. Chem. Phys.* ***121***, 9423

[3] M Brauner, J S Briggs and H Klar 1989, *J. Phys.B: At.Mol. Opt. Phys.* ***22***, 2265

[4] B Senger 1988, *Z. Phys. D****9***, 79

[5] D J Lynch et al 1976, *J. Chem. Phys.* ***64***, 2616

email : mpurkait 2007@rediffmail.com

IOP Conf. Series: Journal of Physics: Conf. Series **875** (2017) 102002 doi:10.1088/1742-6596/875/11/102002

Interference in fast bare ions colliding with diatomic molecules

S. Halder[1], A. Mondal[2] and M. Purkait[3]

Department of Physics, Ramakrishna Mission Residential College, Narendrapur, Kolkata-700103, India

Keywords : Ion-molecule collision, charge-transfer, cross sections

Synopsis Total as well as differential cross sections for electron capture in collisions of bare ions with diatomic molecules have been investigated at incident projectile energies ranging from 20 to 1000 keV/amu in the framework of distorted wave (DW) approximation. The present DW calculation is the simplest possible quantum calculation for this system. The present differential cross sections (DCS) are in good agreement with the shape of the observed angular dependence at forward angles.

Electron capture from molecular targets differs from capture from atoms because of the multi-center configuration of molecules, At high impact energies, a diatomic molecule may be regarded as a target where the incoming-projectile wave fronts scatter from both atomic centres providing interference patterns in the outgoing waves. This idea was recognized by Tuan and Gerjuoy [1] who first calculated electron-capture cross-sections from molecular hydrogen by protons at high collision velocities. In 2008, Schmidt et al [2] reported on strong interference effects in fast $H^+ + H_2$ collisions when only one electron is transferred to a fast H^+ ion to form a fast neutral H atom. Later, Misra et al investigated the first experimental observation of Young-type interference effects in a two electron transfer process in $He^{2+} + H_2$ collision at 1.2 and 2 MeV respectively. In the present theoretical investigation, we have focused our attention on single transfer of diatomic molecules by the impact of bare ions in the incident energy range 50-5000 keV. The differential cross-sections for electron capture, averaged over all orientations of the internuclear axis $\vec{\rho}$ of the hydrogen molecule is given in the DW approximation by

$$\frac{d\sigma_M}{d\Omega} = 2\left(\frac{\mu_i\mu_f}{4\pi^2}\frac{k_f}{k_i}\right)\frac{1}{4\pi}|T_{if}|^2 d\Omega_\rho.$$

he factor of 2 in the right hand side occurs because either of the two molecular electrons may be involved in the capture process. The transition amplitude is given by

$$T_{if} = \langle\psi_f^-|V_i|\psi_i\rangle.$$

The initial and final states are given by

$$\psi_i = e^{i\vec{k}_i\cdot\vec{r}_i}\phi_M(\vec{r}_1,\vec{r}_2,\vec{\rho}),$$

$$\psi_f^- = e^{i\vec{k}_f\cdot\vec{R}_f}\phi_I(\vec{r}_2,\vec{\rho})\phi_{nlm}(\vec{s})C_{t-e_1},$$

where $\vec{k}_i(\vec{k}_f)$ is the initial (final) momenta. ϕ_M, ϕ_I and ϕ_{nlm} are the wavefunction for H_2, H_2^+ and hydrogen like projectile respectively. Here C_{t-e_1} represent the interaction between the target ion and the captured electron. Here we consider the electron e_1 in any one of the atom in the molecule is active which is captured by the projectile and the other electron e_2 is passive in the other atom of the molecule (H_2^+). We will represent theoretical results for single capture cross sections for $H^+ - H_2/He^{2+} - H_2$ collisions in the intermediate to high energy range and will compared them with the available experimental dada.

The author gratefully acknowledges the financial support from the Council of Scientific & Industrial Research (CSIR), New Delhi, India through project No: 03/1366/16/EMR-II.

References

[1] T F Tuan and E Gerjuoy 1960 *Phys. Rev.* ***117*** 756

[2] H T Schmidt et al 2008 *Phys. Rev. Lett.* ***101*** 083201

[3] D Misra et al 2009 *J. Phys. Rev. Lett.* ***102*** 153201

[1] $E - mail$: *halder.saheb58@gmail.com*
[2] $E - mail$: *abhoy.mondal5@gmail.com*
[3] $E - mail$: *mpurkait 2007@rediffmail.com*

IOP Conf. Series: Journal of Physics: Conf. Series **875** (2017) 102003 doi:10.1088/1742-6596/875/11/102003

Anisotropic two-body dissociation by highly charged ion impact

Jyoti Rajput*[1], A. Agnihotri†, A. Cassimi†, X. Fléchard‡, S. Guillous†, W. Iskandar§, A. Méry†, J. Rangama†, C. P. Safvan¶[2]

* Department of Physics and Astrophysics, University of Delhi, Delhi - 110007, India
† CIMAP, CEA-CNRS-ENSICAEN-UNICAEN, Normandie Université, BP5133, F-14050 Caen Cedex 04, France
‡ Normandie Université, ENSICAEN, UNICAEN, CNRS/IN2P3, LPC Caen, 14000 Caen, France
§ Lawrence Berkley National Laboratory, Berkeley, CA-94720, USA
¶ Inter-University Accelerator Centre, Aruna Asaf Ali Marg, New Delhi - 110067, India

Synopsis The angular distribution of fragment emission with respect to the incoming beam direction is measured. In case of two body breakup of di-cationic molecular ions, such as D_2O^{2+} and $C_2H_2^{2+}$, a strong anisotropy is observed when one of the fragments is an H^+ or D^+. These molecular ions were produced by impact of slow highly charged ions on neutral species. The technique of recoil ion momentum spectroscopy was used to measure the complete momentum vectors of all fragments in coincidence.

From the studies on fragmentation of charged molecular ions, it is becoming increasingly apparent that the emission of dissociation fragments is anisotropic in nature. In case of laser induced fragmentation, anisotropy had been observed with reference to the laser polarization axis [1] while for fragmentation induced by charged particles, the anisotropy is with respect to the incoming projectile beam [2]. Many of the studies reported in literature have focussed on diatomic molecules and results for polyatomic molecules are rather scarce.

We carried out experiments to address the anistropic emission of fragment ions in two-body dissociation of few hydrogen (or deuteron) containing polyatomic molecules, such as D_2O^{2+} and $C_2H_2^{2+}$, induced by impact of highly charged ions. The molecules considered for this study have very different geometries. The experiments were carried out at the Low Energy Ion Beam Facility (LEIBF) of Inter-University Accelerator Centre, Delhi, India and also at the ARIBE-GANIL facility in Caen, France. At both these facilities an electron cyclotron resonance (ECR) ion source was used to extract beams of highly charged ions. At LEIBF a recoil ion momentum spectrometer was used to measure the momenta of all ions in coincidence with the target gas being supplied by an effusive gas jet. At ARIBE-GANIL a COLTRIMS setup was used where the cold target molecules were produced using supersonic expansion. The facility at GANIL was also equipped with a post collision charge state analyzer which enabled us to separate events coming from pure ionization from the ones involving capture of one or two electrons.

The results of our experiments show strong anisotropic effects in the emission of recoil ion fragments with respect to the incoming projectile beam direction. These effects are common to all the molecules considered during this investigation. As an example, the angular distribution of fragments H^+ and C_2H^+ produced from dissociation of $C_2H_2^{2+}$ is shown in the figure below.

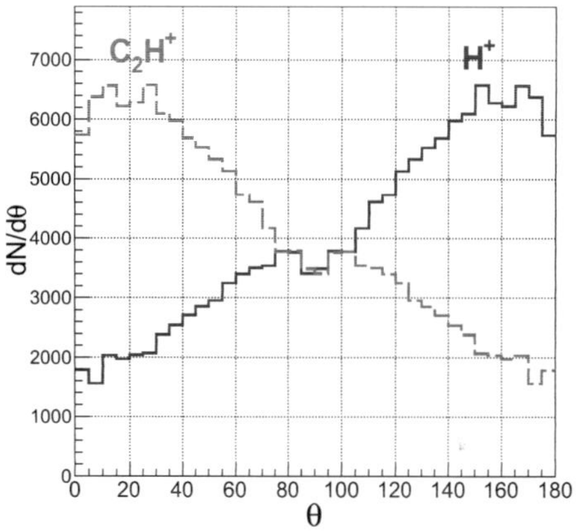

Figure 1. Fragment angular distribution for two body breakup of $C_2H_2^{2+}$. Angle θ represents the angle between the fragment momentum vector and direction of the incoming projectile beam.

The authors thank Indo-French Centre for Promotion of Advanced Research (IFCPAR/CEFIPRA) for supporting this collaboration (4604-2). JR thanks Department of Science and Technology, Government of India (SR/FTP/PS-098/2011) for financial support.

References

[1] D. Pavičić *et al.* 2007 *Phys. Rev. Lett.* **98** 243001

[2] J. Rajput and C. P. Safvan 2014 *J. Chem. Phys.* **141** 164313

[1]E-mail: jrajput.du@gmail.com
[2]E-mail: cp.safvan@gmail.com

IOP Conf. Series: Journal of Physics: Conf. Series **875** (2017) 102004 doi:10.1088/1742-6596/875/11/102004

Dissociation dynamics of NH_3 under slow highly charged ion impact

C. P. Safvan[*1], **T. Sairam**[*], **and Pragya Bhatt**[*]

[*] Inter University Accelerator Center, Aruna Asaf Ali Marg, New Delhi 110067, India

Synopsis The effect of energy deposited by slow highly charged ions on the dynamics of dissociation of NH_3 molecule is studied.

The study of dissociation of multiply charged molecules is important not only for fundamental understanding of collision physics but also for practical applications in astrophysics [1] and plasma physics [2]. Collisions between highly charged ions and neutral molecules result in highly reactive and unstable molecular ions which further dissociate into either charged or neutral atomic/molecular species. These dissociated fragments if detected in coincidence can provide information about the dynamical properties of their precursor molecular ion.

We have used the technique of recoil ion momentum spectroscopy (RIMS) [3] to understand the dynamics of dissociation of the ammonia molecule when it is bombarded by 400 keV Ar^{8+} ions. These Argon ions are produced using an electron cyclotron resonance ion source. The recoil ions and electrons generated in these collisions are separated by a homogeneous electric field. The recoils ions are detected by a micro channel plate coupled with a position sensitive detector; the ejected electrons are detected by a channel electron multiplier and they serve as the reference for the ion arrival time. We have measured the momenta of fragment ions generated in the dissociation of NH_3^{q+} (q=2,3); the momentum of undetected particle is obtained by applying the principle of conservation of momentum.

Neutral NH_3 molecule in its ground state has a trigonal pyramidal geometry C_{3v}, however, its singly and doubly charged states may have C_{3v} as well as planar geometry D_{3h} [4]. We have obtained the correlation between the momentum vectors of the fragments of NH_3^{q+} to find out what happens when ammonia is multiply ionized. We have measured the kinetic energy release (KER) for different dissociation channels of NH_3^{q+}. The KER value measured for a given precursor molecular ion indicates which state was accessed in an ionizing event. Under highly charged ion impact, a number of states may be accessed and therefore a broad distribution of KER is obtained in our measurements.

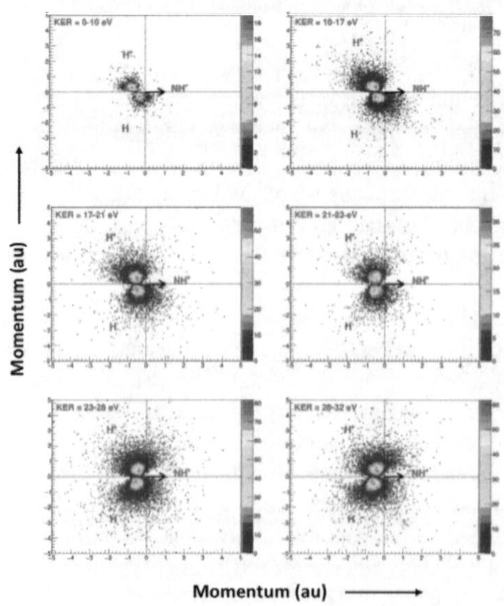

Figure 1. Momentum map obtained for dissociation of NH_3^{2+} into the channel H^0+H^++NH^+.

The variation of angles between the fragment momenta with KER (see, Figure 1) is studied [4]. It is observed that the angular dependence among the fragments is affected by the variation in the KER value. This implies that different multiply charged ionic states of ammonia may have different geometries before dissociation into a particular fragmentation channel.

References

[1] T. E. Cravens 2002 *Science* **296** 1042

[2] R. Janev, 2001, *Atomic and Plasma-Material Interaction Data for Fusion* **9** 1

[3] J. Ullrich, 1997, *J. Phys. B: At. Mol. Opt. Phys.* **30** 2917

[4] P. Bhatt *et al.*, 2017 *to be communicated*

[1]E-mail: safvan@iuac.res.in

IOP Conf. Series: Journal of Physics: Conf. Series **875** (2017) 102005 doi:10.1088/1742-6596/875/11/102005

Bond rearrangement in NH_3 under slow highly charged ion impact

Pragya Bhatt*[1], T. Sairam*, and C. P. Safvan*

* Inter University Accelerator Center, Aruna Asaf Ali Marg, New Delhi 110067, India

Synopsis Ultra fast bond rearrangement in ammonia is observed under slow highly charged ion impact.

The migration of protons is an important process from the point of view of isomerization of molecules during chemical reactions. A number of reports are available in the literature on the formation of H_2^+ and H_3^+ ions as these are an essential facet in the reactions through which many molecules, like, H_2O, CH_4, NH_3 and several other hydrocarbons are thought to be formed in the astrophysical environment. Recently the rearrangement of H-atoms in NH_3 has been reported by few workers [1, 2]

In the present experiment [3], the low energy highly charged Argon ions are generated from an electron cyclotron resonance source and are made to collide with a gas jet of NH_3 under single collision condition. The recoil ions and ejected electrons are extracted in opposite directions by applying a homogeneous electric field; these are detected by a micro channel plate and a channel electron multiplier, respectively. The position information of the recoil ions is obtained by a delay line anode. The momentum of an ion is obtained by measuring its time of flight from collision center to the detector and its position simultaneously. Triple ion coincidence technique is employed to study the dynamical properties of the precursor ion NH_3^{q+} (q=3-5) giving rise to three fragment ions. Only the fragmentation channels involving rearrangement of hydrogen atoms are discussed here.

We have observed that the dissociation of NH_3^{q+} (q=3-5) occurs concertedly where the H^+, H_2^+ and N^{p+} (p=1-3) ions fly out of plane. Further, the bond rearrangement, though having a low cross section, is faster in comparison to the Coulomb explosion. The rearrangement process is observed in case of multiply charged NH_3 also which is contrary to earlier reports where rearrangement is suggested to occur for up to doubly charged parent ions only. The kinetic energy release value for the dissociation of NH_3^{2+} is found to be consistent with the earlier theoretical calculations and the measurements.

To understand the structure of dissociating NH_3^{2+}, the *ab initio* calculations at the unrestricted Hartree Fock level under adiabatic approximation are performed (see, Figure 1). An interesting result is observed that the dissociation channels involving rearrangement of H-atoms originate from the doubly charged NH_3 only when it has a planar structure. However, the neutral NH_3 has a trigonal pyramidal geometry in its ground state.

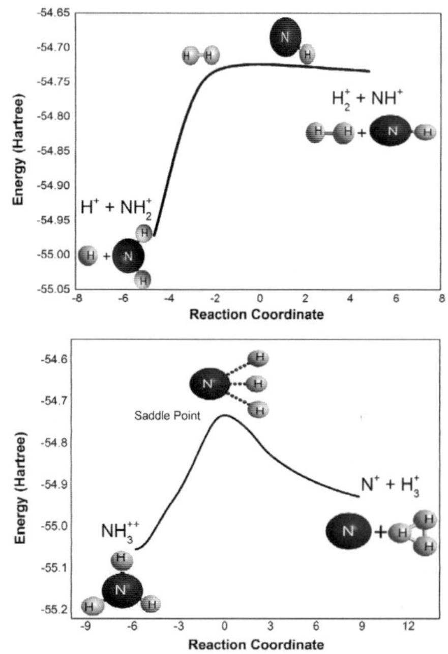

Figure 1. Reaction Path for the dissociation of NH_3^{2+} into bond rearrangement channels considering its initial geometry to be planar.

References

[1] B. Jochim *et* al., 2009 *P*hys. Chem. Chem. Phys. **42** 091002

[2] N. Walsh *e*t al., 2015 *P*hys. Chem. Chem. Phys. **17** 18944

[3] P. Bhatt *e*t al., 2016 *communicated*

[1]E-mail: pbpragya@gmail.com

IOP Conf. Series: Journal of Physics: Conf. Series **875** (2017) 102006 doi:10.1088/1742-6596/875/11/102006

Three-body dissociation of OCS^{3+} : separating sequential and concerted pathways

Herendra Kumar[*1], **Pragya Bhatt**[†] , **C. P. Safvan**[†] **and Jyoti Rajput**[*2]

[*] Department of Physics and Astrophysics of University of Delhi, Delhi-110007, India.
[†] Inter University Accelerator Centre, Aruna Asaf Ali Marg, New Delhi-110067, India.

Synopsis Three body fragmentation pathways of OCS^{3+} molecular ion leading to detection of $C^+ + O^+ + S^+$ in coincidence have been studied using the technique of recoil ion momentum spectroscopy. This fragmentation can occur either via a one step concerted process or a two step sequential process. In the first step of the sequential pathway either CO^{2+} or CS^{2+} is formed. By separating the events coming from the two processes, branching ratios are determined and the possible electronic states associated with the intermediate CO^{2+} or CS^{2+} have been identified.

Studying many-body fragmentation of poly-atomic molecules is a big challenge to both theory and experiment. One of the questions in the field which seeks resolution is the separation of events based on concerted or sequential mechanism. Even in a simple case of a tri-atomic molecule, the three-body dynamics can be very complicated. Working towards the same goal we have studied the three-body dissociation pathways of a simple tri-atomic molecule OCS.

The experiment was conducted at the Low Energy Ion Beam Facility (LEIBF) of Inter University Accelerator Centre (IUAC), New Delhi, India in an ion-molecule collision setup. A beam of Xe^{9+} ions at an energy of 200 kV/q, extracted from an electron cyclotron resonance ion source was made to interact with neutral OCS molecules effusing from a needle. All the fragmentation products resulting from the interaction were collected using a multi-hit recoil ion momentum spectrometer. More details of the experimental setup can be found in reference [1].

The three-body breakup of OCS^{3+} molecular ion, which is the focus of the present work, can be broadly categorized in terms of concerted and sequential mechanisms. In the case of concerted fragmentation process , the reaction products are generated in one step,

$OCS^{3+} \rightarrow C^+ + O^+ + S^+$

while the sequential breakup is a two step process,

$OCS^{3+} \rightarrow CO^{2+} + S^+ \rightarrow C^+ + O^+ + S^+$

or

$OCS^{3+} \rightarrow CS^{2+} + O^+ \rightarrow C^+ + O^+ + S^+$

Since the final products in both these cases are the same set of three ions in coincidence, it becomes difficult to separate events coming from the two different mechanisms. A new method was recently developed in the group of *Prof. Itzik Ben-Itzhak* (JRM Laboratory, Kansas State University, USA) which enables the separation of events based on the fragmentation mechanism. The method is based on analysing the three-body fragmentation data set in two different frames of reference, one in the centre of mass frame of OCS^{3+} and other in the frame of reference of the intermediate molecular ion, CO^{2+} or CS^{2+}. Upon following the procedure, the kinetic energy release spectra associated with the breakup of the intermediate ion is obtained which enables identification of possible electronic states of the intermediate populated during the first step of the sequential process. The kinetic energy release spectra of the CO^{2+}intermediate is shown in the figure below. Some of the electronic states [2, 3] which may be populated are also marked. Further, using this new method, we have determined the branching ratio of all the fragmentation pathways observed in our data.

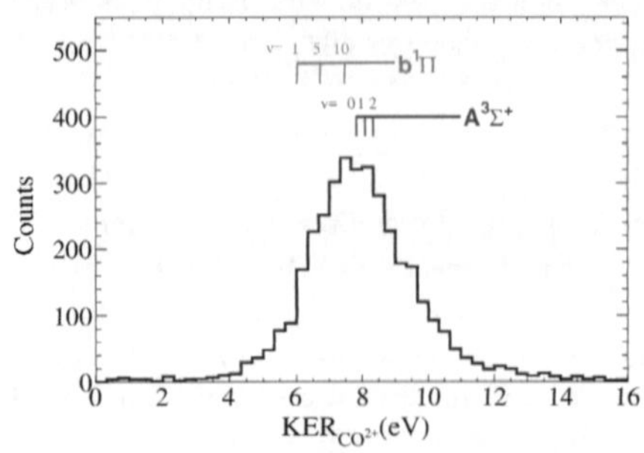

Figure 1. Kinetic energy release spectrum associated with the second step of sequential breakup of OCS^{3+} via CO^{2+} as intermediate.

References

[1] Kumar *et al.* 2014 *J. Mass Spect.* **374** 44

[2] M. Lundqvist *et al.* 1995 *Phys. Rev. Lett.*, **75** 1058

[3] F. Mrugala 2008 *J. Chem. Phys.* **129** 064314

[1]E-mail: harendraamu@gmail.com
[2]E-mail: jrajput.du@gmail.com

IOP Conf. Series: Journal of Physics: Conf. Series **875** (2017) 102007 doi:10.1088/1742-6596/875/11/102007

Stability of dimer and trimer of Naphthalene studied in electrostatic storage Mini-Ring.

S. Martin[*1], L. Chen[*], J. Bernard[*], G. Montagne*, C. Joblin [†]and A. Cassimi[£]

[*] Université de Lyon, F-69622, Lyon, France Université Lyon 1, Villeurbanne CNRS, UMR 5306, ILM
† Université de Toulouse, UPS-OMP, IRAP, Toulouse, France
[£]CIMAP, GANIL Université de CAEN, Bd. H. Becquerel, Caen, France

Synopsis We report studies on stabilities of dimer and trimer of polycyclic aromatic hydrocarbon PAH in a small electrostatic storage device. Different nslaser energies were used at different time delays as the probe to determine the dissociation energy. The effect of dehydrogenation on the stability is also studied.

We present measurements on the stabilities of the dimer and trimer of PAH studied with the MINIRING [1]. Stabilities of Dimer-nH (n=1-4) of Naphthalene are also studied.

Dimer and Trimer cations of PAH were produced on the Nanogan ECR (Electron Cyclotron Resonance) ions source and accelerated up to 12 keV. The production of dimers and dehydrogenated dimers of Naphthalene was strongly depending on the source conditions such as injected naphtalene gas pressure and the 10 GHz HF power. Typically a 10^{-5} mbar pressure and 0.6W HF power were used to produce intact dimers. Lower pressure promoted dehydrogenated dimers. The studied molecules were stored in the electrostatic storage ring "Mini-Ring" for up to 100 ms. "Natural" decays were studied by measuring the neutral fragments emitted by the stored molecules. A position sensitive de-tector was used to measure the position and time of the neutral fragment emitted at each turn. For the intact dimer of naphthalene, we observed a small spot on the multichannel plate indicating a small Kinetic Energy Release (KER) of the fragmentation reaction. This is in good agreement with a "Van Der Waals" bind-ing leading to a fragmentation of the dimers in two monomers . In contrast, the dehydrogenat-ed dimer shows a broad spot indicating the emission of a small fragment (H or C_2H_2) by a "quasi covalent" dimer. Natural decays (e.g., figure 1) and decays induced by the laser were also studied. Simulations for the intact dimer of naphthalene using a simple RRKM model are in good agreement with experimental decays. The cooling measurements of the dimers and trimers of napthalene will also be presented.

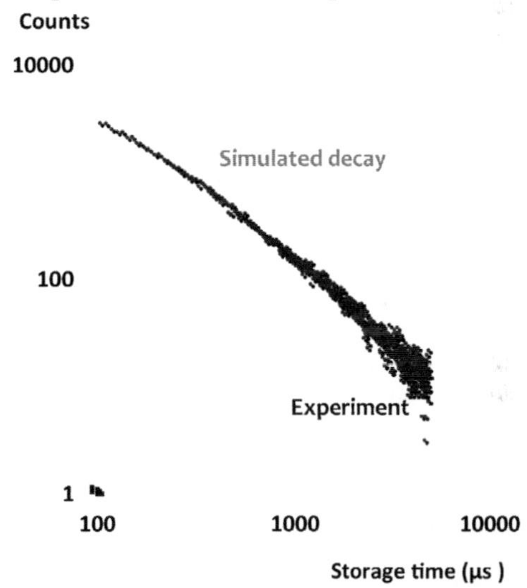

Figure 1. Experimental and simulated decays for dimer-Naphthalene using low dissociation energy of 0.7 eV. This measurement is in agreement with the Van der Waals nature of the binding of the dimer of Naphthalene.

References

[1] S. Martin *et al* 2013 *Phys. Rev. Lett.* **110** 063003

E-mail: smartin@univ-lyon1.fr

IOP Conf. Series: Journal of Physics: Conf. Series **875** (2017) 102008 doi:10.1088/1742-6596/875/11/102008

Ab initio study of proton collisions with BeH

I. Rabadán *[1], **L.Méndez** *, **J. W. Gao** †‡, **Y. Wu** †, **J. G. Wang** †

* Departamento de Química, Universidad Autónoma de Madrid, 28049 Madrid, Spain
† Institute of Applied Physics and Computational Mathematics, 10088 Beijing, China
‡ Laboratoire de Chimie Physique Matière et Rayonnement, UMR 7614, UPMC Univeristé Paris 6, Paris, France

Synopsis We present calculations of electron capture cross sections in proton collisions with BeH molecules in the energy range 25 eV/u $< E <$25 keV/u.

A key issue in the design of future fusion tokamaks is the choice of plasma facing materials. In ITER, beryllium will be the plasma facing component of the main chamber. An ITER-like-wall is currently installed in the Joint European Torus (JET) and several experiments have already been carried out in order to understand the erosion of the first wall and the material migration. Some of these experiments [1] have pointed out the importance of the chemical assisted physical sputtering mechanism that leads to the release of BeD molecules, which are then ionized and dissociated by collision with the main plasma components.

In this work, we present a theoretical study of the electron capture (EC) reaction

$$H^+ + BeH(^2\Sigma^+) \rightarrow H + BeH^+ \quad (1)$$

Orientation-averaged total cross sections are obtained by expanding the scattering wave function in a basis set of molecular electronic states of the H_2Be^+ molecule. Our treatment assumes rectilinear projectile trajectories and uses the sudden approximation for molecular rotation and vibration. The molecular wave functions have been obtained with a 3-electron full configuration interaction (CI) method, with the 1s electrons of Be frozen. As an illustration, we show in Fig. 1 a cut of the potential energy surfaces (lines) for a nuclear geometry in which the BeH target is fixed at the equilibrium distance of its electronic ground state ($\rho_e = 2.568$ bohr). One can note the good agreement with the energies calculated with a multi-reference single- and double-excitation CI method [2] (solid symbols).

We show in Fig. 2 the orientation-averaged EC cross sections calculated within the Franck-Condon approximation ($\rho = \rho_e$). Here, we note that the transitions to the EC channels that dissociate into H(1s)+$BeH^+(2^1\Sigma^+, 1^3\Pi)$ (3-state calculation), furnish the main mechanism of reaction (1). The importance of including rotational couplings, and not only radial ones, in the calculation is also emphasized in the figure. In addition, SC-AOCC calculations [3] will also be performed for higher energies in the future. The relatively large cross section for this reaction indicates that it could be relevant in plasma modelling.

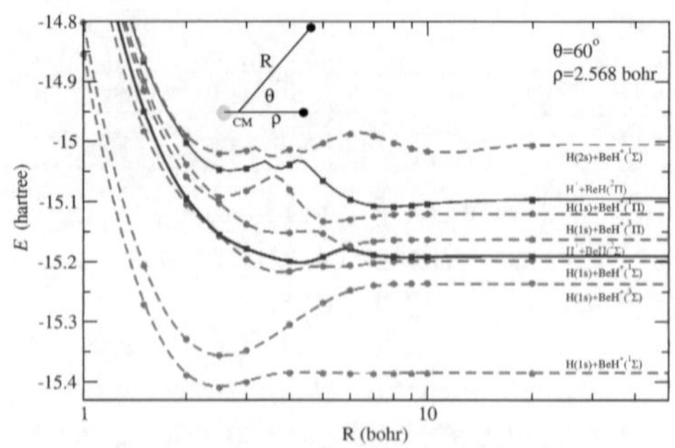

Figure 1. Potential energy curves of H_2Be^+ for the nuclear geometry shown in the figure.

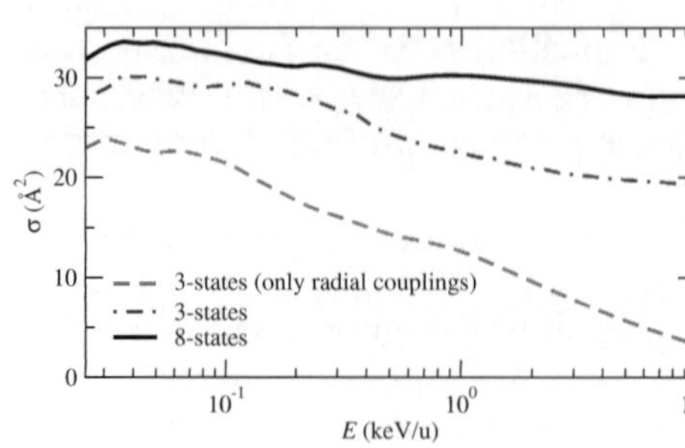

Figure 2. Orientation-averaged EC total cross section for reaction (1) as a function of the collision energy.

This work has been partially supported by a project UAM-Santander de Cooperación Interuniversitaria con Asia and Ministerio de Economía y Competitividad (Spain) (ENE2014-52432-R).

References

[1] S. Brezinsek *et al.* 2015 *Nuclear Fusion* **55** 063021

[2] R. J. Buenker and R. A. Phillips 1985 *J. Mol. Struct. THEOCHEM* **123** 291

[3] N. Sisourat *et al.* 2011 *Phys. Rev. A* **84**, 052722

[1] E-mail: ismanuel.rabadan@uam.es

IOP Conf. Series: Journal of Physics: Conf. Series **875** (2017) 102009 doi:10.1088/1742-6596/875/11/102009

Proton impact fragmentation of water molecules

P.N. Terekhin*, † 1, M.A. Quinto*, J.M. Monti*, O.A. Fojón*, R.D. Rivarola*

* Instituto de Física Rosario (CONICET-UNR), Bv 27 de Febrero, 2000 Rosario, Argentina
† NRC Kurchatov Institute, Kurchatov Sq. 1, 123182 Moscow, Russia

Synopsis The Continuum Distorted Wave-Eikonal Initial State approximation for impact of H^+ on H_2O molecules is used to calculate absolute cross sections for multiple electron removal processes at intermediate and high collision energies. Resulting fragmentation patterns are also matter of study. Results are compared with previous calculations and available experimental data.

Interaction of charged particles with molecular targets is of great interest in a number of areas of research such as nuclear fusion [1], plasma physics [2] and others. Particularly, collisions of ions with H_2O are of great importance in radiotherapy [3]. The ionization of water molecules produce different charged species, which can damage the surrounding cells. Therefore, it is important to analyze pure ionization, capture and transfer-ionization reactions to study fragmentation processes.

We present theoretical calculations of fragmentation cross sections based on the model proposed in the work [4] for water molecules interacting with H^+ impact. In the present work, cross sections are obtained in the framework of the three-body Continuum Distorted Wave-Eikonal Initial State approximation (3B-CDW-EIS) [5]. The initial wavefunctions of the active electrons bound to a particular water molecular orbital are described employing the complete neglect of the differential overlap (CNDO) approximation [6]. A trinomial distribution analysis has been employed to compute exclusive probabilities using the independent electron (IEL) model. A unitarization procedure is employed to avoid overestimations of 3B-CDW-EIS impact parameter probabilities.

From the comparison with the available theoretical and recent experimental results, we conclude that exclusive probabilities are required for a reliable description of the fragmentation yields (see Figure 1).

The developed approach may be used to investigate fragmentation patterns of more complex molecules.

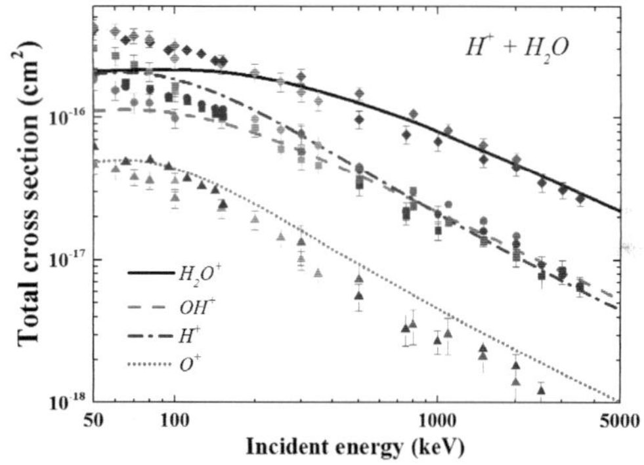

Figure 1. Fragmentation cross sections as a function of the incident energy for H^+ impacting on H_2O.

References

[1] S. Yu Simon *et al.* 2007 *Nucl. Fusion* **47** 721-727

[2] I. Murakami, J. Yan, H. Sato, M. Kimura, R.K. Janev, T. Kato 2008 *At. Data Nucl. Data Tables* **94** 161-222

[3] D. Schardt, T. Elsässer and D. Schulz-Ertner 2010 *Rev. Mod. Phys.* **82** 383-425

[4] M. Murakami, T. Kirchner, M. Horbatsch, H.J. Lüdde 2012 *Phys. Rev. A* **85** 052704

[5] P.D. Fainstein, V.H. Ponce and R.D. Rivarola 1991 *Phys. B: At. Mol. Opt. Phys.* **24** 3091-3119

[6] J. Pople, D. Santry and G. Segal 1965 *J. Chem. Phys.* **43** S136-S151

[1] E-mail: P.N.Terekhin@yandex.ru

IOP Conf. Series: Journal of Physics: Conf. Series **875** (2017) 102010 doi:10.1088/1742-6596/875/11/102010

Multiple electron processes from H_2O by He^{2+} and Li^{3+} impact

P.N. Terekhin[*, †, 1], **M.A. Quinto**[*], **J.M. Monti**[*], **O.A. Fojón**[*], **R.D. Rivarola**[*]

[*] Instituto de Física Rosario (CONICET-UNR), Bv 27 de Febrero, 2000 Rosario, Argentina
[†] NRC Kurchatov Institute, Kurchatov Sq. 1, 123182 Moscow, Russia

Synopsis We employ the Continuum Distorted Wave-Eikonal Initial State approximation for He^{2+} and Li^{3+} ions impacting on water vapor targets to study multiple electron removal processes at the intermediate to high projectile energy range. Transition probabilities and absolute cross sections for pure ionization, capture and transfer-ionization channels were calculated. Results are compared with previous calculations and recent experimental data.

Investigation of single and multiple-electron removal processes occurring between fully stripped ions and H_2O molecules is of fundamental interest in many areas such as nuclear fusion [1], plasma physics [2] and others. In particular, it is of fundamental importance in radiotherapy [3] enabling the understanding of the basic governing mechanisms involved. A large variety of final projectile and target charge states may be produced during these processes. Consequently, calculations of cross sections of pure ionization, capture and transfer-ionization reactions are essential to determine the main features of multiple electron processes.

He^{2+} and Li^{3+} beams are chosen in this work. The three-body Continuum Distorted Wave-Eikonal Initial State approximation (3B-CDW-EIS) [4] is used to calculate transition probabilities (capture and ionization) as a function of the impact parameter and absolute cross sections for the considered collisions. The initial wavefunctions of the active electrons bound to a particular water molecular orbital are described employing the complete neglect of the differential overlap (CNDO) approximation [5]. A trinomial distribution analysis has been employed to compute exclusive probabilities using the independent electron (IEL) model, where electron correlation is neglected [6]. A unitarization procedure is employed to avoid overestimations of 3B-CDW-EIS impact parameter probabilities.

From the comparison with the available theoretical and recent experimental results, we conclude that inclusive probabilities are required for a reliable description of the processes of interest for single-electron removal processes (see Figure 1) and exclusive probability analysis for multiple-electron ones.

The developed method for calculation of one-electron ionization and capture probabilities allows to study multiple electron processes for complex targets in particular such as macromolecules of DNA and RNA to model scenarios for the radiobiological consequences of the impact of charged energetic particles on those macromolecules.

[1] E-mail: P.N.Terekhin@yandex.ru

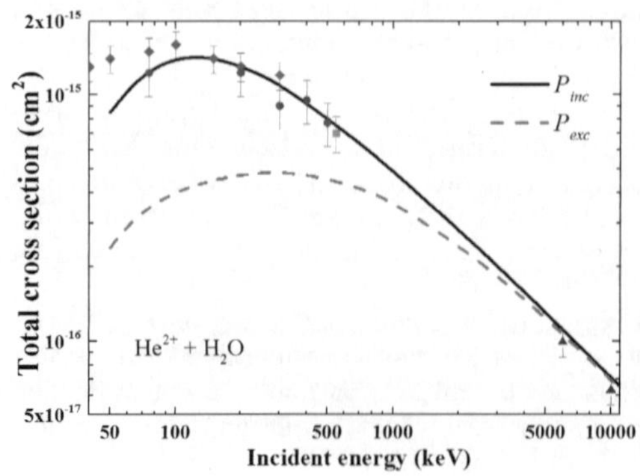

Figure 1. Total single ionization cross sections as a function of the incident energy for He^{2+} impacting on H_2O.

References

[1] S. Yu Simon *et al.* 2007 *Nucl. Fusion* **47** 721-727

[2] I. Murakami, J. Yan, H. Sato, M. Kimura, R.K. Janev, T. Kato 2008 *At. Data Nucl. Data Tables* **94** 161-222

[3] D. Schardt, T. Elsässer and D. Schulz-Ertner 2010 *Rev. Mod. Phys.* **82** 383-425

[4] P.D. Fainstein, V.H. Ponce and R.D. Rivarola 1991 *Phys. B: At. Mol. Opt. Phys.* **24** 3091-3119

[5] J. Pople, D. Santry and G. Segal 1965 *J. Chem. Phys.* **43** S136-S151

[6] J. Bradley, R. J. S Lee, M. McCartney and D. S. F. Crothers 2004 *J. Phys. B: At. Mol. Opt. Phys.* **37** 3723-373

IOP Conf. Series: Journal of Physics: Conf. Series **875** (2017) 102011 doi:10.1088/1742-6596/875/11/102011

Electron emission from water and neucleobases with radio-sensitizer

Lokesh C Tribedi* [1]**, M. Roy Chowdhury***, **S. Bhattacharjee***, **C. Bagdia***, **A. Mandal***, **D. Misra***, **C. Champion**†, **J. Monti**‡, **R.D. Rivarola**‡

*Tata Institute of Fundamental Research, Colaba, Mumbai 400005, India
†Universite´ Bordeaux 1, CNRS/IN2P3, CENBG, Chemin duSolarium, BP120, 33175 Gradignan, France
‡Institute de Fisica Rosario (CONICET-UNR), Universidad Nacional de Rosario, 2000 Rosario, Argentina

Synopsis We report a comprehensive study of the double differential e-emission cross sections from water, adenine, uracil and bromouracil in collisions with fast highly charged C, O and Si ions ~MeV/u i.e. at the Bragg peak. A large enhancement in e-DDCS for bromouracil (a radiosensitizer) over adenine and uracil has been observed which can not be explained by electron statistics, rather in terms of collective shape resonance or Auger cascade in Br. Large set of TCS data provides a scaling law in terms of v_p and q_p, The agreement with the QM model calculations (CDW-EIS, CB1) for such large molecules is encouraging. We have also measured the K-ionization and K-K e-transfer TCS for adenine.

The secondary electron emission from nucleobases and water is an important parameter to estimate the radiation damage caused by the fast ions, at the Bragg peak of energy loss, in the context of hadron-therapy. However, a realistic estimation of collision cross sections, the energy loss in such a solid or liquid medium needs much more input. The angular distribution of electron double differential cross sections (DDCS) and total ionization cross section (TCS) as well as the recoil-ion fragmentation spectrum provide crucial inputs towards the mechanism of such ionization process. The presence of a radio-sensitizer, which enhances the e-emission, helps to reduce the radiation dosage. Therefore the study of the influence of such sensitizers in a nucleobase has been a topic of current study. But there is no quantitative study of the exact enhancement so far.

A recently installed ECR based ion-accelerator on 400 kV deck and the existing 14 MV Pelletron tandem accelerator in TIFR, Mumbai are being used to investigate many body effects on collisions C_{60}, PAH, water and nucleobases [1-8]. The collision energy of C, O and Si ions vary between ~100 keV-100 MeV. A series of experiments are carried out with for uracil, adenine and bromouracil targets from a hot-oven vapor source. The experiment with water target involves a vapor target. An electron spectrometer is used for the measurement of low energy electron DDCS whereas the recoil-ion TOF setup is used for TCS measurement and fragmentation. Low energy electrons in the range of 1-600 eV are detected which also include the C, N and O K-LL Auger e-emission. The dramatically large forward backward asymmetry in electron emission for uracil [4,6,7] and adenine indicates a size and multi-center effect. An unusually large enhancement in e-emission from, the radio-sensitizing molecule, bromouracil has been quantitatively measured which goes beyond the electron statistics. Different mechanisms in terms of collective shape resonance and Auger cascade decay are possible candidates for an explanation. The K-K electron transfer and the K-ionization for nucleobase molecule are reported for the Ist time.

The CDW-EIS model provides an excellent agreement in contrast to earlier observed deviation for uracil molecule. The TCS data for water, uracil and adenine are found to follow a scaling behavior which is different from q^2-law (q=charge state) which will provide an useful input for the model calculation for radiation damage. Some relevant examples will be presented, based on our recent work.

References

[1] Biswas *et al.* 2015 *Phys. Rev. A* **92** 060701 (R)
[2] A. Kelkar *et al* 2015 **92** 052708
[3] Bhattacharjee *et al* 2016 *J. Phys. B* **49** 065202
[4] Agnihotri *et al* 2013 *Phys Rev A* **87** 032716
[5] Nandi *et al* 2013 *Phys Rev A* **87** 052710
[6] L. Tribedi *et al* 2012 *Eur Phys J D* **66** 303
[7] Agnihotri *et al* 2012 *Phys Rev A* **85** 032711
[8] S. Kasthurirangan *et al* 2013 *PRL* **111** 243201

[1] E-mail: lokesh@tifr.res.in
and ltribedi@gmail.com

Comparison between anion and cation emission from CH_4 molecules colliding with 10.5-keV C^+ ions: fragment-energy aspects

J.-Y. Chesnel*[1], Z. Juhász†[2], E. Lattouf*, B. A. Huber*, S. T. S. Kovács†, P. Herczku†, E. Bene†, V. Vizcaino*, A. Méry*, J.-C. Poully*, J. Rangama*, J. A. Tanis*,§, and B. Sulik†

* Centre de Recherche sur les Ions, les Matériaux et la Photonique (CIMAP), 6 Bd Mar. Juin, F-14000 Caen, France
† Institute for Nuclear Research, Hungarian Academy of Sciences (MTA Atomki), H-4001 Debrecen, Hungary
§ Department of Physics, Western Michigan University, Kalamazoo, Michigan 49008 USA

Synopsis It is shown that both positive and negative ions can be ejected from gas-phase methane molecules when bombarded with positive ions at keV energies. The kinetic energy distribution of the negatively charged fragments shows strong similarities with that of the ejected cations. However, in a hard binary collision between the C^+ projectile and an H atom, the formation of H^- ions is more endothermic than the formation of H^+ ions.

We show that both cations (H^+, C^+, CH_x^+) and anions (H^-, C^-) are emitted from methane molecules when bombarded by positive ions at an impact energy of a few keV. The experiment was performed at GANIL in Caen, France. A 10.5-keV C^+ ion beam crossed an effusive gas jet of CH_4. The collision-induced ionic fragments were selected according to their kinetic energy by means of an electrostatic analyzer. To avoid electron contamination of the anion spectra, the electrons and anions were magnetically separated so that only anions could reach the detector [1].

Figure 1 shows the energy distribution of the anions and cations ejected at the forward angle of 70° (with respect to the beam direction) in 10.5-keV C^+ + CH_4 collisions. The anion yield is relatively large, more than 3% of the cation yield, and the total cross section for anion formation is estimated to be higher than 10^{-18} cm^2. The main component of each spectrum is a broad structure, slowly decreasing with energy. This component results from soft many-body processes involving large impact parameters. On the other hand, at angles below 90°, pronounced peaks are observed at higher emission energies (Fig. 1). As recently observed for other collision systems [1-3], these peaks are due to recoil ions formed in hard binary collisions occurring at small impact parameters.

The kinetic energy distribution of the ejected anions shows strong similarities with that of the cations. This finding suggests a statistical picture in which the final charge state distribution of the emitted H centers barely depends on how closely the atomic centers approach each other during the collision. However, the binary peak due to H^- ions is centered at a lower energy than the peak due to H^+ ions with an energy difference of 20–30 eV in the 40°–70° angular range. This result shows that the formation of H^- ions in hard binary collisions is more endothermic than that of H^+ fragments.

This work was supported by the Hungarian National Science Foundation OTKA (Grant No. K109440) and the French-Hungarian CNRS-MTA Cooperation (ANION project).

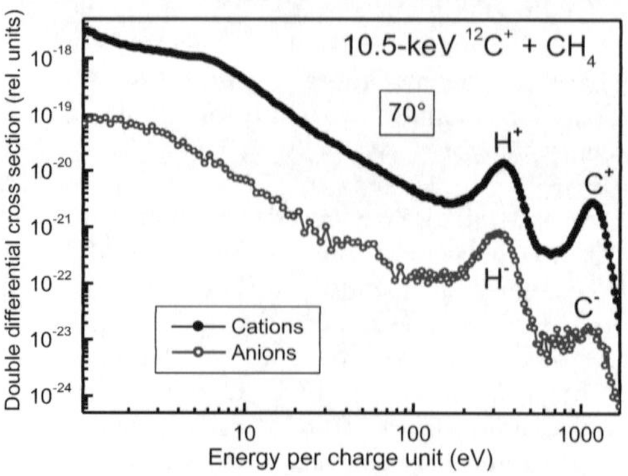

Figure 1. Energy distributions of anions and cations emitted at 70° in 10.5-keV C^+ + CH_4 collisions.

References

[1] J-Y Chesnel *et al* 2015 *Phys Rev A* **91** 060701(R)
[2] Z. Juhász *et al* 2013 *Phys. Rev. A* **87** 032718
[3] E. Lattouf *et al* 2014 *Phys. Rev. A* **89** 062721

[1]E-mail: ean-yves.chesnel@ensicaen.fr
[2]E-mail: zjuhasz@atomki.hu

IOP Conf. Series: Journal of Physics: Conf. Series **875** (2017) 102014 doi:10.1088/1742-6596/875/11/102014

Vibrational radiative cooling of isolated C_4^- and C_6^-

Ryuta Suzuki*[1], Naoko Kono*, Reito Andou§, Takeshi Furukawa*, Jun Matsumoto§, Toshiyuki Azuma†*, Hajime Tanuma*, Klavs Hansen‡¶, Haruo Shiromaru§,

* Department of Physics, Tokyo Metropolitan University, Hachioji, Tokyo 192-0397, Japan,
§ Department of Chemistry, Tokyo Metropolitan University, Hachioji, Tokyo 192-0397, Japan,
† AMO Physics Laboratory, RIKEN, Wako, Saitama 351-0198, Japan,
‡ Department of Physics, University of Gothenburg, 41296 Gothenburg, Sweden,
¶ Tianjin International Center of Nanoparticles and Nanosystems, Tianjin University, Tianjin 300072, P.R.China.

Synopsis We have observed the storage-time dependence of the laser-induced electron detachment yields for the linear carbon cluster anions C_4^- and C_6^- stored in an electrostatic ion storage ring (TMU E-ring) with several laser wavelengths. The electron detachment yield reflects the decrease of the internal energy of the stored ions. We succeeded in obtaining the time evolution of the excitation energy distribution below the first dipole-allowed electronic excited energy E_{ex} for C_4^- and C_6^-, which was well consistent to the behavior explained by vibrational IR-radiative cooling.

Radiative cooling of isolated molecular ions has recently been recognized as an important factor in the creation and stabilization of interstellar molecular anions. The radiative cooling process occurs both by IR photon emission with vibrational transition and/or visible-photon emission, referred to as recurrent fluorescence (RF) [1]. The RF was experimentally measured for the molecular ions, C_4^- and C_6^-, equipped with the low-lying electronic excited state through the observation of the characteristic fast cooling [1–3]. The RF mechanism is valid only when the internal energy is larger than that of the electronic excited state E_{ex}. Once the stored ions are cooled down below this energy, further cooling should proceed slowly by the normal vibrational cooling. In this study, we focused on this slow process, which is observable only by storing the ions for the longer time than the typical vibrational cooling time scale of approximately tens ms in an electrostatic ion storage ring (the TMU E-ring).

We irradiated the stored ions with a tunable visible optical parametric oscillator (OPO) laser to induce the statistical thermally-driven delayed detachment. The neutralized products were detected by a microchannel plate (MCP) detector. The observed laser induced neutral yields reflects the population in a particular internal energy region E_0 after laser absorption of the energy of $nh\nu$ where n is the number of photon absorbed, and thus $E_0 - nh\nu$ ($= E_{obs}$) before photo-excitation. This energy window E_0 is defined by the decay times determined by the experimental configuration (E_0 = 3.88 eV for C_4^-, 4.49 eV for C_6^-), as demonstrated for these anions with internal energies above E_{ex} [1,2].

Measuring the laser induced neutral yields as a function of laser firing time after storage at each sampling energy E_{obs} = 0.43 -1.06 eV, we observed the time evolution of the excitation energy distribution during the cooling processes via the vibrational transitions at internal energies below E_{ex} (1.34 eV for C_4^-, 1.16 eV for C_6^-). The neutral yields increased and then decreased with time in the tens ms range as the result of the decrease in the excitation energy distribution. Every sampling energy E_{obs} was smaller than E_{ex}, and the neutral yields varied with much longer time than the RF cooling time scale of approximately 0.1 ms. This result is consistent with the behavior explained by vibrational cooling. At higher sampling energy E_{obs}, the peak of the neutral yields appeared earlier. From time difference of these peaks, we estimated the vibrational cooling rate. These were reasonably reproduced by simulation.

References

[1] G. Ito *et al.* 2014 *Phys. Rev. Lett.* **112** 183001

[2] N. Kono *et al.* 2015 *Phys. Chem. Chem. Phys.* **17** 24732

[3] Y. Ebara *et al.* 2016 *Phys. Rev. Lett.* **117** 133004

[1]E-mail: suzuki-ryuta@ed.tmu.ac.jp

IOP Conf. Series: Journal of Physics: Conf. Series **875** (2017) 102015 doi:10.1088/1742-6596/875/11/102015

Electron-transfer studies in potassium collisions with tetrachloromethane

Khrystyna Regeta[*1], **Tiago Cunha**[*], **Filipe Ferreira da Silva**[*], **Gustavo García**[#], **Paulo Limão-Vieira**[*2¶]

[*] Laboratório de Colisões Atómicas e Moleculares, CEFITEC, Departamento de Física, FCT, Universidade NOVA de Lisboa, 2829-516 Caparica, Portugal

[#] Instituto de Física Fundamental, Consejo Superior de Investigaciones Científicas (CSIC), Serrano 113-bis, 28006 Madrid, Spain

Synopsis We have studied electron transfer process in collisions of neutral tetrachloromethane molecular beam and potassium atom beam yielding negative ion formation. The fragment anions formed in this collision process have been assigned and compared against previously reported DEA (dissociative electron attachment) measurements.

Tetrachloromethane, CCl_4, has been widely used as a chemical cleaning agent, in polymerization processes as well as a feed gas in plasma processing. CCl_4 has also been widely used in research investigations due to its high cross section for electron attachment (*s*-wave) at virtually no electron energy, i.e. a shape resonance at ~ 0 eV [1,2], well-attuned to set the electron energy scale and energy resolution in such experiments.

The experimental setup consists of a crossed molecular beam experiment with two chambers (more details were given in Ref. [3]): one where the neutral K beam is produced in a charge exchange chamber and the second where the collision with the molecular target beam takes place. Anionic products formed after electron transfer process from K to the target molecule are extracted and mass analysed into a reflectron time-of-flight (TOF) spectrometer.

We report here time-of-flight mass spectra of anions formed in collisions of potassium atoms with tetrachloromethane molecules as a function of the lab frame collision energy. A TOF mass spectrum at 200 eV is shown in Figure 1. The most relevant fragment anions produced are: Cl^-, Cl_2^-, CCl_2^- and CCl_3^-. The dominant fragment anion, Cl^-, appears also in DEA [1,4] experiments, where its formation requires breaking a single bond.

We performed measurements at different collision energies and found that signal intensity is higher at lower collision energies. This is reminiscent of longer collision times which allows delaying autodetachment in detriment to dissociation. Such is only possible due to the stabilizing nature of the potassium cation in the vicinity of the temporary negative ion. Other fragment anions are considerably less intense, which we also found in DEA experiments [1,4].

Currently, we are working on the implementation of a K^+ beam energy loss analyser, which will give us valuable information on the accessed resonant states of the target molecules.

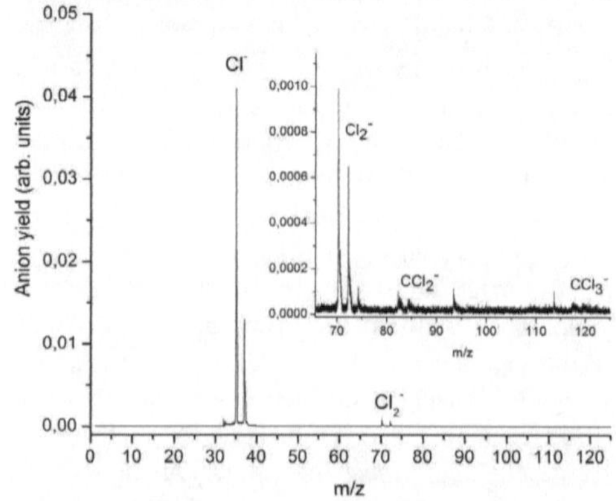

Figure 1. TOF mass spectrum of negative ion formation in K/ CCl_4 electron transfer collisions at 200 eV lab frame.

Further details will be given at the meeting.

References

[1] S. Matejcik, A Kiendler, A. Stamatovic, T.D. Märk 1995 *Int. J. Mass Spectrom. Ion Process.* **149/150** 311

[2] T. Oster, A. Khün and E. Illenberger 1989 *Int. J. Mass Spectrom. Ion Process.* **89** 1

[3] R. Antunes PhD thesis 2011 "The Role of Halouracils in Radiotherapy Studied by Electron Transfer in Atom-Molecule Collision Experiments", *Universidade Nova de Lisboa*

[4] E. Illenberger 1982 *Ber. Bunsenges. Phys. Chem.* **86** 252

[1]E-mail: k.regeta@fct.unl.pt

[2]E-mail: plimaovieira@fct.unl.pt

IOP Conf. Series: Journal of Physics: Conf. Series **875** (2017) 102016 doi:10.1088/1742-6596/875/11/102016

Detection of recurrent fluorescence photons emitted from C_4^-

Mai Yoshida* [1], **Takeshi Furukawa*** [2], **Jun Matsumoto†, Hajime Tanuma*, Toshiyuki Azuma‡*, Haruo Shiromaru†, and Klavs Hansen§¶**

* Department of Physics, Tokyo Metropolitan University, Hachioji, Tokyo 192-0397, Japan
† Department of Chemistry, Tokyo Metropolitan University, Hachioji, Tokyo 192-0397, Japan
‡ AMO Physics Laboratory, RIKEN, 2-1 Hirosawa, Wako, Saitama 351-0198, Japan
§ Department of Physics, University of Gothenburg 41296 Gothenburg, Sweden
¶Tianjin International Center of Nanoparticles and Nanosystems, Tianjin University, China

Synopsis We detected recurrent fluorescence (RF) in the region of visible light emitted from small carbon cluster anions C_4^- and C_6^- stored in an electrostatic ion storage ring. The observed RF decay profile for C_4^- is consistent to our simulation under the assignment that the RF is associated with the $C^2\Pi_g$-$X^2\Pi_u$ transition. In addition, observation of the RF at two different wavelengths suggests that the RF band is rather broad.

Cooling dynamics of isolated molecules plays an important role in molecular physics, photochemistry and molecular evolution in universe. The major cooling process of isolated molecules has been considered to be infrared radiation associated with vibrational transition. It is usually slow and of the order of ms or longer in the time scale.

For the last few years, a much faster pathway of radiative cooling associated with electronic transition has been confirmed experimentally for some isolated molecules, for example, chain-form small carbon cluster anions C_4^- and C_6^- [1,2] and polycyclic hydrocarbon (PAH) cations [3]. It is the visible-photon emission process via the electronic transitions after inverse internal conversion, converting the vibrational energy to the electronic energy. This process is referred to as recurrent fluorescence (RF) or Poincaré fluorescence. At first, the RF process has been studied not by detecting the emitted photon, but by observing the statistical delayed detachment or dissociation process of the molecular ions, which provides the information on the evolution of the internal energy and the cooling rate: the unusually fast cooling is a signature of the RF [1-3].

Recently we have succeeded in direct measurements of the RF photons from the C_6^- stored in an ion storage ring [4]. We observed energy-resolved photons by employing a photomultiplier tube combined with a bandpass filter-I (CWL: 607 nm ($h\nu$ = 2.04 eV), FWHM: 35nm) suitable for 2.04eV photons corresponding to the $C^2\Pi_g$ - $X^2\Pi_u$ electronic transition of C_6^-.

In this report, we show new results of the RF photons from C_4^-. Hot anions produced in the ion source, from C_2^- to C_6^-, were simultaneously stored in the ring. The time profile of the photon intensities detected by the photomultiplier tube, i.e., the synchronized periodic structure to the ion revolution, contributed to exclude uncorrelated background signals.

As is the case for the C_6^- anions, we detected the significant number of the 2.72 eV RF photons associated with the $C^2\Pi_u$ - $X^2\Pi_g$ electronic transition of C_4^- anions by using a bandpass filter-II (CWL: 460 nm ($h\nu$ = 2.69 eV), FWHM: 7 nm), and confirmed that the decay profile of the fluorescence is consistent with the simulated evolution of the level population based on the detailed balance theory. However, to our surprise, we found that the substantial amount of the RF photons for the C_4^- anions are detected with a bandpass filter–I.

The obtained experimental evidence is so far limited, however, this observation suggests that the RF band may be equipped with a non-negligible tail component to the longer wavelength side. This behavior is expected to be a crucial clue to clarify vibrational structures both on the high- and low-energy sides of the band origin of the RF spectrum, and they may imply transitions from the initial vibronic state to other states accompanying the vibrational excitation or deexcitation for several modes [4]

References

[1] G. Ito *et al.* 2014 *PRL* **112** 183001; N. Kono *et al.* 2015 *PCCP* **17** 24732
[2] V. Chandrasekaran *et al.* 2014 *JPCL* **5** 4078
[3] S. Martin *et al.* 2013 *PRL* **110** 063003; S. Martin *et al.* 2015 *PRA* **92** 053425
[4] Y. Ebara *et al.* 2016 *PRL* **117** 133004

[1] E-mail: yoshida-mai3@tmu.ac.jp
[2] E-mail: takeshi@tmu.ac.jp

IOP Conf. Series: Journal of Physics: Conf. Series **875** (2017) 102017 doi:10.1088/1742-6596/875/11/102017

Multiple ionization and dissociation of ethylene induced by collision of Xe^{9+}

K. Yokokawa*, J. Matsumoto*, H. Shiromaru*[1], P. Bhatt†, H. Kumar‡, C. P. Safvan†[2]

* Department of Chemistry, Tokyo Metropolitan University, 1-1 Minami-Osawa, Hachioji 192-0397, Japan
†Inter-University Accelerator Centre, Aruna Asaf Ali Marg, New Delhi 110067, India
‡Department of Physics and Astrophysics, University of Delhi, Delhi 110007, India

Synopsis Multiple ionization and subsequent ion-ion dissociation of ethylene was studied by using intermediate energy of Xe^{9+}. Delayed dissociation of $C_2H_4^{2+} \rightarrow C_2H_3^+ + H^+$ is discussed.

Collision-induced multiple ionization of organic molecules sometimes leads to characteristic reactions such as hydrogen migration and efficient formation of H_3^+ [1,2]. As for ethylene, which is the smallest planar hydrocarbon, multiple ionization by collision of Ar^{8+} at low-energy region, where electron-capture is dominant, was studied focusing attention onto the fragmentation scheme [3]. In the present study, collisional multiple ionization of ethylene was studied at considerably higher energy region, where electron emission from the target also works.

Experiments were conducted at the beamline in the Low Energy Ion Beam Facility of the Inter-University Accelerator Centre, using 2.25 MeV Xe^{9+} beam [4]. A schematic drawing of the setup is shown in Fig. 1. The target ethylene was introduced as an effusive beam, and the emitted electrons triggered the multi-hit position sensitive TOF measurement of the recoil ions.

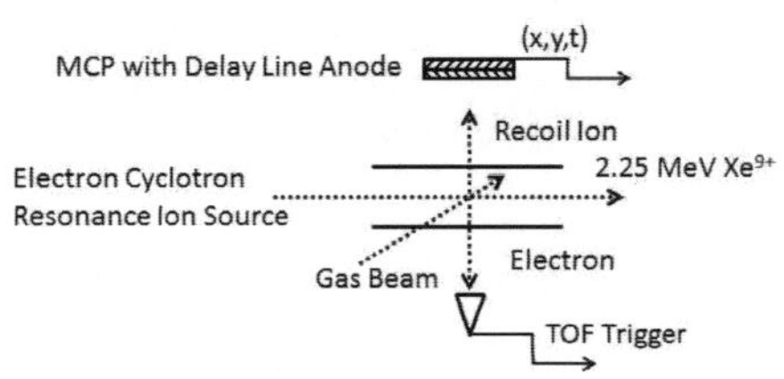

Figure 1. Schematic view of the apparatus. The details are given in [4].

A double-hit time-of-flight coincidence map, around the area for the heavy fragment ions, m/q = 24 – 28, with the light ones, m/q = 1 – 3, is shown in Fig. 2, in which the channels are represented as $[C_2H_3^+,H^+]$ for example. The $[C_2H^+,H_3^+]$ island is very weak but clearly above the background level. The diagonal line starting from the island $[C_2H_3^+,H^+]$ indicates delayed dissociation takes place. While the decay profile slightly deviates from a single exponential function, the lifetime of the metastable $C_2H_4^{2+}$ is estimated to be an order of 100 ns. For other channels, delayed reactions were not observed. By further analysis of the velocity vectors of the fragment ion pairs, the kinetic energy release distribution and angular distribution with respect to the projectile beam direction will be discussed.

This work was partly supported by TMU travel fund for students (KY).

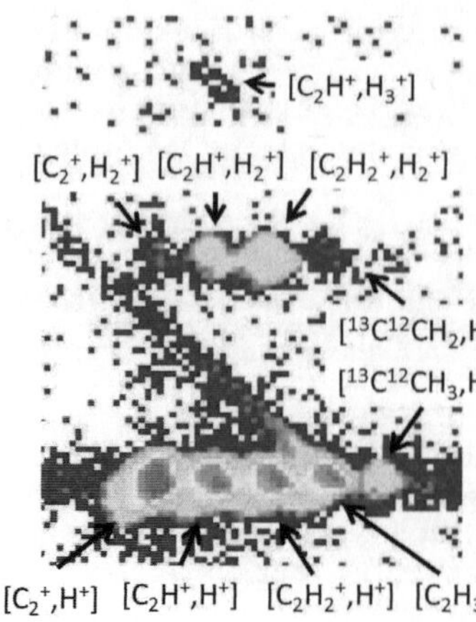

Figure 2. TOF coincidence map for $[C_2H_{4-n}^+(n=1{\sim}4), H^+]$, $[C_2H_{4-n}^+(n=2{\sim}4), H_2^+]$, and $[C_2H^+, H_3^+]$ channels. The color indicates the intensity in the log scale. The diagonal line is due to the delayed dissociation.

References

[1] S. De *et al.* 2006 *Phys. Rev.Lett.* **97** 213201
[2] K. Hayakawa *et al.* 2011 *J. Phys. B: At. Mol. Opt. Phys.* **44** 075207
[3] G. Veshapidze *et al.* 2004 *J. Phys. B: At. Mol. Phys.* **37** 2969
[4] A. Kumar *et al.* 2014 *Int. J. Mass Spectrom.* **374** 44

[1] E-mail: shiromaru-haruo@tmu.ac.jp
[2] E-mail: safvan@iuac.res.in

IOP Conf. Series: Journal of Physics: Conf. Series **875** (2017) 102018 doi:10.1088/1742-6596/875/11/102018

Cold inelastic collisions of He(^{1}S) with the smallest astrophysical anion observed, CN$^-$($^1\Sigma^+$): an accurate quantum dynamical study.

Carelli F.[*1], **Boese A.D.**[†2], **Gianturco F.A.**[*3], **Wester R.**[*4]

[*]Institute of Ion and Applied Physics, University of Innsbruck, Technickerstrasse 25, A-6020 Innsbruck, Austria
[†]Institute of Chemistry, Karl-Franzens University of Graz, Heinrichstrasse 28, A-8010, Graz, Austria

Synopsis Accurate state-to-state rovibrational inelastic collisional cross sections involving the smallest molecular anion astronomically detected with He(^{1}S) based on two recently computed energy surfaces will be presented. A detailed comparison between the collisional findings provided either using a high-level of accuracy in the potential energy surface or taking advantage of the full-dimensional approach will be discussed.

In dense and dark regions of the interstellar medium, two-body inelastic collisions involving p-/o-H_2($^1\Sigma_g^+$) and He(^{1}S) as projectiles are known to be crucial. They affect in fact the populations of the internal rovitational levels involved in the radiative transitions by which a given species can be observed and its abundance inferred. If collisions with He(^{1}S) are not as important as those involving p-/o-H_2($^1\Sigma_g^+$) on the basis of their column densities ratio, the former however provide an interesting choice for testing theory and supporting experimental measurements [1, 2].

CN$^-$($^1\Sigma^+$) is the smallest detected interstellar anion and to support the observations as well as to help to predict its abundance in interstellar regions other than the molecular envelope IRC+10216 where it has been observed, accurate inelastic collisional rate coefficients with the He(^{1}S) and p-/o-H_2($^1\Sigma_g^+$) are essential [3].

Two potential energy surfaces for the ground electronic state of the CN$^-$-He anionic van der Waals complex have been recently constructed [4]. The first describes the complex in the rigid-rotor approximation at the CCSD(T)/aug-cc-pV{5,6}Z (cp corrected) + △CCSDT/cc-pV{T,Q}Z + △CCSDT(Q)/cc-pVTZ + rel. CCSD(T)/aug-cc-pV{T,Q}Z-DK + core-valence CCSD(T)/aug-cc-pwCV{T,Q}Z level and shows a general accuracy of about 0.2/0.4 cm^{-1}. The second one is full-dimensional at the CCSD(T)/aug-cc-pV{5,6}Z (cp corrected) level and is averaged on the lowest anionic diatom vibrational wave functions. Assuming the system CN$^-$-He as a prototype for polar negatively charged van der Waals complexes, a detailed comparison between the inelastic collisional findings obtained using either the full-dimensional approach or very high level of accuracy will be presented and discussed.

References

[1] M. Costes, C. Naulin 2016 *Chemical Science* **7** 2462

[2] A. Faure, P. Jankowski, T. Stoecklin, K. Szalewicz 2016 *Nature Scientific Reports* **6** 28449

[3] M. Agundez, J. Cernicharo et al. 2010 *Astron. & Astrophys.* **517** L2

[4] F. Carelli et al., in preparation

[1]E-mail: Fabio.Carelli@uibk.ac.at
[2]E-mail: Adrian_Daniel.Boese@uni-graz.at
[3]E-mail: Francesco.Gianturco@uibk.ac.at
[4]E-mail: Roland.Wester@uibk.ac.at

Probing anion resonances in FeO^-: a species of astrophysical relevance

Roby Chacko[†1], **Shreyak Banhatti**[†2], **A. K. Gupta**[*3], **and G. Aravind**[†4]

[†] Department of Physics, Indian Institute of Technology Madras, Chennai - 600 036, India
[*] Nuclear Physics Division, Bhabha Atomic Research Centre, Mumbai 400 085, India

Synopsis We report the first experimental probe on resonance states in FeO anion, a molecule of astrophysical relevance. Excitation of FeO^- through collision with argon was performed to measure the kinetic energy release in the centre of mass frame. Results show access of two anion resonances upon collisional excitation. We discuss our evaluation of these two resonances from kinetic energy release in the centre of mass frame. The astrophysical importance of these anion resonances will be presented.

FeO is the first iron-bearing molecule discovered in the inter-stellar medium (ISM) [1], [2]. Depletion of refractory elements in the ISM motivates the search and study of iron-bearing ISM molecules. Recently, another iron-bearing molecule, FeCN was unambiguously identified in the envelope of carbon-rich asymptotic giant branch star, IRC+10216 [3]. Although ISM anions were predicted in 1981 by Herbst *et. al.* [4], only in 2006 the first ISM anion C_6H^- was detected [5]. This was followed by recent detections of CN^-, C_3N^-, C_4H^-, C_6H^-, C_8H^- and C_5N^-. Anion resonances are states embedded in the detachment continuum and are short-lived states. They are formidable to study theoretically due to high electron correlation. Hence experimental investigations such as dissociative electron attachment [6], collisional or photoexcitation become important to probe these resonances. In the present work, we have probed FeO^- which is detected in its neutral state in the ISM. Resonance states of FeO^- can be accessed by electron attachment to FeO in the ISM and would result in autodetachment or dissociation. Anion resonances thus critically determine the ISM reactions and relative abundances of various ISM species.

FeO^- were produced in Cs-sputter source and were accelerated to 15keV and eventually collided with Argon in a collision cell. The collisionally excited anion dissociates and the anionic daughters scattered in the forward and backward directions were only detected using an energy analyser. Two daughter anions were detected in our experiments as shown in the figure. The kinetic energy(KE) released in the center of mass frame was evaluated from the lab frame KE release. The energy released in the center of mass frame for both the daughter anions were different, indicating access of two different anionic states in this experiment. In our recent work on FeC anions we have observed two anion resonances [7]. We have evaluated the ground state of the FeO^- and its vibrational wavefunctions using LEVEL program [7]. We employ reflection method, on the CM frame energy release to deduce the two anion resonances accessed in our experiment. Our results are of paramount importance in studying the disscociation dynamics involving iron containing ISM molecules. The important implications of our results to astronomy will be discussed in the presentation.

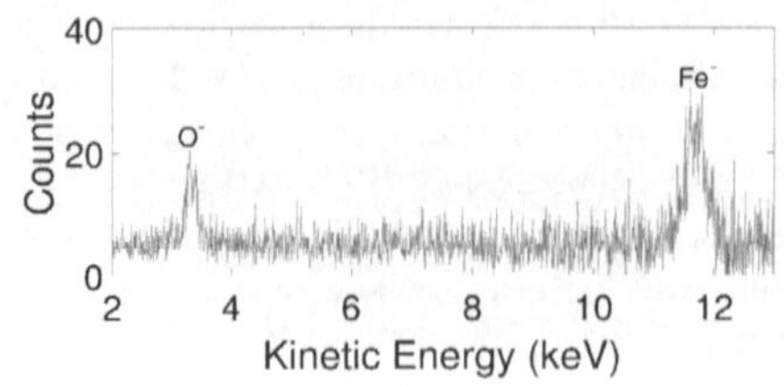

Figure 1. Daughter-ion peaks recorded for the CID of FeC^- anion

References

[1] C. M. Walmsley *et al.* 2002 *Astrophys. J.* **566** L109
[2] R. S. Furuya *et al.* 2003 *Astr. Astrophys.* **409** L21
[3] L. N. Zack *et al.* 2011 *Astrophys. J. Lett.* **733** L36
[4] E. Herbst 1981 *Nature* **289** 656
[5] M. C. McCarthy *et al.* 2006 *Astrophys. J. Lett.* **652** L141
[6] V. S. Prabhudesai *et al.* 2005 *Phys. Rev. Lett.* **95** 143202
[7] M. Nrisimhamurty *et al.* 2016 *Astrophys. J* **833** 269

[1]E-mail: robychako@gmail.com
[2]E-mail: shreyak.rocker@gmail.com
[3]E-mail: anit@tifr.res.in
[4]E-mail: garavind@iitm.ac.in

IOP Conf. Series: Journal of Physics: Conf. Series **875** (2017) 102020 doi:10.1088/1742-6596/875/11/102020

Reactions of O^- with D_2 at low temperatures 10 – 300 K

Radek Plašil*[1], Thuy Dung Tran*, Štěpán Roučka*, Pavol Jusko*, Dmytro Mulin*, Illia Zymak*, Petr Dohnal*, and Juraj Glosík*

* Department of Surface and Plasma Science, Faculty of Mathematics and Physics, Charles University in Prague, 180 00 Prague, Czech Republic

Synopsis The reaction of O^- anions with molecular deuterium D_2 has been studied experimentally using a cryogenic 22-pole radiofrequency ion trap. There were observed two reaction channels. In the associative detachment D_2O is formed and for atom transfer formation $OD^- + D$ was observed. The rate coefficients of the reactions have been determined at temperatures ranging from 10 K to 300 K.

The reaction of O^- with D_2 has two exothermic channels corresponding to associative detachment and deuterium atom transfer

$$O^- + D_2 \longrightarrow D_2O + e^-, \qquad \Delta H = -3.65 \text{ eV},$$

$$O^- + D_2 \longrightarrow OD^- + D, \qquad \Delta H = -0.28 \text{ eV}.$$

Energetically possible are also reactions leading to the formation of metastable D_2O^- anion in three body or radiative association. However these processes have not been detected experimentally. Studies of gas-phase processes involving water and especially those leading to isotopic fractionation [1] is essential for understanding of the water formation in the Universe.

The reactions were studied using a 22-pole radiofrequency ion trap. It was mounted on a cryo-cooler in an ultra-high vacuum system. The measuring procedure was based on iterative filling of the trap with a well-defined number of primary ions O^- where they react with molecular deuterium. The content of the trap were analyzed after chosen times using a quadrupole mass spectrometer and a micro-channel plate detector. Further detail may be found in [2] and references therein.

We studied previously analogous reactions involving O^- and H_2. At first we measured electron energy spectrum originating from associative detachment [3] at 300 K. Later on we studied both reactions using the 22-pole ion trap [4].

Very interesting dependence on temperature is shown on Figure 1. Up to now the associative detachment has been measured mostly at temperatures higher than 300 K and the observed reaction rate coefficient was about ⅓ of the Langevin rate. It is explained by the fact that trajectories on only one of three electronic surfaces of the collision system can lead directly to autodetachment.

At lower temperatures we observed unexpectedly high value of the associative detachment rate coefficient. Theoretical model in our paper [4] for the reaction $O^- + H_2$ explained this behavior by rearrangement of the collisional system in a shallow local minimum. The increase of the reaction rate coefficient at low temperatures is in qualitative agreement with the theoretical predictions. The presented data will be compared with classical trajectory simulation model. An isotopic effect will be evaluated.

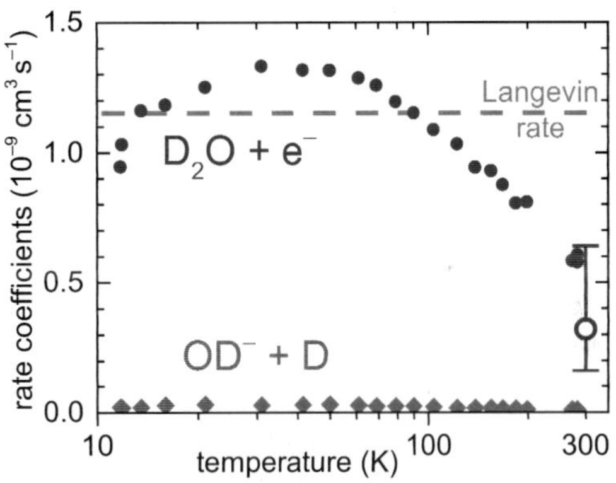

Figure 1. Measured temperature dependencies of the reaction rate coefficients for both channels of $O^- + D_2$ reaction. The open circle indicates results of our previous experiment [4].

We thank the TU Chemnitz, the DFG and prof. D. Gerlich for lending us the apparatus. Supported by GACR 15-15077S and GACR 17-18067S.

References

[1] L. I. Cleeves *et al.* 2014 *Science* **345** 1590
[2] D. Gerlich and G. Borodi 2009 *Faraday Discussions* **142** 57
[3] P. Jusko *et al.* 2013 *Int. J. Mass Spectr.* **352** 19
[4] P. Jusko *et al.* 2015 *J. Chem. Phys.* **142** 014304

[1] E-mail: Radek.Plasil@mff.cuni.cz

Charge dependence of fragmentation process induced by ion collisions with furan molecule

E. Erdmann*[1], M. Łabuda*, S. Díaz-Tendero†, N.F. Aguirre†, M. Alcamí†

*Department of Theoretical Physics and Quantum Information, Gdańsk University of Technology, Narutowicza 11/12, 80-233 Gdańsk, Poland
†Departamento de Química, Módulo 13, Universidad Autónoma de Madrid, 28049, Madrid, Spain

Synopsis The goal of this work is to describe the system evolution after ion-molecule interaction. We combine different quantum chemistry and statistical mechanics approaches in order to give extended description of the process. Herein we report on a recent study of the fragmentation mechanism of neutral, singly- and doubly-ionized furan molecule in the gas phase.

Furan belongs to the family of ring structures often taken as an elementary analogue system of deoxyribose, fundamental component of the DNA backbone chain. Therefore, it can serve as a model system for track simulations in biological medium. Transfer of energy to the environment of the DNA initiates a sequence of events depending on the quantity of energy being transferred and electronic structure of the surrounding matter. Extensive theoretical and experimental research has already focused on understanding the fundamental reactions occurring in biological matter after irradiation. However, as the mechanism of formation of various products is still unclear, in this work we aim at extending and complementing previous studies.

In this communication, we present a merged approach of Density Functional Theory (DFT) quantum chemistry methods with statistical Monte Carlo technique. Firstly, we have performed molecular dynamics simulations using the Atom-Centered Density Matrix Propagation (ADMP) method [1] for the internal energy deposited to the system varying from 5 to 30 eV. Extensive statistical analysis of the obtained trajectories allowed us to identify most abundant dissociation channels. The dynamics of the neutral system is dominated by unimolecular opening of the furan ring and hydrogen transfer. In the case of singly-ionized furan the most abundant channels consist of production of formyl radical (CHO) and cyclopropenyl cation ($C_3H_3^+$) as well as $C_3H_4O^+$ cation and hydrogen atom for higher energies. Doubly-ionized furan fragmentation follows a dissociation path with one channel clearly distinguishable as most probable (67% of the trajectories for the energy of 14 eV). This channel leads to production of two fragments: formyl cation (CHO^+) and cyclopropenyl cation ($C_3H_3^+$) by direct cleavage of the furan ring.

Snapshots of the most probable reaction channel for singly- and doubly-ionized furan are shown in Fig. 1. Subsequently, in order to gain complementary energetic and structural information on the fragmentation mechanism, we have performed calculations of the critical points on the potential energy surface of the most apparent channels appearing in the dynamical calculations.

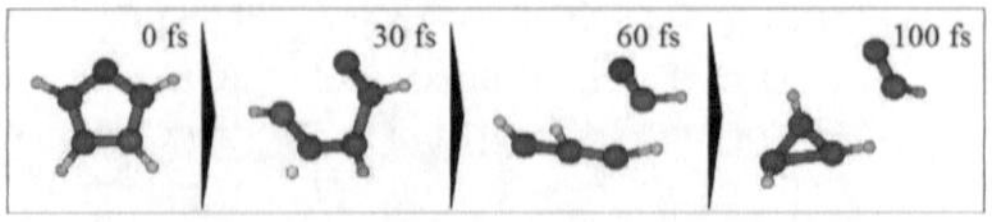

Figure 1. Mechanism of fragmentation to $C_3H_3^+$/CHO and $C_3H_3^+$/CHO^+ channels.

The statistical Microcanonical Metropolis Monte Carlo (M_3C) method [2,3] has been employed as an alternative approach to understanding the fragmentation process. The M_3C technique provides fragmentation probabilities as a function of the excitation energy. Comparison of our preliminary results with the experimental data [4] leads to the conclusion that M_3C method is a promising theoretical tool for reproducing the mass spectrum of the studied system.

Acknowledgments: We acknowledge financial support from XLIC COST Action, Erasmus Plus programme (M.Ł.) and GUT. Calculations have been performed at Wroclaw Centre for Networking and Supercomputing and Centro de Computación Científica-UAM in Madrid.

References

[1] H. B. Schlegel *et al.* (2001)J. Chem. Phys. **114,** 9758
[2] S. Díaz-Tendero *et al.* (2005) Phys. Rev. A **71,** 033202
[3] N.F. Aguirre *et al.* (to be submitted)
[4] M. Dampc *et al.* (2015) J. Phys. B: At. Mol. Opt. Phys. **48,** 165202

[1]E-mail: eerdmann@mif.pg.gda.pl

IOP Conf. Series: Journal of Physics: Conf. Series **875** (2017) 102022 doi:10.1088/1742-6596/875/11/102022

Excitation and fragmentation in high velocity C_nN^+ - He collisions

T.Mahajan[†], T.Id Barkach[*], N.F.Aguirre[¤], M.Alcami[¤], M.Bonnin[†], M.Chabot[*], S.Diaz-Tendero[¤], F.Geslin[*], T.Hamelin[*], F.Hammache[*], C.Illescas[¤], A.Jallat[*], A.Jorge[¤], T.Launoy[§], T.K.C.Le[†], A.LePadellec[‡], F.Martin[¤], A.Meyer[*], L.Perrot[*], T.Pino[†], B.Pons[#], N. de Séréville[*], K.Béroff[†1]

[†] Institut des Sciences Moléculaires d'Orsay, CNRS- Univ.Paris-Sud F-91406 Orsay, France
[*] Institut de Physique Nucléaire d'Orsay, CNRS- Univ.Paris-Sud F-91405 Orsay, France
[¤] Departamento de Quimica Universidad Autonoma de Madrid, 28049 Madrid, Spain
[§] Laboratoire de Chimie Quantique et Photophysique Univ Libre de Bruxelles, CP160/09 1050 Bruxelles, Belgium
[‡] Institut de Recherche en Astrophysique et Planétologie, CNRS-INP Univ. Toulouse 3 F-31028 Toulouse, France
[#] CELIA Univ.Bordeaux CNRS UMR 5107 CEA 351 Cours de la Libération, 33405 Talence, France

Synopsis: We will present measurements and modeling for two aspects of the C_nN^+ - He collisions (n=1-3, v=2.25 a.u) : cross sections for electronic excitation processes and fragmentation branching ratios for the excited and ionized C_nN^{q+} molecules produced in the collision (q=-1,0,1,2-5).

The study of molecule-atom collisions is a difficult topic, both from the experimental and theoretical points of view. In the high velocity regime, mostly small molecular systems have been studied [1]. On the other hand, the so-called Independent Atom and Electron (IAE) model was applied recently with reasonable success to C_n^+-He, Ar systems with state of the art CTMC and SCAOCC P(b) probabilities [2]. We will test this approach again in this work.

Fragmentation of the excited molecular system is another topic of interest. In high velocity collisions (τ_{coll}~10^{-16}s) it occurs well after the excitation and can be treated separately. The MMMC approach and its new more general version M3C [3] is dedicated to treatment of statistical fragmentation. The systems studied here belong typically to this class of fragmentation [4]. We will ultimately compare our experimental fragmentation branching ratios (BR) to predictions of this statistical approach.

Experiments have been performed at the Tandem accelerator in Orsay with beams of C_nN^+ molecular ions (n=1-3) of constant velocity v=2.25 a.u colliding with helium atoms. The setup is identical to the one described in [5] (see also Jallat et al, this conference). Briefly the setup allows to reconstruct, from fragments complete collection and identification in charge and mass, the charge q of the projectile after the collision, signature of the process. An example is given in Table 1 for the case of double electron capture (q=-1) in the C_2N^+ - He collision where contribution of various channels to the {C_2N^-} production is reported.

We will present two types of results. First experimental cross sections for various electronic processes will be presented and compared to predictions of the IAE + CTMC calculations. These calculations will use structure calculations for C_nN^+ systems that we performed. Second, fragmentation BR for C_nN^{q+} species with q=-1,0,1,2,3-5 will be presented. In addition to the fundamental aspects discussed before, these BR are also of interest in astrochemistry as already pointed out [6]. Note that C_nN species, in their neutral and anionic forms, have been detected in interstellar medium [7] and planetary atmospheres [8].

Table 1 Measured fragmentation BR of {C_2N^-} species produced by double electron capture in the collision C_2N^+ - He (v=2.25 a.u).

Channel	Exp. BR	Error
C_2N^-	0.40	0.04
CN^- + C	0.51	0.06
C^- + CN	0.07	0.02
C_2^- + N	≤0.01	
C^- + C + N	0.020	0.012

References

[1] H.Luna et al Phys. Rev. A 2016 **93** 052705
[2] G.Labaigt et al JPB 48, 075201 2015
[3] N.F. Aguirre et al JCTC 2017 *to appear*
[4] G. Martinet et al Phys. Rev. Lett. 2004 **93** 063401
[5] T. Launoy et al Phys. Rev. A 2017 **95** 022711
[6] M. Chabot et al ApJ 2013 **771**:90
[7] M.Agundez et al Chem.Tev 2013 **113** 8710
[8] V.Vuitton et al Planet. Space Sc. 2009 **57** 1558

[1] E-mail: karine.beroff@u-psud.fr

Development of transferable water-halide potentials

Raúl Rodríguez-Segundo*[1], Daniel J. Arismendi-Arrieta*[2], Rita Prosmiti*[3]

* Institute of Fundamental Physics (IFF-CSIC), Serrano 123, 28006 Madrid, Spain

Synopsis The behavior of ions in aqueous environment is of fundamental interest in physics and chemistry. A topic of research is focused on the modeling of the ion-ion, ion-water, and water-water interactions. So far, the water-water interactions has been widely studied what has lead to a variety of water potential models of increasing complexity. However, how ions change the water network is an open debate in the literature. Thus, the development of a generalized representation for the ion-water interactions coupled to available reliable water models provide a rational route to follow for studying ion-hydrates from finite-size cluster systems to condensed phase.

Despite experimental and theoretical efforts to provide a direct molecular-level understanding of hydration of ions in different environments, from gas to condensed phases, investigations have been constrained by the limited predictive abilities of the empirical models, the high computational cost for correlated *abinitio* methods, and the intrinsic deficiencies of common functionals in describing weak interactions. A fully *ab initio* approach to these appears, at first glance, daunting. However, progress has been made, with the most advanced *ab initio*-based approaches clearly being a step forward combining computational efficiency with fitting accuracy.

Thus, the first step to achieve it to adopt a bottom-up approach starting from the study of small to medium size clusters, where valuable information has been acquired by IR experimental studies and *ab initio* computations. We focus on the representation of the interactions based on first-principle data of solvated monovalent (halide X= F^-, Cl^-, Br^-, I^-) ions [1]. The new ion-water potential energy surfaces (PESs) are constructed by an extended version of the Thole type model (TTM) to describe halide ion-water interactions in conjunction with different TTM-based water model potentials. So, $V_{tot} = V_{W_n}^{inter} + V_{W}^{intra} + V_{i-TTM}^{inter}$, where $V_{W_n}^{inter}$ corresponds to MB-pol, TTM4, TTM3 and TTM2 water models, and V_{W}^{intra} the PS water monomer PES. The new $V_{i-TTM}^{inter} = V_{TTM,elec} + V_{TTM,ind} + \sum_{i=1}^{n}(V_{rep}(X,i) + V_{disp}(X,i)$ PESs count with four contributions: electrostatic, induction, repulsion and dispersion interaction energy terms, with n being the number of surrounding water molecules. Thus, the potential functions explicitly count with polarization terms, include angular anisotropy in short- and long-range terms by adjusting to *ab initio* energies and polarizabilities from explicitly correlated CCSD(T)-F12/cc-pVQZ-F12/aug-cc-pVQZ-PP and CCSD(T)/t-aug-cc-pV5Z/t-aug-cc-pV5Z-PP calculations, respectively, while dispersion coefficients are obtained from DFT/LC-ωPBE/aug-cc-pVTZ calculations using the XDM model (see Figure 1). The quality of the generalized V_{tot} PESs and their performance are checked by comparing to computationally expensive *ab initio* CCSD(T)-F12/DF-MP2 calculations on optimized structures and energies of larger ion-water (with n up to 8) systems. In addition, DFT approaches have also employed, and in a systematic way the performance of various functionals with and without dispersion corrections is investigated. [2, 3]

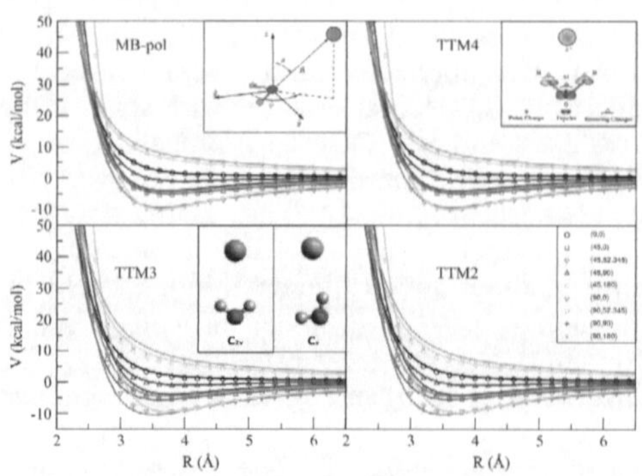

Figure 1. Potential curves (solid lines) and corresponding CCSD(T)-F12 reference data (symbols) calculated for different (θ, ϕ) orientations of the $I^-(H_2O)$ dimers employing the indicated TTM water models. In the insert panels we display the coordinate system used (upper/left), the charge and dipole-moment positions for i-TTM4/MB-pol models (upper/right), and the C_s and C_{2v} minimum and saddle-point geometries of $I^-(H_2O)$ (lower/left).

References

[1] D. J. Arismendi-Arrieta *et al.* 2015 *J. Phys. Conf. Ser.* **635** 032009

[2] D. J. Arismendi-Arrieta *et al.* 2016 *J. Phys. Chem. B* **120** 1822

[3] R. Rodríguez *et al.* 2017 *in preparation*

[1]E-mail: raul@iff.csic.es
[2]E-mail: d.arismendi@iff.csic.es
[3]E-mail: rita@iff.csic.es

IOP Conf. Series: Journal of Physics: Conf. Series **875** (2017) 102024 doi:10.1088/1742-6596/875/11/102024

K-K electron transfer and K-ionization in fast ion collisions with adenine and water using KLL Auger electron technique

Chandan Bagdia*, S. Bhattacharjee*, M. Roychowdhury*, A. Mandal*, Lokesh C Tribedi* [1]

* Department of Atomic and Nuclear Physics, Tata Institute of Fundamental Research, Colaba, Mumbai – 400005, India.

Synopsis In this report, we present a study of K-shell ionization, state selective K-to-K electron capture cross sections from C and N atoms in the adenine and other gas molecules using K-LL Auger electron technique, which is applied for first time. Energy dependence of K-ionization and K-K electron capture has also been measured. The e-transfer cross section is found to be substantial compared to the K-ionization. Total σ_{KK} and σ_{KI} for the entire molecule was deduced.

In hadron therapy, DNA/RNA damages as a result of energy loss of projectile in bio-molecular system. Energy loss of projectile can take place via different processes like electron emission, electron capture from target. Study of damage of DNA/RNA or water via electron emission upon ion impact is hot topic in recent time [1-4]. Electron capture is a fundamental process in ion-molecule collision. We report here the first measurement of state selective, 1s-1s electron capture cross section from constituent atoms in nucleobase (adenine) molecule ($C_5H_5N_5$). The 1s-ionization (KI) cross section has also been measured.

For the first time, by measuring the projectile charge state dependence of KLL Auger electron emission, the K-K e-capture cross section has been obtained. Experiments have been performed using effusive vapor jet of adenine molecules. Electrons emitted upon ion impact has been measured using hemispherical electron energy analyzer mounted on turntable inside high vacuum chamber. Data has been collected at 150^0 from beam direction and then total cross section was calculated by assuming isotropic distribution of Auger electron emission. In some cases whole angular distribution was measured. Energy dependence of cross section has been measured for C, O and Si projectile of energy 2-5 MeV/u. Measurements for N_2 and CO_2 have been performed in flooded chamber experiment.

The data obtained from adenine has been compared with those from gas molecules (N_2 and CO_2). The KI cross sections, obtained using no-K-vacancy projectiles, has been compared with the ECPSSR (Energy loss (E) and Coulomb deflection (C) perturbed stationary states (PSS) and relativistic corrections (R)) calculation and a good agreement has been observed (see Figure). The σ_{KK}, derived from the KLL yield with bare or H-like projectiles, shows almost energy independence for C and O ions. The σ_{KI} however tends to reduce with energy and well reproduced by the ECPSSR. Detailed analysis and results will be presented.

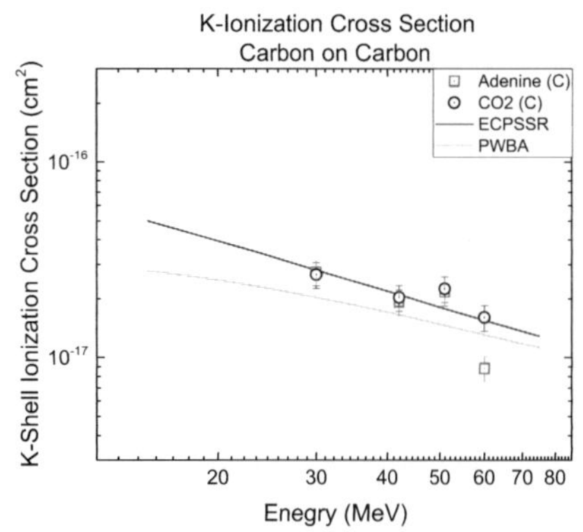

Figure 1. Energy dependence of KI for C on C from adenine and CO_2 along with ECPSSR and PWBA.

References

[1]A. Agnihotri *et al.* 2013 *Phys. Rev. A* **87** 032716
[2]A. Agnihotri *et al.* 2012 *Phys. Rev. A* **85** 032711
[3]A. Agnihotri *et al. 2013 J. Phys. B* **46** 185201
[4]S. Bhattacharjee *et al. 2016 J. Phys. B* **49** 065202

[1] E-mail: lokesh@tifr.res.in; ltribedi@gmail.com

IOP Conf. Series: Journal of Physics: Conf. Series **875** (2017) 102025 doi:10.1088/1742-6596/875/11/102025

Conformation analysis and semiclassical dynamics study of charge exchange process induced by collision of C^{2+} ions with tetrahydrofuran

Ewa Erdmann* and Marta Łabuda[1]*

* Faculty of Applied Physics and Mathematics, Gdańsk University of Technology, Narutowicza 11/12, 80233 Gdańsk, Poland

Synopsis We present the results of recent investigations of charge transfer processes in the collisions of C^{2+} ions with heterocyclic molecules of tetrahydrofuran.

Charge transfer (CT) in atomic and molecular collisions has been intensively studied by us in the past decade. [1-3] The importance of this process is evident in different fields of science such as chemistry, material science, astrophysics, laser research, biophysics and many others. For instance, the understanding of the CT mechanism is essential to address radiation damage in DNA. Important damage is due to secondary particles: low-energy electrons, radicals, or singly and multiply charged ions generated along the track after the interaction of the ionizing radiation with the biological medium. In this regard, we present a complete theoretical study of a fundamental CT process occurring in collision between multiply charged carbon ions and the tetrahydrofuran molecule. Tetrahydrofuran (THF, C_4H_8O) heterocyclic molecule is a model molecule which can be considered as a simple analog of the deoxyribose sugar and also structural unit for the DNA or RNA building blocks. A detailed analysis of the electronic structure of the THF molecule by employing several accurate quantum chemistry approaches is presented to determine the geometry of the molecule and to define the most important orbitals which are included in the process. The planar ($\theta = 0°$) and perpendicular ($\theta = 90°$) attacks of the C^{2+} projectile on the THF target molecule (being in envelope C_{2v} and twisted C_s symmetry orientation) were investigated using *ab initio* potential energy curves and non-adiabatic couplings calculated along internuclear distance *R* followed by a semi-classical collisional treatment with the impact energy range from eV to keV. Single and double electron capture channels have been identified in the potential energy curves. A strong delocalization of the π electrons of the THF ring towards the C^{2+} ion leading to the creation of a chemical bond between the oxygen atom and the colliding ion has been observed.

The presence of the cluster of the $[C^{+}\text{-}C_4H_8O]^{+}$ temporary formed during the interaction has been noticed, which confirms the experimental prediction [4] for the C^{+}- THF system. Another important conclusion of our work concerns the calculated values of CT cross sections, which strongly depend on the orientation of the THF molecule. It has been pointed out that change of the symmetry of the THF from C_{2v} to C_s in the collision with C^{2+} ions leads to the significant increase of the cross sections values even by three orders of magnitude (Figure 1).

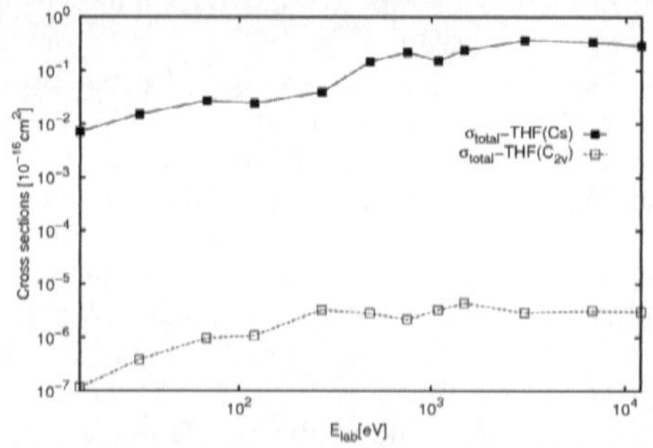

Figure 1. Averaged charge transfer cross sections for different orientations of the C^{2+} projectile toward the THF molecule at C_{2v} and C_s symmetry.

Taking into account the dependence between the fragmentation and electron capture processes, this would suggest that the CT process seems to be the most efficient and playing the most significant role when the carbon ion is colliding the THF in C_s symmetry with displaced orientation of $\alpha = 23°$ along C-O bond. Our work is a basis to study the fragmentation processes induced by the collision of ions and protons with the biologically relevant biomolecules and their analogs which we plan to present in future.

We thank the COST Action XLIC, Erasmus Plus Programme (M.Ł.) for financial support. Calculations were done at the Wrocław Centre for Net-working and Supercomputing and Academic Computer Centre in Gdańsk.

References

[1] M. C. Bacchus-Montabonel, M. Łabuda, Y. Tergiman, J .Sienkiewicz, 2005 Phys. Rev. A **72** 052706
[2] M. Łabuda, J. González-Vázquez, L. González 2010 Phys. Chem. Chem. Phys. **12** 5439
[3] M. Łabuda, J. González-Vázquez, F. Martin and L. González. 2012 Chemical Physics **400** 165
[4] T. Wąsowicz, B. Pranszke 2015 J. Phys. Chem. A **119** 581

[1] E-mail: marta.labuda@pg.gda.pl

Benchmarking the performance of density-functional-based approaches on intermolecular interactions of Helium–water complexes

M. Blanco de Paz[† 1], **D. J. Arismendi Arrieta**[†2], **R. Prosmiti**[† 3]

[†] Institute of Fundamental Physics (IFF-CSIC), Serrano 123, 28006 Madrid, Spain

Synopsis The He-water systems remain relatively unexplored. As only under extreme conditions, noble gases can be made to react to produce compounds, we investigated the characteristics of the interaction among He atom and water ice channels or clathrate cavities from first-principles calculations. The ability of DFT-based methods to describe such weak interactions might lead to further insights on noble-gas inclusion compounds.

In this work we studied structural properties and energetics of various water containing systems, from filled ice structures (e.g. horizontal and perpendicular I_{II} and I_h) to cavities of clathrate hydrates (e.g. 5^{12}, $5^{12}6$, etc) [1]. Such compounds are crystalline solids with an ice-like hydrogen bonded network that trap atoms or molecules, and in our case the "guest" atom in the clathrate cavity or in the ice channels is the He; the lightest noble gas atom. The predominant intramolecular interactions in such complexes are the hydrogen bonding between water molecules, and van der Waals (vdW) forces between the He and water molecules. Thus, *ab initio* technology have been used to investigate the guest-host interactions. Electronic structure calculations were carried out at MP2/CCSD(T) levels of theory and serve as the reference data. In turn, DFT and dispersion-corrected (DFT-D) [2] methods are then employed to describe the intermolecular interactions of such weakly bound systems. The main disadvantage of DFT-based approaches is the poorly description of hydrogen bond and vdW interactions. Therefore, we carried out calculations, choosing sets of dimers formed by one He atom and increasing number of water complexes. Given the delicate balance between the interaction energy components present on these systems, we aim to evaluate DFT and DFT-D functionals from the "simplest system" formed by one water molecule and govern exclusively by vdW forces; followed by the study of the "fundamental units" present in all ice-like frameworks (such as pentamers and hexamers), where hydrogen bonding start to have relevance on the interaction; and finally, by analyzing the interactions in the "building blocks" of ice channels and clathrate cavities.

For such purpose, we proposed [3] a protocol based on the analysis of: a) the relative error on a single optimized configuration. As we found that this is not a conclusive filter, we also calculated b) the interaction energy curve along a relevant coordinate for each type of He-water system. We found that non-local correlation effects can be recovered only with functionals of the higher rungs of Jacob's ladder, while dispersion corrections should be included once wisely and well tested, as vdW forces are fundamentals on the description of intermediate and long range interactions. In this sense, we observed that by increasing the number of water molecules, or hydrogen bonds, the DFT functionals which keep the same interaction energy curve profile as the reference one, are: TPSS, B3LYP, B2PLYP, mPW2PLYP, CAM-B3LYP, LC-wPBE and wB97X (see Figure 1). Besides, in most of the functionals used (including GGA, meta-GGA, hybrid-GGA, hydrid-meta-GGA, double-hybrid, range-separated hydrid), the part of dispersion is better recovered by including D3BJ dispersive term. Futhermore, considering the computational cost, the functionals with best relation between computational efficiency and accuracy are: LC-wPBE and wB97XD.

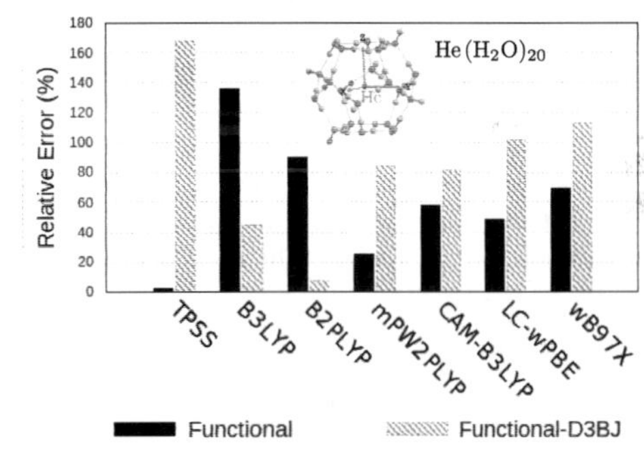

Figure 1. Performance of the indicated DFT and DFT-D functionals on a single MP2-optimized structure of 5^{12} He-clathrate-like cage (see inset panel).

References

[1] E. D. Sloan, 2003 *Nature* **426** 353-363

[2] S. Grimme *et al.* 2016 *Chem. Rev* **116** 5105

[3] M. Blanco de Paz *et al.* 2017 *in preparation*

[1]E-mail: maria.blanco@iff.csic.es
[2]E-mail: d.arismendi@iff.csic.es
[3]E-mail: rita@iff.csic.es

Electron emission in ionization of bromouracil by fast bare carbon ions

Madhusree Roy Chowdhury*, S. Bhattacharjee*, C. Bagadia* and Lokesh C. Tribedi* [1]

* Tata Institute of Fundamental Research, Colaba, Mumbai 400005, India

Synopsis Double differential cross sections (DDCS) of the secondary electrons emitted from a vapor jet of bromouracil in collision with 42 MeV bare carbon ions have been measured for different forward and backward emission angles over a wide energy range. The DDCS for bromouracil were compared with that of uracil measure earlier. An unusually large increase in electron emission cross section have been observed for bromouracil.

Hadron therapy using mainly protons and carbon ions have gained immense importance in cancer treatment in last several years. The advantage of ions compared to photons lies in their selective dose depth penetration, which results in negligible damage to the healthy cells surrounding the tumors. The low energy secondary electrons produced along the ion track causes strand breaks of the DNA/RNA [1]. The efficiency of damage thus depends on the amount of secondary electrons produced during irradiation by ions.

Radiosensitizers are agents well known for increasing the killing efficiency of malignant cells in presence of ion beams. Inserting a high-Z element (Au, Pt, Gd) into the DNA and further exposure to ion beams, results in the production of cascade of low energy electrons which cause strand breaks of DNA/RNA. Several breaks when strongly localized results in cell death, preventing the malignant cells from multiplying further in the body. Earlier studies have shown that with the insertion of nanoparticle, the amount of single and double strand breaks increases in DNA [2]. Halouracils are also known to exhibit radiosensitizing effects. These molecules are commercially available, where, the hydrogen at the 5th place of carbon in uracil, is replaced by a halogen. Different experiments and theoretical studies are being performed for halouracils, most of which deal with the different fragmentation channels of such molecules.

We present a quantitative measurement of the amount of enhancement in low energy electron production from bromouracil by comparing it with the existing data of uracil [3]. The DDCS of electrons emitted from bromouracil in collision with 42 MeV bare carbon ions have been measured using a hemispherical electrostatic analyzer followed by a channel electron multiplier for different angles ranging between 20^0 and 160^0. The vapor jet of bromouracil was obtained by heating the powder in an oven for approximately 24 hrs. The heating was done slowly to avoid any clogging in the jet nozzle. A quartz crystal based thickness monitor was used to monitor the rate of deposition of the vapor to ensure a uniform flow. Extreme care was taken for the measurement of low energy electrons, which could otherwise get affected by the field created due to the static charge build up by the deposition of the vapor inside the chamber. A comparison of the DDCS at each individual angle between uracil and bromouracil showed an increase in electron production from bromouracil for all angles. This large enhancement in the electron production cannot be attributed only by scaling with the total number of electrons in bromouracil compared to that in uracil. The production of low energy electrons from bromouracil may be explained by the mechanism of plasmon excitation or Auger cascade, though a clear understanding is not available.

The present measurement for bromouracil is the first of it's kind and provides an important information about the enhancement in low energy electron production, explaining the radiosensitizing properties of halouracils.

[1]E-mail: lokesh@tifr.res.in
and ltribedi@gmail.com

References

[1] B. Boudaiffa *et al.* 2000 *Science* **287** 1658
[2] E. Porcel *et al.* 2010 *Nanotech.* **21** 085103
[3] A. N. Agnihotri *et al.* 2013 *PRA* **87** 032716
[4] A. N. Agnihotri *et. al.* 2013 *J. Phys. B* **46** 185201
[5] A. N. Agnihotri *et. al.* 2012 *PRA* **85** 032711

IOP Conf. Series: Journal of Physics: Conf. Series **875** (2017) 102028 doi:10.1088/1742-6596/875/11/102028

Electron emission from water molecule in collisions with fast highly charged C, O and Si-ions and scaling law

S. Bhattacharjee* [1], **C. D. Bagdia***, **M. R. Chowdhury***, **J. Monti**†, **R. D. Rivarola**†, **D. Misra***, **L. C. Tribedi*** [2]

* Tata Institute of Fundamental Research, Homi Bhabha Road, Mumbai 400005

† Institute de Fisica Rosario (CONICET-UNR), Universidad Nacional de Rosario, 2000 Rosario, Argentina

Synopsis Double differential cross sections (DDCS) of vapor phase water molecules by similar velocity C^{6+}, O^{8+} and Si^{13+} ion impact have been measured by the electron spectroscopy technique. The total ionization cross sections (TCS) were obtained by numerical integration of the DDCS results. All the existing data have been plotted after scaling. Gradual deviation of the TCS from the first Born predicted q_p^2 –scaling has been observed.

Study of electron emission from water molecule is especially relevant because of its importance in fundamental problems in radiation biology and radiation therapy [1]. So it is important to study differential cross section of emitted electrons from water molecule for interaction with highly charged high-velocity projectiles.

Several experiments have been carried out to study ionization of biological molecules bombarded with high-velocity ions. But when it comes to the measurement of differential cross sections, literature is not so vast and most of the existing results are for light projectiles like e^-, H^+ or $He^{1+,2+}$.

In the present work, we have measured the double differential cross section (DDCS) of emitted electrons for the collision of 4 MeV/u C^{6+}, Si^{13+} and 2.00 - 3.75 MeV/u O^{8+} ion beams with the water vapor target. The DDCS measurements were done using electron spectroscopy technique- based on a rotatable hemispherical e.s. analyzer to detect low energy electron in an angle resolved manner. We have compared our experimental results with the CDW-EIS model (both prior and post versions) calculations. The comparison shows an overall good agreement with the data for C and O for which perturbation strength (S_P) is low. For H-like Si ions for which the S_p is larger the model overestimated the experimental results. Another interesting aspect is the better agreement for O-projectile compared to the C-data wheras the C-projectile is used in hadron therapy.

Using the experimental DDCS data we have, by integration, obtained the single differential cross section (SDCS) and the total ionization cross section (TCS). In order to obtain a scaling law for the TCS data we have plotted the scaled total cross section (TCS/q_p^2) as a function of the projectile velocity. We have compared our present results with other existing results (like H^+ [2-4], He^{2+} [5,6] and C^{6+} [7]) including the 6.0-MeV/u C^{6+} data [6]. Existing data of light projectiles like H^+, He^+ or He^{2+} show a good agreement with the q_p^2 dependence. But clear deviation can be seen for C^{6+}, O^{8+} or Si^{13+}. These data suggests a q_p^n dependence where n is found to be 1.7 ± 0.1 which is different from that obtained for uracil earlier, for which the n was found to be 1.5.

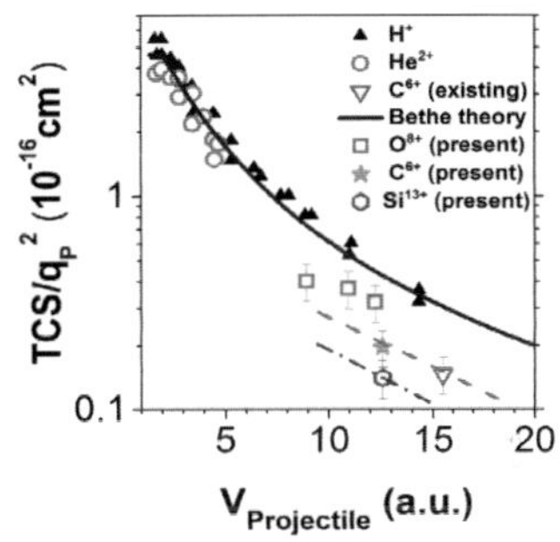

Figure 1. The projectile velocity dependence of scaled TCS i.e. TCS/q_p^2 for different projectiles.

References

[1] L H Toburen 1998 *Radiat. Environ. Biophys.* **37** 221-233
[2] M A Bolorizadeh *et al.* 1986 *Phys. Rev. A* **33** 888
[3] M Rudd *et al.* 1985 *Rev. Mod. Phys.* **64** 441
[4] M Rudd *et al.* 1985 *Phys. Rev. A* **31** 492
[5] L H Toburen *et al.* 1980 *Radiat. Res.* **82** 27-44
[6] M Rudd *et al.* 1985 *Phys. Rev. A* **32** 2128
[7] D Ohsawa *et al.* 2009 *NIM B* **267** 781-790
[8] S Bhattacharjee *et al.* (Submitted to PRA)
[9] S Bhattacharjee *et al.* 2016 *J. Phys. B* **49** 065202

[1] E-mail: shamik127@gmail.com

[2] E-mail: lokesh@tifr.res.in and ltribedi@gmail.com

IOP Conf. Series: Journal of Physics: Conf. Series **875** (2017) 102029 doi:10.1088/1742-6596/875/11/102029

Bare C-ion impact ionization of adenine molecules : DDCS and TCS measurements

S. Bhattacharjee* [1], C. D. Bagdia*, M. R. Chowdhury*, J. Monti† , R. D. Rivarola†, L. C. Tribedi* [2]

* Tata Institute of Fundamental Research, Homi Bhabha Road, Mumbai 400005
† Institute de Fisica Rosario (CONICET-UNR), Universidad Nacional de Rosario, 2000 Rosario, Argentina,

Synopsis Double differential cros sections (DDCS) for vapour phase adenine molecule under the impact of 3.5-MeV/u C^{6+} ions have been measured. The measured DDCS values were compared against the those for other nucleobases, such as, uracil, as well as the CDW-EIS model calculations. An excellent agreement of the experimental result with the theoretical model was observed for both the energy and angular distributions.

Collision of high velocity projectiles with atoms or molecules produces a large number of secondary electrons, near the end of the projectiles trajectory i.e. the Bragg Peak (BP) region. These secondary electrons can further initiate ionization events. Study of energy distribution of these secondary electrons from biological molecules is particularly important, because of its growing importance in radiation damage and radiation therapy[1]. Because of this reason, an increasing number of experimental studies have been devoted recently to study ionization of biological molecules [1-5].

Radiation therapy using proton beam is the most widely used technique. Carbon projectiles are also quite commonly used for this purpose since it offers more confinement of the radiation dose region, thereby minimizing the damage caused to the cells near the infected region.

In present work we have studied the double differential cross sections (DDCS) of the emitted secondary electrons from vapour phase adenine under the impact of 3.5-MeV/u C^{6+} projectile. DDCS study of adenine molecule is extremely important to understand DNA/RNA damage of cells in collisions with high velocity ions. The 3.5-MeV/u C^{6+} beam was generated using the Pelletron Accelerator facility at TIFR, Mumbai. DDCS spectrum was obtained using an electro static hemispherical e-analyzer. A heated oven at a temperature of $\sim 180^0$ C was used along with a quartz crystal thickness monitor. To make sure about the steady flow of vapour the oven was heated very slowly i.e. in about 24 hours. Fig 1(a,b) shows a typical absolute DDCS of emitted secondary electrons for two different emission angles. The solid lines in the figure represent the CDW-EIS model calculation for the 3.5-MeV/u C^{6+} projectile. The present results show a fairly good agreement with the theoretical model.

The cross section results for adenine were compared with the results of a previous experiment for uracil [4] target or even a bromouracil target at the same velocity C^{6+} projectile. The ratio of DDCS and TCS (total ionization cross section) values for adenine and uracil were obtained and the ratios were shown to vary from the expected value based on the electron numbers.

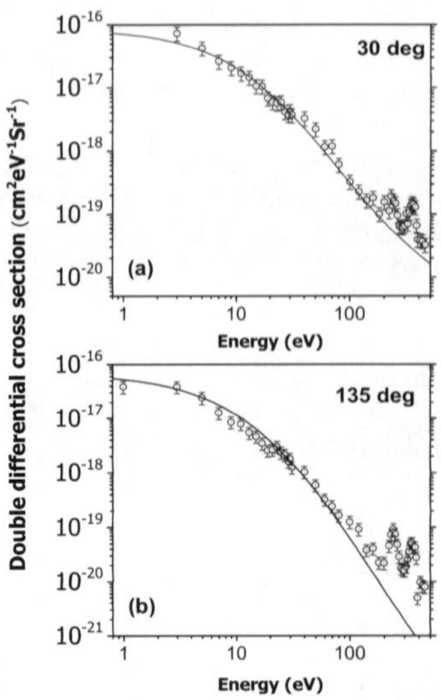

Figure 1. DDCS of emitted e^- for 3.5-MeV/u C^{6+} on adenine target.

References

[1] J. Tabet *et al.* 2010 *Phys. Rev. A* **82** 022703
[2] J. Tabet *et al.* 2010 *Phys. Rev. A* **81** 012711
[3] Y. Iriki *et al.* 2011 *Phys. Rev. A* **84** 032704
[4] Agnihotri *et al.* 2013 *J. Phys. B* **46** 185201
[5] A. Itoh *et al.* 2013 *Phys. Rev. A* **88** 052711

[1] E-mail: shamik127@gmail.com

[2] E-mail: lokesh@tifr.res.in and ltribedi@gmail.com

IOP Conf. Series: Journal of Physics: Conf. Series **875** (2017) 102031 doi:10.1088/1742-6596/875/11/102031

Stereodynamics of asymmetric ion-pair formation in collisions of highly-charged ions with rare gas dimers

Tomoko Ohyama-Yamaguchi[*1] **and Atsushi Ichimura**[†2]

[*] Tokyo Metropolitan College of Industrial Technology, Shinagawa, Tokyo 140-0011, Japan
[†] Institute of Space and Astronautical Science, JAXA, Sagamihara, Kanagawa 252-5210, Japan

Synopsis Stereodyanamical effects are analyzed for multiple ionization of rare gas dimers by slow highly charged ions using the three-center Coulombic over-barrier model previously developed by the present authors.

More than ten years ago, we proposed a three-center Coulombic over-barrier model to describe sequential multiple ionization of a rare gas dimer BC collided by a slow highly charged ion A^{q+} [1]. In more recent works [2, 3], we modified the model so as to introduce the effect of partial screening during a collision for non-active target atomic site (either B or C) and also for projectile site (A) in respective steps of electron removal. The model predicts the population distribution over charge-states (Q,Q') of dissociating ion pairs just after the collision. Measured result [4, 5] of the ion pair distribution in $Ar^{9+}+Ar_2$ collisisons was reasonably reproduced in the model by taking a screening parameter as $s = 0.3 \sim 0.4$ [3].

In the present work, stereodynamical effects are examined in relation to the screening effect. We have calculated the ion pair formation cross sections as a function of the orientation angle, $\cos\theta = \hat{\boldsymbol{d}}\cdot\hat{\boldsymbol{v}}$, where $\boldsymbol{d}$ denotes a molecular axis vector from C to B with $\boldsymbol{v}$ being the projectile beam velocity. In addition, to obtain a physical insight more clearly, we introduce a pair of atomic impact parameters b_B and b_C respectively defined as the vertical distances from sites B and C to the incident trajectory (see Fig.1). Hence, we distinguish the *near* and *far* sites; thereby discuss the ion-pair formation cross section $\sigma(Q_{near}, Q_{far})$. The near and far sites would be determined through the measurement of momentum transfer.

Figure 2 shows the angular dependence of the cross sections $\sigma(Q_{near}, Q_{far})$ for $(Q_{near}, Q_{far}) = (2,1)$, (1,2), (3,2), and (2,3). It is seen from the figure that the $(2,1)$ population overwhelms $(1,2)$, and similarly $(3,2)$ overwhelms $(2,3)$. All the curves in the figure are symmetric with respect to $\cos\theta$. We see two maxima in $(2,1)$ and $(3,2)$ cross sections, and two cusps in $(1,2)$ and $(2,3)$ at $\cos\theta = \pm 1$. These behaviors come from the geometry of saddle point formation in the three-center Coulombic potential [2, 3].

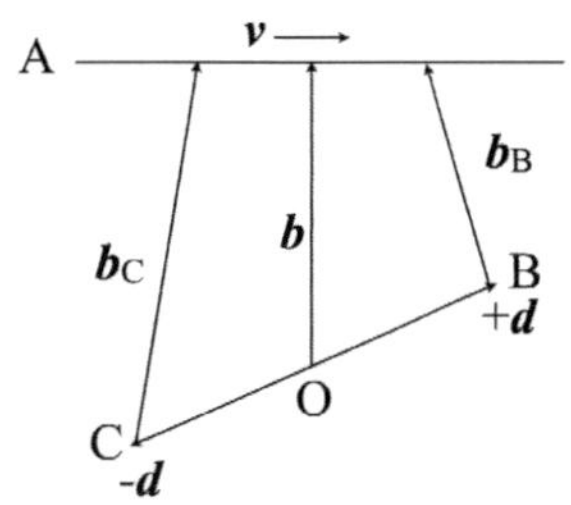

Figure 1. Atomic impact parameters b_B for $A^{q+}+B$ and b_C for $A^{q+}+C$ with molecular impact parameter b for $A^{q+}+BC$.

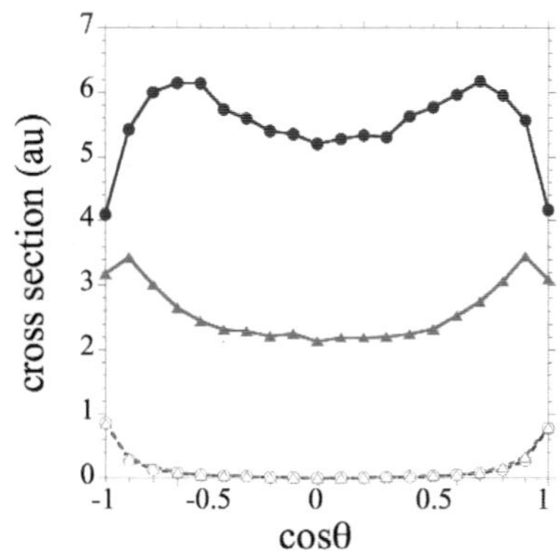

Figure 2. Angular dependence of the ion-pair formation cross sections $\sigma(Q_{near}, Q_{far})$ in $A^{9+}+Ar_2$ collisions; those for (2,1) are indicated by closed circles (●), those for (1,2) by open circles (○), those for (3,2) by closed triangles (▲), and those for (2,3) by open triangles (△).

References

[1] T. Ohyama-Yamaguchi and A. Ichimura 2003 *Nucl. Instrum. Methods B* **205** 620

[2] T. Ohyama-Yamaguchi and A. Ichimura 2013 *Physica Scripta* T **156** 014043

[3] T. Ohyama-Yamaguchi and A. Ichimura 2015 *J. Phys. Conf. Ser.* **635** 032101

[4] Matsumoto *et al.* 2010 *J. Phys. Rev. Lett.* **105** 263202

[5] Iskandar *et al.* 2014 *Phys. Rev. Lett.* **113** 143201

[1]E-mail: yamaguti@metro-cit.ac.jp
[2]E-mail: ichimura@isas.jaxa.jp

IOP Conf. Series: Journal of Physics: Conf. Series **875** (2017) 102032 doi:10.1088/1742-6596/875/11/102032

Orientation-dependent dissociation dynamics of H_2O_2

M. Nrisimhamurty, C. Bagdia, L. C. Tribedi, and D. Misra[1]

Department of Nuclear and Atomic Physics
Tata Institute of Fundamental Research, Dr. Homi Bhabha Road, Colaba, Mumbai-400005, INDIA

Synopsis The ion-induced fragmentation dynamics and the angular anisotropy effects in collisions of 1 MeV Ar^{8+} ions with H_2O_2 is studied.

Multiple ionization and subsequent fragmentation of molecules by energetic highly-charged ions is of fundamental importance, but remained elusive for quite a long time due to their multi-electronic nature and complexity involved during ion-induced breakup. Few decades ago, the advent of the state-of-the-art recoil-ion momentum spectroscopy (RIMS) [1] technique brought an upsurge in the ion-induced fragmentation studies. Moreover, it provided robust kinematically-complete information about the various dissociation channels involved, in terms of kinetic energy release, momentum distributions and angular correlations. Not only that, RIMS also aided in the development of next generation theoretical methodologies to unravel the ion-induced dissociation dynamics. With the help of momentum-imaging technique, dissociation dynamics of a plethora of di- and tri-atomic molecules have been investigated, but many polyatomic molecules are still unexplored. Here, we report on the ion-impact induced dissociation studies of H_2O_2.

Hydrogen peroxide (H_2O_2), a non-planar molecule, is a subject of interest due to its significance in the atmospheric chemistry of ozone. H_2O_2 is also found to be one of the main oxidants that decomposes sulfur dioxide (SO_2) in cloud droplets which eventually leads to acid rains. Besides being an important oxidant H_2O_2, along with HCHO, plays a decisive role in the behavior of oxidation power and self cleaning capacity of the atmosphere [2, 3]. Dissociation studies on H_2O_2 is scanty because it is highly unstable and slowly decomposes into oxygen and water. This could be also a reason for the formation of water on the dust grains in the interstellar medium and is supported by the recent identification of interstellar hydrogen peroxide [4]. In addition, hydrogen peroxide plays an important role in understanding the radiation damage of biological matter [2]. Hence, current observations on dissociation dynamics of H_2O_2 are of immense importance.

In the present work, 1 MeV Ar^{8+} ions are obtained from the electron cyclotron resonance ion accelerator (ECRIA) at TIFR, Mumbai. Electrons that are created during the fragmentation are detected by a channel electron multiplier detector which serve as the start signal. Ion fragments are projected onto a position- and time-sensitive detector [5]. From the coincidence observations, we found two dissociation channels for two-body decay of H_2O_2 whereas H_2O_2 fragments into H^+, H^+ and O_2^+ in a three-body dissociation process. Interestingly, a signature of angular anisotropy is observed as shown in the Fig. 1. This could be attributed to post-collisional effects. Thus, these results are of paramount importance in the context of orientation and alignment dependent dissociation dynamics.

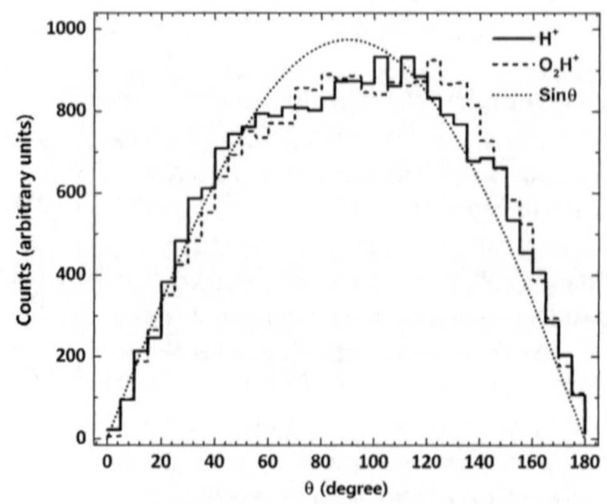

Figure 1. Angle between the individual momentum vectors and beam direction for the dissociation channel $H_2O_2^{2+} \longrightarrow H^+ + O_2H^+$. Solid and dashed line corresponds to H^+ and O_2H^+ fragments, respectively. The dotted curve corresponds to isotropic distribution ($\sin\theta$).

References

[1] J. Ullrich *et al.* 1997 *J. Phys. B* **30** 2917

[2] D. Nandi *et al.* 2003 *Chem. Phys. Lett.* **373** 454

[3] T. Klippel *et al.* 2011 *Atmos. Chem. Phys.* **11** 4391

[4] F. Du *et al.* 2012 *Astron. Astrophys.* **538** A91

[5] A. Khan *et al.* 2015 *Rev. Sci. Instrum.* **86** 043105

[1]E-mail: dmisra@tifr.res.in

IOP Conf. Series: Journal of Physics: Conf. Series **875** (2017) 102033 doi:10.1088/1742-6596/875/11/102033

Theoretical study for the exchange reactions of H' + NH → NH' + H and D + NH → ND + H

Xian-Fang Yue* [1]

* Department of Physics and Information Engineering, Jining University, Qufu 273155, China

Synopsis In this work, the author has investigated the rotational integral cross sections and product angular distributions of the title reactions. The results are compared with previous work, and large differences were found and compared.

NH and NH_2 radicals are of great importance in nitrogen chemistry in flames. Therefore, the title reactions have been studied in both experimental [1-2] and theoretical [3-4] aspects. Qu *et al.* [1] measured the rate coefficients of the reaction $NH(X^3\Sigma^-)$ + $D(^2S)$ in a quasistatic laser-flash photolysis, laser-induced fluorescence system. Li *et al.* [3] carried out the exact quantum mechanical (EQM) and statistical quantum mechanical (SQM) calculations using a quantum wavepacket method. In the present work, the author has carried out the investigation of the title reactions by employing the quasi-classical trajectory (QCT) method.

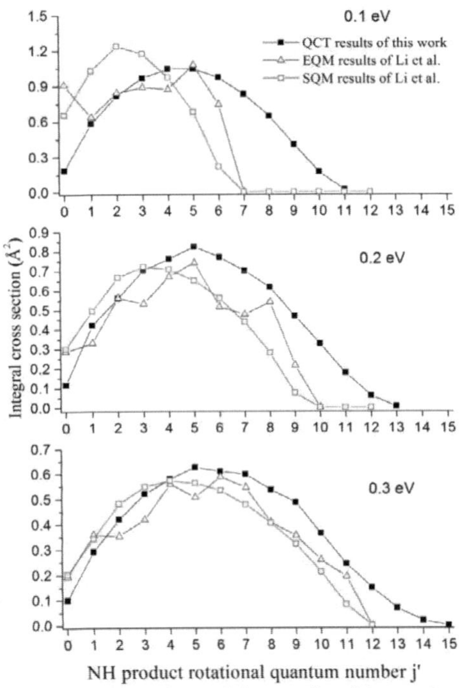

Figure 1. Comparison of the rotational integral cross sections for the H' + NH (v=0, j=0) → NH' (v', j') + H reaction at three collision energies obtained with the QCT method by this work and the EQM and SQM methods by Li *et al.* [3].

As shown in Figure 1, all of the nascent NH product from the reaction H' + NH (v=0, j=0) → NH' (v', j') + H are in the ground vibrational state. However, their rotational state distribution is inverted. These are in consistent with previous EQM and SQM results studied by Li et al. [3]. At the collision energy of 0.1 eV, the ground rotational state distribution of the QCT calculations is lower than that of both the EQM and SQM [3]. This may be due to the quantum effect. At collision energies of the 0.2 and 0.3 eV, the product rotational state distributions of the QCT results are in good agreement with the EQM and SQM [3], meaning that the quantum effect becomes weak in high collision energies.

The product vibrational and rotational state distribution demonstrate that the title reaction via an indirect reaction mechanism. The reaction mechanism is also can be deduced through characters of the differential cross section (DCS). The DCS of the H' + NH (v=0, j=0) → NH' (v', j') + H reaction shows the backward and forward distributions, which indicates that the reaction process undergoes a longer time. That is to say, the title reaction is indirect.

The vector correlations between the product and reactant are also calculated. The results indicate that the product rotational angular momentum $\boldsymbol{j'}$ is not only aligned, but also oriented along the negative direction of y-axis.

References

[1] Z.-W. Qu *et al.* 2005 *J. Chem. Phys.* **122** 204313
[2] H. Umemoto et al. 1997 *J. Chem. Phys.* **106** 4985
[3] Z. Li *et al.* 2009 *J. Chem. Phys.* **131** 1243133
[4] S. Y. Lin *et al.* 2006 *Phys. Rev. A.* **74** 022703

[1] E-mail: xfyuejnu@163.com

IOP Conf. Series: Journal of Physics: Conf. Series **875** (2017) 102034 doi:10.1088/1742-6596/875/11/102034

Rotational energy transfer in collisions of ammonia with rare gas atoms and H_2

Jérôme Loreau*[1] and Ad van der Avoird†

* Service de Chimie Quantique et Photophysique, Université Libre de Bruxelles (ULB), Brussels, Belgium
† Theoretical Chemistry, Institute for Molecules and Materials, Radboud University, Nijmegen, The Netherlands

Synopsis We investigate inelastic collisions of ammonia NH_3 and its isotopologue ND_3 with rare gas atoms or molecular hydrogen. We use state of the art *ab initio* potential energy surfaces to study the low energy dynamics of the collision by means of the quantum-mechanical close-coupling method. We discuss the experimental implications of our work.

Collisions of ammonia with rare gas atoms and H_2 have been the subject of a large number of studies, both theoretical and experimental, mainly focusing on rotationally inelastic scattering in an astrophysical context. Indeed, NH_3 is often used as a probe of the physical conditions in various interstellar environments [1], and accurate values for the cross sections and rate constants for NH_3-He and NH_3-H_2 collisions are required in order to interpret observations. Recent experimental advances have renewed interest in these systems by allowing measurements of quantum-state-resolved integral and differential cross sections in crossed beam experiments [2, 3, 4], which can be directly compared to theoretical predictions based on accurate potential energy surfaces and quantum-mechanical scattering calculations. The same techniques, combined with a Stark-decelerated beam of NH_3, will allow the measurement of integral and differential cross sections at very low collision energies, where the cross section is dominated by resonances that are very sensitive to the details of the potential energy surface [5]. Finally, inelastic collisions of NH_3 with He and Ne are also of interest in order to understand how the rotational cooling of ammonia in a buffer gas cell proceeds [6,7].

In this context, we have recently investigated the rotationally-inelastic scattering of NH_3 and ND_3 by rare gas atoms or H_2 using new potential energy surfaces obtained by means of the coupled cluster method with single, double and perturbative triple excitations (CCSD(T)) [1, 8]. The close-coupling method is employed to calculate integral and differential cross sections at low collision energy (up to 300 cm^{-1}) corresponding to rotationally inelastic scattering. We compare the theoretical predictions with the experimental data and discuss the implications for the sympathetic cooling of ammonia in collisions with rare gas atoms. We also demonstrate the importance of explicitly taking into account the umbrella (inversion) motion of NH_3 in order to obtain accurate scattering cross sections in the cold regime.

References

[1] S. Maret, A. Faure, E. Scifoni, and L. Wiesenfeld 2009 *Mon. Not. R. Astron. Soc.* **399** 425

[2] O. Tkac, A.K. Saha, J. Onvlee, C.-H. Yang, G. S., C. K. Bishwakarma, S.Y.T. van de Meerakker, A. van der Avoird, D.H. Parker, and A. J. Orr-Ewing 2014 *Phys. Chem. Chem. Phys.* **16** 477

[3] O. Tkac, A.K. Saha, J. Loreau, D.H. Parker, A. van der Avoird, and A. J. Orr-Ewing 2015 *J. Phys. Chem. A* **119** 5979

[4] O. Tkac, A. K. Saha, J. Loreau, Q. Ma, P. J. Dagdigian, D. H. Parker, A. van der Avoird, and A. J. Orr-Ewing 2015 *Mol. Phys.* **113** 3925

[5] Q. Ma, A. van der Avoird, J. Loreau, M. Alexander, S. Y. T. van de Meerakker, and P. J. Dagdigian 2015 *J. Chem. Phys.* **143** 044312

[6] O. Schullian, J. Loreau, N. Vaeck, A. van der Avoird, B. R. Heazlewood, C. J. Rennick, and T. P. Softley 2015 *Mol. Phys.* **113** 3972

[7] M. J. Doppelbauer, O. Schullian, J. Loreau, N. Vaeck, A. van der Avoird, C. J. Rennick, T. P. Softley, and B. R. Heazlewood 2017 *J. Chem. Phys.* **146** 044302

[8] J. Loreau and A. van der Avoird 2015 *J. Chem. Phys.* **143** 184303

[1]E-mail: jloreau@ulb.ac.be

Improved model for the interference effects in the ionization of H_2 by fast ions

L. Nagy[1] **and L. Czipa**

Faculty of Physics, Babeş-Bolyai University, str. Kogălniceanu nr. 1, 400084 Cluj, Romania

Synopsis A more precise description of the interference effects in the ionization of H_2 by fast ions is presented. The interference is due to the coherent emission of the electron waves from the vicinity of the two nuclei. The obtained results improve the previous ones mainly for perpendicular ejection of the electron relative to the projectile trajectory, where the experimentally observed secondary oscillations are reproduced.

Interference effects due to the two-center character of the target were observed in the ejected electron spectrum for the ionization of H_2 by fast ions more than a decade ago [1]. The oscillations observed in the differential cross section ratio H_2/2H as a function of electron velocity were analyzed theoretically using a simple plane wave description of the final state [2]. A dependence on the ejection angle of the interference pattern was predicted, which in principle was confirmed experimentally [3]. Other theoretical investigations were also published [4,5]. However, some discrepancies remained between the theoretical and experimental cross section ratios, mainly for 90^0 ejection.

In the present work we improve the previous calculations [2]. The two atomic amplitudes are calculated with a higher precision, and added up coherently. The obtained interference term has a more complex behavior, than predicted before. Our results are compared on Figure 1 to the previous theoretical cross section ratios, to the analytical formula deduced in [2] and to the experimental data. For the small ejection angle the obtained oscillations are in very good agreement with the experimental data and the CDW-EIS calculations [4], but have a higher frequency than the analytical formula's prediction. The main novelty of the present description may be observed at 90^0 – the model reproduces the secondary oscillations observed in the experiments, while the plane wave (PW) model [2] and the analytical formula predict a nearly constant ratio and the CDW-EIS calculations [5] give an increasing ratio as a function of velocity. It is somehow surprising the appearance of these secondary oscillations in a first-order approach. We explain this behavior with the modulation of the oscillations caused by the different distances between the detector and the nuclei by the oscillations originated from the different timing of the momentum transfer to the electron from the vicinity of the two nuclei. The disagreement between the present theory and the experiments for small electron velocities is due to the second-order interactions with the nuclei and to electron correlations, which are not taken into account in the present description.

[1]E-mail: lnagy@phys.ubbcluj.ro

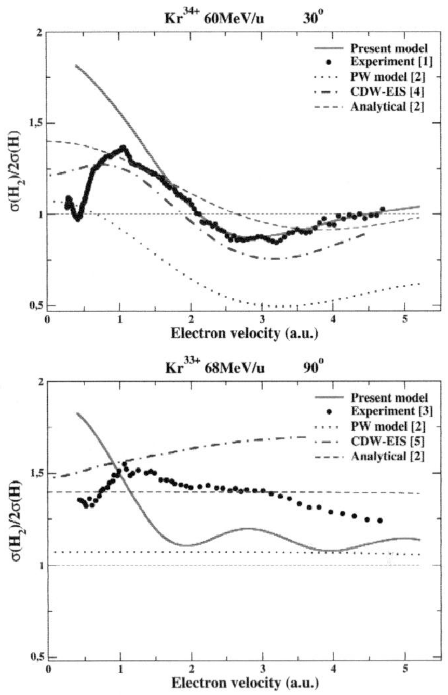

Figure 1. Calculated and experimental H_2/2H differential cross section ratios as a function of ejected electron velocity for different ejection angles.

References

[1] N. Stolterfoht *et al.* 2001 *Phys. Rev. Lett.* **87** 023201

[2] L. Nagy *et al.* 2002 *J. Phys. B: At. Mol. Opt. Phys.* **35** L453

[3] N. Stolterfoht *et al.* 2003 *Phys. Rev. A* **67** 030702(R)

[4] M.E. Galassi *et al.* 2002 *Phys. Rev. A* **66** 052705

[5] M.E. Galassi, R.D. Rivarola and P.D. Fainstein 2004 *Phys. Rev. A* **70** 032721

IOP Conf. Series: Journal of Physics: Conf. Series **875** (2017) 102036 doi:10.1088/1742-6596/875/11/102036

Angular distributions in two body breakup of OCS^{q+} ions

Herendra Kumar*[1], Pragya Bhatt¶, C. P. Safvan¶ and Jyoti Rajput*[2]

* Department of Physics and Astrophysics of University of Delhi, Delhi-110007, India.
¶ Inter University Accelerator Centre, Aruna Asaf Ali Marg, New Delhi-110067, India.

Synopsis Several two-body dissociation channels of multiply ionized OCS molecules generated by Xe^{9+} impact are studied for the angular distributions of fragment ions with respect to the incident projectile direction. We have used the technique of recoil ion momentum spectroscopy (RIMS) for measuring momenta of the fragment ions in coincidence.

The angular distribution of the correlated dissociation fragments with respect to the beam direction has been a topic of interest for both experimentalists and theoreticians. This is of fundamental importance in various areas of science and technology (see, e.g. [1]). In the past decades, many experimental methods have been used to study the anisotropy in the angular distributions of fragments of multiply ionized molecules formed by impact of highly charged ions (e.g. [2] and references therein).

To study the angular distribution of fragmented ions, we performed an experiment in which a beam of multiply charged ions 1.8 MeV Xe^{9+} are made to collide with neutral OCS molecules. This experiment was performed at the Low Energy Ion Beam Facility (LEIBF) of the Inter University Accelerator Centre (IUAC), New Delhi, India. At the collision point one or more electron(s) are released and the molecule gets multiply ionized (i.e. OCS^{q+}). The technique of multi-hit RIMS, using an electron signal as a start and employing a position sensitive microchannel plate (MCP) detector was used to measure the time and position information of the recoil ions generated upon fragmentation of OCS^{q+}. The three-dimensional momentum vectors are derived from the measured time of flight and position information of the detected fragment ions (for details see [3]).

The dissociation of OCS^{q+} molecular ions can have many possible two body fragmentation channels. According to charge on fragment ions, we classify dissociation channels into two categories: charge symmetric e.g. (CO^+, S^+) and charge asymmetric e.g. (CO^+, S^{2+}). These channels are studied for angular distribution of their momentum vectors with respect to projectile ion beam. For an isotropic distribution of the fragment ions, the angular distribution is expected to follow a $sin\theta$ distribution. The angular distributions of fragment ions resulting from symmetric channels match well with $sin\theta$ distribution, however for the asymmetric case, the angular distributions of fragment ions deviate from $sin\theta$ distribution. A deviation from the sine distribution indicates anisotropy. The anisotropy of (CO^+, S^{2+}) asymmetric channel varies with its measured kinetic energy release (KER). This KER distribution ranges upto about 28 eV. If we divide the KER range into three equal parts, the maximum anisotropy is shown for KER condition over lowest range i.e. 0 to 9 eV (see figure 1). Further, we note that the fragment ion of higher charge of asymmetric channel is emitted preferentially in the backward to the beam direction. For example, in (CO^+, S^{2+}) channel, the S^{2+} is observed in the backward direction.

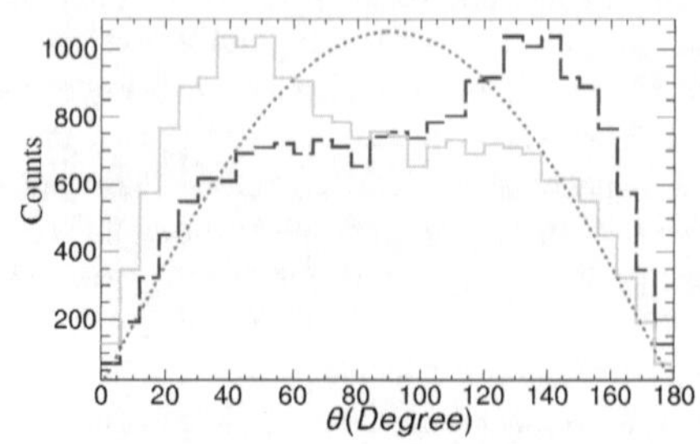

Figure 1. Angular distributions for breakup of OCS^{3+} into CO^++S^{2+}. θ is the angle between momentum vector of fragment ion and direction of incoming projectile. The dashed (blue) line and solid (green) line indicate the angular distribution of S^{2+} and CO^+ fragments, respectively. The dotted red line is for the $sin\theta$ distribution. This plot is under the condition $0 \leq$ KER ≤ 9 eV.

References

[1] Z. D. Pesic *et al.* 2009 *J. Phys. B: At. Mol. Opt. Phys.* **42** 235202

[2] B. Siegmann *et al.* 2003 *Nucl. Instr. and Meth. in Phys. Res.B* **205** 629-633

[3] Kumar *et al.* 2014 *J. Mass Spect.* **374** 44

[1]E-mail: harendraamu@gmail.com
[2]E-mail: jrajput.du@gmail.com

IOP Conf. Series: Journal of Physics: Conf. Series **875** (2017) 102037 doi:10.1088/1742-6596/875/11/102037

Investigation of fragmentation dynamics of nitrogen dimers by collisions with highly charged ions

X. L. Zhu* [1]**, S. Yan*, W. T. Feng*, D. L. Guo*, Y. Gao*, S. F. Zhang*, Z. K. Huang*,†, H. B. Wang*,†, D. B. Qian*, D. M. Zhao*, D. P. Dong*,†, and X. Ma**† [2]

* Institute of Modern Physics, Chinese Academy of Sciences, Lanzhou 730000, China
† University of Chinese Academy of Sciences, Beijing 100049, China

Synopsis The fragmentation dynamics of nitrogen molecule dimers have been studied by using highly charged Ne^{8+} ions at impact energy of 1.0 MeV. For its three-body fragmentation, three fragmentation mechanisms have been identified by analyzing Dalitz plot and Newton diagram of the momenta of three fragment ions. The sequential dissociation of covalent first breakdown is quite surprising in contrast to the fragmentation of triatomic molecule.

The multiple-ionization of clusters induced by highly charged ions may result in multiple bond breakage. If the target is nitrogen molecule dimer, which offers a combination of covalent bond and van der Waals bond [1], the opportunity to learn about the competition among different bonding schemes in the fragmentation process may be provided.

In an intuitive consideration, the covalent bond energy is three orders larger than the van der Waals bond energy, so the breakage of the latter one will be likely to happen in the cluster ion fragmentation. Does this assumption always true? In the present work, we investigated the three-body fragmentations of nitrogen dimer induced by 1 MeV Ne^{8+} ions by employing Reaction Microscopes in Lanzhou [2]. By detecting all fragment ions in coincidence, we obtained the momenta of each ion in the 3-body fragmentation of $(N_2)_2^{3+}$ ions. Through the corresponding Dalitz plot and Newton diagram of fragment ions momenta, contributions from three fragmentation mechanisms were clearly observed: (a) one step concerted fragmentation, in which the covalent bond N-N breaks up simultaneously with the breakup of the van der Waals bond N_2-N_2. (b) van der Waals-first-covalent-second sequential dissociation: the van der Waals bond breaks down in the first step, and a stable $(N_2)^+$ molecular ion is emitted, leaving a rotating metastable $(N_2)^{2+}$ ion behind. The $(N_2)^{2+}$ ion fragments in a second step after a time delay; (c) covalent -first-van der Waals -second sequential dissociation: the covalent bond breaks up firstly, and induces the production of $N^+/(N_3)^{2+}$ ion pair, and leaves the $(N_3)^{2+}$ ion in a metastable state. In the second step, the transient ion $(N_3)^{2+}$ fragments into N^+ and $(N_2)^+$ ions through Coulomb explosion. The measured Newton diagram of three-body fragmentation is shown in Figure 1. The type (c) of sequential dissociation is surprising; it suggests the covalent bond may be more easily to fragment in some cases. It may be due to a potential energy minimum existing in the potential energy surface, which gives the priorities to the formation of metastable $(N_3)^{2+}$ than that of $(N_2)^{2+}$ [3].

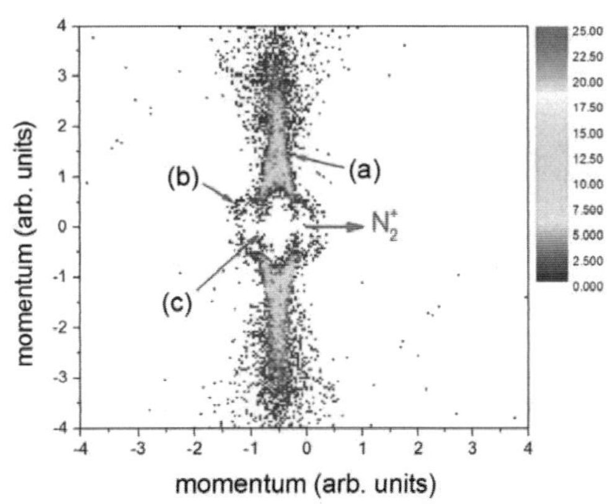

Figure 1. The Newton diagram for three-body fragmentation, $(N_2)_2^{3+} \rightarrow N^+ + N^+ + N_2^+$.

This work was partially supported by the NSFC of China under Grants Nos. 11274317 and 10979007.

References

[1] F. Trinter *et al.* 2014 *Nature* **505** 664
[2] X. Ma *et al.* 2011 *Phys. Rev. A* **83** 052707
[3] Paolo Tosi, *et al.* 1999 *Phys. Rev. Lett.* **82** 450

[1] E-mail: zhuxiaolong@impcas.ac.cn
[2] E-mail: x.ma@impcas.ac.cn

Charge equilibration times for slow highly charged ions in single layer graphene

E. Gruber*, R. A. Wilhelm†, J. Schwestka*, V. Smejkal*, R. Kozubek‡, A. V. Krasheninnikov†, M. Schleberger‡, S. Facsko†, and F. Aumayr* [1]

* TU Wien, Institute of Applied Physics, 1040 Vienna, Austria, EU
† Helmholtz-Zentrum Dresden-Rossendorf (HZDR), 01328 Dresden, Germany, EU
‡ Universität Duisburg-Essen, 47048 Duisburg, Germany, EU

Synopsis We report on charge exchange and energy loss measurements for slow highly charged Xe^{q+} ($q \leqslant 35$) ions after transmission through a single layer of freestanding graphene. Surprisingly short charge equilibration times of only a few femtoseconds are found, which cannot be explained within currently available models.

Highly charged ions can be used as a tool to measure the short-time response of 2D materials to an extremely large, local external field. A charge state of q = 35 implies a local electric field strength of 1.8×10^{11} Vm^{-1} at a distance of 5 Å from it (corresponding to laser power densities of $\sim 10^{17}$ Wcm^{-2}). Previous work on scattering of HCI from solid surfaces or their transmission through freestanding carbon membranes reported unexpectedly large charge capture within 5 - 30fs [1 and refs. therein]. Recently we took the final step and presented results for the ultimately thin carbon target, a freestanding single layer of graphene [2]. Measurements of the charge state and energy of the transmitted ions surprisingly show that a large number of electrons is extracted from a small surface area (see figure 1).

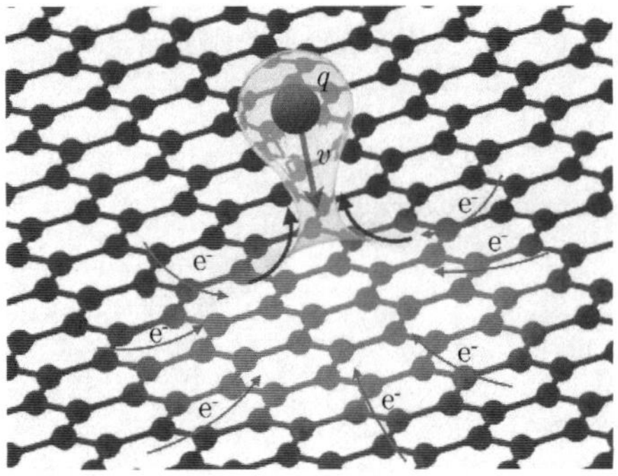

Figure 1. The strong electric field of a highly charged ion is able to capture dozens of electrons from the single layer graphene within a few femtoseconds [2].

Charge equilibration times are derived from the mean exit charge state in dependence of the incident charge state and ion velocity. We find that Xe^{q+} ions equilibrate within a time of only a few fs (see figure 2).

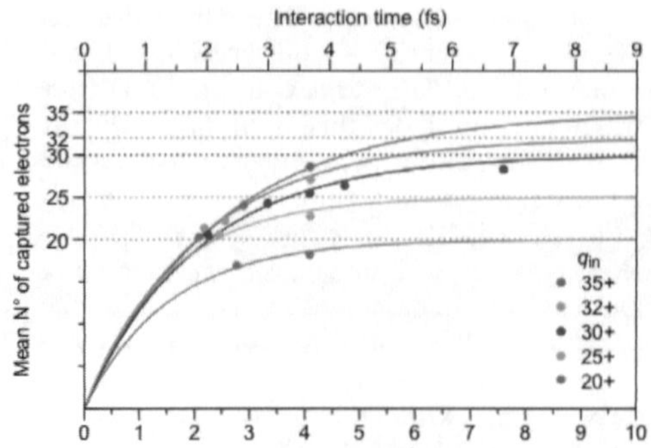

Figure 2. Number of captured (and stabilized) electrons after transmission of highly charged Xe ions (incident charge state q_{in} = 20-35) through a single layer graphene sheet as a function of the inverse projectile velocity [2].

Our current understanding of HCI-solid interaction is not able to explain an almost complete de-excitation of the projectiles within the limited time span of the interaction with a 2D target. This strongly points to a new mechanism, which will be presented at the conference.

References

[1] R.A. Wilhelm *et al.* 2014 *Phys. Rev. Lett.* **112**, 153201
[2] E. Gruber *et al.* 2016 *Nature Comm.* **7**, 13948

[1] E-mail: aumayr@iap.tuwien.ac.at

van der Waals effects in GIFAD for light atoms on insulating surfaces

G. A. Bocan*[1], J. D. Fuhr*, M. S. Gravielle†

* CONICET and Centro Atómico Bariloche, S.C. de Bariloche, Argentina

† Instituto de Astronomía y Física del Espacio (IAFE, CONICET-UBA), Ciudad de Buenos Aires, Argentina

Synopsis We theoretically address grazing incidence fast atom diffraction (GIFAD) for light atoms on insulating alkali-halide surfaces. We combine a description of the projectile-surface interaction obtained from density functional theory (DFT) calculations with a semiquantum treatment of the dynamics. We obtain simulated intensity profiles and compare them with reported experimental GIFAD patterns. For different systems and incidence conditions we focus on the relevance of van der Waals (vdW) interactions on the diffraction patterns.

Grazing incidence fast atom diffraction (GIFAD) is an extraordinarily sensitive phenomenon. First reported for light atoms on wide band-gap insulating surfaces [1, 2], it has since been observed for a variety of systems, becoming a powerful surface analysis technique as well as a most useful tool for testing potential energy surfaces (PESs).

Achieving an appropriate theoretical description of GIFAD is a challenging task. A projectile-surface potential which includes the key features of the interaction, and a scattering dynamics representation which retains the quantum character of the process are necessary ingredients for attaining good accord with experiments. In this work we accomplish this by combining a description of the projectile-surface interaction by means of a three-dimensional PES, built from high-precision density functional theory calculations; with a semiquantum treatment of the elastic scattering process within the surface initial-value representation (SIVR) [3].

GIFAD theoretical studies that include van der Waals (vdW) interactions are still scarce. However, for certain systems and incidence conditions the role of vdW could be significant. In a recent article, we reported on an important vdW effect on GIFAD for the H/LiF(001) system [4]. It is thus the aim of the present work to further discuss those results and to explore similar systems such as, for example, He/KCl(001) and H/KCl(001).

We consider different systems, incidence channels and perpendicular impact energies. Our results show visible vdW effects on the diffraction patterns provided i) the projectile have a relatively high polarizability, ii) there be no polarization effects along the channel and iii) perpendicular impact energy values be low enough so that the projectile be reflected relatively far from the surface.

These results position GIFAD as a highly sensitive quality check for the various approaches to vdW within DFT, and thus as a promising tool in the development of more accurate descriptions of these interactions.

[1]E-mail: gisela.bocan@cab.cnea.gov.ar

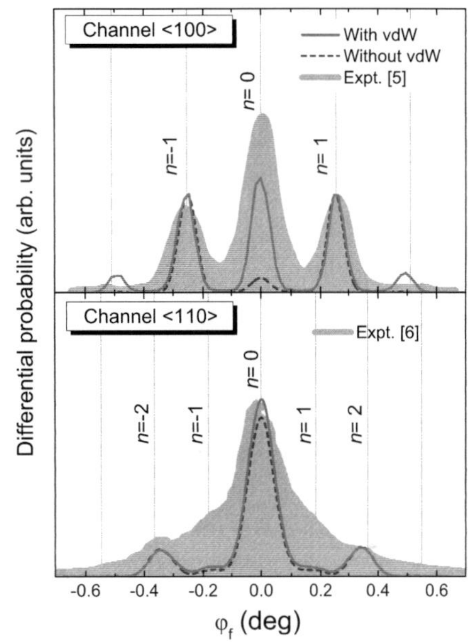

Figure 1. GIFAD for H/LiF(001). Impact energy $E_{tot} = 1$ keV. Incidence angle θ_i relative to surface plane. TOP: $\theta_i = 0.976°$, BOTTOM: $\theta_i = 1.356°$.

References

[1] A. Schüller *et al.* 2007 *Phys. Rev. Lett.* **98** 016103

[2] P. Rousseau *et al.* 2007 *Phys. Rev. Lett.* **98** 016104

[3] M. S. Gravielle *et al.* 2014 *Phys. Rev. A* **90** 052718

[4] G. A. Bocan *et al.* 2016 *Phys. Rev. Lett.* **94** 022711

[5] A. S. Muzas *et al.* 2016 *Nucl. Instr. Meth. B* **382** 49

[6] P. Rousseau *et al.* 2008 *J. Phys. Conf. Ser.* **133** 012013

IOP Conf. Series: Journal of Physics: Conf. Series **875** (2017) 112003 doi:10.1088/1742-6596/875/12/112003

Energy loss due to plasmon excitation by the impact of charged particles on solid surfaces: Beyond a standard approximation

J. L. Gervasoni[*†‡], **R. O. Barrachina**[*†‡1], **S. Segui**[*‡], **F. Navarrete**[*†‡], **N. Arista**[*†]

[*] Centro Atómico Bariloche (CNEA). Av. Bustillo 9500, 8400 S. C. de Bariloche, Argentina.
[†] Instituto Balseiro (CNEA - UNCuyo). Av. Bustillo 9500, 8400 S. C. de Bariloche, Argentina.
[‡] Consejo Nacional de Investigaciones Científicas y Técnicas (CONICET), Argentina.

Synopsis We demonstrate that a standard approximation in the integration conditions of the transferred momentum, which is frequently employed for the study of the interaction of charged projectiles with condensed matter, might lead to inconsistent results. We propose a method for avoiding these shortcomings. Finally we exemplify this issue by studying the excitation of bulk plasmons by a particle penetrating a semi-infinite material at an oblique trajectory.

The increasing improvement of spectroscopic techniques based on the interaction of charged particles with condensed matter leads to a continuous revision and refining of the theoretical models used to describe this process. In this framework, the study of plasmon excitation is of great relevance for the characterization of surfaces, thin films and nanomaterials. In this communication, we reevaluate a standard approximation in the integration conditions of the transferred momentum [1], which has been usually applied in the literature for more than forty years. We show that this approximation might lead to inconsistent results as, for instance, an spurious dependence of the excitation of bulk plasmons on the incidence angle θ for processes occurring deep inside the solid (see, e.g. [2], fig. 7a). Here we propose a method for avoiding these shortcomings of the standard approach, by employing the same integration conditions along every direction of the transferred momentum.

We exemplify this method by evaluating the energy loss rate dW/dt due to bulk plasmon excitations for a particle of charge Z and velocity $\mathbf{v}$ entering into a semi-infinite material of a given dielectric function $\varepsilon(\mathbf{k},\omega)$. We define θ as the angle between $\mathbf{v}$ and the normal to the planar interfase. For the sake of simplicity, we employ a specular reflection model and an extended pseudo-medium method [3], being the generalization to more sophisticated theories straightforward. The electric potential is obtained through Poisson's equation in a Fourier transformed space, and its time derivative leads to the energy loss rate, from which we single out the bulk contribution [4].

In figure 1 we show dW/dt at very large distance from the planar surface, well inside an aluminium sample, i.e. with a plasma frequency $\omega_p = 0.55$ a.u. We clearly see that our calculation leads to an energy loss rate which is independent of the angle, as expected. On the other hand, the use of the standard approximation leads to a spurious dependence with θ. A similar analysis for the production of surface plasmons is under way.

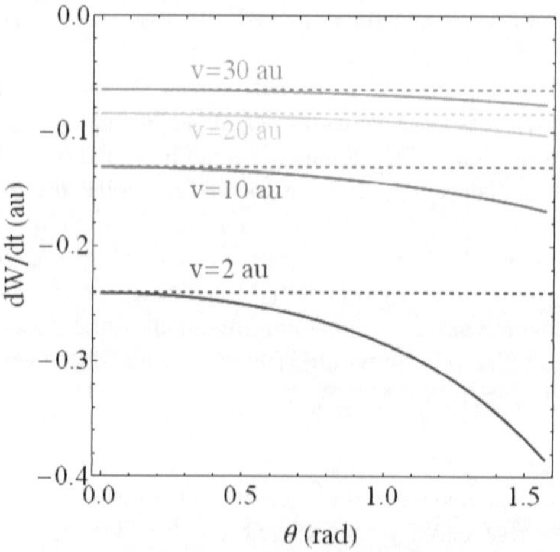

Figure 1. Energy loss rate dW/dt due to bulk plasmon excitation for a particle of charge $Z = 1$ inside Aluminium, as a function of the incidence angle θ for various values of the velocity v. The present calculation (broken line) is compared with the standard model (full line), as described in the text.

References

[1] G. D. Mahan 1973 *Phys. Status Solidi B* **55** 703

[2] C. Denton *et al.* 1998 *Phys. Rev. A* **57** 4498

[3] J. L. Gervasoni, N. R. Arista 1992 *Surf. Sci.* **260** 329

[4] J. L. Gervasoni *et al* 2015 *Nucl. Instr. Meth. B* **354** 197

[1]E-mail: barra@cab.cnea.gov.ar

IOP Conf. Series: Journal of Physics: Conf. Series **875** (2017) 112004 doi:10.1088/1742-6596/875/12/112004

Development of laser target sight-on system based on multiple transmission through a tapered glass capillary for ion microbeam irradiation

Mikiko Koushima*, Tokihiro Ikeda†,* [1], Mitsuyoshi Matsubara*, Takafumi Masuyama*, Tatsuya Minowa*, and Wei-Guo Jin*

* Department of Physics, Toho University, 2-2-1 Miyama, Funabashi, Chiba 274-8510 Japan
† RIKEN Nishina Center for Accelerator-Based Science, 2-1 Hirosawa, Wako, Saitama 351-0198 Japan

Synopsis An estimation method of micro-meter sized laser profile created by the laser beam transmission through a tapered glass capillary is developed for the target sight-on system of ion microbeam irradiation to living cells. In our system, both the laser and the ions are designed to be transmitted through the same capillary so that the laser spot is surely on the ion track. By selecting the excitation light of a fluorescent protein in cell nucleus, the fluorescent volume can be confirmed as the targeted area.

A method to irradiate single cells with a microbeam of a few MeV ions extracted from a tapered glass capillary with thin end-window has been developed [1]. One of the advantages of this method is that the projectile ion can stop inside the cell employing the short range, which is a few to 100 μm in water for MeV-H or He ions. Therefore, micron-scale pin-point damage can be realized with the maximized stopping power called as Bragg peak.

Depending on the irradiation angle defined as the angle between the capillary axis and the objective lens axis, an assist to align the target on the beam axis is necessary. We propose a laser target sight-on system, where the capillary will be used for both the ion and the laser transmissions. The laser micro-spot created by the capillary coincides with the area to be irradiated. When the wavelength (λ) of the laser is selected as the same λ of the excitation light of the fluorescent dye or protein at a local area in the cell, the selectivity of the irradiation will become better.

In order to develop the sight-on system, we first demonstrated a single fluorescent bead being aimed. Figure 1(a) shows that a single bead with a diameter of 24 μm in a 1D-fluorescent bead array yielded fluorescence (peak λ = 508 nm) by a micro-meter sized excitation light (λ = 488 nm) created by a capillary at Toho University. The photo was taken through a band pass filter with the center λ of 510 nm and the width of 20 nm so that the high intensity of the excitation light was suppressed by 10^{-6} times. Figure 1(b) is a laser beam profile at 1 mm- downstream from the capillary outlet. The profile was visualized by a thin layer of fluorescent beads of 2 μm in diameter formed on a microscope slide (Fig. 1(c).) The diameters of the rings in Fig. 1(b) were extracted automatically by home-made software and compared with those of Fraunhofer diffraction known as ring pattern by parallel light transmitted through a small aperture. The profile was also measured by a power-meter to obtain the density distribution in a laser spot. We estimated the spot size of the practical area and the maximum power density. The spot size possibly depends on the capillary shape. The obtained data as well as a simulation [2] will be used for the feedback to fabricate the better capillary shape.

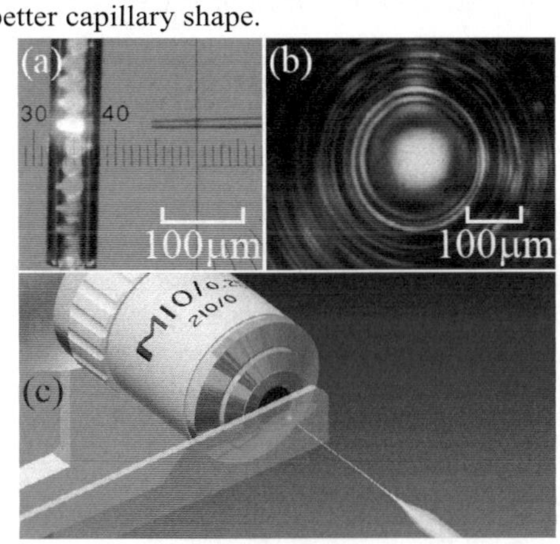

Figure 1. (a) A fluorescent micro-bead was spot-lighted by an excitation light from a glass capillary. (b) A beam profile photo on a fluorescent screen. (c) The screen on a microscope slide was between the capillary tip and the objective lens.

References

[1] V. Mäckel *et al.* 2014 *Rev. Sci. Instrum.* **85** 014302
[2] W-G. Jin *et al.* 2015 *J. Phys. Soc. Jpn.* **84** 114301

[1] E-mail: tokihiro@riken.jp

IOP Conf. Series: Journal of Physics: Conf. Series **875** (2017) 112005 doi:10.1088/1742-6596/875/12/112005

XMCD going ultra-cold: Experiments at 100 mK and 7 T

Ivan Baev* [1], **Torben Beeck***, **M. Benedetta Casu**†, **Michael Martins*** **and Wilfried Wurth***‡

* Physics Department, University of Hamburg, 22761 Hamburg, Germany
† Institute of Physical Chemistry, University of Tübingen, 72074 Tübingen, Germany
‡ DESY Photon Science, 22761 Hamburg, Germany

Synopsis First experiments along with the general capabilities of the TTXMCD endstation at P04, PETRA III, DESY are presented. The TTXMCD endstation allows measuring the magnetic properties of solids, thin films, molecules or atomic adsorbates at ultra-cold temperatures in an element specific way. At such temperatures, electron-electron interaction phenomena are studied, such as the Kondo effect for Co on Cu(100).

The investigation of magnetic properties by means of the X-Ray Magnetic Circular Dichroism (XMCD) has proven to be a valuable tool especially when it comes down to site specific effects. A wide range of samples that need high magnetic fields and ultra-cold temperatures, such as molecular magnets, clusters and other dilute magnetic species are difficult to study directly in the gas phase. The new TTXMCD endstation [1] therefore aims at a solid state sample environment with a well-defined temperature and magnetic field control.

Cobalt adatoms on crystalline copper surfaces have already been studied locally with STM. The Kondo transition temperatures are determined to be ~ 50 K for Cu(111) [2] and ~ 80 K for Cu(100) [3]. Our measurements show, that even below the Kondo transition temperature the electron screening is strongly affected by the exact temperature. The XMCD also allows distinguishing between the spin and orbital d-band magnetic momenta, which are inaccessible with STM based experiments.

The Kondo effect affects both: spin and orbital momenta, which are screened by roughly one order of magnitude on copper surfaces in respect to the atomic value and to similar experiments of adatoms on metallic surfaces. We also found that the spin and orbital momenta behave significantly different for (100) and (111) surfaces. Furthermore, the spin screening strength is increased at ultra-cold temperatures. This is given by the μ_l / μ_s ratio, which rises from ~0.3 at 30 K to ~1.3 at 100 mK for Co/Cu(100). The presented experiments represent a great benchmark for electronic structure calculations in the field of strongly interacting electrons.

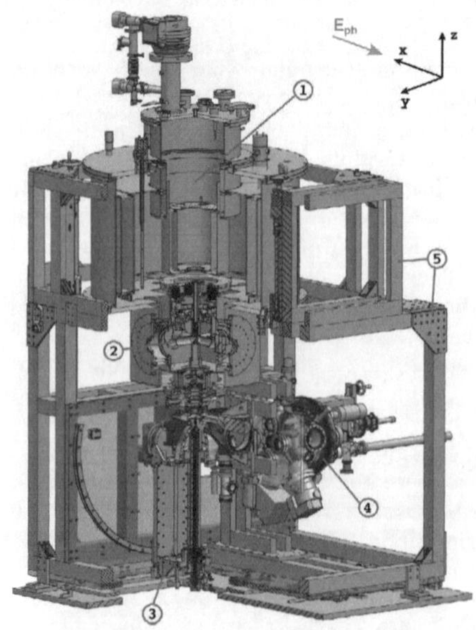

Figure 1. CAD drawing of the TTXMCD endstation from [1]: It contains a dilution refrigerator ①, a preparation chamber ④ and a vector magnet ②.

Additionally, experiments on metal free organic molecules (derivatives of the Blatter radical) are performed at 100 mK. A radical carries a magnetic momentum in terms of an unpaired electron. Our experiments show the feasibility of K-shell related magnetism studies via XMCD. This opens up experimental possibilities for a huge class of important molecules.

This work is funded by the DFG via the collaboration research center SFB 668.

References

[1] T. Beeck *et al.* 2016 *Rev. Sci. Instrum.* **87** 045116
[2] L. Vitali *et al.* 2008 *Phys. Rev. Lett.* **101** 216802
[3] N. Néel *et al.* 2007 *Phys. Rev. Lett.* **98** 016801

[1] E-mail: ivan.baev@desy.de

IOP Conf. Series: Journal of Physics: Conf. Series **875** (2017) 112006 doi:10.1088/1742-6596/875/12/112006

Energy deposition around swift proton and carbon ion tracks in biomaterials

Rafael Garcia-Molina *[1], **Maurizio Dapor** †[2], **Pablo de Vera** ‡[3], **Isabel Abril** §[4]

* Departamento de Física - CIOyN, Universidad de Murcia, E-30100 Murcia, Spain.
† European Centre for Theoretical Studies in Nuclear Physics and Related Areas (ECT*), Bruno Kessler Foundation, and Trento Institute for Fundamental Physics and Applications (INFN-TIFPA), I-38123 Trento, Italy.
‡ School of Mathematics and Physics, Queen's University Belfast, BT7 1NN Belfast, Northern Ireland, UK.
§ Departament de Física Aplicada, Universitat d'Alacant, E-03080 Alacant, Spain

Synopsis Hadron therapy is a modern cancer treatment based on the interaction of proton or heavier ion beams with living tissue, whose purpose is the destruction of the malignant tumor cells producing minimal effects on the surrounding healthy tissue. To study the physical basis of the Relative Biological Effectiveness (RBE) of different projectiles, such as protons or carbon ions, we calculate the radial distribution of energy deposited by the secondary electrons generated in biomaterials by these ions at characteristic energies around the Bragg peak. This is done by means of the simulation code SEED, which follows in detail the motion and interactions of the secondary electrons as well as the subsequent electron cascade.

The interaction of energetic ions with biological targets lies in the basis of hadron therapy, which is a promising technique for oncological treatment due to its higher precision in energy delivery to the tumor cells, whereas sparing most of the healthy tissues surrounding the sensitive region [1].

The use of energetic ion beams for cancer treatment, suggested more than sixty years ago and being widely exploited nowadays, is based on the sharp and well-located dose maximum (the so-called Bragg peak) appearing at the end of their depth-dose curve. The advantages of ion beams as compared to other radiations lie mainly on their higher relative biological effectiveness (RBE), as well as in the precision with which the tumor can be irradiated, with minor side effects to the surrounding healthy regions.

A precise knowledge of the energy loss by swift ions in targets of biological interest (mainly liquid water, but also other constituents of living tissues, such as cell components…) is mandatory to suitably model the depth-dose profiles. This stage in the energy delivery by the ion beam to the target is successfully described by the SEICS code [2].

There are other important issues concerning the biodamage induced by ion bombardment. Among other questions, it is important to know the spatial distribution of the energy deposited around the ion's track by the secondary electrons generated by the projectile [3,4], as well as the subsequent electron cascade, which can propagate appreciable amounts of energy relatively far away from the ion's path.

[1] E-mail: rgm@um.es
[2] E-mail: dapor@ectstar.eu
[3] E-mail: p.devera@qub.ac.uk
[4] E-mail: ias@ua.es

Given the importance of low energy electrons in producing biodamage [5], it is essential to know in detail the spatial and energy distribution of electrons around the tracks of different types of projectiles (proton or carbon ions), in relevant biological targets for representative energies around the Bragg peak.

For this purpose, we use the SEED code [6], which provides a detailed description of the elastic and inelastic processes in condensed phase materials. The whole cascade of secondary electrons is followed, because any truncation or cut off underestimates the secondary electron emission yield. Also, at very low electron energies, trapping phenomena due to electron-polaron interactions and electron-phonon interactions are considered.

By analyzing the radial distribution of energy deposited by the generated secondary electrons around the tracks of these projectiles in liquid water and other targets of biological interests, we hope to assess the physical basis of the higher RBE of carbon ion beams as compared to proton beams.

References

[1] M. Durante, J. S. Loeffler 2010 *Nature Rev. Clin. Oncol.* 7 37
[2] R. Garcia-Molina *et al.* 2011 *Phys. Med. Biol.* **56** 6475
[3] P. de Vera *et al.* 2013 *Phys. Rev. Lett.* **110** 148104
[4] P. de Vera *et al.* 2015 *Phys. Rev. Lett.* **114** 018101
[5] B. Boudaïffa *et al.* 2000 *Science* **287** 1658
[6] M. Dapor 2017 *Transport of Energetic Electrons in Solids: Computer Simulation with Applications to Materials Analysis and Characterization*, 2nd ed., Springer Tracts in Modern Physics, Springer

IOP Conf. Series: Journal of Physics: Conf. Series **875** (2017) 112007 doi:10.1088/1742-6596/875/12/112007

Simulation of the energy loss of proton beams interacting with few layer graphene foils

Juan José Esteve-Paredes *[1], Jorge E. Valdés †[2], Jaime Sánchez-Claros *[3], Isabel Abril ‡[4], Rafael Garcia-Molina *[5]

* Departamento de Física - CIOyN, Universidad de Murcia, E-30100 Murcia, Spain.
† Departamento de Física, Universidad Técnica Federico Santa María, Valparaíso 2390123, Chile.
‡ Departament de Física Aplicada, Universitat d'Alacant, E-03080 Alacant, Spain.

Synopsis We have simulated the passage of proton beams, having energies $E < 10$ keV, through graphene targets, considering the case of few layer foils. For this purpose and for comparison, we implement Monte Carlo and deterministic semi-classical approaches to describe the interaction of protons with carbon atoms in graphene structures. Our results show that the energy loss through a single layer graphene foil is smaller than recent *ab initio* calculations based on time dependent density functional theory (TD-DFT). Besides, for $E > 6$ keV the energy loss into a graphene target composed of n layers is n times the energy loss through a single layer graphene foil.

The interaction of proton beams, having velocities $v < 1$ au ($E < 25$ keV), with graphene targets has been the subject of recent research [1-3]. In all these studies, the energy loss ΔE into a single layer of graphene practically follows a linear dependence with the projectile velocity v, although slight differences in the absolute value of ΔE appear for each calculation.

In this work, we obtain the energy loss by simulating the passage of protons through multilayered graphene targets (n<10 layers). For that purpose, we simulate particle trajectories inside the solid to obtain information on the energy losses parameters. Newton equations of motion of the particle under the influence of the forces due to the target nuclei and core electrons are solved assuming a screened type interaction potential. Furthermore, a quantum dissipative force resulting in an energy loss due to the electron-hole excitations of valence electrons is included using a non-linear model based on density functional formalism [4].

We show in figure 1 the calculated energy loss of a proton beam through a single layer graphene foil. It can be seen than in all cases the energy loss practically follows a linear dependence with the projectile velocity, except at the lowest velocities. However, our values for ΔE are clearly smaller than the ones obtained by methods based in TD-DFT [1-3].

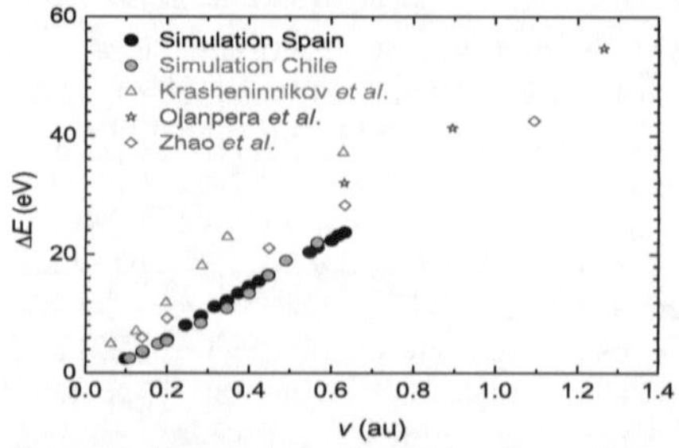

Figure 1. Energy loss of a proton beam through a single layer of graphene.

We have also simulated the passage of protons through graphene foils made of several layers. These results, presented in figure 2, clearly show that ΔE is proportional to the number of layers, for $v > 0.5$ au (i.e., $E > 6$ keV).

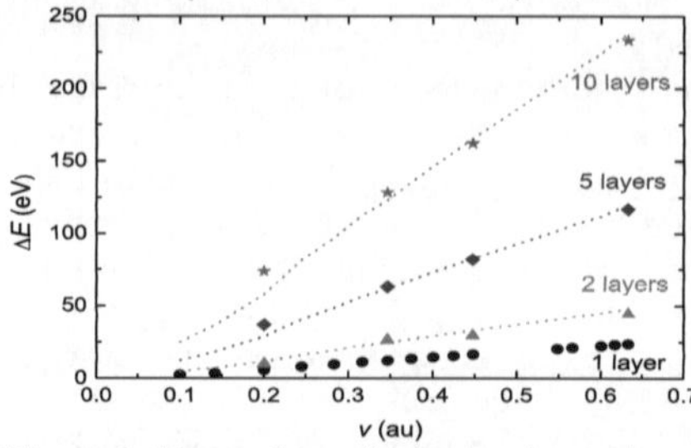

Figure 2. Energy loss of a proton beam through a few layers graphene foil: simulations (symbols) and $n \times \Delta E$ of a single layer (lines).

[1] E-mail: juanjose.esteve@um.es
[2] E-mail: jorge.valdes@usm.cl
[3] E-mail: js.claros29@gmail.com
[4] E-mail: ias@ua.es
[5] E-mail: rgm@um.es

References

[1] A. V. Krasheninnikov *et al.* 2007 *Phys. Rev. Lett.* **99** 016104
[2] A. Ojanperä *et al.* 2014 *Phys. Rev. B* **89** 035120
[3] S. Zhao *et al.* 2015 *J. Phys.: Condens. Matter* **27** 025401
[4] J. E. Valdés *et al.* 2013 *Carbon* **52** 137

Active discharging method for stable sub-micron sized beams of slow highly charged ions using tapered glass capillary with electrodes

Tokihiro Ikeda* [1], **Takao M. Kojima** *, **Yoshio Natsume** †, **Jun Kimura** ‡, **and Tomoko Abe** *

* RIKEN Nishina Center for Accelerator-Based Science, 2-1 Hirosawa, Wako, Saitama 351-0198 Japan
† Electron Tube Division, Hamamatsu Photonics KK, 314-5 Shimokanzo, Iwata, Shizuoka 438-0193 Japan
‡ Tokyo Branch Office, Hamamatsu Photonics KK, 3-8-21 Toranomon, Minato, Tokyo 105-0001 Japan

Synopsis Stable transmission of a 0.75 μm-sized Ar^{8+} beam of 10–100-keV energy through glass tandem optics was achieved. The optics consisted of a 4-electrode tapered capillary and a sub-μm-sized tapered capillary tip, arranged in series. To actively discharge the capillary, and thus obtain stable transmission, an electrode was applied with a mechanical relay closing at a 0.01–0.1 Hz frequency.

Slow (10–100 keV) highly-charged ion (HCI) beams have been widely used for processing and analysis of materials. In order to use the slow HCI in microscopic analysis and modification, small and stable micro-HCI beam generators are required. In 2006, we demonstrated a simple and convenient method for producing microbeams using a single tapered glass capillary with an outlet diameter of 24 μm. The beam transportation is based on self-organized charge patches on the inner surface of the capillary, which are induced by the incident beam itself [1]. However, transmitted microbeams sometimes become unstable and their transmission can be blocked or suddenly increased.

In order to solve the problem, a tandem glass-made optics consisting of a 4-electrode capillary, from Hamamatsu Photonics K.K. (Fig.1(a)), and a tapered capillary tip with a sub-μm sized outlet (Figs. 1(b)) was developed. The larger capillary had 4 ring-electrodes on its outer surface. The unstable transmission was expected to be due to too much charge-up around the position marked by a circle *S* in Fig.1(a). The outlet surface of the larger capillary and the inlet surface of the smaller one touched each other and were connected to *electrode-A*, so that the excess charge is removed toward the ground level periodically through a mechanical relay (active discharging, Fig.1(c)). The relay was controlled by TTL signals with a frequency $f_{TTL} \sim 0.1$ Hz [2]. Figure 1(d) shows the number of transmitted ions per second $N_{transmit}$ for a 104-keV Ar^{8+} beam at the Slow Highly Charged Ion Facility in RIKEN. During a period (3050-3350s) without relay operation, the transmission became gradually blocked and eventually stopped. When the TTL sending restarted, the transmission also started again and remained constant until the input beam stopped.

When only a 4-electrode capillary was used, the density of the extracted beam became up to 14 times larger than that of the input beam. A transmission efficiency up to 70% was observed with good reproducibility when the electrodes were biased with appropriate voltages.

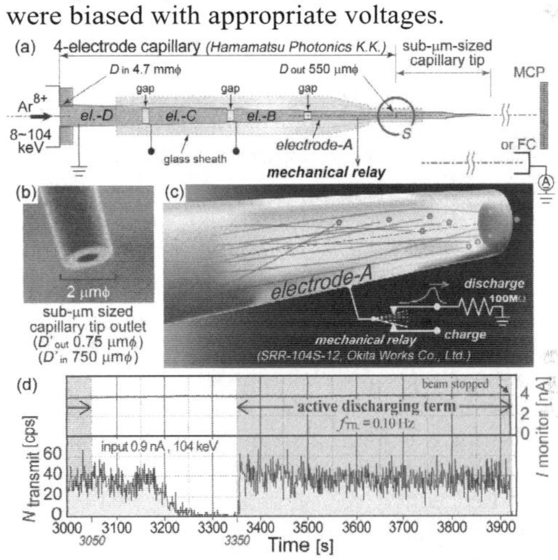

Figure 1. (a) A 4-electrode capillary (larger capillary) and a sub-μm-sized capillary tip (smaller one) were in series. (b) A magnified view of the tip outlet. (c) Charge flow around a circle *S* in Fig. 1(a). (d) The time evolution of the transmission with or without the active discharging. Regarding the stable $I_{monitor}$ (red curve) measured at an entrance aperture of the experimental chamber, input current to the larger capillary was estimated to be stable.

References

[1] T. Ikeda *et al.* 2006 *Appl. Phys. Lett.* **89** 163502
[2] T. Ikeda *et al.* 2016 *Appl. Phys. Lett.* **109** 133501

[1] E-mail: tokihiro@riken.jp

M-X-ray emission in interaction of slow highly charged Xe^{q+} ions (q=26-40) with metallic foils

Ł. Jabłoński*, D. Banaś*, J. Braziewicz*, J. Czub*, P. Jagodziński†, A. Kubala-Kukuś*, D. Sobota*, I. Stabrawa* and M. Pajek*[1]

* Institute of Physics, Jan Kochanowski University, 25-406 Kielce, Poland
† Department of Mathematics and Physics, Kielce University of Technology, 25-314 Kielce, Poland

Synopsis Emission of M-X-rays in interaction of slow highly charged Xe^{q+} ions, delivered by electron beam ion source (EBIS), with metallic surfaces was studied. Measured M-X-ray spectra for different charge states give access to interpret both the dynamics of neutralization of highly charged ions at surfaces and relaxation processes of hollow atoms formed in this process. The measured spectra are interpreted in terms of the performed atomic structure calculations using the GRASP code.

In interaction of slow highly charged ions (HCI) with matter the so-called "hollow atoms" are formed in the process of fast neutralization of HCI at the surface [1,2]. In these exotic atomic states a large part of the electrons are in high Rydberg levels while the inner shells remain empty. The X-rays and Auger electrons emitted in deexcitation of hollow atoms carry information about formation and the structure of hollow atoms as well as their relaxation processes. In particular, the observed X-rays from hollow atoms give access to identify the mutivacancy configurations corresponding to a moment of X-ray emission.

In this work we report on the measurements of X-rays emitted in interaction of ~3 keV×q Xe^{q+} ions (q=26-40) with metallic foils of Be, Au, Ti and Ta. The ion beams were produced in the EBIS-A facility [3,4] consisting of the electron beam ion trap (EBIT), ion extraction and beam transport system, analyzing dipole magnet and UHV experimental chamber equipped with a 5-axis sample manipulator. The X-rays were measured with the XFlash silicon drift detector (SDD) having a resolution of about 80 eV in 1-2 keV photon energy range studied.

The X-rays measured for different charge states of Xe^{q+} ions were interpreted as the M-x-ray nl-3l transitions (where n≥4) corresponding to different multivacancy configurations, including thus both x-ray satellites and hypersatellites. The energies of these transitions were calculated using the multiconfiguration Dirac-Fock (MCDF) method using the GRASP code [5].

In order to interpret the measured spectra, the experimental broadening of spectral lines have been taken into account. The observed intensities of nl-3l transitions, including their satellite structures, yielded information about population of electrons in high n-states. In this way an initial population of electrons as appears in ion neutralization process, followed by the Auger and radiative deexcitation cascade was probed experimentally. Consequently, the predictions of the classical over-the-barrier model (OBM) describing the neutralization of HCI at surfaces were tested experimentally. For Xe^{26+} ions with no M-shell vacancies expected the observed M X-rays may indicate more complex electronic structure of these ions or the internal dielectronic excitation process [6].

Acknowledgments

The equipment was purchased thanks to financial support of the European Regional Development Fund in the framework of the Polish Innovative Economy Operational Program (Contact No. WNP-POIG.02.02.00-26-023/08).

References

[1] H. Winter and F. Aumayr 1999 *J. Phys. B* **58** R39
[2] J.-P. Briand *et al.* 1990 *Phys. Rev. Lett.* **65** 159
[3] G. Zschornack *et al.* 2008 *Rev. Sci. Instrum.* **79** 02A703
[4] D. Banaś *et al.* 2010 *J. Instrum.* **5** C09005
[5] P. Jönsson *et al.* 2013 *Comp. Phys. Comm.* **184** 21972203
[6] R. Schuch *et al.* 1993 *Phys. Rev. Lett.* **70** 1073

[1] E-mail: pajek@ujk.edu.pl

Detection of circular Rydberg states in lifetime measurements

Gaurav Sharma* [1] **and Nitin K. Puri*** [2]

* Dept. of Applied Physics, Delhi Technological University, New Delhi-110042, India

Synopsis The possible reason for the non detection of the circular Rydberg state during the earlier life time measurements is discussed. It is concluded that for populating and detecting a CRS the energy of the ion, Plasmon frequency of the target, position of measurement and step size between two consecutive steps near the last cascade position plays an important role.

Circular Rydberg states (CRS) have much longer radiative lifetimes than low orbital angular momentum states and are therefore candidates for high-fidelity gate operations [1] and determining the relativistic and the quantum electrodynamic effects, the fundamental constants such as the nuclear mass. The electron in CRS decays through the yrast chain following the electric dipole selection rules. The 2p state, the last candidate of the cascade that leads to emission of Ly-α emission takes place in the H-like heavy ions, is populated very late. Each delayed Ly-α transition originating from certain CRS appears as a hump at a particular delay time. Recently these states has been observed in lifetime measurement [2] in a beam-foil experiment which are supported theoretically [3].

For the formation of particular Rydberg state with principal quantum number n of an ion with core charge Z; it is possible to predict the ionic velocity v>vc [4] such that $vc=wp*(1.43n+ 0.64k)^2/Z$ where wp is the Plasmon frequency of the solid and n and k are the principal and angular quantum numbers. This estimation of the velocity is plotted in fig.1 suggesting that a much higher energy is required to populate low l states compared to CRS. We have examined some earlier lifetime measurement [2,5–8] to understand the non detection of CRS. Table 1 represents some experimental data with the possible CRS populated and the measurements taken during the experiments. It is clear from the table that in most of the experiments the data point was not taken up to the desired cascade position or the step size of the measurements was too large to observe the delayed decay. The crude extract of the analysis realizes that the combination of three parameters play a key role for the detection of CRS through lifetime measurement. First is the velocity of the incident particle and Plasmon frequency for populating a particular (n,l); second is the measurement position should be around the delay interval of last cascade. Another is the precise measurement with small intervals near the last cascade delay interval with step size of the measurement less than or equal to the lifetime of the last candidate of the chain i.e. 2p or 3p, for the detection of CRS.

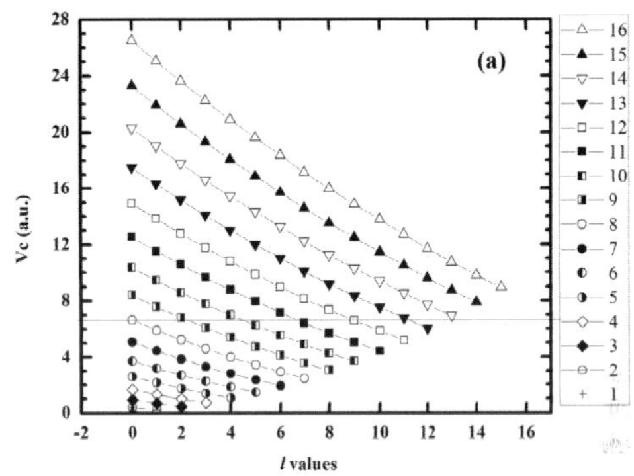

Figure 1. Graphical representation of velocity required to populate a (n,l) level for 125MeV S ions. Horizontal line represents velocity corresponding to 125MeV and the states crossing the line are the possible excitations.

Energy,Species	Nc	d (mm)	M(mm)	Comment
125MeV S [5]	14	3.45	2,20	Δd large
2530 MeV Kr, H-like Kr [6]	35	17 ns	0.2ns	Delay not sufficient
700 MeV Kr He-like Kr [7]	25	108.78	10	
163 MeV Ni He-like Ni [8]	17	18.17	10.5	
164 MeV Fe H-like Fe [2]	17	21.56	36	Observed

Table1. Experimental lifetime data with the measured and possible CRS positions. (Nc is the possible principle quantum number of CRS populated d is corresponding delay, Δd is the step size between two consecutive measurements, and M is experimental measured points)

References

[1] M Saffman et *al.* 2010 *Rev.Mod. Phys.* **82** 2313
[2] T. Nandi 2008 *Astrophys. Journal,* **673** L103
[3] N.N. Nedeljković et *al.* 2014 *J. Quant. Spectrosc. Radiat. Transf.* **134** 46
[4] J.Muller and J.Burgdorfer 1991 *Phys.Rev.A* **43** 11
[5] H-D.Betz et.*al*1983 Targets *Phys.Rev.Lett.***50** 34
[6] S.Cheng et *al.* 1993 *Am. Inst. Phys.* **274** 394
[7] S.Cheng et *al.* 1994 *Phys. Rev. A* **50** 2197
[8] T.Nandi et *al.* 2004 *J. Phys. B: At.Mol Opt. Phys.* **37** 703

[1]E-mail: gaurav.kmc891101@gmail.com
[2]E-mail: nitin.phy@dce.edu

Charge-patch enhanced surface scattering in the transmission of hundred-keV proton through tapered glass capillary

J. X. Shao[1], A. X. Yang, B. H. Zhu, X. M. Chen[2]*

School of Nuclear Science and Technology, Lanzhou University, Lanzhou, China

Synopsis Based on our previous work for proton transmission in nanocapillary, the simulation of hundred-keV protons through tapered-capillary was conducted. The deposited charge patch, small-angle scattering under surface as well as surface scattering were considered. We found, only the charge patch could not make the transmission effective. The charge-patch-assisted surface scattering make the density be enhanced.

Taper glass capillary has been attracting many attention as a focusing device for different ions. For keV-energy highly charged ions, the focusing ratio is about 1-7 [1], where the deposited charge patch is dominant. For MeV ions, it can be as large as 100 [2], by the small-angle scattering of incident ions. For hundred-keV protons, the focusing ratio can be 6 and decreased as E^{-1} [3]. But the scenario of hundred-keV ion transmission through tapered-capillary is not clear.

Based on our previous work [4], the simulation of hundred-keV protons through tapered-capillary was conducted. We want to identify the primary reason for the distinct focusing ratio of ions with different energies ranging from keV to hundreds of keV and even MeV. The relative size of capillary is same to Ref. [3], where the ratio of outer to inner diameter is 0.33. In simulation, the charge patch, small-angle scattering and surface scattering are considered.

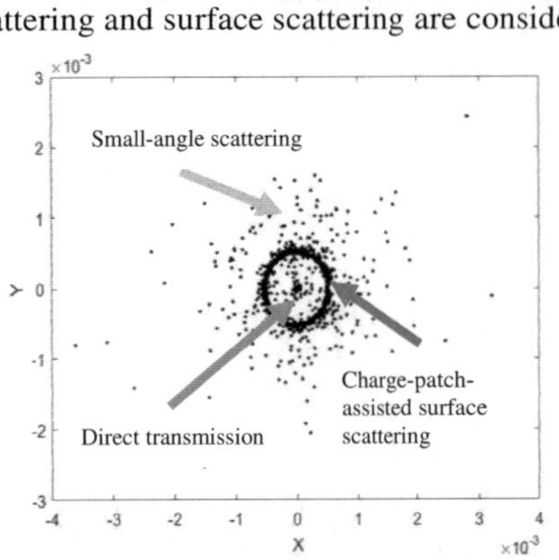

Figure 1. Position of transmission proton of 100 keV for incident angle along the axis.

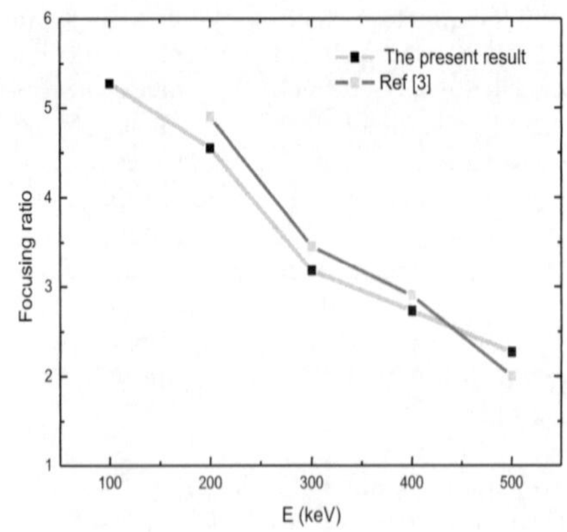

Figure 2. Focusing ratio changed with energy

Figure 1 is the position of proton outer of capillary. The charge-patch-assisted surface scattering is dominant in the three process. In figure 2, the focusing ratio changed with the incident energy are plotted and compared with experimental results [3]. The focusing ratio decreases with the incident energy.

References

[1] T. Schweigler et al. 2011 *Nucl. Instrum. Methods Phys. B* **269** 1253

[2] J. Hasegawa et al. 2011 *J. App. Phys.* **110** 044913

[3] K. A. Vokhmyanina et al. 2006 *J. Phys. A: Math. Gen.* **39** 4775

[4] G. Y. Wang et al. 2015 *Scientific Report* **5** 15169

[1]E-mail: shaojx@lzu.edu.cn

[2]E-mail: chenxm@lzu.edu.cn

IOP Conf. Series: Journal of Physics: Conf. Series **875** (2017) 112012 doi:10.1088/1742-6596/875/12/112012

Polyanionic metal clusters: How to make 'em and to break 'em

Lutz Schweikhard* [1]

* University of Greifswald, Institute of Physics, Felix-Hausdorff-Str. 6, 17489 Greifswald, Germany

Synopsis Polyanionic metal clusters are produced by electron attachment in both Paul and Penning traps. After size and charge-state selection, the cluster properties are further investigated by various methods including photo-dissociation. Depending on the particular cluster species various decay modes are observed.

Clusters of a few up to a few hundred atoms bridge the gap between single atoms and the bulk phases of matter. Their properties are governed by their geometric and electronic structure. There is a competition between both aspects and depending on the material one or the other aspect dominates. Insight can be gained by varying both the cluster size, i.e. the number of constituents and thus their spatial arrangements, as well as their charge state – and thus the occupation of the electronic levels.

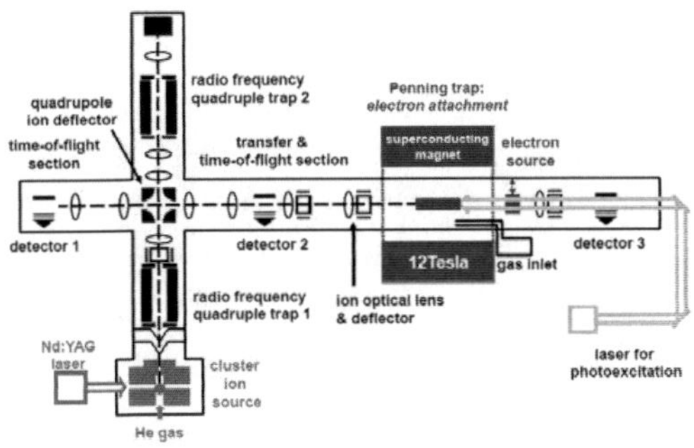

Figure 1. Schematic overview of ClusterTrap setup.

While there is a long tradition of the investigation of multiply charged cationic systems, the production of multiply negatively charged clusters, i.e. the attachment of further excess electron is more challenging. A successful general approach was found at CusterTrap [1] (Fig. 1) by subjecting mono-anionic clusters to an electron bath in a Penning trap [2,3]. Gold clusters, e.g., of charge states up to z = -6, i.e. with six excess electrons [1] and, recently, aluminum clusters even up to z = -11 have been observed. The corresponding appearance sizes, i.e. minimum clusters sizes, were studied in detail [4].

In addition, polyanionic clusters have been probed with respect to their stability and decay processes upon excitation, e.g. by laser irradiation. The decay pathways include electron emission and evaporation of neutral atoms. They depend on the cluster sizes under investigation [5]. Recently, new decay pathways were found, where larger fragments (either neutral or charged) break off (Fig. 2).

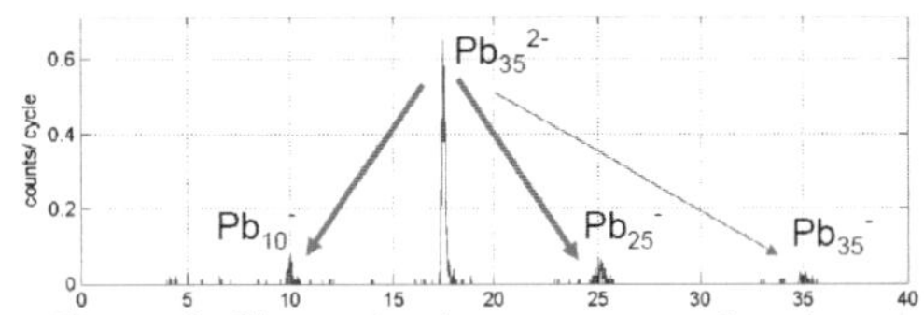

Figure 2. Cluster abundance spectrum after photo-excitation of size-selected dianionic lead clusters..

On the technical side, the production of polyanionic clusters by attachment of further excess electrons has been extended to Paul traps, i.e. rf traps that operate without a magnetic field (which, in the case of the high-mass clusters has to be provided by a superconducting magnet). To this end, a "3-state digital ion trap" has been developed [6] for the electron-beam irradiation of stored clusters in the field-free time windows of the guiding field [1].

Furthermore, the on-going investigations of polyanionic metal clusters include time-resolved measurements of delayed photodissociation, which have already proven very valuable in the case of mono-cations [7,8].

References

[1] F. Martinez *et al.* 2014 *Int. J. Mass Spectrom.* **365-366** 266
[2] A. Herlert *et al.* 1999 *Physica Scripta* **T80** 200
[3] C. Yannouleas *et al.* 2001 *Phys. Rev. Lett.* **86** 2996
[4] F. Martinez *et al.* 2015 *J. Phys. Chem. C* **119** 10949
[5] A. Herlert *et al.* 2012 *New J. Phys.* **14** 055015
[6] S. Bandelow *et al.* 2013 *Int. J. Mass Spectrom.* **353** 49
[7] U. Hild *et al.* 1998 *Phys. Rev. A* **57** 2786
[8] M. Vogel *et al.* 2001 *Phys. Rev. Lett.* **87** 013401

[1] E-mail: lschweik@physik.uni-greifswald.de

IOP Conf. Series: Journal of Physics: Conf. Series **875** (2017) 112013 doi:10.1088/1742-6596/875/12/112013

Electric-field noise and carbon diffusion on Au(110)

E. Kim,[†1] A. Safavi-Naini,[*] D. A. Hite,[¶] K. S. McKay,[¶] D. P. Pappas,[¶] P. F. Weck,[‡] and H. R. Sadeghpour[§]

[†]Department of Physics and Astronomy, University of Nevada, Las Vegas, NV 89154-4002
[*]JILA, 440 University Avenue, Boulder, Colorado 80302
[¶]NIST, 325 Broadway, Boulder, Colorado 80305
[‡]Sandia National Laboratories, P.O. Box 5800, Albuquerque, New Mexico 87185-0779
[§]ITAMP, Harvard-Smithsonian Center for Astrophysics, Cambridge, Massachusetts 02138

Synopsis: The decoherence of trapped-ion quantum gates due to heating of their motional modes is a fundamen-tal science and engineering problem. This heating is attributed to electric-field noise arising from the trap-electrode surfaces. We have investigated the origin of this noise by focusing on the diffusion of carbon-containing adsorbates on the Au(110) surface. We show how the carbon adatom diffusion on the gold surface changes the energy landscape, and how the adatom dipole moment varies with the diffusive motion. A simple model for the diffusion noise, which varies quadratically with the variation of the dipole moment, predicts a noise spectrum, in accord with the measured values.

Trapped ions are a promising platform for demonstrating coherent operations for quantum information applications [1]. However, heating of their motional modes remains a major obstacle to continued progress [2]. In particular, motional heating caused by electric-field noise originating from the trap-electrode surfaces has proven to be a difficult problem to mitigate, ever since it was first observed more than two decades ago. This decoherence mechanism scales strongly with the inverse of the distance separating the ion from the nearest electrode, and, therefore, is a barrier to scalability through miniaturization. The origin of this noise source has been suspected to arise from surface processes, based on experimental evidence on scaling with ion-electrode distance [3], electrode temperature [4], and spectral-density frequency dependence [5–6]. However, the exact mechanism that gives rise to electric-field noise at the location of the ion still remains elusive.

In this work, we have investigated diffusion of a carbon adatom on gold surfaces to gain better understanding of electric-field noise. Electron spectroscopy indicates that carbon is a dominant contaminant on gold trap-electrode surfaces [7], and scanning tunneling microscopy (STM) images (see Fig. 1) reveal ordered Au(110)-(2×1) like structures after trap electrodes have been treated with ion bombardment. Density functional theory (DFT) calculations provide the first detailed values for the energy and dipole landscapes of the adsorbed carbon atoms on Au(110), and are used subsequently in an analytical derivation of noise due to classical diffusion. The model shows how the electric-field noise varies in a non-monotonic fashion as a function of the degree of carbon-adatom coverage. The electric-field noise spectral density is a function of the variation in the adatom dipole moment, the surface diffusion constant, and patch size with different work functions. Using realistic parameters, the theory predicts an electric-field noise spectral density consistent with experimental measurement.

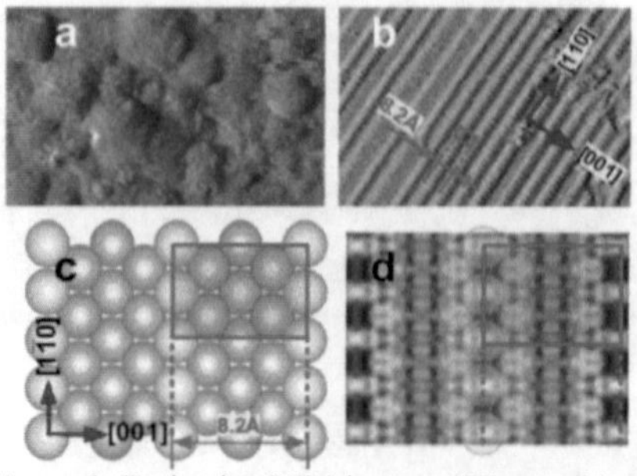

Figure 1. Derivative STM images (17 nm × 10 nm) of (a) untreated and (b) treated ion-trap surfaces by ion bombardment; (c) Top view of the Au(110)-(2 × 1) reconstructed structure and (d) its simulated STM image.

References

[1] C. Monroe and J. Kim 2013 *Science* **339** 1164
[2] D. A. Hite *et al.* 2013 *MRS Bull.* **38** 826
[3] L. Deslauriers *et al.* 2006 *Phys. Rev. Lett.* **97** 103007
[4] J. Labaziewicz *et al.* 2008 *Phys. Rev. Lett.* **101** 180602
[5] R. J. Epstein *et al.* 2007 *Phys. Rev. A* **76**, 033411
[6] D. T. C. Allcock et al. 2011 New J. Phys. **13** 123023
[7] D. A. Hite *et al.* 2012 *Phys. Rev. Lett.* **109** 103001

[1]E-mail: kimej@physics.unlv.edu

IOP Conf. Series: Journal of Physics: Conf. Series **875** (2017) 112014 doi:10.1088/1742-6596/875/12/112014

Rb adsorbate-induced negative electron affinity on quartz

J. A. Sedlacek,[*] E. Kim,[†,1] S. T. Rittenhouse,[¶‡] P. F. Weck,[§] H. R. Sadeghpour,[¤] and J. P. Shaffer[*]

[*]Homer L. Dodge Department of Physics and Astronomy, The University of Oklahoma, Norman, OK 73019, USA
[†]Department of Physics and Astronomy, University of Nevada Las Vegas, Las Vegas, NV 89154, USA
[¶]Dpartment of Physics and Astronomy, Western Washington University, Bellingham, WA 98225, USA
[‡]Department of Physics, The United States Naval Academy, Annapolis, MD 21402, USA
[§]Sandia National Laboratories, P.O. Box 5800, Albuquerque, New Mexico 87185-0779, USA
[¤]ITAMP, Harvard-Smithsonian Center for Astrophysics, Cambridge, Massachusetts 02138, USA

Synopsis We have investigated Rb adsorbates on the SiO_2 (0001) surface. Using Rydberg atom electromagnetically induced transparency, we investigate the electric fields resulting from Rb adsorbed on the quartz surface, and measure the activation energy of the Rb adsorbates. We show that the adsorbed Rb induces a negative electron affinity (NEA) on the quartz surface. The NEA surface allows low energy electrons to bind to the surface and cancel the electric field from the Rb adsorbates. Our results are important for integrating Rydberg atoms into hybrid quantum systems and the fundamental study of atom-surface interactions, as well as applications for electrons bound to a surface.

Due to recent technological advances in fabrication and trapping, hybrid quantum systems (HQS) consisting of atoms and surfaces, as well as electrons and surfaces, are fast emerging as ideal platforms for a diverse range of studies in quantum control, quantum simulation and computing, strongly correlated systems and microscopic probes of surfaces [1–3]. Miniaturization of chip surfaces is necessary to achieve large platform scalability, but decoherence and noise emerge as serious challenges as feature sizes shrink [4–6]. Mitigating noise is a fundamental step in realizing the full potential of HQSs.

In this work, we show that adsorption of Rb atoms on a quartz (0001) surface, contrary to prevailing assumption, can reduce the E-field near the surface, using spin-polarized density functional theory (DFT). On the surface of quartz, the Rb atom is bound to two oxygen atoms. Total-energy calculations for the quartz (0001) surface with various Rb coverage indicate that the lowest bound state for one monolayer (ML) has a binding energy of E_b = 0.35 eV. For the lower experimentally investigated coverages, our DFT calculations show an increase of E_b by about 1.4. The calculated E_b is comparable in magnitude with the measured activation energy, E_a, and is consistent with the expectation $E_b \leq E_a$ [7]. We demonstrate, by appealing to theoretical arguments and ab initio calculations, that the reduction in E-field is caused by a transformation of the quartz into a negative electron affinity (NEA) surface via adsorption of Rb atoms on the surface (see Figure 1). The repulsion on quartz occurs because the surface vacuum level dips below the conduction band minimum. We find that the binding of electrons to the surface substantially reduces the E-field above the surface. Reducing E-fields on a quartz surface by making quartz a NEA surface by Rb adsorption is a promising pathway for coupling Rydberg atoms to surfaces.

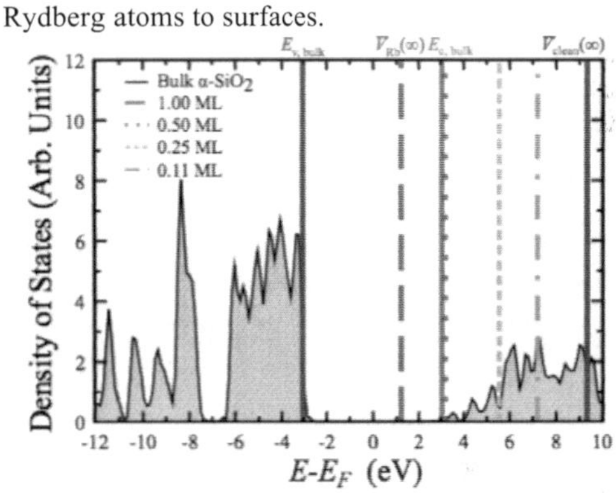

Figure 1. Density of states of bulk α-quartz. The Fermi level, E_F is at E = 0. The valence band maximum, $E_{v,bulk}$, conduction band minimum $E_{c,bulk}$ of bulk α-quartz, and the vacuum levels of the SiO_2 (0001) surface without and with Rb adsorbates, respectively, $V_{clean}(\infty)$ and $V_{Rb}(\infty)$ are labelled.

References

[1] G. Kurizki *et al.* 2015 *Proc. Natl. Acad. Sci.* **112** 3866
[2] Z.-L. Xiang *et al.* 2013 *Rev. Mod. Phys.* **85** 623
[3] C. Dong *et al.* 2015 *National Science Review* **2** 510
[4] R. O. Behunin *et al.* 2012 *Phys. Rev. A* **86**, 052509
[5] M. Brownnutt *et al.* 2015 *Rev. Mod. Phys.* **87** 1419
[6] J. D. Carter and J. D. D. Martin 2013 *Phys. Rev. A* **88** 043429
[7] L. J. Brillson 2010 *Surfaces and Interfaces of Electronic Materials*, Vol. **7** (John Wiley & Sons)

[1]E-mail: kimej@physics.unlv.edu

IOP Conf. Series: Journal of Physics: Conf. Series **875** (2017) 112015 doi:10.1088/1742-6596/875/12/112015

Density-functional approximations on CO_2@sI clathrate hydrate interactions

Daniel J. Arismendi-Arrieta[*1], **Álvaro Valdés**[†2], **Rita Prosmiti** [*3]

[*] Institute of Fundamental Physics (IFF-CSIC), CSIC, Serrano 123, 28006 Madrid, Spain
[†] Departamento de Física, Universidad Nacional de Colombia, Calle 26, Cra 39, Edificio 404, Bogotá, Colombia

Synopsis Clathrate hydrates have unique physical and chemical properties and considering that the stability of such systems is strongly related with its guest. In the last decades a renewed interest on these substances have grown up due to their potential industrial applications, with gas hydrate research becoming a joint field of various engineering disciplines and fundamental sciences.

So far, clathrate hydrates have attracted the interest of both experimentalists and theorists [1], focused mainly on two fundamental problems, a) at which conditions (pressure, temperature, concentration) gas hydrate may be formed or decomposed, and b) what structure type of gas hydrate could be obtained at these conditions. Although many characteristics of clathrate hydrates are well known experimentally, other relevant to microscopic aspects, such as accurate estimations through modelling is a scientific challenge. Related with this, computational simulations have been usually limited by the accuracy of the models in describing guest-host interactions, while currently first-principles and *ab initio* calculations are being used to determine molecular scale parameters (interaction force-fields, structural variables, spectra), that are essential for understanding fundamental aspects of hydrate formation, stability and thermodynamics.

Such facts, motivated us to evaluate the description of the guest-host CO_2@sI clathrate hydrate interactions from wave-function (WFT) and density function theory (DFT) electronic structure calculations. Computed couple-cluster or MP2 reference interaction energies for a representative set of configurations are used to assess the accuracy of modern DFT approximations (empirical and nonempirical), including the effect of semilocal approach, exact-exchange/correlation admixture, range-separation, as well as dispersion-correction terms [2]. In this work, results are presented for the T and D cages of the CO_2@sI clathrate hydrate (see Figure 1) obtained from twelve functionals available or recently implemented in *Gaussian* package [3]. Each density functional approximation (DFA) was evaluated following a protocol of complementary strategies [4] (see Figure 1). First, an extended benchmark dataset was introduced by performing reference explicitly correlated CCSD(T)-F12/DF-MP2 calculations, and sampling configurations in near-equilibrium and representative stretched regions of the potential; not, as traditionally, only equilibrium configurations. In turn, error analysis was carried out for each DFA for both the minimum interaction energies, as well as the remaining set of energies, following by scans of the potential along relevant orientations, complemented by correlation plots for a complete exploration of the systems configurational space. Finally, comparisons with semiempirical, as well as *ab initio*-based effective pairwise models commonly used as a first approximation in molecular simulations, indicate that for microscopic considerations on the interactions further refinements in such analytical representations should be incorporated in these analytical representations.

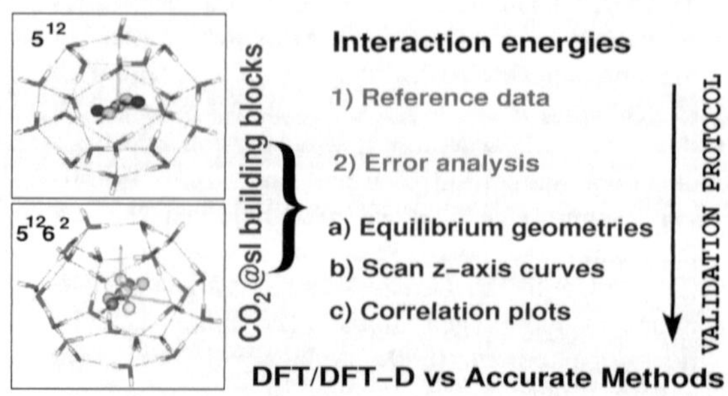

Figure 1. Protocol used to validate the performance of different DFAs with and without dispersion corrections for the sI building blocks, $CO_2(H_2O)_{20,24}$, interactions.

References

[1] E. D. Sloan *et al.* 2007 *Clathrate Hydrates of Natural Gases* **CRC Press 3rd Edition**

[2] M.G. Medvedev *et al.* 2017 *Science* **355** 49

[3] M.J. Frisch *et al.* 2016 *Gaussian 09*; 2004 *Gaussian 03* Gaussian, Inc., Wallingford CT.

[4] D. J. Arismendi-Arrieta *et al.* 2017 *to be submitted*

[1]E-mail: d.arismendi@iff.csic.es
[2]E-mail: avaldesl@unal.edu.co
[3]E-mail: rita@iff.csic.es

IOP Conf. Series: Journal of Physics: Conf. Series **875** (2017) 122001 doi:10.1088/1742-6596/875/13/122001

Description of an elastic scattering process in the pilot-wave formulation

D. C. Cordeiro Ballesteros*, M. Feole*, C. Collet†1, F. Navarrete*2, R. O. Barrachina*3,

* Centro Atómico Bariloche and Instituto Balseiro, Comisión Nacional de Energía Atómica and Universidad Nacional de Cuyo, Av. Bustillo 9500, 8400 Bariloche, Argentina

† Ecole Nationale Supérieure d'Ingénieurs de Caen, 6 Boulevard Maréchal Juin, 14000 Caen, France

Synopsis We apply De Broglie - Bohm theory to study the elastic collision of a particle by a force center in a stationary formulation. This approach allows us to clarify certain characteristics of the collision process, such as the occurrence of quantum vortices. It also provides an adequate framework for analyzing contextuality effects produced by the degree of coherence of the projectile's beam.

In 1926 Max Born [1] attempted to perform a quantum description of an atomic collision; not in terms of Schrödinger's or Heisenberg's formalisms but of De Broglie's pilot–wave formulation, arguing that it could provide a more satisfactory framework for that study. But this research program did not come to fruition until recently, when it was revalued as an advantageous option for the numerical calculation of various collision processes [2]. It is in this context that we developed a general stationary formulation of the scattering process in terms of De Broglie - Bohm formalism [3]. We evaluate the quantum trajectories for wave packets that are scattered by a dispersion center. In contrast to a previous analysis [4], we do not restrict our study to head-on collisions, but consider that the wave packets might have different impact parameters b uniformly distributed along a collimation region. This realistic description of the scattering process allows us to overcome the inconsistencies of a standard approach, where the incoming projectiles are described by plane-waves [5]; and by doing so it becomes particularly adequate for analyzing the effects produced by the characteristics of the projectile's beam.

In figure 1 we show the quantum trajectories for the dispersion of an off-centre non-spreading Gaussian wave packet with a well-defined forward impulse p by a model potential corresponding to an isotropic differential cross section. For the sake of simplicity we assume a two-dimensional system, being the generalization to the 3D case straightforward. Let us note that, since quantum trajectories cannot cross each other, only those in the outermost sides of the incoming flux can be scattered. The appearance of quantum vortices is evident in the figure. It can be demonstrated that, instead of being an exception, they represent a necessary ingredient of any scattering process. Finally, the cross section can be obtained by integrating the outgoing flux on the impact parameter b. The standard result is recovered for an infinite collimation region.

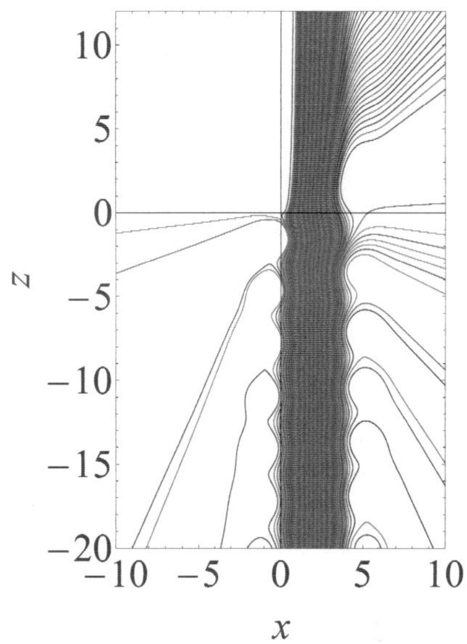

Figure 1. Quantum trajectories for the dispersion of a wave packet of width $\sigma = 1$, impulse $p = 1$ and impact parameter $b = 2$ a.u. by a force center at $(x,z) = (0,0)$.

References

[1] M. Born 1929 *Zeitschrift für Physik* **38** 803

[2] R. E. Wyatt 2005 *Quantum Dynamics with Trajectories: Introduction to Quantum Hydrodynamics* (Springer)

[3] P. R. Holland 1995 *The Quantum Theory of Motion* (Cambridge University Press)

[4] J. Stenson and A. Stetz 2013 *Eur. J. Phys.* **34** 1199

[5] M. Meisinger 2004 *Coulomb Scattering in Bohmian Mechanics* (1) (Universität Innsbruck)

1E-mail: clement.collet@ecole.ensicaen.fr
2E-mail: navarrete@cab.cnea.gov.ar
3E-mail: barra@cab.cnea.gov.ar

IOP Conf. Series: Journal of Physics: Conf. Series **875** (2017) 122002 doi:10.1088/1742-6596/875/13/122002

Model for plasma jet-driven magneto-inertial fusion

Sergei V. Ryzhkov* [1], **Victor V. Kuzenov***†‡ [2]

[*] Department of Thermal Physics, Bauman Moscow State Technical University, Moscow 105005, Russia
[†] N.L. Dukhov All-Russian Research Institute of Automatics (State Corporation «ROSATOM»), Moscow 127055
[‡] A.Yu. Ishlinsky Institute for Problems in Mechanics RAS, Moscow 119526, Russia

Synopsis Mathematical models are developed to model the implosion dynamics for plasma jets-magnetized target system [1-9]. Transport (diffusion, thermal, radiative) processes of atoms, ions, electrons, molecules, photons in a strong magnetic field are taken into account.

We assume that the plasma is quasi-neutral, and use the average ion model to determine Z_i (when it is assumed that all ions of chemical components "i" have the same value Z_i at each point in the computational domain). This can be justified if the distribution of the different degree of ionization is sufficiently narrow.

The average ion charge Z_i (t) can change over time and we can use the formula:

$$\frac{DZ_i}{Dt} = Z_i\left(\nu_I - \nu_{Ph} - \nu_{T.R}\right).$$

The following inelastic processes (electron impact ionization ν_I, photorecombination ν_{Ph} and triple recombination $\nu_{T.R}$) in the one-component case are taken into consideration in calculation:

$$\nu_I = 10^{-7} n_i \left[\frac{I_H}{I(Z_i)}\right]^{3/2} \exp\left[-\frac{I(Z_i)}{kT_e}\right]\left[\frac{I(Z_i)}{kT_e}\right]^{1/2}\left[1+\frac{I(Z_i)}{kT_e}\right]^{-1}$$

$$\nu_{Ph} = 6\times\lambda\times10^{-14} n_i \left[\frac{I(Z_i)}{I_H}\right]^{1/2}\left[\frac{I(Z_i)}{kT_e}\right]^{3/2}\left[1+\frac{I(Z_i)}{kT_e}\right]^{-1}$$

$$\nu_{T.R} = 3{,}3\times10^{-31} Z_i n_i^2 \left[\frac{I_H}{I(Z_i)}\right]^{3/2}\left[\frac{I_H}{kT_e}\right]^{3/2}\left[\frac{I(Z_i)}{kT_e}\right]^{1/2}\left[1+\frac{I(Z_i)}{kT_e}\right]^{-1}$$

where $I_H = 13{,}6$ eV is the ionization potential of hydrogen, $I(Z_i)$ is the ionization energy of the average ion, T_e is the electron temperature (K), $\lambda = 1 \div 5$, $n_e = Z_i n_i$, $n_i = \rho/M_i$ are electron and ion densities (cm^{-3}).

It is known, that plasma dynamical processes are strongly affected by irradiation, if plasma temperature achieve 1 eV. In that case, the gas dynamic fields of thermal physical variables may be calculated only with taking into account the radiation fields. In this work the radiation transfer equation is used in the form of multi-group diffusion approach:

$$\frac{1}{J}\frac{\partial\left(Jq_{i\xi}\right)}{\partial\xi} + \frac{1}{J}\frac{\partial\left(Jq_{i\eta}\right)}{\partial\eta} + \chi_i c U_i = 4\chi_i\sigma_i T^4,$$

$$\frac{c}{3}\frac{\partial U_i}{\partial\xi} + \chi_i q_{i\xi} = 0, \quad \frac{c}{3}\frac{\partial U_i}{\partial\eta} + \chi_i q_{i\eta} = 0,$$

where $U_i(y, z, t)$ is the radiation power density in i-th spectral group, χ_i is the spectral absorption coefficient. Besides of the method mentioned above, the discrete ordinates method has been used in the work, which gets an opportunity to solve the radiation transfer equation on tetrahedral non–structured mesh. Modified alternatively triangular three-layered iterative scheme is applied for the solution of radiation transport equations, where the time step selected via conjugate directions method.

The turbulent viscous coefficient $\mu_\Sigma^{r,z}$ is calculated by using of the Bussinesk hypothesis, when effective viscous is $\mu_\Sigma^{r,z} = \mu_m^{r,z} + \mu_t$, where $\mu_m^{r,z}$ is a molecular viscosity coefficient, taking into account the atomic-molecular collision processes and the presence of a magnetic field in the plasma, and μ_t is a turbulent one, determined from two-parameter turbulence model (Coakley model). The mathematical model and test results are presented and discussed.

This research has been supported by the Ministry of Education and Science of the Russian Federation (Project No. 13.5240.2017/БЧ).

References

[1] H. Nagatomo *et al.* 2016 *J. Phys. Conf. Ser.* **717** 012041
[2] V.V. Kuzenov *et al.* 2013 *Probl. At. Sci. Technol.* 4 (86) 103
[3] S.V. Ryzhkov *et al.* 2013 *Fusion Sci. Technol.* **63** (1T) 135
[4] V.Kuzenov *et al.* 2013 *Probl. At. Sci. Tech.* **1** 12
[5] S. Ryzhkov 2010 *Probl. At. Sci. Technol.* **4** 105
[6] S.A. Belkov *et al.* 1997 *JETP* **84** 272
[7] V.V. Kuzenov *et al.* 2016 *Bull. Russ. Acad. Sci.: Phys.* **80** 598
[8] I.Yu. Kostyukov *et al.* 2011 *Plasma Phys. Rep.* **37** 1092
[9] V.V. Kuzenov *et al.* 2017 *J. Phys. Conf. Ser.* **815** 012024

[1] E-mail: vik.kuzenov@gmail.com P

[2] E-mail: svryzhkov@bmstu.ru

IOP Conf. Series: Journal of Physics: Conf. Series **875** (2017) 122003 doi:10.1088/1742-6596/875/13/122003

Magic wavelengths of Ca^+ ion for linearly and circularly polarized light

Jun Jiang*[1], Li Jiang*, Xia Wang*, Peter Shaw†, Deng-Hong Zhang*, Lu-You Xie* and Chen Zhong Dong*[2]

*Key laboratory of Atomic and Molecular Physics & Functional Materials of Gansu Province, College of Physics and Electronic Engineering, Northwest Normal University, Lanzhou, 730070, China
†School of Engineering, Charles Darwin University, Darwin NT 0909, Australia

Synopsis The dynamic dipole polarizabilities of the low-lying states of Ca^+ ion for linearly and circularly polarized light are calculated by using relativistic configuration interaction plus core polarization (RCICP) approach. Present magic wavelengths for linearly polarized light agree with the available results excellently. Additional magic wavelengths have been found for circularly polarized light, which is very useful for magnetic-sublevel selective trapping.

The magic wavelength, at which the two levels of transitions have same AC stark shifts, was introduced in Refs. [1, 2]. It has been of great interest in ultraprecise optical lattice clocks and the state-insensitive quantum engineering [3, 4]. For circularly polarized light, the dynamic polarizability has vector and tensor component. The dynamic polarizability is different for different magnetic sublevels of atomic states.

In this work, the dynamic dipole polarizabilities of the $4s$, $4p_j$ and $3d_j$ states of Ca^+ ion for linearly and circularly polarized light are calculated by using relativistic configuration interaction plus core polarization (RCICP) approach. The basic strategy of the model is to partition the atom into valence and core electrons. The orbitals of the core are written as linear combinations of S-spinors which can be treated as relativistic generations of Slater-type orbitals. The orbitals of valence are written as linear combinations of L-spinors and S-spinors. L-spinors can be treated as relativistic generalizations of the Laguerre- type orbitals.

Fig.1 shows the dynamic polarizabilities of $4s$ and $4p_{1/2}$ states for the linearly (A=0) and left handed (A=-1) polarized light. In Fig.1 (a), three magic wavelengths are found, which are in good agreement with available results [5, 6]. Fig.1 (b) gives the dynamic polarizabilities for A = -1 for $4s_{m=-1/2}$ and $4p_{1/2m=-1/2}$ states. There are only two magic wavelengths are found for $4s$ – $4p_{1/2}$ transition. There is no magic wavelength between $4p_{1/2}$ – $5s$ and $4p_{1/2}$ – $4d_{3/2}$ transition energy. The magic wavelength 395.5410 nm is very close to the magic wavelength 395.1788 nm for $4s$ - $4p_{1/2}$ transition for A=0. Another magic wavelength 778.3726 nm occurs between $4p_{1/2}$ – $4s$ and $4p_{1/2}$ –$3d_{3/2}$ transition energy. It has 87 nm difference with the magic wavelength 691.2444 nm for linearly polarized light.

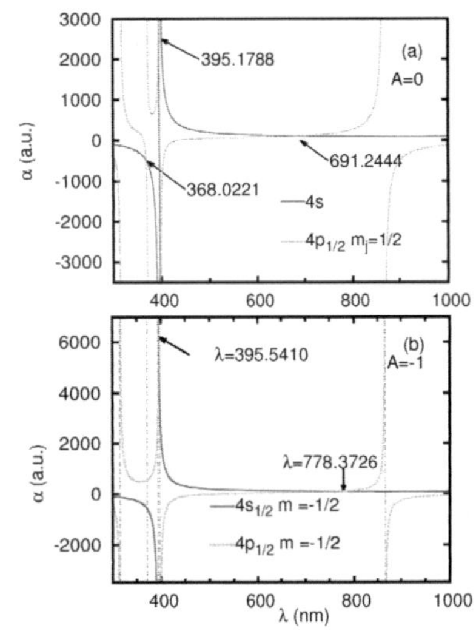

Fig. 1. Dynamic polarizabilities (in au) for the $4s$ - $4p_{1/2}$ of Ca^+ for the linearly (A= 0) and left handed (A=-1) polarized light. Magic wavelengths are identified by arrows.

This work was supported by National Natural Science Foundation of China (Grant No. 11564036, U1332206, 11464042, U1330117, U1331122, U153014) , the Young Teachers Scientific Research Ability Promotion Plan of Northwest Normal University (NWNU-LKQN-15-3).

References

[1] J. Ye *et al.*1999 *Phys Rev. Lett.* 83,4987
[2] H. Katori *et al.*1999 *Phys Rev. Lett.* 82,1116
[3]G. Wilpers *et al.* 2007 *Metrologia* 44,146
[4]N. Lundblad *et al.* 2010 *Phys Rev.A* 81,031611
[5]J.Kaur *et al.2015 Phys Rev.A 92,031402*
[6]Y.B. Tang *et al.2013 Phys Rev.A 87.042517*

[1] E-mail: phyjiang@yeah.net
[2] E-mail: dongcz@nwnu.edu.cn

IOP Conf. Series: Journal of Physics: Conf. Series **875** (2017) 122004 doi:10.1088/1742-6596/875/13/122004

Exploring metastable decay dynamics of polycyclic aromatic nitrogen containing hydrocarbons upon HCN evaporation

P. K. Najeeb[*1], **T. Sairam**[*], **M. V. Vinitha**[*], **Anudit Kala**[*], **S. Vig**[*], **Pragya Bhatt**[†], **C.P. Safvan**[†] **and U. Kadhane**[*]

[*] Department of Physics, Indian Institute of Space Science and Technology, Thiruvananthapuram-695547, India.
[†] Inter University Accelerator Centre, Aruna Asaf Ali Marg, New Delhi-110067, India.

Synopsis : Dicationic states of isoquinoline fragment is explored. A tail feature is observed in the coincidence map showing that $C_8H_6^{2+}$ has a metastable state. Using time of flight difference spectrum life time has been estimated to be $\sim$75ns.

Polycyclic Aromatic Nitrogen containing Hydrocarbons (PANHs) are known to be stable. In the interstellar regions photo or thermal processing may lead to dissociation of these species[1]. Earlier studies have focused on understanding the product formation from neutral and mono-cationic PANHs. However, the product formation from dicationic PANHs (present in interstellar) species has received far less attention. Moreover, understanding the dissociation of these metastable states may give us a hint regarding how the association reactions could have progressed to form these intact dications. Thus, to understand the mechanisms involved in the metastable dicationic species, simplest of PANHs quinoline ($C_9H_7N^+$) (and its isomer isoquinoline ($C_9H_7N^+$)) have been chosen. Even though, loss of C_2H_2 or HCN is feasible from these species, it has been shown that HCN is energetically favorable. In this article, we report metastable intermediate states that could be involved in the formation of $C_7H_3^+$, which belongs to $C_{2n+1}H_3$ class and unlike their analogues C_3H_3 and C_5H_3 less explored. We report $C_7H_3^+$ from quinoline and isoquinoline through an intermediate $C_8H_6^{2+}$ which possess a metastable state.

$$quinoline/isoquinoline \rightarrow C_8H_6^{2+} + HCN \quad (1)$$
$$C_8H_6^{2+} \rightarrow CH_3^+ + C_7H_3^+ \quad (2)$$

Earlier reports have obtained $C_8H_6^{2+}$ in the time of flight spectrum [2], since only a single hit information is obtained, it was attributed to be a stable species. Field et al.[3] have shown that the intensity variation of the difference of the time of flight(ToF) of the two ions in the tail region can give us an quantitative estimate regarding the lifetimes of the metastable species as shown in eqn.3.

$$I(t_2-t_1) = Aexp(-\frac{m1+m2}{m1-m2}\frac{t_2-t_1}{2\tau}) \quad (3)$$

Here, m1 and m2 denotes the mass of the two fragments the metastable state is dissociating into, τ denotes the lifetime. It should be noted that the $C_7H_3^+$ molecule and its isomers are linear in their ground state. Moreover, removal of CH_3^+ is not energetically favored from the ring structures of $C_8H_6^{2+}$. The experiment was performed using ECR ion source based low energy ion beam facility (LEIBF) at Inter-University Accelerator Centre (IUAC), New Delhi, India [4].

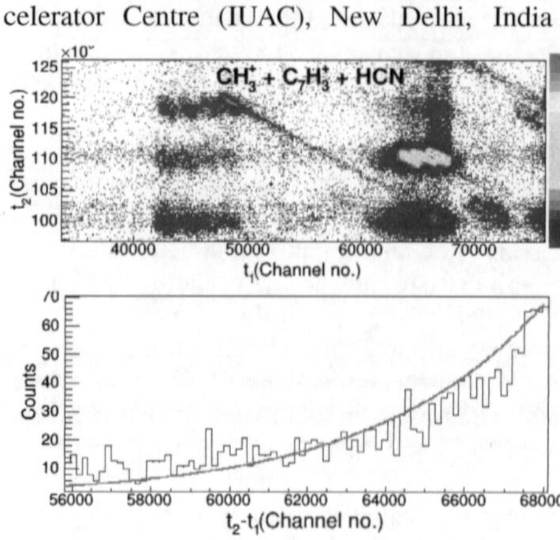

Figure 1. Top:The tail structure shown is formed when lifetime of parent $C_8H_6^{2+}$ has lifetime comparble to the time of flight of the fragments. Bottom: Shows the experimental intensity distribution as the function of ToF difference. This function is fit (red line) with eqn.3. giving lifetime (τ) of $C_8H_6^{2+}$ as $\sim$=75ns.

References

[1] De Haas A.J. *et al.* 2017 *Physical Chemistry Chemical Physics* **19**, 2974.

[2] Roithova J. *et al.* 2007 *Physical Chemistry Chemical Physics* **9**, 731.

[3] Field T,A. and Eland J.H. 1993 *Chemical Physics Letters* **11**, 4.

[4] Kumar A. *et al.* 2014 *International Journal of Mass Spectrometry* **374**, 44.

[1]E-mail: najeeb.12@iist.ac.in